AN INTRODUCTION TO LTE

AN INTRODUCTION TO LTE

LTE, LTE-ADVANCED, SAE, VoLTE AND 4G MOBILE COMMUNICATIONS

Second Edition

Christopher Cox

Director, Chris Cox Communications Ltd, UK

This edition first published 2014

Registered office
John Wiley & Sons Ltd, The Atrium, Southern Gate, Chichester, West Sussex, PO19 8SQ, United Kingdom

For details of our global editorial offices, for customer services and for information about how to apply for permission to reuse the copyright material in this book please see our website at www.wiley.com.

Library of Congress Cataloging-in-Publication Data

Cox, Christopher (Christopher Ian), 1965-
An introduction to LTE LTE, LTE-advanced, SAE, VoLTE and 4G mobile communications / Christopher Cox.
pages cm
Includes index.
ISBN 978-1-118-81803-9 (cloth)
1. Long-Term Evolution (Telecommunications) 2. Mobile communication systems – Standards. I. Title.
TK5103.48325.C693 2014
621.3845′6 – dc23

2014007432

A catalogue record for this book is available from the British Library.

ISBN:9781118818039

Set in 10/12 Times by Laserwords Private Limited, Chennai, India

To my nieces, Louise and Zoe

Contents

Preface

This book is about the world's dominant 4G mobile telecommunication system, LTE.

In writing the book, my aim has been to give the reader a concise, system level introduction to the technology that LTE uses. The book covers the whole of the system, both the techniques used for radio communication between the base station and the mobile phone, and the techniques used to transfer data and signalling messages across the network. I have avoided going into excessive detail, which is more appropriate for specialized treatments of individual topics and for the LTE specifications themselves. Instead, I hope that the reader will come away from this book with a sound understanding of the system and of the way in which its different components interact. The reader will then be able to tackle the more advanced books and the specifications with confidence.

The target audience is twofold. Firstly, I hope that the book will be valuable for engineers who are working on LTE, notably those who are transferring from other technologies such as GSM, UMTS and cdma2000, those who are experts in one part of LTE but who want to understand the system as a whole and those who are new to mobile telecommunications altogether. Secondly, the book should give a valuable overview to those who are working in non technical roles, such as project managers, marketing executives and intellectual property consultants.

Structurally, the book has four parts. The first part lays out the foundations that the reader will need in the remainder of the book. Chapter 1 is an introduction, which relates LTE to earlier mobile telecommunication systems and lays out its requirements and key technical features. Chapter 2 covers the architecture of the system, notably the hardware components and communication protocols that it contains and its use of radio spectrum. Chapter 3 reviews the radio transmission techniques that LTE has inherited from earlier mobile telecommunication systems, while Chapters 4 and 5 describe the more recent techniques of orthogonal frequency division multiple access and multiple input multiple output antennas.

The second part of the book covers the air interface of LTE. Chapter 6 is a high level description of the air interface, while Chapter 7 relates the low level procedures that a mobile phone uses when it switches on, to discover the LTE base stations that are nearby. Chapter 8 covers the low level procedures that the base station and mobile phone use to transmit and receive information, while Chapter 9 covers a specific procedure, random access, by which the mobile phone can contact a base station without prior scheduling. Chapter 10 covers the higher level parts of the air interface, namely the medium access control, radio link control and packet data convergence protocols.

The third part covers the signalling procedures that govern how a mobile phone behaves. In Chapter 11, we describe the high level procedures that a mobile phone uses when it switches on, to register itself with the network and establish communications with the outside world. Chapter 12 covers the security procedures used by LTE, while Chapter 13 covers the procedures that manage the quality of service and charging characteristics of a data stream. Chapter 14 describes the mobility management procedures that the network uses to keep track of the mobile's location. Chapter 15 describes how LTE inter-operates with the earlier technologies of GSM and UMTS, while Chapter 16 discusses inter-operation with other technologies such as wireless local area networks and cdma2000. Chapter 17 covers the self-configuration and self-optimization capabilities of LTE.

The final part covers more specialized topics. Chapters 18, 19 and 20 describe the enhancements that have been made to LTE in later releases of the specifications, notably an enhanced version of the technology that is known as LTE-Advanced. Chapters 21 and 22 cover the two most important solutions for the delivery of voice calls to LTE devices, namely circuit switched fallback and the IP multimedia subsystem. Finally, Chapter 23 reviews the performance of LTE and discusses the techniques that are used to estimate the coverage and capacity of an LTE network.

LTE has a large number of acronyms, and it is hard to talk about the subject without using them. However, they can make the material appear unnecessarily impenetrable to a newcomer, so I have aimed to keep the use of acronyms to a reasonable minimum, often preferring the full name or a colloquial one. There is a full list of abbreviations in the introductory material and new terms are highlighted using italics throughout the text.

I have also endeavoured to keep the book's mathematical content to the minimum needed to understand the system. The LTE air interface makes extensive use of complex numbers, Fourier transforms and matrix algebra, but the reader will not require any prior knowledge of these in order to understand the book. We do make limited use of complex numbers in Chapters 3 and 4 to illustrate our discussion of modulation, and introduce Fourier transforms and matrices in subsections of Chapters 4 and 5 to cover the more advanced aspects of orthogonal frequency division multiple access and multiple antennas. Readers can, however, skip this material without detracting from their overall appreciation of the subject.

Acknowledgements

Many people have given me assistance, support and advice during the creation of this book. I am especially grateful to Liz Wingett, Susan Barclay, Sophia Travis, Sandra Grayson, Mark Hammond and the rest of the publishing team at John Wiley & Sons, Ltd for the expert knowledge and gentle encouragement that they have supplied throughout the production process.

I am indebted to Michael Salmon and Geoff Varrall for encouraging me to write the first edition of this book and to the publishing team at Wiley for requesting a second. The advice and feedback I have received while preparing the manuscript have been invaluable and have given me many opportunities to correct errors and improve the material. In this respect, I would particularly like to thank Jeff Cartwright, Joseph Hoy, Julian Nolan, Michael Salmon, Mohammad Anas, Obi Chiemeka, Pete Doherty, Les Granfield, Karl van Heeswijk, Kit Kilgour and Paul Mason. I am especially indebted to Nicola Rivers, for her support and encouragement throughout the preparation of the second edition. Naturally, the responsibility for any remaining errors or omissions in the text, and for any lack of clarity in the explanations, is entirely my own.

Much of my knowledge of the more detailed aspects of LTE, notably of circuit switched fallback and the IP multimedia subsystem, has been gathered while delivering courses on behalf of various training providers. I am indebted to the directors and staff of Imagicom, Informa Telecoms Academy, Wray Castle and Mpirical, for the support and learning opportunities that they have provided to me. I would also like to extend my thanks to the delegates who have attended my training courses on LTE. Their questions and corrections have extended my knowledge of the subject, while their feedback has regularly suggested ways to explain topics more effectively.

Several diagrams in this book have been reproduced from the technical specifications for LTE, with permission from the European Telecommunications Standards Institute (ETSI), © 2013, 2012, 2011, 2010, 2006. 3GPP™ TSs and TRs are the property of ARIB, ATIS, CCSA, ETSI, TTA and TTC who jointly own the copyright for them. They are subject to further modifications and are therefore provided to you 'as is' for information purposes only. Further use is strictly prohibited.

Analysys Mason Limited kindly supplied the market research data underlying the illustrations of network traffic and operator revenue in Figures 1.6 and 21.1. I would like to extend my appreciation to Hilary Bailey, Morgan Mullooly, Terry Norman and James Allen for providing this information. The measurements of network traffic in Figure 1.5 and the subscription data underlying Figures 1.9 and 1.10 are by Ericsson, and I am grateful to Elin Pettersson and Svante Bergqvist for making these available.

List of Abbreviations

16-QAM	16 quadrature amplitude modulation
1G	First generation
1xRTT	1x radio transmission technology
2G	Second generation
3G	Third generation
3GPP	Third Generation Partnership Project
3GPP2	Third Generation Partnership Project 2
4G	Fourth generation
64-QAM	64 quadrature amplitude modulation
AAA	Authentication, authorization and accounting
ABMF	Account balance management function
ABS	Almost blank subframe
ACK	Positive acknowledgement
ACM	Address complete message
ADC	Analogue to digital converter
AES	Advanced Encryption Standard
AF	Application function/Assured forwarding
AKA	Authentication and key agreement
AM	Acknowledged mode
AMBR	Aggregate maximum bit rate
AMR	Adaptive multi rate
AMR-WB	Wideband adaptive multi rate
ANDSF	Access network discovery and selection function
ANM	Answer message
API	Application programming interface
APN	Access point name
APN-AMBR	Per APN aggregate maximum bit rate
ARIB	Association of Radio Industries and Businesses
ARP	Allocation and retention priority
ARQ	Automatic repeat request
AS	Access stratum/Application server
ASME	Access security management entity
ATCF	Access transfer control function
ATGW	Access transfer gateway
ATIS	Alliance for Telecommunications Industry Solutions

AuC	Authentication centre
AVP	Attribute value pair/Audio visual profile
AWS	Advanced Wireless Services
B2BUA	Back to back user agent
BBERF	Bearer binding and event reporting function
BBF	Bearer binding function
BCCH	Broadcast control channel
BCH	Broadcast channel
BD	Billing domain
BE	Best effort
BGCF	Breakout gateway control function
BICC	Bearer independent call control
BM-SC	Broadcast/multicast service centre
BPSK	Binary phase shift keying
BSC	Base station controller
BSR	Buffer status report
BSSAP+	Base station subsystem application part plus
BSSGP	Base station system GPRS protocol
BTS	Base transceiver station
CA	Carrier aggregation
CAMEL	Customized applications for mobile network enhanced logic
CBC	Cell broadcast centre
CBS	Cell broadcast service
CC	Call control/Component carrier
CCCH	Common control channel
CCE	Control channel element
CCO	Cell change order
CCSA	China Communications Standards Association
CDF	Charging data function
CDMA	Code division multiple access
CDR	Charging data record
CFI	Control format indicator
CGF	Charging gateway function
CIF	Carrier indicator field
CLI	Calling line identification
CM	Connection management
CMAS	Commercial mobile alert system
C-MSISDN	Correlation mobile subscriber ISDN number
CoMP	Coordinated multi-point transmission and reception
COST	European Cooperation in Science and Technology
CP	Cyclic prefix
CQI	Channel quality indicator
CRC	Cyclic redundancy check
C-RNTI	Cell radio network temporary identifier
CS	Circuit switched

CS/CB	Coordinated scheduling and beamforming
CSCF	Call session control function
CSFB	Circuit switched fallback
CSG	Closed subscriber group
CSI	Channel state information
CS-MGW	Circuit switched media gateway
CTF	Charging trigger function
D2D	Device to device
DAC	Digital-to-analogue converter
dB	Decibel
dBi	Decibels relative to an isotropic antenna
dBm	Decibels relative to one milliwatt
DCCH	Dedicated control channel
DCI	Downlink control information
DeNB	Donor evolved Node B
DFT	Discrete Fourier transform
DFT-S-OFDMA	Discrete Fourier transform spread OFDMA
DHCP	Dynamic host configuration protocol
DiffServ	Differentiated services
DL	Downlink
DL-SCH	Downlink shared channel
DNS	Domain name server
DPS	Dynamic point selection
DRS	Demodulation reference signal
DRVCC	Dual radio voice call continuity
DRX	Discontinuous reception
DSCP	Differentiated services code point
DSL	Digital subscriber line
DSMIP	Dual-stack mobile IP
DTCH	Dedicated traffic channel
DTM	Dual transfer mode
DTMF	Dual tone multi-frequency
EAG	Explicit array gain
eAN	Evolved access network
EAP	Extensible authentication protocol
EATF	Emergency access transfer function
ECGI	E-UTRAN cell global identifier
ECI	E-UTRAN cell identity
ECM	EPS connection management
ECN	Explicit congestion notification
E-CSCF	Emergency call session control function
EDGE	Enhanced Data Rates for GSM Evolution
EEA	EPS encryption algorithm
EF	Expedited forwarding
eHRPD	Evolved high rate packet data

EIA	EPS integrity algorithm
EICIC	Enhanced inter cell interference coordination
EIR	Equipment identity register
EIRP	Equivalent isotropic radiated power
eMBMS	Evolved MBMS
EMM	EPS mobility management
eNB	Evolved Node B
EPC	Evolved packet core
ePCF	Evolved packet control function
EPDCCH	Enhanced physical downlink control channel
ePDG	Evolved packet data gateway
EPRE	Energy per resource element
EPS	Evolved packet system
E-RAB	Evolved radio access bearer
ERF	Event reporting function
ESM	EPS session management
E-SMLC	Evolved serving mobile location centre
ESP	Encapsulating security payload
ETSI	European Telecommunications Standards Institute
ETWS	Earthquake and tsunami warning system
E-UTRAN	Evolved UMTS terrestrial radio access network
EV-DO	Evolution data optimized
FCC	Federal Communications Commission
FDD	Frequency division duplex
FDMA	Frequency division multiple access
FD-MIMO	Full-dimension MIMO
FFT	Fast Fourier transform
FTP	File transfer protocol
GBR	Guaranteed bit rate
GCP	Gateway control protocol
GERAN	GSM EDGE radio access network
GGSN	Gateway GPRS support node
GMLC	Gateway mobile location centre
GMM	GPRS mobility management
GNSS	Global navigation satellite system
GP	Guard period
GPRS	General Packet Radio Service
GPS	Global Positioning System
GRE	Generic routing encapsulation
GRX	GPRS roaming exchange
GSM	Global System for Mobile Communications
GSMA	GSM Association
GTP	GPRS tunnelling protocol
GTP-C	GPRS tunnelling protocol control part
GTP-U	GPRS tunnelling protocol user part

GUMMEI	Globally unique MME identifier
GUTI	Globally unique temporary identity
HARQ	Hybrid ARQ
HeNB	Home evolved Node B
HI	Hybrid ARQ indicator
HLR	Home location register
H-PCRF	Home policy and charging rules function
HRPD	High rate packet data
HSDPA	High speed downlink packet access
HSGW	HRPD serving gateway
HSPA	High speed packet access
HSS	Home subscriber server
HSUPA	High-speed uplink packet access
HTTP	Hypertext transfer protocol
I	In phase
IAM	Initial address message
IARI	IMS application reference identifier
IBCF	Interconnection border control function
ICIC	Inter-cell interference coordination
ICS	IMS centralized services
I-CSCF	Interrogating call session control function
ICSI	IMS communication service identifier
IDC	In device coexistence
IEEE	Institute of Electrical and Electronics Engineers
IETF	Internet Engineering Task Force
iFC	Initial filter criteria
IFOM	IP flow mobility
II-NNI	Inter IMS network to network interface
IKE	Internet key exchange
IMEI	International mobile equipment identity
IM-MGW	IMS media gateway
IMPI	IP multimedia private identity
IMPU	IP multimedia public identity
IMS	IP multimedia subsystem
IMS-ALG	IMS application level gateway
IMSI	International mobile subscriber identity
IM-SSF	IP multimedia service switching function
IMT	International Mobile Telecommunications
IP	Internet protocol
IP-CAN	IP connectivity access network
IPSec	IP security
IP-SM-GW	IP short message gateway
IPv4	Internet protocol version 4
IPv6	Internet protocol version 6
IPX	IP packet exchange

IRL	Isotropic receive level
ISDN	Integrated services digital network
ISI	Inter symbol interference
ISIM	IP multimedia services identity module
ISR	Idle mode signalling reduction
ISRP	Intersystem routing policy
ISUP	ISDN user part
ITU	International Telecommunication Union
IWF	Interworking function
JP	Joint processing
JR	Joint reception
JT	Joint transmission
LA	Location area
LBS	Location-based services
LCS	Location services
LCS-AP	LCS application protocol
LDAP	Lightweight directory access protocol
LGW	Local gateway
LIPA	Local IP access
LIR	Location info request
LPP	LTE positioning protocol
LRF	Location retrieval function
LTE	Long term evolution
LTE-A	LTE-Advanced
M2M	Machine to machine
MAC	Medium access control
MAP	Mobile application part
MAPCON	Multi access PDN connectivity
MAR	Multimedia authentication request
MBMS	Multimedia broadcast/multicast service
MBMS-GW	MBMS gateway
MBR	Maximum bit rate
MBSFN	Multicast/broadcast over a single frequency network
MCC	Mobile country code
MCCH	Multicast control channel
MCE	Multicell/multicast coordination entity
MCH	Multicast channel
MDT	Minimization of drive tests
ME	Mobile equipment
MEGACO	Media gateway control
MeNB	Master evolved Node B
MGCF	Media gateway control function
MGL	Measurement gap length
MGRP	Measurement gap repetition period
MGW	Media gateway

MIB	Master information block
MIMO	Multiple input multiple output
MIP	Mobile IP
MM	Mobility management
MME	Mobility management entity
MMEC	MME code
MMEGI	MME group identity
MMEI	MME identifier
MMSE	Minimum mean square error
MMTel	Multimedia telephony service
MNC	Mobile network code
MO	Management object
MOS	Mean opinion score
MPLS	Multiprotocol label switching
MRB	Media resource broker
MRF	Multimedia resource function
MRFC	Multimedia resource function controller
MRFP	Multimedia resource function processor
M-RNTI	MBMS radio network temporary identifier
MSC	Mobile switching centre
MSISDN	Mobile subscriber ISDN number
MSK	Master session key
MSRP	Message session relay protocol
MT	Mobile termination
MTC	Machine-type communications
MTC-IWF	Machine-type communications interworking function
MTCH	Multicast traffic channel
M-TMSI	M temporary mobile subscriber identity
MTSI	Multimedia telephony service for IMS
MU-MIMO	Multiple user MIMO
NACC	Network-assisted cell change
NACK	Negative acknowledgement
NAI	Network access identifier
NAP-ID	Network access provider identity
NAS	Non-access stratum
NAT	Network address translation
NH	Next hop
NMO	Network mode of operation
OCF	Online charging function
OCS	Online charging system
OMA	Open Mobile Alliance
OFCS	Offline charging system
OFDM	Orthogonal frequency division multiplexing
OFDMA	Orthogonal frequency division multiple access
OSA	Open service access

OSI	Open systems interconnection
OTDOA	Observed time difference of arrival
OUI	Organizational unique identifier
PAPR	Peak-to-average power ratio
PBCH	Physical broadcast channel
PBR	Prioritized bit rate
PCC	Policy and charging control
PCCH	Paging control channel
PCEF	Policy and charging enforcement function
PCell	Primary cell
PCFICH	Physical control format indicator channel
PCH	Paging channel
PCRF	Policy and charging rules function
PCS	Personal Communications Service
P-CSCF	Proxy call session control function
PDCCH	Physical downlink control channel
PDCP	Packet data convergence protocol
PDN	Packet data network
PDP	Packet data protocol
PDSCH	Physical downlink shared channel
PDU	Protocol data unit
PESQ	Perceptual evaluation of speech quality
P-GW	Packet data network gateway
PHB	Per hop behaviour
PHICH	Physical hybrid ARQ indicator channel
PL	Path loss/Propagation loss
PLMN	Public land mobile network
PLMN-ID	Public land mobile network identity
PMCH	Physical multicast channel
PMD	Pseudonym mediation device
PMI	Precoding matrix indicator
PMIP	Proxy mobile IP
PoC	Push to talk over cellular
POLQA	Perceptual objective listening quality assessment
PPR	Privacy profile register
PRACH	Physical random access channel
PRACK	Provisional response acknowledgement
PRB	Physical resource block
P-RNTI	Paging radio network temporary identifier
ProSe	Proximity services
PS	Packet switched
PSAP	Public safety answering point
PSS	Primary synchronization signal
PSTN	Public switched telephone network
P-TMSI	Packet temporary mobile subscriber identity
PUCCH	Physical uplink control channel

PUSCH	Physical uplink shared channel
PWS	Public warning system
Q	Quadrature
QAM	Quadrature amplitude modulation
QCI	QoS class identifier
QoS	Quality of service
QPSK	Quadrature phase shift keying
RA	Routing area
RACH	Random access channel
RADIUS	Remote authentication dial in user service
RANAP	Radio access network application part
RA-RNTI	Random access radio network temporary identifier
RB	Resource block
RBG	Resource block group
RCS	Rich communication services
RE	Resource element
REG	Resource element group
RF	Radio frequency/Rating function
RFC	Request for comments
RI	Rank indication
RIM	Radio access network information management
RLC	Radio link control
RLF	Radio link failure
RN	Relay node
RNC	Radio network controller
RNTI	Radio network temporary identifier
ROHC	Robust header compression
R-PDCCH	Relay physical downlink control channel
RRC	Radio resource control
RRH	Remote radio head
RS	Reference signal
RSCP	Received signal code power
RSRP	Reference signal received power
RSRQ	Reference signal received quality
RSSI	Received signal strength indicator
RTCP	RTP control protocol
RTP	Real time transport protocol
S1-AP	S1 application protocol
SAE	System architecture evolution
SaMOG	S2a mobility based on GTP
SAR	Server assignment request
SC	Service centre
SCC-AS	Service centralization and continuity application server
SCell	Secondary cell
SC-FDMA	Single-carrier frequency division multiple access

SCS	Service capability server
S-CSCF	Serving call session control function
SCTP	Stream control transmission protocol
SDF	Service data flow
SDP	Session description protocol
SDU	Service data unit
SEG	Secure gateway
SeNB	Slave evolved Node B
SFN	System frame number
SGsAP	SGs application protocol
SGSN	Serving GPRS support node
S-GW	Serving gateway
SIB	System information block
SID	Silence information descriptor
SIM	Subscriber identity module
SINR	Signal-to-interference plus noise ratio
SIP	Session initiation protocol
SIPTO	Selective IP traffic offload
SI-RNTI	System information radio network temporary identifier
SLF	Subscription locator function
SM	Session management
SMS	Short message service
SMS-GMSC	SMS gateway MSC
SMS-IWMSC	SMS interworking MSC
SMTP	Simple mail transfer protocol
SNR	Subscribe notifications request
SOAP	Simple object access protocol
SON	Self-optimizing network/Self organizing network
SPR	Subscription profile repository
SPS	Semi persistent scheduling
SPT	Service point trigger
SR	Scheduling request
SRB	Signalling radio bearer
SRS	Sounding reference signal
SRVCC	Single radio voice call continuity
SS	Supplementary service
SS7	Signalling system 7
SSID	Service set identifier
SSS	Secondary synchronization signal
S-TMSI	S temporary mobile subscriber identity
STN-SR	Session transfer number single radio
SU-MIMO	Single-user MIMO
SVD	Singular value decomposition
TA	Timing advance/Tracking area
TAC	Tracking area code
TAI	Tracking area identity

TCP	Transmission control protocol
TDD	Time division duplex
TDMA	Time division multiple access
TD-SCDMA	Time division synchronous code division multiple access
TE	Terminal equipment
TEID	Tunnel endpoint identifier
TETRA	Terrestrial Trunked Radio
TFT	Traffic flow template
THIG	Topology hiding inter network gateway
TM	Transparent mode
TMSI	Temporary mobile subscriber identity
TPC	Transmit power control
TR	Technical report
TrGW	Transition gateway
TS	Technical specification
TTA	Telecommunications Technology Association
TTC	Telecommunication Technology Committee
TTI	Transmission time interval
UA	User agent
UAR	User authorization request
UCI	Uplink control information
UDP	User datagram protocol
UDR	User data repository/User data request
UE	User equipment
UE-AMBR	Per UE aggregate maximum bit rate
UICC	Universal integrated circuit card
UL	Uplink
UL-SCH	Uplink shared channel
UM	Unacknowledged mode
UMB	Ultra Mobile Broadband
UMTS	Universal Mobile Telecommunication System
URI	Uniform resource identifier
USIM	Universal subscriber identity module
USSD	Unstructured supplementary service data
UTDOA	Uplink time difference of arrival
UTRAN	UMTS terrestrial radio access network
VANC	VoLGA access network controller
VLR	Visitor location register
VoIP	Voice over IP
VoLGA	Voice over LTE via generic access
VoLTE	Voice over LTE
V-PCRF	Visited policy and charging rules function
VRB	Virtual resource block
vSRVCC	Single radio video call continuity
WCDMA	Wideband code division multiple access

WCS	Wireless Communications Service
WiMAX	Worldwide Interoperability for Microwave Access
WINNER	Wireless World Initiative New Radio
X2-AP	X2 application protocol
XCAP	XML configuration access protocol
XML	Extensible markup language
ZUC	Zu Chongzhi

1

Introduction

Our first chapter puts LTE into its historical context, and lays out its requirements and key technical features. We begin by reviewing the architectures of UMTS and GSM, and by introducing some of the terminology that the two systems use. We then summarize the history of mobile telecommunication systems, discuss the issues that have driven the development of LTE and show how UMTS has evolved first into LTE and then into an enhanced version known as LTE-Advanced. The chapter closes by reviewing the standardization process for LTE.

1.1 Architectural Review of UMTS and GSM

1.1.1 High-Level Architecture

LTE was designed by a collaboration of national and regional telecommunications standards bodies known as the *Third Generation Partnership Project* (3GPP) [1] and is known in full as 3GPP *Long-Term Evolution*. LTE evolved from an earlier 3GPP system known as the *Universal Mobile Telecommunication System* (UMTS), which in turn evolved from the *Global System for Mobile Communications* (GSM). To put LTE into context, we will begin by reviewing the architectures of UMTS and GSM, and by introducing some of the important terminology.

A mobile phone network is officially known as a *public land mobile network* (PLMN), and is run by a *network operator* such as Vodafone or Verizon. UMTS and GSM share a common network architecture, which is shown in Figure 1.1. There are three main components, namely the core network, the radio access network and the mobile phone.

The *core network* contains two domains. The *circuit switched* (CS) domain transports phone calls across the geographical region that the network operator is covering, in the same way as a traditional fixed-line telecommunication system. It communicates with the *public switched telephone network* (PSTN) so that users can make calls to land lines and with the circuit switched domains of other network operators. The *packet switched* (PS) domain transports data streams, such as web pages and emails, between the user and external *packet data networks* (PDNs) such as the internet.

The two domains transport their information in very different ways. The CS domain uses a technique known as *circuit switching*, in which it sets aside a dedicated two-way connection

---

An Introduction to LTE: LTE, LTE-Advanced, SAE, VoLTE and 4G Mobile Communications, Second Edition.
Christopher Cox.

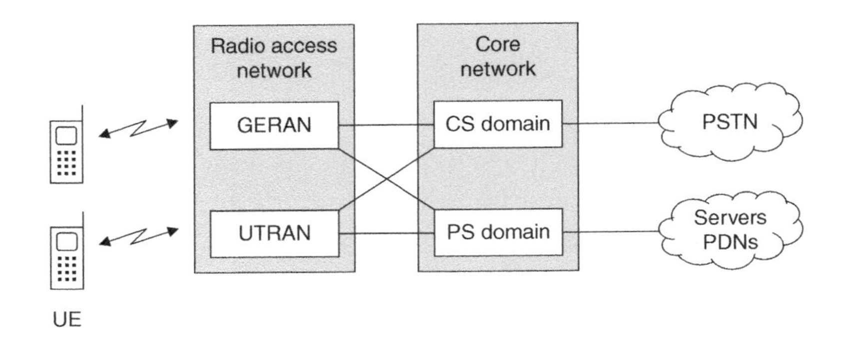

Figure 1.1 High-level architecture of UMTS and GSM

for each individual phone call so that it can transport the information with a constant data rate and minimal delay. This technique is effective, but is rather inefficient: the connection has enough capacity to handle the worst-case scenario in which both users are speaking at the same time, but is usually over-dimensioned. Furthermore, it is inappropriate for data transfers, in which the data rate can vary widely.

To deal with the problem, the PS domain uses a different technique, known as *packet switching*. In this technique, a data stream is divided into packets, each of which is labelled with the address of the required destination device. Within the network, *routers* read the address labels of the incoming data packets and forward them towards the corresponding destinations. The network's resources are shared amongst all the users, so the technique is more efficient than circuit switching. However, delays can result if too many devices try to transmit at the same time, a situation that is familiar from the operation of the internet.

The *radio access network* handles the core network's radio communications with the user. In Figure 1.1, there are actually two separate radio access networks, namely the *GSM EDGE radio access network* (GERAN) and the *UMTS terrestrial radio access network* (UTRAN). These use the different radio communication techniques of GSM and UMTS, but share a common core network between them.

The user's device is known officially as the *user equipment* (UE) and colloquially as the *mobile*. It communicates with the radio access network over the *air interface*, also known as the *radio interface*. The direction from network to mobile is known as the *downlink* (DL) or *forward link* and the direction from mobile to network is known as the *uplink* (UL) or *reverse link*.

A mobile can work outside the coverage area of its network operator by using the resources from two public land mobile networks: the *visited network*, where the mobile is located and the operator's *home network*. This situation is known as *roaming*.

1.1.2 Architecture of the Radio Access Network

Figure 1.2 shows the radio access network of UMTS. The most important component is the *base station*, which in UMTS is officially known as the *Node B*. Each base station has one or more sets of antennas, through which it communicates with the mobiles in one or more *sectors*. As shown in the diagram, a typical base station uses three sets of antennas to

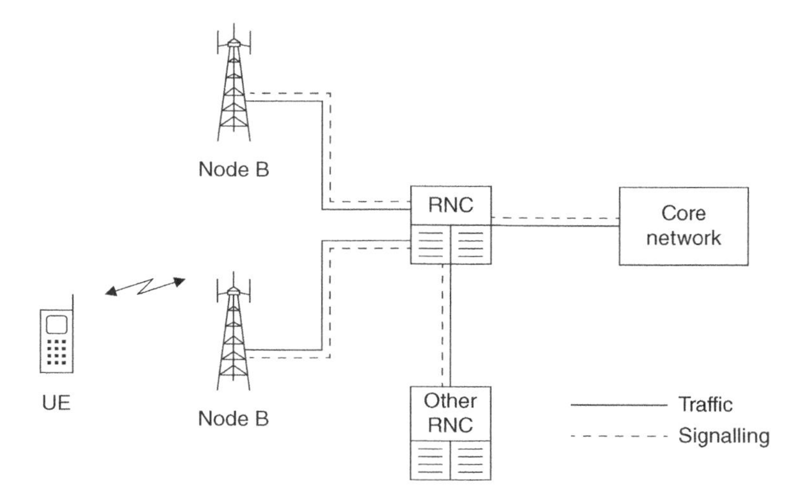

Figure 1.2 Architecture of the UMTS terrestrial radio access network

control three sectors, each of which spans an arc of 120°. In a medium-sized country like the United Kingdom, a typical mobile phone network might contain several thousand base stations altogether.

The word *cell* can be used in two different ways [2]. In Europe, a cell is usually the same thing as a sector, but in the United States, it usually means the group of sectors that a single base station controls. We will stick with the European convention throughout this book, so that the words cell and sector mean the same thing.

Each cell has a limited size, which is determined by the maximum range at which the receiver can successfully hear the transmitter. It also has a limited capacity, which is the maximum combined data rate of all the mobiles in the cell. These limits lead to the existence of several types of cell. *Macrocells* provide wide-area coverage in rural areas or suburbs and have a size of a few kilometres. *Microcells* have a size of a few hundred metres and provide a greater collective capacity that is suitable for densely populated urban areas. *Picocells* are used in large indoor environments such as offices or shopping centres and are a few tens of metres across. Finally, subscribers can buy *home base stations* to install in their own homes. These control *femtocells*, which are a few metres across.

Looking more closely at the air interface, each mobile and base station transmits on a certain radio frequency, which is known as the *carrier frequency*. Around that carrier frequency, it occupies a certain amount of frequency spectrum, known as the *bandwidth*. For example, a mobile might transmit with a carrier frequency of 1960 MHz and a bandwidth of 10 MHz, in which case its transmissions would occupy a frequency range from 1955 to 1965 MHz.

The air interface has to segregate the base stations' transmissions from those of the mobiles, to ensure that they do not interfere. UMTS can do this in two ways. When using *frequency division duplex* (FDD), the base stations transmit on one carrier frequency and the mobiles on another. When using *time division duplex* (TDD), the base stations and mobiles transmit on the same carrier frequency, but at different times. The air interface also has to segregate the different base stations and mobiles from each other. We will see the techniques that it uses in Chapters 3 and 4.

When a mobile moves from one part of the network to another, it has to stop communicating with one cell and start communicating with the next cell along. This process can be carried out using two different techniques, namely *handover* for mobiles that are actively communicating with the network and *cell reselection* for mobiles that are on standby. In UMTS, an active mobile can actually communicate with more than one cell at a time, in a state known as *soft handover*.

The base stations are grouped together by devices known as *radio network controllers* (RNCs). These have two main tasks. Firstly, they pass the user's voice information and data packets between the base stations and the core network. Secondly, they control a mobile's radio communications by means of signalling messages that are invisible to the user, for example by telling a mobile to hand over from one cell to another. A typical network might contain a few tens of radio network controllers, each of which controls a few hundred base stations.

The GSM radio access network has a similar design, although the base station is known as a *base transceiver station* (BTS) and the controller is known as a *base station controller* (BSC). If a mobile supports both GSM and UMTS, then the network can hand it over between the two radio access networks, in a process known as an *inter-system handover*. This can be invaluable if a mobile moves outside the coverage area of UMTS, and into a region that is covered by GSM alone.

In Figure 1.2, we have shown the user's traffic in solid lines and the network's signalling messages in dashed lines. We will stick with this convention throughout the book.

1.1.3 Architecture of the Core Network

Figure 1.3 shows the internal architecture of the core network. In the circuit switched domain, *media gateways* (MGWs) route phone calls from one part of the network to another, while *mobile switching centre* (MSC) *servers* handle the signalling messages that set up, manage

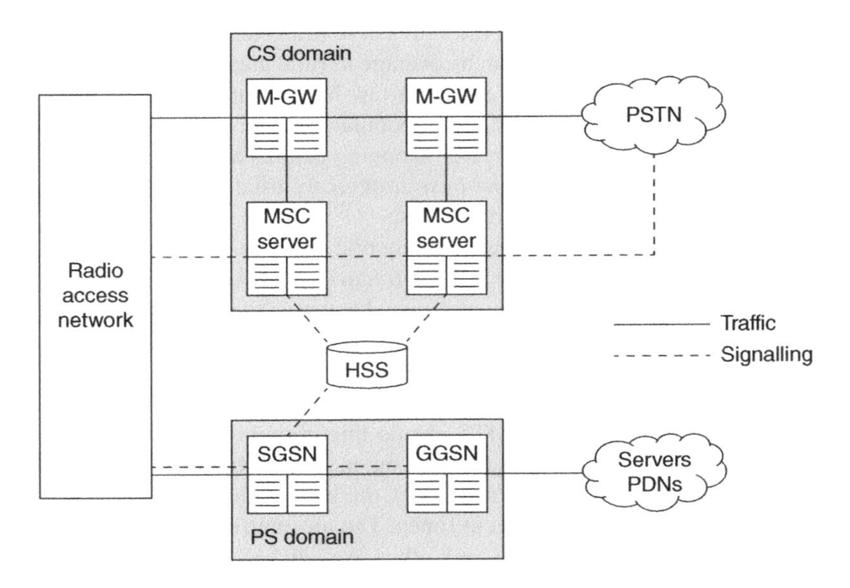

Figure 1.3 Architecture of the core networks of UMTS and GSM

and tear down the phone calls. They respectively handle the traffic and signalling functions of two earlier devices, known as the mobile switching centre and the *visitor location register* (VLR). A typical network might just contain a few of each device.

In the packet switched domain, *gateway GPRS support nodes* (GGSNs) act as interfaces to servers and packet data networks in the outside world. *Serving GPRS support nodes* (SGSNs) route data between the base stations and the GGSNs, and handle the signalling messages that set up, manage and tear down the data streams. Once again, a typical network might just contain a few of each device.

The *home subscriber server* (HSS) is a central database that contains information about all the network operator's subscribers and is shared between the two network domains. It amalgamates the functions of two earlier components, which were known as the *home location register* (HLR) and the *authentication centre* (AuC).

1.1.4 Communication Protocols

In common with other communication systems, UMTS and GSM transfer information using hardware and software *protocols*. The best way to illustrate these is actually through the protocols used by the internet. These protocols are designed by the *Internet Engineering Task Force* (IETF) and are grouped into various numbered *layers*, each of which handles one aspect of the transmission and reception process. The usual grouping follows a seven layer model known as the *Open Systems Interconnection* (OSI) model.

As an example (see Figure 1.4), let us suppose that a web server is sending information to a user's browser. In the first step, an *application layer* protocol, in this case the *hypertext transfer protocol* (HTTP), receives information from the server's application software, and passes it to the next layer down by representing it in a way that the user's application layer will eventually be able to understand. Other application layer protocols include the *simple mail transfer protocol* (SMTP) and the *file transfer protocol* (FTP).

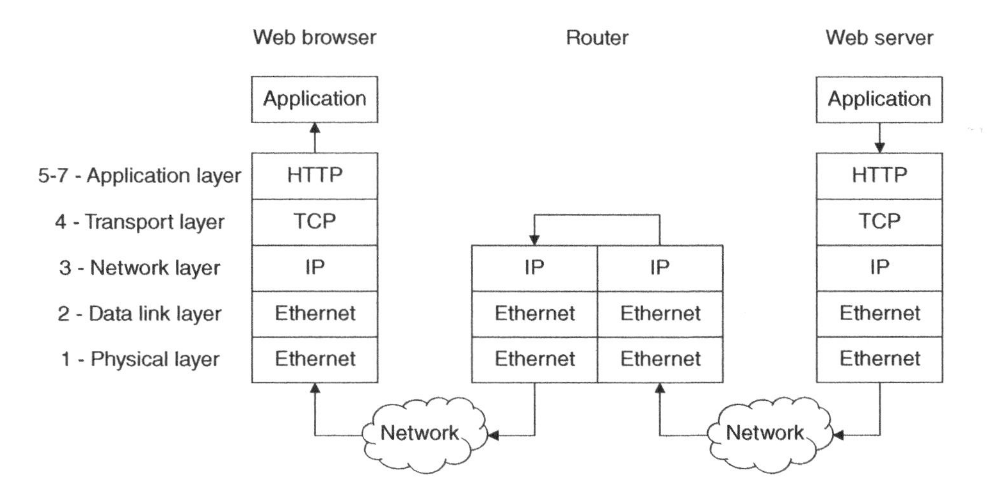

Figure 1.4 Examples of the communication protocols used by the internet, showing their mapping onto the layers of the OSI model

The *transport layer* manages the end-to-end data transmission. There are two main protocols. The *transmission control protocol* (TCP) re-transmits a packet from end to end if it does not arrive correctly, and is suitable for data such as web pages and emails that have to be received reliably. The *user datagram protocol* (UDP) sends the packet without any re-transmission and is suitable for data such as real time voice or video for which timely arrival is more important.

In the *network layer*, the *internet protocol* (IP) sends packets on the correct route from source to destination, using the IP address of the destination device. The process is handled by the intervening routers, which inspect the destination IP addresses by implementing just the lowest three layers of the protocol stack. The *data link layer* manages the transmission of packets from one device to the next, for example by re-transmitting a packet across a single interface if it does not arrive correctly. Finally, the *physical layer* deals with the actual transmission details; for example, by setting the voltage of the transmitted signal. The internet can use any suitable protocols for the data link and physical layers, such as *Ethernet*.

At each level of the transmitter's stack, a protocol receives a data packet from the protocol above in the form of a *service data unit* (SDU). It processes the packet, adds a header to describe the processing it has carried out, and outputs the result as a *protocol data unit* (PDU). This immediately becomes the incoming service data unit of the next protocol down. The process continues until the packet reaches the bottom of the protocol stack, at which point it is transmitted. The receiver reverses the process, using the headers to help it undo the effect of the transmitter's processing.

This technique is used throughout the radio access and core networks of UMTS and GSM. We will not consider their protocols in any detail at this stage; instead, we will go straight to the protocols used by LTE as part of Chapter 2.

1.2 History of Mobile Telecommunication Systems

1.2.1 From 1G to 3G

Mobile telecommunication systems were first introduced in the early 1980s. The *first generation* (1G) systems used analogue communication techniques, which were similar to those used by a traditional analogue radio. The individual cells were large and the systems did not use the available radio spectrum efficiently, so their capacity was by today's standards very small. The mobile devices were large and expensive and were marketed almost exclusively at business users.

Mobile telecommunications took off as a consumer product with the introduction of *second generation* (2G) systems in the early 1990s. These systems were the first to use digital technology, which permitted a more efficient use of the radio spectrum and the introduction of smaller, cheaper devices. They were originally designed just for voice, but were later enhanced to support instant messaging through the *Short Message Service* (SMS). The most popular 2G system was the Global System for Mobile Communications (GSM), which was originally designed as a pan-European technology, but which later became popular throughout the world. Also notable was *IS-95*, otherwise known as *cdmaOne*, which was designed by Qualcomm, and which became the dominant 2G system in the United States.

The success of 2G communication systems came at the same time as the early growth of the internet. It was natural for network operators to bring the two concepts together, by allowing users to download data onto mobile devices. To do this, so-called 2.5G systems built on the original ideas from 2G, by introducing the core network's packet switched domain and by modifying the air interface so that it could handle data as well as voice. The *General Packet Radio Service* (GPRS) incorporated these techniques into GSM, while IS-95 was developed into a system known as *IS-95B*.

At the same time, the data rates available over the internet were progressively increasing. To mirror this, designers first improved the performance of 2G systems using techniques such as *Enhanced Data Rates for GSM Evolution* (EDGE) and then introduced more powerful *third generation* (3G) systems in the years after 2000. 3G systems use different techniques for radio transmission and reception from their 2G predecessors, which increases the peak data rates that they can handle and which makes still more efficient use of the available radio spectrum.

Unfortunately, early 3G systems were excessively hyped and their performance did not at first live up to expectations. Because of this, 3G only took off properly after the introduction of 3.5G systems around 2005. In these systems, the air interface includes extra optimizations that are targeted at data applications, which increase the average rate at which a user can upload or download information, at the expense of introducing greater variability into the data rate and the arrival time.

1.2.2 Third Generation Systems

The world's dominant 3G system is the Universal Mobile Telecommunication System (UMTS). UMTS was developed from GSM by completely changing the technology used on the air interface, while keeping the core network almost unchanged. The system was later enhanced for data applications, by introducing the 3.5G technologies of *high-speed downlink packet access* (HSDPA) and *high-speed uplink packet access* (HSUPA), which are collectively known as *high-speed packet access* (HSPA).

The UMTS air interface has two slightly different implementations. *Wideband code division multiple access* (WCDMA) is the version that was originally specified, and the one that is currently used through most of the world. *Time division synchronous code division multiple access* (TD-SCDMA) is a derivative of WCDMA, which is also known as the low chip rate option of UMTS TDD mode. TD-SCDMA was developed in China, to minimize the country's dependence on Western technology and on royalty payments to Western companies. It is deployed by one of China's three 3G operators, China Mobile.

There are two main technical differences between these implementations. Firstly, WCDMA usually segregates the base stations' and mobiles' transmissions by means of frequency division duplex, while TD-SCDMA uses time division duplex. Secondly, WCDMA uses a wide bandwidth of 5 MHz, while TD-SCDMA uses a smaller value of 1.6 MHz.

cdma2000 was developed from IS-95 and is mainly used in North America. The original 3G technology was known as cdma2000 *1x radio transmission technology* (1xRTT). It was subsequently enhanced to a 3.5G system with two alternative names, cdma2000 *high-rate packet data* (HRPD) or *evolution data optimized* (EV-DO), which uses similar techniques to high-speed packet access. The specifications for IS-95 and cdma2000 are produced by a similar collaboration to 3GPP, which is known as the *Third Generation Partnership Project 2* (3GPP2) [3].

There are three main technical differences between the air interfaces of cdma2000 and UMTS. Firstly, cdma2000 uses a bandwidth of 1.25 MHz. Secondly, cdma2000 is backwards compatible with IS-95, in the sense that IS-95 mobiles can communicate with cdma2000 base stations and vice-versa, whereas UMTS is not backwards compatible with GSM. Thirdly, cdma2000 segregates voice and optimized data onto different carrier frequencies, whereas UMTS allows them to share the same one. The first two issues hindered the penetration of WCDMA into the North American market, where there were few allocations of bandwidths as wide as 5 MHz and there were a large number of legacy IS-95 devices.

The final 3G technology is *Worldwide Interoperability for Microwave Access* (WiMAX). This was developed by the *Institute of Electrical and Electronics Engineers* under IEEE standard 802.16 and has a very different history from other 3G systems. The original specification (IEEE 802.16–2001) was for a system that delivered data over point-to-point microwave links instead of fixed cables. A later revision, known as *fixed WiMAX* (IEEE 802.16–2004), supported point-to-multipoint communications between an omni-directional base station and a number of fixed devices. A further amendment, known as *mobile WiMAX* (IEEE 802.16e), allowed the devices to move and to hand over their communications from one base station to another. Once these capabilities were all in place, WiMAX started to look like any other 3G communication system, albeit one that had been optimized for data from the very beginning.

1.3 The Need for LTE

1.3.1 The Growth of Mobile Data

For many years, voice calls dominated the traffic in mobile telecommunication networks. The growth of mobile data was initially slow, but in the years leading up to 2010 its use started to increase dramatically. To illustrate this, Figure 1.5 shows measurements by Ericsson of the total traffic being handled by networks throughout the world, in petabytes (million gigabytes) per month [4]. The figure covers the period from 2007 to 2013, during which time the amount of data traffic increased by a factor of over 500.

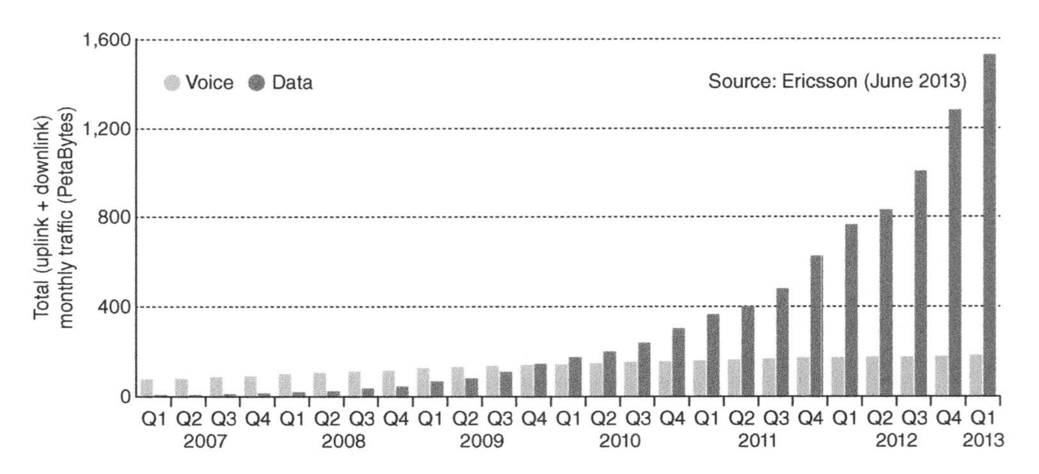

Figure 1.5 Measurements of voice and data traffic in worldwide mobile telecommunication networks, in the period from 2007 to 2013. Source: Ericsson mobility report, June 2013

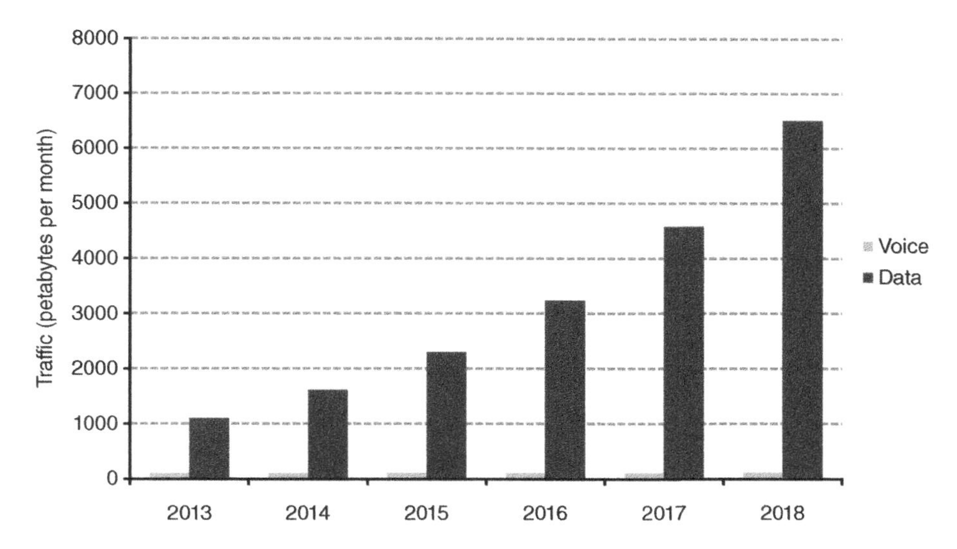

Figure 1.6 Forecasts of voice and data traffic in worldwide mobile telecommunication networks, in the period from 2013 to 2018. Data supplied by Analysys Mason

This trend is set to continue. For example, Figure 1.6 shows forecasts by Analysys Mason of the growth of mobile traffic in the period from 2013 to 2018. Note the difference in the vertical scales of the two diagrams.

In part, this growth was driven by the increased availability of 3.5G communication technologies. More important, however, was the introduction of the Apple iPhone in 2007, followed by devices based on Google's Android operating system from 2008. These smartphones were more attractive and user-friendly than their predecessors and were designed to support the creation of applications by third party developers. The result was an explosion in the number and use of mobile applications, which is reflected in the diagrams. As a contributory factor, network operators had previously tried to encourage the growth of mobile data by the introduction of flat rate charging schemes that permitted unlimited data downloads. That led to a situation where neither developers nor users were motivated to limit their data consumption.

As a result of these issues, 2G and 3G networks started to become congested in the years around 2010, leading to a requirement to increase network capacity. In the next section, we review the limits on the capacity of a mobile communication system and show how such capacity growth can be achieved.

1.3.2 Capacity of a Mobile Telecommunication System

In 1948, Claude Shannon discovered a theoretical limit on the data rate that can be achieved from any communication system [5]. We will write it in its simplest form, as follows:

$$C = B\log_2(1 + \text{SINR}) \tag{1.1}$$

Here, SINR is the *signal-to-interference plus noise ratio*, in other words the power at the receiver due to the required signal, divided by the power due to noise and interference. B is the bandwidth of the communication system in Hz, and C is the *channel capacity* in bits s^{-1}. It is theoretically possible for a communication system to send data from a transmitter to a receiver without any errors at all, provided that the data rate is less than the channel capacity. In a mobile communication system, C is the maximum data rate that one cell can handle and equals the combined data rate of all the mobiles in the cell.

The results are shown in Figure 1.7, using bandwidths of 5, 10 and 20 MHz. The vertical axis shows the channel capacity in million bits per second (Mbps), while the horizontal axis shows the signal-to-interference plus noise ratio in decibels (dB):

$$\mathrm{SINR(dB)} = 10\log_{10}(\mathrm{SINR}) \tag{1.2}$$

1.3.3 Increasing the System Capacity

There are three main ways to increase the capacity of a mobile communication system, which we can understand by inspection of Equation 1.1 and Figure 1.7. The first, and the most important, is the use of smaller cells. In a cellular network, the channel capacity is the maximum data rate that a single cell can handle. By building extra base stations and reducing the size of each cell, we can increase the capacity of a network, essentially by using many duplicate copies of Equation 1.1.

The second technique is to increase the bandwidth. Radio spectrum is managed by the *International Telecommunication Union* (ITU) and by regional and national regulators, and the increasing use of mobile telecommunications has led to the increasing allocation of spectrum to 2G and 3G systems. However, there is only a finite amount of radio spectrum available and it is also required by applications as diverse as military communications and radio astronomy. There are therefore limits as to how far this process can go.

The third technique is to improve the communication technology that we are using. This brings several benefits: it lets us approach ever closer to the theoretical channel capacity,

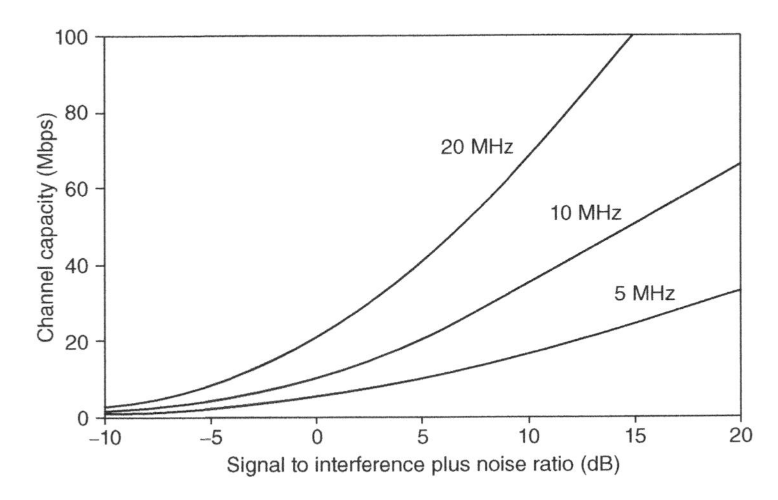

Figure 1.7 Shannon capacity of a communication system, in bandwidths of 5, 10 and 20 MHz

and it lets us exploit the higher SINR and greater bandwidth that are made available by the other changes above. This progressive improvement in communication technology has been an ongoing theme in the development of mobile telecommunications and is the main reason for the introduction of LTE.

1.3.4 Additional Motivations

Three other issues have driven the move to LTE. Firstly, a 2G or 3G operator has to maintain two core networks: the circuit switched domain for voice, and the packet switched domain for data. Provided that the network is not too congested, however, it is also possible to transport voice calls over packet switched networks using techniques such as *voice over IP* (VoIP). By doing this, operators can move everything to the packet switched domain, and can reduce both their capital and operational expenditure.

In a related issue, 3G networks introduce delays of the order of 100 ms for data applications, in transferring data packets between network elements and across the air interface. This is barely acceptable for voice and causes great difficulties for more demanding applications such as real-time interactive games. Thus a second driver is the wish to reduce the end-to-end delay, or *latency*, in the network.

Thirdly, the specifications for UMTS and GSM have become increasingly complex over the years, due to the need to add new features to the system while maintaining backwards compatibility with earlier devices. A fresh start aids the task of the designers, by letting them improve the performance of the system without the need to support legacy devices.

1.4 From UMTS to LTE

1.4.1 High-Level Architecture of LTE

In 2004, 3GPP began a study into the long term evolution of UMTS. The aim was to keep 3GPP's mobile communication systems competitive over timescales of 10 years and beyond, by delivering the high data rates and low latencies that future users would require. Figure 1.8 shows the resulting architecture and the way in which that architecture developed from that of UMTS.

In the new architecture, the *evolved packet core* (EPC) is a direct replacement for the packet switched domain of UMTS and GSM. There is no equivalent to the circuit switched domain, which allows LTE to be optimized for the delivery of data traffic, but implies that voice calls have to be handled using other techniques that are introduced below. The *evolved UMTS terrestrial radio access network* (E-UTRAN) handles the EPC's radio communications with the mobile, so is a direct replacement for the UTRAN. The mobile is still known as the user equipment, though its internal operation is very different from before.

The new architecture was designed as part of two 3GPP work items, namely *system architecture evolution* (SAE), which covered the core network, and *long-term evolution* (LTE), which covered the radio access network, air interface and mobile. Officially, the whole system is known as the *evolved packet system* (EPS), while the acronym LTE refers only to the evolution of the air interface. Despite this official usage, LTE has become a colloquial name for the whole system, and is regularly used in this way by 3GPP. We will use LTE in this colloquial way throughout the book.

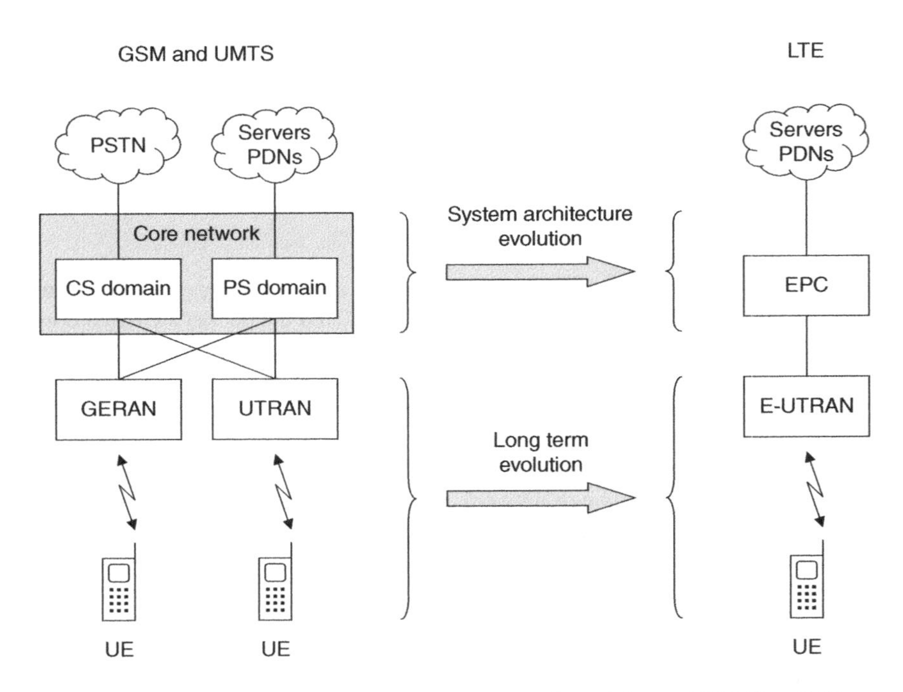

Figure 1.8 Evolution of the system architecture from GSM and UMTS to LTE

1.4.2 Long-Term Evolution

The main output of the study into long-term evolution was a requirements specification for the air interface [6], in which the most important requirements were as follows.

LTE was required to deliver a peak data rate of 100 Mbps in the downlink and 50 Mbps in the uplink. This requirement was exceeded in the eventual system, which delivers peak data rates of 300 Mbps and 75 Mbps respectively. For comparison, the peak data rate of WCDMA, in Release 6 of the 3GPP specifications, is 14 Mbps in the downlink and 5.7 Mbps in the uplink. (We will discuss the different specification releases at the end of the chapter.)

It cannot be stressed too strongly, however, that these peak data rates can only be reached in idealized conditions, and are wholly unachievable in any realistic scenario. A better measure is the *spectral efficiency*, which expresses the typical capacity of one cell per unit bandwidth. LTE was required to support a spectral efficiency three to four times greater than that of Release 6 WCDMA in the downlink and two to three times greater in the uplink.

Latency is another important issue, particularly for time-critical applications such as voice and interactive games. There are two aspects to this. Firstly, the requirements state that the time taken for data to travel between the mobile phone and the fixed network should be less than 5 ms, provided that the air interface is uncongested. Secondly, we will see in Chapter 2 that mobile phones can operate in two states: an active state in which they are communicating with the network and a low-power standby state. The requirements state that a phone should switch from standby to the active state, after an intervention from the user, in less than 100 ms.

There are also requirements on coverage and mobility. LTE is optimized for cell sizes up to 5 km, works with degraded performance up to 30 km and supports cell sizes of up to 100 km. It is also optimized for mobile speeds up to 15 km h^{-1}, works with high performance up to 120 km h^{-1} and supports speeds of up to 350 km h^{-1}. Finally, LTE is designed to work with a variety of different bandwidths, which range from 1.4 MHz up to a maximum of 20 MHz.

The requirements specification ultimately led to a detailed design for the LTE air interface, which we will cover in Chapters 3–10. For the benefit of those familiar with other systems, Table 1.1 summarizes its key technical features, and compares them with those of WCDMA.

1.4.3 System Architecture Evolution

The main output of the study into system architecture evolution was a requirements specification for the fixed network [7], in which the most important requirements were as follows.

The evolved packet core routes packets using the Internet Protocol (IP) and supports devices that are using IP version 4, IP version 6 or dual stack IP version 4/version 6. In addition, the EPC provides users with always-on connectivity to the outside world, by setting up a basic IP connection for a device when it switches on and maintaining that connection until it switches off. This is different from the behaviour of UMTS and GSM, in which the network only sets up an IP connection on request and tears that connection down when it is no longer required.

Unlike the internet, the EPC contains mechanisms to specify and control the data rate, error rate and delay that a data stream will receive. There is no explicit requirement on the maximum time required for data to travel across the EPC, but the relevant specification suggests a user plane latency of 10 ms for a non-roaming mobile, increasing to 50 ms in a typical roaming scenario [8]. To calculate the total delay, we have to add the earlier figure for the delay across the air interface, giving a typical delay in a non-roaming scenario of around 20 ms.

The EPC is also required to support inter-system handovers between LTE and earlier 2G and 3G technologies. These cover not only UMTS and GSM, but also non-3GPP systems such as cdma2000 and WiMAX.

Table 1.1 Key features of the air interfaces of WCDMA and LTE

Feature	WCDMA	LTE	Chapter
Multiple access scheme	WCDMA	OFDMA and SC-FDMA	4
Frequency re-use	100%	Flexible	4
Use of MIMO antennas	From Release 7	Yes	5
Bandwidth	5 MHz	1.4, 3, 5, 10, 15 or 20 MHz	6
Frame duration	10 ms	10 ms	6
Transmission time interval	2 or 10 ms	1 ms	6
Modes of operation	FDD and TDD	FDD and TDD	6
Uplink timing advance	Not required	Required	6
Transport channels	Dedicated and shared	Shared	6
Uplink power control	Fast	Slow	8

Table 1.2 Key features of the radio access networks of UMTS and LTE

Feature	UMTS	LTE	Chapter
Radio access network components	Node B, RNC	eNB	2
RRC protocol states	CELL_DCH, CELL_FACH, CELL_PCH, URA_PCH, RRC_IDLE	RRC_CONNECTED, RRC_IDLE	2
Handovers	Soft and hard	Hard	14
Neighbour lists	Always required	Not required	14

Table 1.3 Key features of the core networks of UMTS and LTE

Feature	UMTS	LTE	Chapter
IP version support	IPv4 and IPv6	IPv4 and IPv6	2
USIM version support	Release 99 USIM onwards	Release 99 USIM onwards	2
Transport mechanisms	Circuit & packet switching	Packet switching	2
CS domain components	MSC server, MGW	n/a	2
PS domain components	SGSN, GGSN	MME, S-GW, P-GW	2
IP connectivity	After registration	During registration	11
Voice and SMS applications	Included	External	21, 22

Tables 1.2 and 1.3 summarize the key features of the radio access network and the evolved packet core, and compare them with the corresponding features of UMTS. We will cover the architectural aspects of the fixed network in Chapter 2 and the operational aspects in Chapters 11–17.

1.4.4 LTE Voice Calls

The evolved packet core is designed as a data pipe that simply transports information to and from the user; it is not concerned with the information content or with the application. This is similar to the behaviour of the internet, which transports packets that originate from any application software, but is different from that of a traditional circuit switched network in which the voice application is an integral part of the system.

Because of this issue, voice applications do not form an integral part of LTE. However, an LTE mobile can still make a voice call using two main techniques. The first is *circuit switched fallback*, in which the network transfers the mobile to a legacy 2G or 3G cell so that the mobile can contact the 2G/3G circuit switched domain. The second is by using the *IP multimedia subsystem* (IMS), an external network that includes the signalling functions needed to set up, manage and tear down a voice over IP call. We will discuss these two techniques in Chapters 21 and 22.

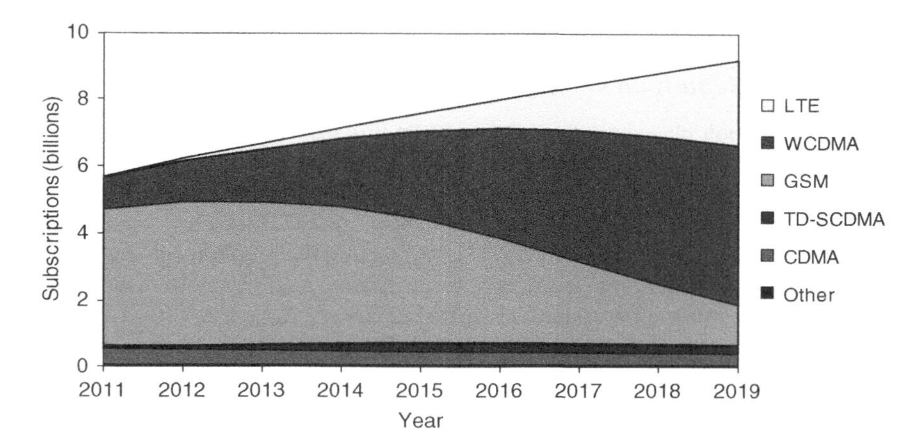

Figure 1.9 Numbers of subscriptions to different mobile communication technologies, with historical data up to 2013 and forecasts thereafter. Source: www.ericsson.com/TET

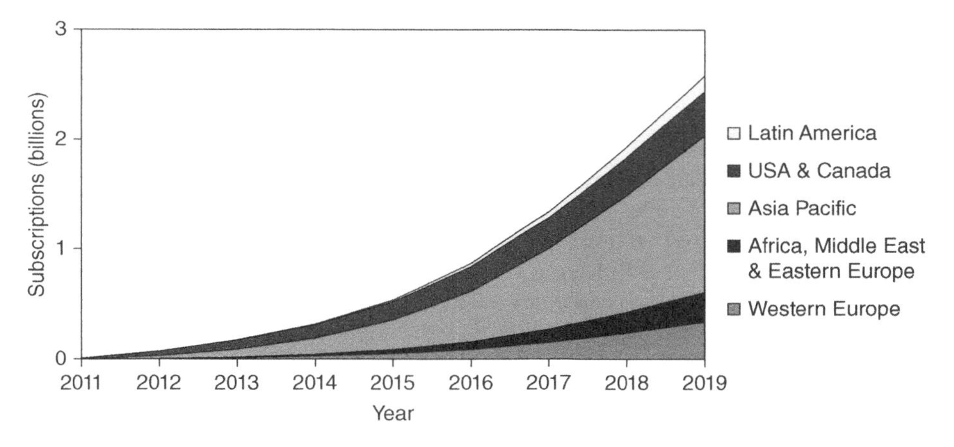

Figure 1.10 Number of subscriptions to LTE in different regions of the world, with historical data up to 2013 and forecasts thereafter. Source: www.ericsson.com/TET

1.4.5 The Growth of LTE

The first LTE networks were launched in Norway and Sweden at the end of 2009. To illustrate the subsequent growth of LTE, Figure 1.9 shows the total number of subscriptions to the most important mobile communication technologies in the period from 2011 to 2019, while Figure 1.10 shows the regional breakdown for the case of LTE. The diagrams are constructed using data published by Ericsson [9] and show historical data up to 2013 and forecasts thereafter. By the end of 2019, LTE is forecast to grow to around 2.5 billion subscribers worldwide.

1.5 From LTE to LTE-Advanced

1.5.1 The ITU Requirements for 4G

The design of LTE took place at the same time as an initiative by the International Telecommunication Union. In the late 1990s, the ITU had helped to drive the development of 3G technologies by publishing a set of requirements for a 3G mobile communication system, under the name *International Mobile Telecommunications* (IMT) *2000*. The 3G systems noted earlier are the main ones currently accepted by the ITU as meeting the requirements for IMT-2000.

The ITU launched a similar process in 2008, by publishing a set of requirements for a *fourth generation* (4G) communication system under the name *IMT-Advanced* [10–12]. According to these requirements, the peak data rate of a compatible system should be at least 600 Mbps on the downlink and 270 Mbps on the uplink, in a bandwidth of 40 MHz. We can see right away that these figures exceed the capabilities of LTE.

1.5.2 Requirements of LTE-Advanced

Driven by the ITU's requirements for IMT-Advanced, 3GPP started to study how to enhance the capabilities of LTE. The main output from the study was a specification for a system known as *LTE-Advanced* [13], in which the main requirements were as follows.

LTE-Advanced was required to deliver a peak data rate of 1000 Mbps in the downlink, and 500 Mbps in the uplink. In practice, the system has been designed so that it can eventually deliver peak data rates of 3000 and 1500 Mbps respectively, using a total bandwidth of 100 MHz that is made from five separate components of 20 MHz each. Note, as before, that these figures are unachievable in any realistic scenario.

The specification also includes targets for the spectrum efficiency in certain test scenarios. Comparison with the corresponding figures for WCDMA [14] implies a spectral efficiency 4.5–7 times greater than that of Release 6 WCDMA on the downlink, and 3.5–6 times greater on the uplink. Finally, LTE-Advanced is designed to be backwards compatible with LTE, in the sense that an LTE mobile can communicate with a base station that is operating LTE-Advanced and vice-versa.

1.5.3 4G Communication Systems

Following the submission and evaluation of proposals, the ITU announced in October 2010 that two systems met the requirements of IMT-Advanced [15]. One system was LTE-Advanced, while the other was an enhanced version of WiMAX under IEEE specification 802.16 m, known as mobile WiMAX 2.0.

Qualcomm had originally intended to develop a 4G successor to cdma2000 under the name *Ultra Mobile Broadband* (UMB). However, this system did not possess two of the advantages that its predecessor had done. Firstly, it was not backwards compatible with cdma2000, in the way that cdma2000 had been with IS-95. Secondly, it was no longer the only system that could operate in the narrow bandwidths that dominated North America, due to the flexible bandwidth support of LTE. Without any pressing reason to do so, no network operator ever announced plans to adopt the technology and the project was dropped in 2008. Instead, most cdma2000 operators decided to switch to LTE.

That left a situation where there were two remaining routes to 4G mobile communications: LTE and WiMAX. Of these, LTE has by far the greater support amongst network operators and equipment manufacturers, to the extent that several WiMAX operators have chosen to migrate their networks over to LTE. Because of this support, LTE is likely to be the world's dominant mobile communication technology for some years to come.

1.5.4 The Meaning of 4G

Originally, the ITU intended that the term 4G should only be used for systems that met the requirements of IMT-Advanced. LTE did not do so and neither did mobile WiMAX 1.0 (IEEE 802.16e). Because of this, the engineering community came to describe these systems as 3.9G. These considerations did not, however, stop the marketing community from describing LTE and mobile WiMAX 1.0 as 4G technologies. Although that description was unwarranted from a performance viewpoint, there was actually some sound logic to it: there is a clear technical transition in the move from UMTS to LTE, which does not exist in the move from LTE to LTE-Advanced.

It was not long before the ITU admitted defeat. In December 2010, the ITU gave its blessing to the use of 4G to describe not only LTE and mobile WiMAX 1.0, but also any other technology with substantially better performance than the early 3G systems [16]. They did not define the words 'substantially better', but that is not an issue for this book: we just need to know that LTE is a 4G mobile communication system.

1.6 The 3GPP Specifications for LTE

The specifications for LTE are produced by the Third Generation Partnership Project, in the same way as the specifications for UMTS and GSM. They are organized into *releases* [17], each of which contains a stable and clearly defined set of features. The use of releases allows equipment manufacturers to build devices using some or all of the features of earlier releases, while 3GPP continues to add new features to the system in a later release. Within each release, the specifications progress through a number of different versions. New functionality can be added to successive versions until the date when the release is frozen, after which the only changes involve refinement of the technical details, corrections and clarifications.

Table 1.4 lists the releases that 3GPP have used since the introduction of UMTS, together with the most important features of each release. Note that the numbering scheme was changed after Release 99, so that later releases are numbered from 4 through to 12.

LTE was first introduced in Release 8, which was frozen in December 2008. This release contains most of the important features of LTE and we will focus on it throughout the early chapters of the book. In specifying Release 8, however, 3GPP omitted some of the less important features of the system. These features were eventually included in Release 9, which we will cover in Chapter 18. Release 10 includes the extra capabilities that are required for LTE-Advanced and will be covered in Chapter 19, while the later enhancements in Releases 11 and 12 will be covered in Chapter 20. 3GPP has also continued to add new features to UMTS throughout Releases 8 to 12. This process allows network operators who stick with UMTS to remain competitive, even while other operators move over to LTE.

Table 1.4 3GPP specification releases for UMTS and LTE

Release	Date frozen	New features
R99	March 2000	WCDMA air interface
R4	March 2001	TD-SCDMA air interface
R5	June 2002	HSDPA, IP multimedia subsystem
R6	March 2005	HSUPA
R7	December 2007	Enhancements to HSPA
R8	December 2008	LTE, SAE
R9	December 2009	Enhancements to LTE and SAE
R10	June 2011	LTE-Advanced
R11	June 2013	Enhancements to LTE-Advanced
R12	September 2014	Enhancements to LTE-Advanced

The specifications are also organized into several *series*, each of which covers a particular component of the system. Table 1.5 summarizes the contents of series 21 to 37, which contain all the specifications for LTE and UMTS, as well as specifications that are common to LTE, UMTS and GSM. (Some other series numbers are used exclusively for GSM.) Within these series, the breakdown among the different systems varies widely. The 36 series is devoted to the techniques that are used for radio transmission and reception in LTE and is an important source of information for this book. In the other series, some specifications are applicable to UMTS alone, some to LTE alone and some to both, so it can be tricky to establish which

Table 1.5 3GPP specification series used by UMTS and LTE

Series	Scope
21	High-level requirements
22	Stage 1 service specifications
23	Stage 2 service and architecture specifications
24	Non-access stratum protocols
25	WCDMA and TD-SCDMA air interfaces and radio access network
26	Codecs
27	Data terminal equipment
28	Tandem free operation of speech codecs
29	Core network protocols
30	Programme management
31	UICC and USIM
32	Operations, administration, maintenance, provisioning and charging
33	Security
34	UE test specifications
35	Security algorithms
36	LTE air interface and radio access network
37	Multiple radio access technologies

specifications are the relevant ones. To help deal with this issue, the book contains references to all the important specifications that we will use.

When written out in full, an example specification number is TS 23.401 v 11.6.0. Here, TS stands for *technical specification*, 23 is the series number and 401 is the number of the specification within that series. 11 is the release number, 6 is the technical version number within that release and the final 0 is an editorial version number that is occasionally incremented for non-technical changes. 3GPP also produces *technical reports*, denoted TR, which are purely informative and have three-digit specification numbers beginning with an 8 or 9.

In a final division, each specification belongs to one of three *stages*. Stage 1 specifications define the service from the user's point of view and lie exclusively in the 22 series. Stage 2 specifications define the system's high-level architecture and operation, and lie mainly (but not exclusively) in the 23 series. Finally, stage 3 specifications define all the functional details. The stage 2 specifications are especially useful for achieving a high-level understanding of the system. The most useful ones for LTE are TS 23.401 [18] and TS 36.300 [19], which respectively cover the evolved packet core and the air interface. There is, however, an important note of caution: these specifications are superseded later on and cannot be relied upon for complete accuracy. Instead, the details should be checked if necessary in the relevant stage 3 specifications.

The individual specifications can be downloaded from 3GPP's specification numbering web page [20] or from their FTP server [21]. The 3GPP website also has summaries of the features that are covered by each individual release [22].

References

1. 3rd Generation Partnership Project (3GPP) (2013) www.3gpp.org (accessed 15 October 2013).
2. 4G Americas (2010) MIMO and Smart Antennas for 3G and 4G Wireless Systems, Section 2, May 2010.
3. 3rd Generation Partnership Project 2 (2012) www.3gpp2.org (accessed 15 October 2013).
4. Ericsson (2013) Ericsson Mobility Report, June 2013 www.ericsson.com/mobility-report (accessed 18 November 2013).
5. Shannon, C.E. (1948) A mathematical theory of communication. *The Bell System Technical Journal*, **27**, 379–428, and 623–656.
6. 3GPP TS 25.913 (2009) Requirements for Evolved UTRA (E-UTRA) and Evolved UTRAN (E-UTRAN), Release 9, December 2009.
7. 3GPP TS 22.278 (2012) Service Requirements for the Evolved Packet System (EPS), Release 11, September 2012.
8. 3GPP TS 23.203 (2013) Policy and Charging Control Architecture, Release 11, Section 6.1.7.2, September 2013.
9. Ericsson (2013) Traffic Exploration Tool, www.ericsson.com/TET (accessed 18 November 2013).
10. International Telecommunication Union (2008) Requirements, Evaluation Criteria and Submission Templates for the Development of IMT-Advanced. ITU report ITU-R M.2133.
11. International Telecommunication Union (2008) Requirements Related to Technical Performance for IMT-Advanced Radio Interface(s). ITU report ITU-R M.2134.
12. International Telecommunication Union (2008) Guidelines for Evaluation of Radio Interface Technologies for IMT-Advanced. ITU report ITU-R M.2135.
13. 3GPP TS 36.913 (2012) Requirements for Further Advancements for Evolved Universal Terrestrial Radio Access (E-UTRA) (LTE-Advanced), Release 11, September 2012.
14. 3GPP TS 25.912 (2012) Feasibility Study for Evolved Universal Terrestrial Radio Access (UTRA) and Universal Terrestrial Radio Access Network (UTRAN), Release 11, Section 13.5, September 2012.
15. International Telecommunication Union (2010) ITU Paves Way for Next-Generation 4G Mobile Technologies, www.itu.int/net/pressoffice/press_releases/2010/40.aspx (accessed 15 October 2013).

16. International Telecommunication Union (2010) ITU World Radiocommunication Seminar Highlights Future Communication Technologies, www.itu.int/net/pressoffice/press_releases/2010/48.aspx (accessed 15 October 2013).
17. 3rd Generation Partnership Project (2013) 3GPP – Releases, www.3gpp.org/releases (accessed 15 October 2013).
18. 3GPP TS 23.401 (2013) General Packet Radio Service (GPRS) Enhancements for Evolved Universal Terrestrial Radio Access Network (E-UTRAN) Access, Release 11, September 2013.
19. 3GPP TS 36.300 (2013) Evolved Universal Terrestrial Radio Access (E-UTRA) and Evolved Universal Terrestrial Radio Access Network (E-UTRAN); Overall Description; Stage 2, Release 11, September 2013.
20. 3rd Generation Partnership Project (2013) 3GPP – Specification Numbering, www.3gpp.org/specification-numbering (accessed 15 October 2013).
21. 3rd Generation Partnership Project (2013) FTP Directory, ftp://ftp.3gpp.org/specs/latest/ (accessed 15 October 2013).
22. 3rd Generation Partnership Project (2013) FTP Directory, ftp://ftp.3gpp.org/Information/WORK_PLAN/Description_Releases/ (accessed 15 October 2013).

2

System Architecture Evolution

This chapter covers the high-level architecture of LTE. We begin by describing the hardware components in an LTE network and by reviewing the software protocols that those components use to communicate. We then look in more detail at the techniques used for data transport in LTE before discussing the state diagrams and the use of radio spectrum. We will leave some more specialized architectural issues until later chapters, notably those related to quality of service, charging and inter-system operation.

Several specifications are relevant to this chapter. TS 23.401 [1] and TS 36.300 [2] are stage 2 specifications that include descriptions of the system architecture, while the relevant stage 3 specifications [3, 4] contain the architectural details. We will also note some other important specifications as we go along.

2.1 High-Level Architecture of LTE

Figure 2.1 reviews the high-level architecture of the evolved packet system (EPS). There are three main components, namely the user equipment (UE), the evolved UMTS terrestrial radio access network (E-UTRAN) and the evolved packet core (EPC). In turn, the evolved packet core communicates with packet data networks in the outside world such as the internet, private corporate networks or the IP multimedia subsystem. The interfaces between the different parts of the system are denoted Uu, S1 and SGi.

The UE, E-UTRAN and EPC each have their own internal architectures and we will now discuss these one by one.

2.2 User Equipment

2.2.1 Architecture of the UE

Figure 2.2 shows the internal architecture of the user equipment [5]. The architecture is identical to the one used by UMTS and GSM.

The actual communication device is known as the *mobile equipment* (ME). In the case of a voice mobile or a smartphone, this is just a single device. However, the mobile equipment

---

An Introduction to LTE: LTE, LTE-Advanced, SAE, VoLTE and 4G Mobile Communications, Second Edition.
Christopher Cox.

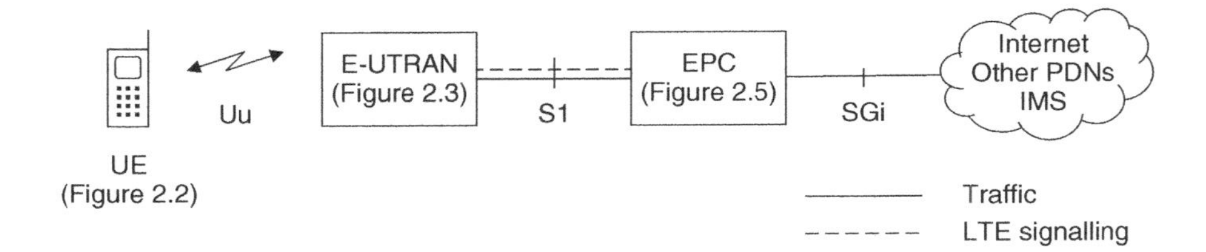

Figure 2.1 High-level architecture of LTE

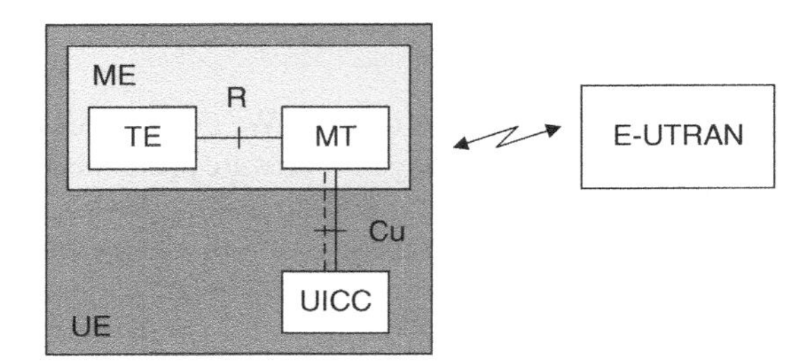

Figure 2.2 Internal architecture of the UE. Source: TS 27.001. Reproduced by permission of ETSI

can also be divided into two components, namely the *mobile termination* (MT), which handles all the communication functions, and the *terminal equipment* (TE), which terminates the data streams. The mobile termination might be a plug-in LTE card for a laptop, for example, in which case the terminal equipment would be the laptop itself.

The *universal integrated circuit card* (UICC) is a smart card, colloquially known as the SIM card. It runs an application known as the *universal subscriber identity module* (USIM) [6], which stores user-specific data such as the user's phone number and home network identity. Some of the data on the USIM can be downloaded from device management servers that are managed by the network operator: we will see some examples shortly. The USIM also carries out various security-related calculations, using secure keys that the smart card stores. LTE supports mobiles that are using a USIM from Release 99 or later, but it does not support the *subscriber identity module* (SIM) that was used by earlier releases of GSM.

In addition, LTE supports mobiles that are using IP version 4 (IPv4), IP version 6 (IPv6) or dual stack IP version 4/version 6. A mobile receives one IP address for every packet data network that it is communicating with; for example, one for the internet and one for any private corporate network. Alternatively, the mobile can receive an IPv4 address as well as an IPv6 address, if the mobile and network both support the two versions of the protocol.

2.2.2 UE Capabilities

Mobiles can have a wide variety of radio capabilities [7, 8]. These cover issues such as the maximum data rate that they can handle, the different types of radio access technology that they support, the carrier frequencies on which they can transmit and receive and the mobile's support for optional features of the LTE specifications. Mobiles pass these capabilities to the

Table 2.1 UE categories

UE category	Release	Maximum # DL bits per ms	Maximum # UL bits per ms	Maximum # DL layers	Maximum # UL layers	Support of UL 64-QAM?
1	R8	10 296	5 160	1	1	No
2	R8	51 024	25 456	2	1	No
3	R8	102 048	51 024	2	1	No
4	R8	150 752	51 024	2	1	No
5	R8	299 552	75 376	4	1	Yes
6	R10	301 504	51 024	4	1	No
7	R10	301 504	102 048	4	2	No
8	R10	2 998 560	1 497 760	8	4	Yes

Source: TS 36.306. Reproduced by permission of ETSI.

radio access network by means of signalling messages, so that the E-UTRAN knows how to control them correctly.

The most important capabilities are grouped together into the *UE category*. As shown in Table 2.1, the UE category mainly covers the maximum data rate with which the mobile can transmit and receive. It also covers some technical issues that are listed in the last three columns of the table, which we will cover in Chapters 3 and 5. Early LTE mobiles were mainly in category 3, so they had maximum data rates on the uplink and downlink of 100 and 50 Mbps respectively. A Release 8 or Release 9 mobile can only attain the maximum theoretical data rates of 300 and 75 Mbps if it lies in category 5.

2.3 Evolved UMTS Terrestrial Radio Access Network

2.3.1 Architecture of the E-UTRAN

The evolved UMTS terrestrial radio access network (E-UTRAN) [9] is illustrated in Figure 2.3. The E-UTRAN handles the radio communications between the mobile and the evolved packet core and just has one component, the *evolved Node B* (eNB).

Each eNB is a base station that controls the mobiles in one or more cells. A mobile communicates with just one base station and one cell at a time, so there is no equivalent of the soft handover state from UMTS. The base station that is communicating with a mobile is known as its *serving eNB*.

The eNB has two main functions. Firstly, the eNB sends radio transmissions to all its mobiles on the downlink and receives transmissions from them on the uplink, using the analogue and digital signal processing functions of the LTE air interface. Secondly, the eNB controls the low-level operation of all its mobiles, by sending them signalling messages such as handover commands that relate to those radio transmissions. In carrying out these functions, the eNB combines the earlier functions of the Node B and the radio network controller, to reduce the latency that arises when the mobile exchanges information with the network.

Each base station is connected to the EPC by means of the S1 interface. It can also be connected to nearby base stations by the X2 interface, which is mainly used for signalling and packet forwarding during handover.

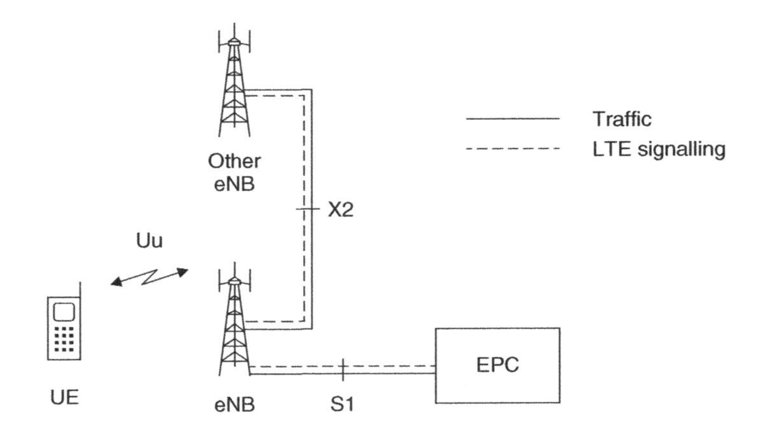

Figure 2.3 Architecture of the evolved UMTS terrestrial radio access network

The X2 interface is optional in two senses. Firstly, communications are only required between nearby base stations that might be involved in handovers, whereas distant base stations do not have to interact. Secondly, the most important X2 communications can also be carried through the evolved packet core using two instances of S1, albeit indirectly and more slowly. Even if it is used, the X2 interface does not have to be configured by hand; instead, a network can set up its X2 interfaces automatically using the self-optimization functions that we discuss in Chapter 17.

2.3.2 Transport Network

Usually, the S1 and X2 interfaces are not direct physical connections. Instead, the information is routed across an underlying IP transport network in the manner shown in Figure 2.4. Each base station and each component of the core network has an IP address, and the underlying

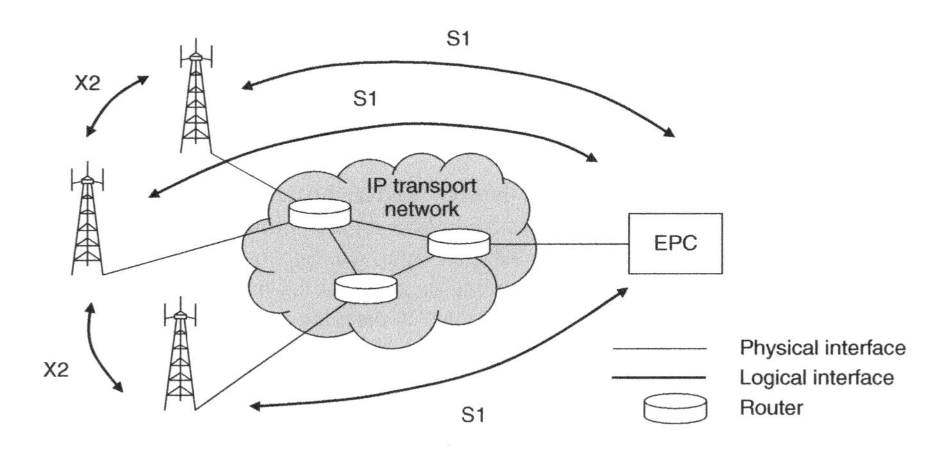

Figure 2.4 Internal architecture of the E-UTRAN transport network

routers use those IP addresses to transport data and signalling messages from one device to another. The S1 and X2 interfaces are better understood as logical relationships, through which the devices know about each other's identities and can exchange information. The evolved packet core uses an IP transport network in the same way, so the same issues will apply to the EPC interfaces that we introduce below.

2.3.3 Small Cells and the Home eNB

We saw in Chapter 1 that operators can greatly increase the capacity of their networks through the progressive use of smaller cells. The smallest example is the *home eNB* (HeNB) [10], which is a base station that a user has purchased to provide femtocell coverage within the home. Home eNBs benefit the user through the provision of better coverage and higher data rates, and also benefit the network operator by taking traffic away from the surrounding macrocells. Their main drawback is cost.

A home eNB belongs to a *closed subscriber group* (CSG), through which it can provide either exclusive or preferential access to mobiles that also belong to the closed subscriber group. A mobile's list of closed subscriber groups is stored by the USIM and can be downloaded from a device management server that is controlled by the network operator. Home eNBs have lower power limitations than normal base stations, can control only one cell and do not support the X2 interface until Release 10.

On the S1 interface, a home eNB can communicate with the evolved packet core either directly or through a device known as a *home eNB gateway* that shields the EPC from the potentially huge numbers of home eNBs. The S1 data and signalling messages are transported by the consumer's internet service provider rather than by the network operator, so they have to be secured more carefully than normal S1 communications.

2.4 Evolved Packet Core

2.4.1 Architecture of the EPC

Figure 2.5 shows the main components of the evolved packet core [11, 12]. We have already seen one component, the home subscriber server (HSS), which is a central database that contains information about all the network operator's subscribers. This is one of the few components of LTE that has been carried forward from UMTS and GSM.

The *packet data network gateway* (P-GW) is the EPC's point of contact with the outside world. Through the SGi interface, each PDN gateway exchanges data with one or more external devices or packet data networks, such as the network operator's servers, the internet or the IP multimedia subsystem. Each packet data network is identified by an *access point name* (APN) [13]. A network operator typically uses a handful of different APNs; for example, one for the internet and one for the IP multimedia subsystem.

Each mobile is assigned to a default PDN gateway when it first switches on, to give it always-on connectivity to a default packet data network such as the internet. Later on, a mobile may be assigned to one or more additional PDN gateways, if it wishes to connect to additional packet data networks such as private corporate networks or the IP multimedia subsystem. Each PDN gateway stays the same throughout the lifetime of the data connection.

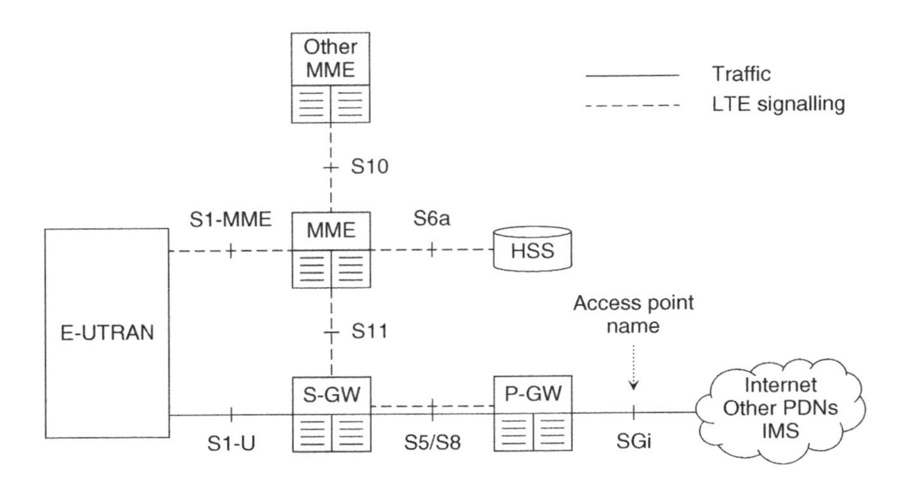

Figure 2.5 Main components of the evolved packet core

The *serving gateway* (S-GW) acts as a high-level router, and forwards data between the base station and the PDN gateway. A typical network might contain a handful of serving gateways, each of which looks after the mobiles in a certain geographical region. Each mobile is assigned to a single serving gateway, but the serving gateway can be changed if the mobile moves sufficiently far.

The *mobility management entity* (MME) controls the high-level operation of the mobile, by sending it signalling messages about issues such as security and the management of data streams that are unrelated to radio communications. As with the serving gateway, a typical network might contain a handful of MMEs, each of which looks after a certain geographical region. Each mobile is assigned to a single MME, which is known as its *serving MME*, but that can be changed if the mobile moves sufficiently far. The MME also controls the other elements of the network, by means of signalling messages that are internal to the EPC.

Comparison with UMTS and GSM shows that the PDN gateway has the same role as the gateway GPRS support node (GGSN), while the serving gateway and MME handle the data routing and signalling functions of the serving GPRS support node (SGSN). Splitting the SGSN in two makes it easier for an operator to scale the network in response to an increased load: the operator can add more serving gateways as the traffic increases, while adding more MMEs to handle an increase in the number of mobiles. To support this split, the S1 interface has two components: the S1-U interface carries traffic for the serving gateway, while the S1-MME interface carries signalling messages for the MME.

The EPC has some other components that were not shown in Figure 2.5. Firstly, the *cell broadcast centre* (CBC) was previously used by UMTS for the rarely implemented *cell broadcast service* (CBS). In LTE, the equipment is re-used for a service known as the *earthquake and tsunami warning system* (ETWS) [14]. Secondly, the *equipment identity register* (EIR) was also inherited from UMTS, and lists the details of lost or stolen mobiles. We will introduce further components later in the book, when we consider the management of quality of service, and the inter-operation between LTE and other mobile communication systems.

2.4.2 Roaming Architecture

Roaming allows users to move outside their network operators' coverage area by using the resources from two different networks. It relies on the existence of a *roaming agreement*, which defines how the operators will share the resulting revenue. The usual architecture is shown in Figure 2.6 [15].

If a user is roaming, then the home subscriber server is always in the home network, while the mobile, E-UTRAN, MME and serving gateway are always in the visited network. The PDN gateway, however, can be in two places. Communications with the internet generally use *home routed traffic*, in which the PDN gateway lies in the home network. By using this architecture, the home network operator can see all the traffic and can charge for it directly, so it only requires a basic roaming agreement with the visited network. The two networks exchange information using an inter-operator backbone known either as the *IP packet exchange* (IPX) or by its older name of *GPRS roaming exchange* (GRX) [16].

Communications with the IP multimedia subsystem generally use *local breakout*, in which the PDN gateway lies in the visited network. This has two important benefits for voice communications: a user can make a local voice call without the traffic travelling back to the home network and can make an emergency call that will be handled by the local emergency services. We will see in Chapter 22 that the IP multimedia subsystem is actually distributed between the home and visited networks, so the home network operator still has visibility of the signalling messages that control the voice call. The HSS indicates whether or not the home network will permit local breakout for each combination of user and APN [17].

The interface between the serving and PDN gateways is known as S5/S8. This has two slightly different implementations, namely S5 if the two devices are in the same network and S8 if they are in different networks, with the main distinction lying in the interfaces' security requirements. For mobiles that are not roaming, the serving and PDN gateways can be integrated into a single device, so that the S5/S8 interface vanishes altogether. This can be useful because of the associated reduction in latency.

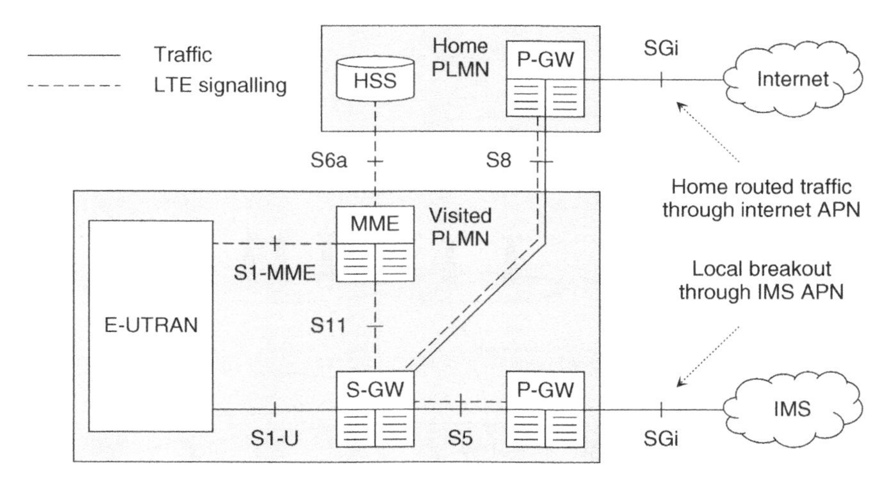

Figure 2.6 Usual architecture of LTE for a roaming mobile that is communicating with the Internet and the IP multimedia subsystem

2.4.3 Network Areas

The EPC is divided into three different types of geographical area [18], which are illustrated in Figure 2.7.

An *MME pool area* is an area through which the mobile can move without a change of serving MME. Every pool area is controlled by one or more MMEs, while every base station is connected to all the MMEs in a pool area by means of the S1-MME interface. Pool areas can also overlap. Typically, a network operator might configure a pool area to cover a large region of the network such as a major city and might add MMEs to the pool as the signalling load in that city increases.

Similarly, an *S-GW service area* is an area served by one or more serving gateways, through which the mobile can move without a change of serving gateway. Every base station is connected to all the serving gateways in a service area by means of the S1-U interface. S-GW service areas do not necessarily correspond to MME pool areas.

MME pool areas and S-GW service areas are both made from smaller, non-overlapping units known as *tracking areas* (TAs). These are used to track the locations of mobiles that are on standby and are similar to the location and routing areas from UMTS and GSM.

2.4.4 Numbering, Addressing and Identification

The components of the network are associated with several different identities [19]. As in previous systems, each network is associated with a *public land mobile network identity* (PLMN-ID). This comprises a three digit *mobile country code* (MCC) and a two or three digit *mobile network code* (MNC). For example, the mobile country code for the UK is 234, while Vodafone's UK network uses a mobile network code of 15.

Each MME has three main identities, which are shown as the shaded parts of Figure 2.8. Each MME pool area is identified using a 16 bit *MME group identity* (MMEGI), while the 8 bit *MME code* (MMEC) uniquely identifies the MME within a pool area. Combining them gives

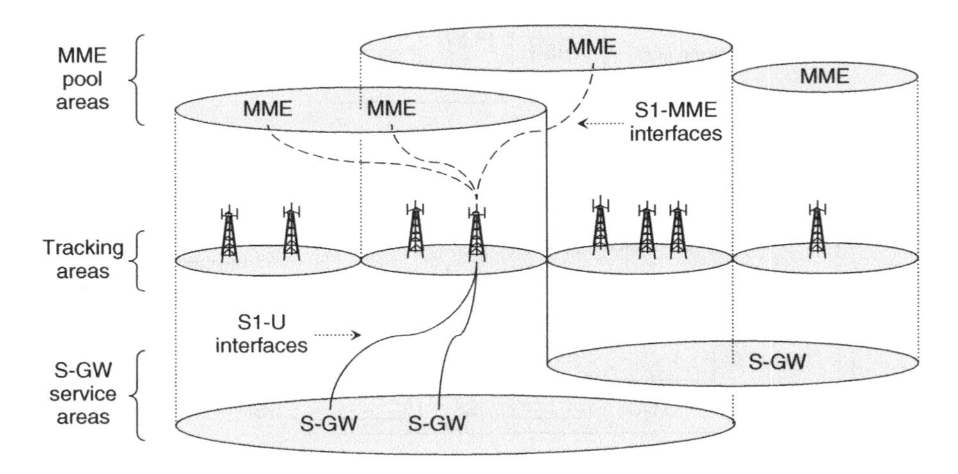

Figure 2.7 Relationship between tracking areas, MME pool areas and S-GW service areas

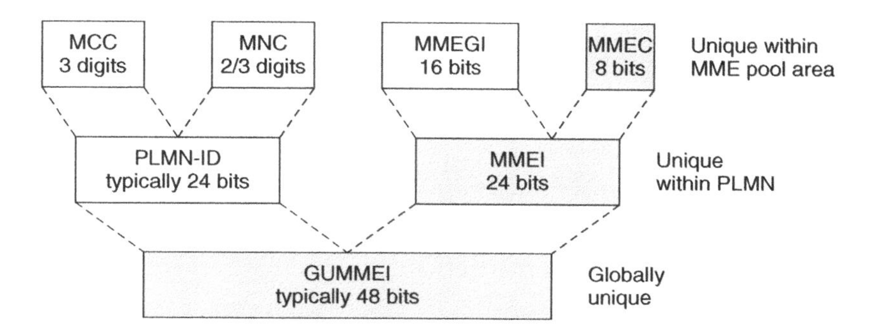

Figure 2.8 Identities used by the MME

the 24 bit *MME identifier* (MMEI), which uniquely identifies the MME within a particular network. By bringing in the network identity, we arrive at the *globally unique MME identifier* (GUMMEI), which identifies an MME anywhere in the world.

Similarly, each tracking area has two main identities. The 16 bit *tracking area code* (TAC) identifies a tracking area within a particular network. Combining this with the network identity gives the globally unique *tracking area identity* (TAI).

Cells have three types of identity. The 28 bit *E-UTRAN cell identity* (ECI) identifies a cell within a particular network, while the *E-UTRAN cell global identifier* (ECGI) identifies a cell anywhere in the world. Also important for the air interface is the *physical cell identity*, which is a number from 0 to 503 that distinguishes a cell from its immediate neighbours.

A mobile is also associated with several different identities. The most important are the *international mobile equipment identity* (IMEI), which is a unique identity for the mobile equipment, and the *international mobile subscriber identity* (IMSI), which is a unique identity for the UICC and the USIM.

The IMSI is one of the quantities that an intruder needs to clone a mobile, so we avoid transmitting it across the air interface wherever possible. Instead, a serving MME identifies each mobile using temporary identities, which it updates at regular intervals. Three types of temporary identity are important, and are shown as the shaded parts of Figure 2.9. The 32 bit *M temporary mobile subscriber identity* (M-TMSI) identifies a mobile to its serving MME.

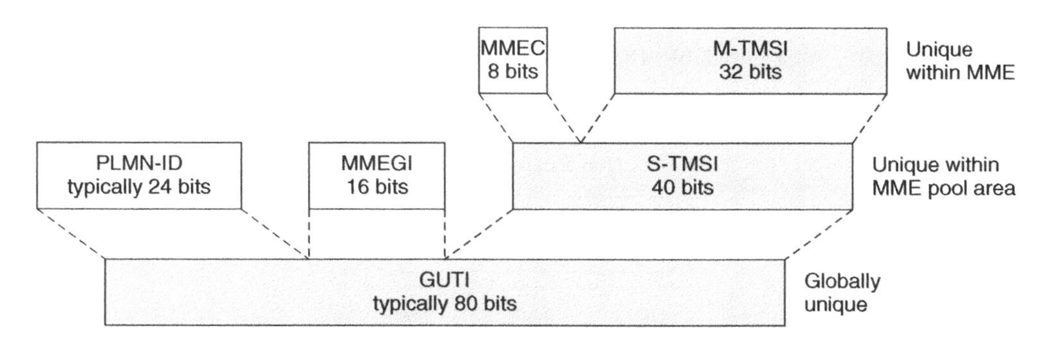

Figure 2.9 Temporary identities used by the mobile

Adding the MME code results in the 40 bit *S temporary mobile subscriber identity* (S-TMSI), which identifies the mobile within an MME pool area. Finally, adding the MME group identity and the PLMN identity results in the most important quantity, the *globally unique temporary identity* (GUTI).

2.5 Communication Protocols

2.5.1 Protocol Model

Each of the interfaces from the previous section is associated with a protocol stack, which the network elements use to exchange data and signalling messages. Figure 2.10 shows the high-level structure of those protocol stacks.

The protocol stack has two planes. Protocols in the *user plane* handle data that are of interest to the user, while protocols in the *control plane* handle signalling messages that are only of interest to the network elements themselves. The protocol stack also has two main layers. The upper layer manipulates information in a way that is specific to LTE, while the lower layer transports information from one point to another. There are no universal names for these layers, but in the E-UTRAN they are known as the *radio network layer* and *transport network layer* respectively.

There are then three types of protocol. *Signalling protocols* define a language by which two devices can exchange signalling messages with each other. *User plane protocols* manipulate the data in the user plane, most often to help route the data within the network. Finally, the underlying *transport protocols* transfer data and signalling messages from one point to another.

On the air interface, there is an extra level of complexity, which is shown in Figure 2.11 [20]. As noted earlier, the MME controls the high-level behaviour of the mobile by sending it signalling messages. However, there is no direct path between the MME and the mobile, through

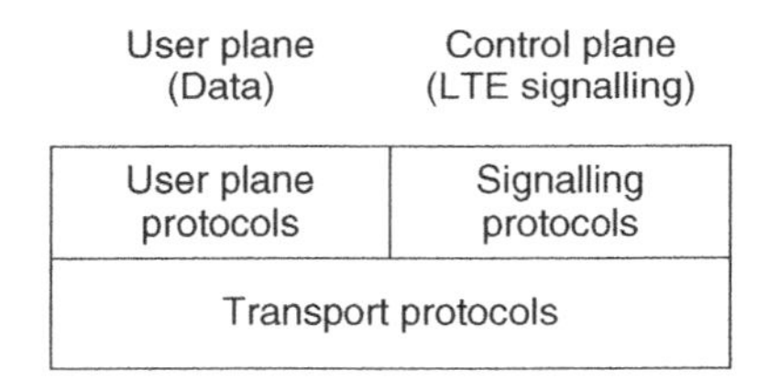

Figure 2.10 High-level protocol architecture of LTE

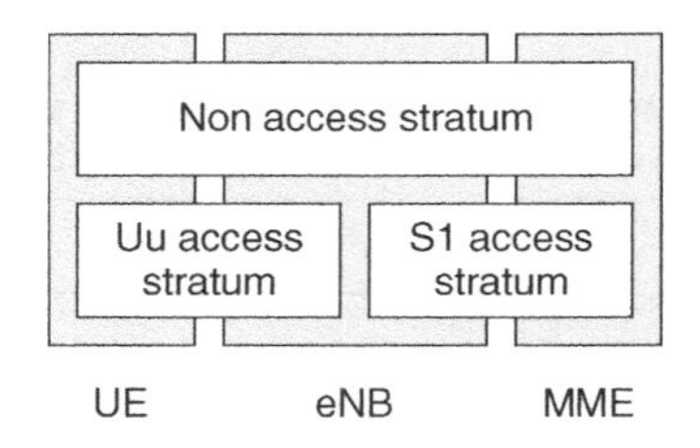

Figure 2.11 Relationship between the access stratum and the non-access stratum on the air interface

which those messages can be transported. To handle this, the air interface is divided into two levels, known as the *access stratum* (AS) and the *non-access stratum* (NAS). The high-level signalling messages lie in the non-access stratum and are transported using the access stratum protocols of the S1 and Uu interfaces.

2.5.2 Air Interface Transport Protocols

The air interface, officially known as the Uu interface, lies between the mobile and the base station. Figure 2.12 shows the air interface's transport protocols. Starting at the bottom, the *air interface physical layer* contains the digital and analogue signal processing functions that the mobile and base station use to send and receive information. The physical layer is described in several specifications that are listed in Chapter 6; the figure shows the most important.

The next three protocols make up the data link layer, layer 2 of the OSI model. The *medium access control* (MAC) protocol [21] carries out low-level control of the physical layer, particularly by scheduling data transmissions between the mobile and the base station. The *radio link control* (RLC) protocol [22] maintains the data link between the two devices, for example by ensuring reliable delivery for data streams that need to arrive correctly. Finally, the *packet data convergence protocol* (PDCP) [23] carries out higher-level transport functions that are related to header compression and security.

2.5.3 Fixed Network Transport Protocols

The interfaces in the fixed network use standard IETF transport protocols, which are shown in Figure 2.13. Each interface is routed across the underlying transport network, so it uses protocols from layers 1 to 4 of the OSI model. At the bottom of the stack, the transport network can use any suitable protocols for layers 1 and 2, such as Ethernet, often supported by another protocol known as *multiprotocol label switching* (MPLS) [24, 25].

Every network element is then associated with an IP address, and the transport network uses the internet protocol (IP) to route information from one network element to another. LTE supports both IP version 4 [26] and IP version 6 [27] for this task. In the evolved packet core,

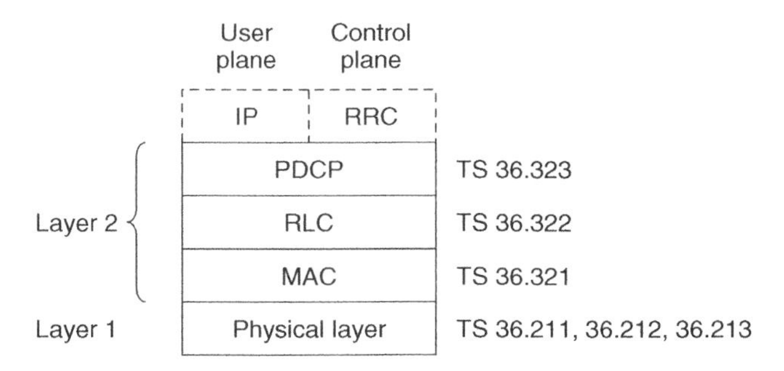

Figure 2.12 Transport protocols used on the air interface. Source: TS 36.300. Reproduced by permission of ETSI

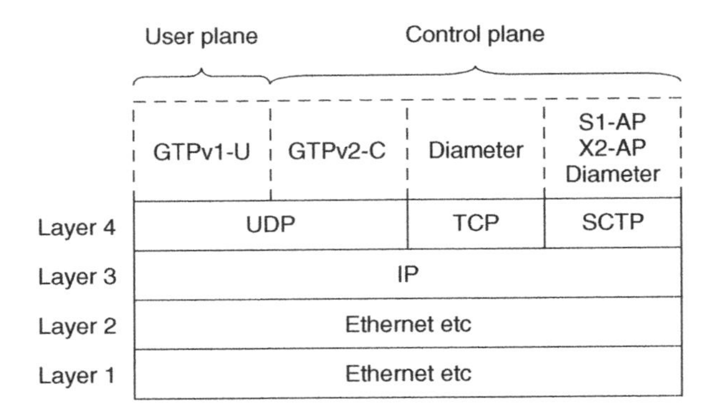

Figure 2.13 Transport protocols used by the fixed network

support of IP version 4 is mandatory and support of version 6 is recommended [28], while the radio access network can use either or both of the two protocols [29, 30].

Above IP, there is a transport layer protocol across the interface between each individual pair of network elements. Three transport protocols are used. The user datagram protocol (UDP) [31] just sends data packets from one network element to another, while the transmission control protocol (TCP) [32] re-transmits packets if they arrive incorrectly. The *stream control transmission protocol* (SCTP) [33] is based on TCP, but includes extra features that make it more suitable for the delivery of signalling messages. The user plane always uses UDP as its transport protocol, to avoid delaying the data. The control plane's choice depends on the overlying signalling protocol, in the manner shown.

2.5.4 User Plane Protocols

The LTE user plane contains mechanisms to forward data correctly between the mobile and the PDN gateway, and to respond quickly to changes in the mobile's location. These mechanisms are implemented by the user plane protocols shown in Figure 2.14. Most of the user plane

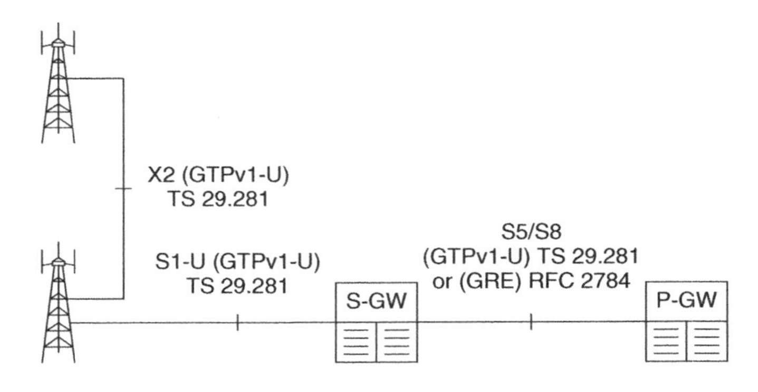

Figure 2.14 User plane protocols used by LTE

interfaces use a 3GPP protocol known as the *GPRS tunnelling protocol user part* (GTP-U) [34]. To be precise, LTE uses version 1 of the protocol, denoted GTPv1-U, along with the 2G and 3G packet switched domains from Release 99. Earlier 2G networks used version 0, which is denoted GTPv0-U. Between the serving gateway and the PDN gateway, the S5/S8 user plane has an alternative implementation. This is based on a standard IETF protocol known as *generic routing encapsulation* (GRE) [35].

GTP-U and GRE forward packets from one network element to another using a technique known as *tunnelling*. The two protocols implement tunnelling in slightly different ways, which we will cover as part of Chapter 13.

2.5.5 Signalling Protocols

LTE uses a large number of signalling protocols, which are shown in Figure 2.15.

On the air interface, the base station controls a mobile's radio communications by means of signalling messages that are written using the *radio resource control* (RRC) protocol [36]. In the radio access network, an MME controls the base stations within its pool area using the *S1 application protocol* (S1-AP) [37], while two base stations can communicate using the *X2 application protocol* (X2-AP) [38].

At the same time, the MME controls a mobile's high-level behaviour using two protocols that lie in the air interface's non-access stratum [39]. These protocols are *EPS session management* (ESM), which controls the data streams through which a mobile communicates with the outside world, and *EPS mobility management* (EMM), which handles internal bookkeeping within the EPC. The network transports EMM and ESM messages by embedding them into lower-level RRC and S1-AP messages and then by using the transport mechanisms of the Uu and S1 interfaces.

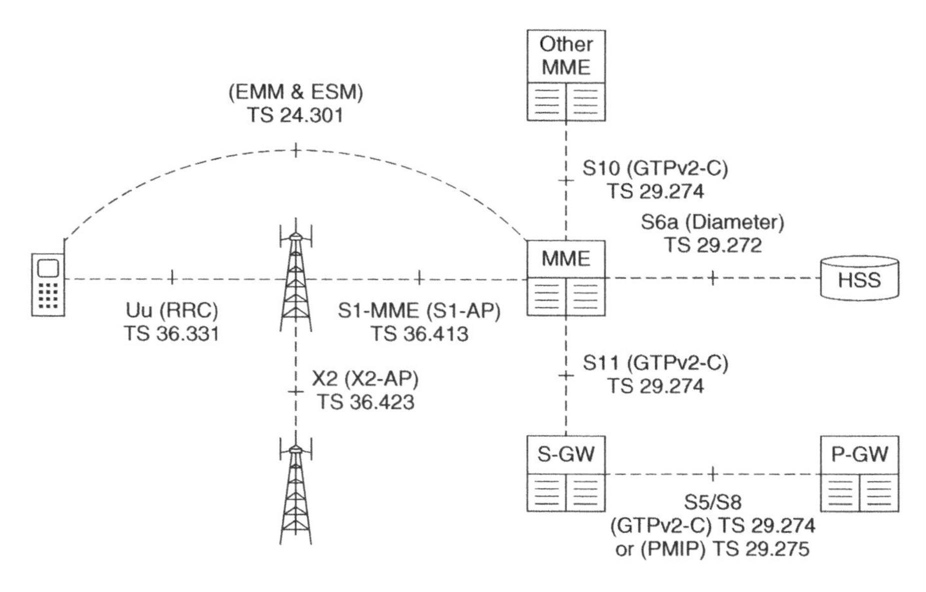

Figure 2.15 Signalling protocols used by LTE

Inside the EPC, the HSS and MME communicate using a protocol based on *Diameter*. The basic Diameter protocol [40] is a standard IETF protocol for authentication, authorization and accounting, which was based on an older protocol known as *Remote Authentication Dial In User Service* (RADIUS) [41, 42]. The basic Diameter protocol can be enhanced for use in specific applications: the implementation of Diameter on the S6a interface [43] is one such application.

Most of the other EPC interfaces use a 3GPP protocol known as the *GPRS tunnelling protocol control part* (GTP-C) [44]. This protocol includes procedures for peer-to-peer communications between the different elements of the EPC, and for managing the GTP-U tunnels that we introduced above. LTE uses version 2 of the protocol, which is denoted GTPv2-C. The 2G and 3G packet switched domains used version 1 of the protocol, GTPv1-C, from Release 99 onwards, while earlier 2G networks implemented GTPv0-C. If the S5/S8 user plane is using GRE, then its control plane uses a signalling protocol known as *proxy mobile IP version 6* (PMIPv6) [45, 46]. PMIPv6 is a standard IETF protocol for the management of packet forwarding, in support of mobile devices such as laptops.

The question then arises of which protocol option to choose on the S5/S8 interface. Operators of legacy 3GPP networks are likely to prefer GTP-U and GTP-C, for consistency with their previous systems and with the other signalling interfaces in the evolved packet core. TS 23.401 describes the system architecture and high level operation of LTE, under the assumption that the S5/S8 interface is using those protocols. We will generally make the same assumption in this book. Operators of non-3GPP networks may prefer GRE and PMIP, which are standard IETF protocols, and which are also used for inter-operation between LTE and non-3GPP technologies. TS 23.402 [47] is a companion specification to TS 23.401, which describes the differences in architecture and operation for a network that uses those protocols.

The signalling protocols introduced above are all binary rather than text-based; each message is a stream of ones and zeros rather than characters, so the messages are short but hard to read. Each message carries various parameters. The 3GPP protocols describe these as *information elements*, but Diameter refers to them as *attribute value pairs* (AVPs) because they combine two pieces of information, namely which parameter is being specified and what its value is.

2.6 Example Signalling Flows

2.6.1 Access Stratum Signalling

Now that we have introduced the network elements and protocol stacks, it is useful to show a few examples of how the different components fit together. Let us first consider an exchange of RRC signalling messages between the mobile and the base station. Figure 2.16 is the message sequence for an RRC procedure known as *UE Capability Transfer* [48]. Here, the serving eNB

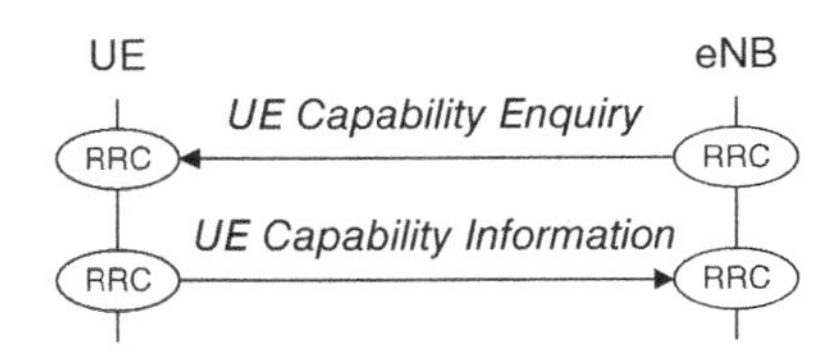

Figure 2.16 UE capability transfer procedure. Source: TS 36.331. Reproduced by permission of ETSI

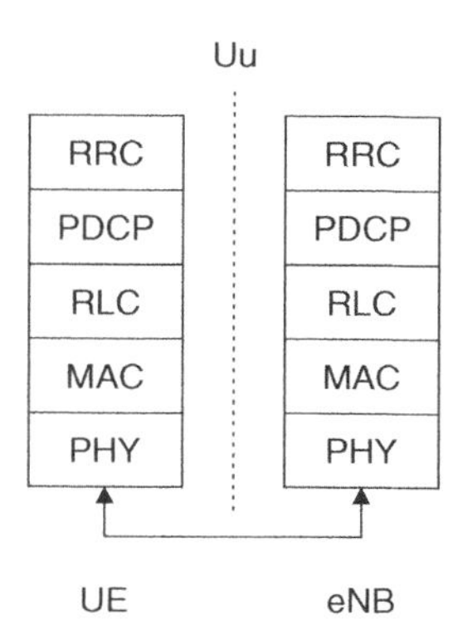

Figure 2.17 Protocol stacks used to exchange RRC signalling messages between the mobile and the base station. Source: TS 36.300. Reproduced by permission of ETSI

wishes to find out the mobile's radio access capabilities, such as the maximum data rate it can handle and the specification release that it conforms to. To do this, the RRC protocol composes a message called *UE Capability Enquiry*, and sends it to the mobile. The mobile responds with an RRC message called *UE Capability Information*, in which it lists the capabilities required.

The corresponding protocol stacks are shown in Figure 2.17. The base station composes its capability enquiry using the RRC protocol, processes it using the PDCP, RLC and MAC and transmits it using the air interface physical layer. The mobile receives the base station's transmission and processes the information by passing it through the same sequence of protocols in reverse. It then reads the enclosed message and composes its reply, which is transmitted and received in exactly the same way.

2.6.2 *Non-Access Stratum Signalling*

The next signalling example is slightly more complex. Figure 2.18a shows the message sequence for an EMM procedure known as a *GUTI reallocation* [49]. Using an EMM *GUTI Reallocation Command*, the MME can give the mobile a new globally unique temporary

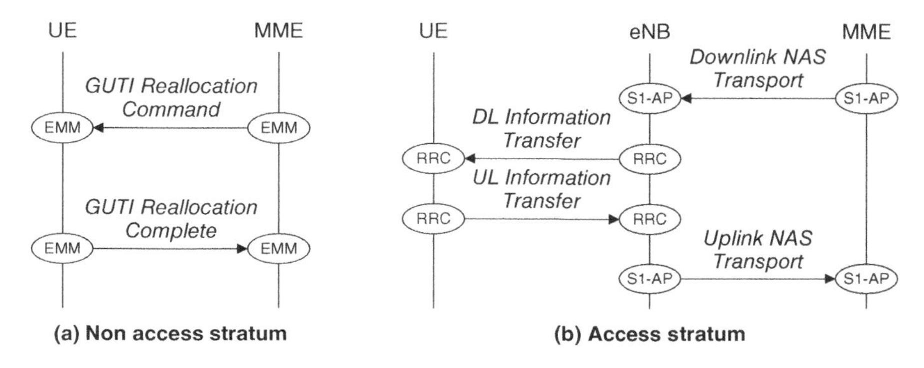

Figure 2.18 GUTI reallocation procedure. (a) Non-access stratum messages. (b) Message transport using the access stratum

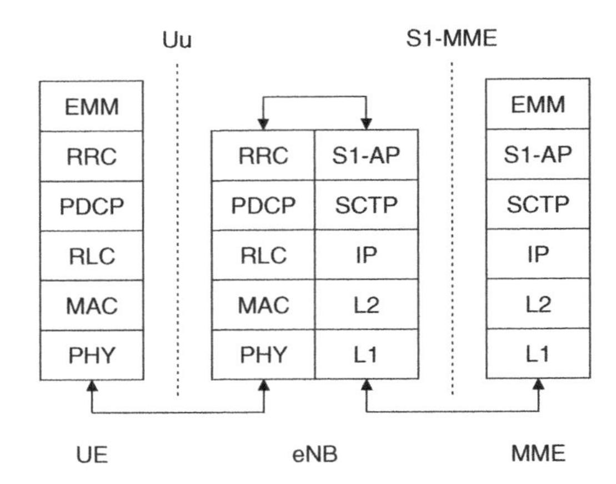

Figure 2.19 Protocol stacks used to exchange non-access stratum signalling messages between the mobile and the MME. Source: TS 23.401. Reproduced by permission of ETSI

identity. In response, the mobile sends the MME an acknowledgement using an EMM *GUTI Reallocation Complete*.

LTE transports these messages by embedding them into S1-AP and RRC messages, as shown in Figure 2.18b. The usual S1-AP messages are known as *Uplink NAS Transport* and *Downlink NAS Transport* [50], while the usual RRC messages are known as *UL Information Transfer* and *DL Information Transfer* [51]. Their sole function is to transport EMM and ESM messages like the ones shown here. However, the network can also transport non-access stratum messages by embedding them into other S1-AP and RRC messages, which can have additional access stratum functions of their own. We will see a few examples later in the book.

Figure 2.19 shows the protocol stacks for this message sequence. The MME writes the *GUTI Reallocation Command* using its EMM protocol, embeds it in the S1-AP *Downlink NAS Transport* message and sends it to the base station using the transport mechanisms of the S1 interface. The base station unwraps the EMM message, embeds it into an RRC *DL Information Transfer* and sends it to the mobile using the air interface protocols that we covered earlier. The mobile reads the message, updates its GUTI and sends an acknowledgement using the same protocol stacks in reverse.

2.7 Bearer Management

2.7.1 The EPS Bearer

LTE transports data packets using the same protocols that are used on the internet. However the transport mechanisms are more complex, because LTE has to address two issues that the internet does not support. The first of these is mobility. On the internet, a device stays connected to the same access point for a significant time and loses its connection to any external servers if the access point is changed. In LTE, a device expects to move from one base station to another, and expects to maintain its connection to external servers when it does so.

The second issue is *quality of service* (QoS), a term that describes the performance of a data stream using parameters such as the guaranteed data rate, maximum error rate and maximum delay. The internet does not offer any QoS guarantees so that, for example, the performance of a VoIP call in a congested network is likely to be poor. In contrast, LTE can offer QoS guarantees and can assign different qualities of service to different data streams and to different users. Depending on the network's exact configuration, for example, a premium LTE user might be able to pay a higher fee so as to guarantee higher quality VoIP calls.

To address these issues, LTE transports data from one part of the system to another using *EPS bearers* [52, 53]. An EPS bearer can be thought of as a bi-directional data pipe, which transfers data on the correct route through the network and with the correct quality of service. The bearer runs between the mobile and the PDN gateway if the S5/S8 interface is based on GTP or between the mobile and the serving gateway if the S5/S8 interface is based on PMIP.

2.7.2 *Default and Dedicated Bearers*

One important issue is the distinction between default and dedicated bearers. The EPC sets up one EPS bearer, known as a *default bearer*, whenever a mobile connects to a packet data network. As shown in Figure 2.20, a mobile receives one default bearer as soon as it registers with the EPC to provide it with always-on connectivity to a default packet data network such as the internet. At the same time, the mobile receives an IP address for it to use when communicating with that network or possibly the combination of an IPv4 address and an IPv6 address. Later on, the mobile can establish connections with other packet data networks, for example, private company networks or the IP multimedia subsystem. If it does so, then it receives an additional default bearer for every network that it connects to, together with an additional IP address.

After connecting to a packet data network and establishing a default bearer, a mobile can also receive one or more *dedicated bearers* that connect it to the same network. This does not lead to the allocation of any new IP addresses; instead, each dedicated bearer shares an IP address with its parent default bearer. A dedicated bearer does, however, have a different quality of

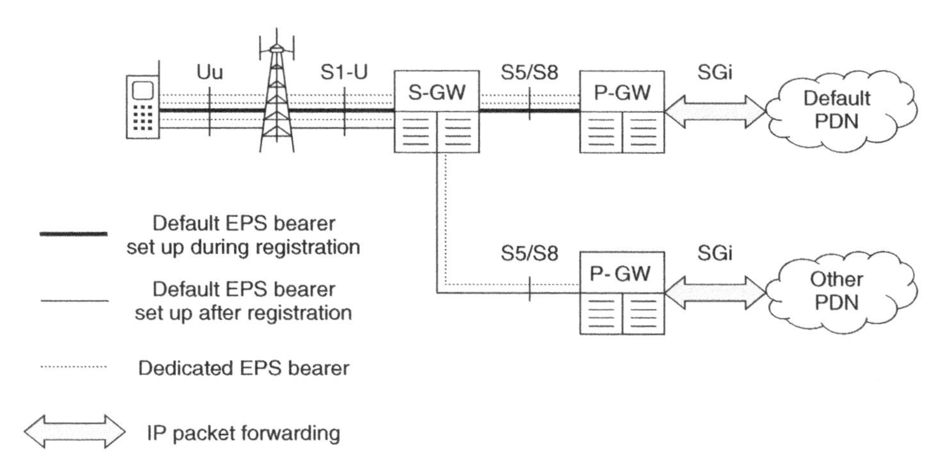

Figure 2.20 Default and dedicated EPS bearers, when using an S5/S8 interface based on GTP

service from the default bearer, such as a guarantee of the minimum long term average data rate. A mobile can have a maximum of 11 EPS bearers [54] to give it connectivity to several networks using several different qualities of service.

2.7.3 Bearer Implementation Using GTP

The EPS bearer spans three different interfaces in the case of a GTP based S5/S8, so it cannot be implemented directly. To deal with this problem (Figure 2.21), the EPS bearer is broken down into three lower-level bearers, namely the *radio bearer*, the *S1 bearer* and the *S5/S8 bearer*. Each of these is itself associated with a set of QoS parameters and receives a share of the EPS bearer's maximum error rate and maximum delay. The combination of a radio bearer and an S1 bearer is sometimes known as an *evolved radio access bearer* (E-RAB).

The radio bearer is then implemented by a suitable configuration of the air interface protocols, while the S1 and S5/S8 bearers are implemented using GTP-U *tunnels*. The resulting protocol stacks are shown in Figure 2.22. To illustrate their operation, let us consider the

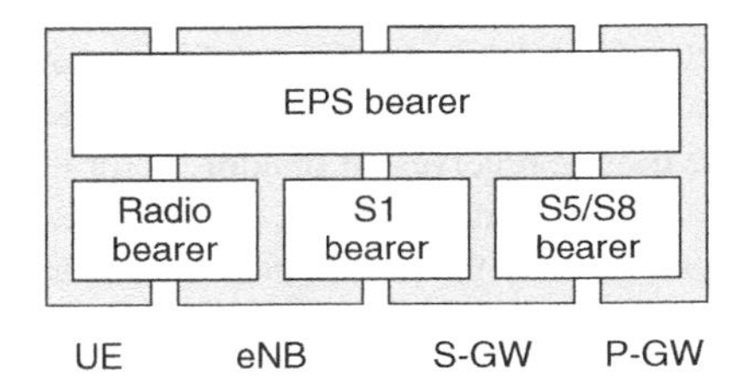

Figure 2.21 Bearer architecture of LTE, when using an S5/S8 interface based on GTP. Source: TS 36.300. Reproduced by permission of ETSI

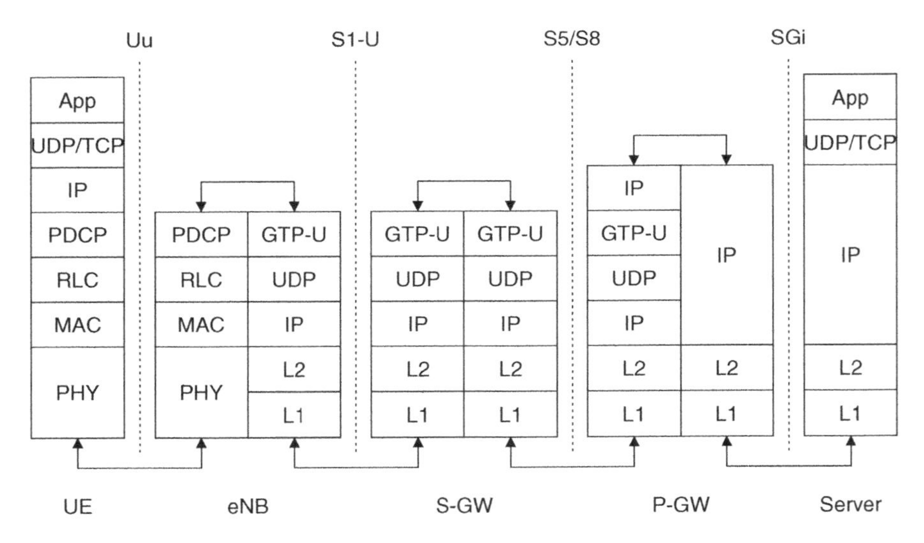

Figure 2.22 Protocol stacks used to exchange data between the mobile and an external server, when using an S5/S8 interface based on GTP. Source: TS 23.401. Reproduced by permission of ETSI

downlink path from the server to the mobile. Initially, a server sends a packet to the mobile using an IP header that includes the mobile's IP address. The packet arrives at the PDN gateway, which inspects that address, identifies the target mobile, and adds a second IP header that is addressed to the mobile's serving gateway. The transport network then uses the serving gateway's IP address to route the packet there. In turn, the serving gateway uses the same mechanism to send the packet to the base station.

The packet forwarding process is supported by the GPRS tunnelling protocol user part (GTP-U), which labels each packet using a 32 bit *tunnel endpoint identifier* (TEID) that identifies the overlying S1 or S5/S8 bearer. By inspecting the tunnel endpoint identifier, the network can distinguish packets that belong to different bearers and can handle them using different qualities of service. We will discuss the techniques that it uses in Chapter 13.

2.7.4 Bearer Implementation Using GRE and PMIP

The GRE protocol also uses tunnels, each of which is identified using a 32 bit *key field* in the GRE packet header. Unlike GTP-C, however, PMIP does not include any signalling messages that can specify the quality of service of a data stream.

If the S5/S8 interface is implemented using GRE and PMIP, then a mobile has only one GRE tunnel on that interface, which handles all the data packets that the mobile is transmitting or receiving without any quality of service guarantees. The EPS bearer only extends as far as the serving gateway [55], but is otherwise implemented in the same way that we have been describing.

2.7.5 Signalling Radio Bearers

LTE uses three special radio bearers, known as *signalling radio bearers* (SRBs), to carry signalling messages between the mobile and the base station [56]. The signalling radio bearers are listed in Table 2.2. Each of them is associated with a specific configuration of the air interface protocols, so that the mobile and base station can agree on how the signalling messages should be transmitted and received.

SRB0 is only used for a few RRC signalling messages, which the mobile and base station use to establish communications in a procedure known as *RRC connection establishment*. Its configuration is very simple and is defined in special RRC messages known as *system information messages*, which the base station broadcasts across the whole of the cell to tell the mobiles about how the cell is configured.

SRB1 is configured using signalling messages that are exchanged on SRB0, at the time when a mobile establishes communications with the radio access network. It is used for all

Table 2.2 Signalling radio bearers

Signalling radio bearer	Configured by	Used by
SRB 0	System information	RRC messages before establishment of SRB 1
SRB 1	RRC message on SRB 0	Subsequent RRC messages NAS messages before establishment of SRB 2
SRB 2	RRC message on SRB 1	Subsequent NAS messages

subsequent RRC messages, and also transports a few EMM and ESM messages that are exchanged prior to the establishment of SRB2. SRB2 is configured using signalling messages that are exchanged on SRB1, at the time when the mobile establishes communications with the evolved packet core. It is used to transport all the remaining EMM and ESM messages.

2.8 State Diagrams

2.8.1 EPS Mobility Management

A mobile's behaviour is defined using three state diagrams [57–59], which describe whether the mobile is registered with the EPC and whether it is active or idle. The first state diagram is the one for EPS mobility management (EMM). It is managed by the EMM protocol in the mobile and the MME, and is shown in Figure 2.23.

The mobile's EMM state depends on whether it is registered with the EPC. In the state EMM-REGISTERED, the mobile is switched on and is registered with a serving MME and a serving gateway. The mobile has an IP address and a default EPS bearer, which gives it always-on connectivity with a default packet data network. In EMM-DEREGISTERED, the mobile is either switched off or out of coverage and has none of these attributes. The mobile moves from EMM-DEREGISTERED to EMM-REGISTERED using the attach procedure that we will cover in Chapter 11 and moves back using the detach procedure.

2.8.2 EPS Connection Management

The second state diagram (Figure 2.24) is for *EPS connection management* (ECM). Once again, these states are managed by the EMM protocol. Each state has two names: TS 23.401 calls them ECM-CONNECTED and ECM-IDLE, while TS 24.301 calls them EMM-CONNECTED and EMM-IDLE. We will use the first of these. The mobile moves from ECM-IDLE to ECM-CONNECTED using the service request procedure that we will cover in Chapter 14 and moves back using a procedure known as S1 release.

The mobile's ECM state depends on whether it is active or on standby, from the viewpoint of the non-access stratum protocols and the EPC. An active mobile is in ECM-CONNECTED state. In this state, the MME knows the mobile's serving eNB, and all the data bearers and signalling radio bearers are in place. Using them, the mobile can freely exchange signalling

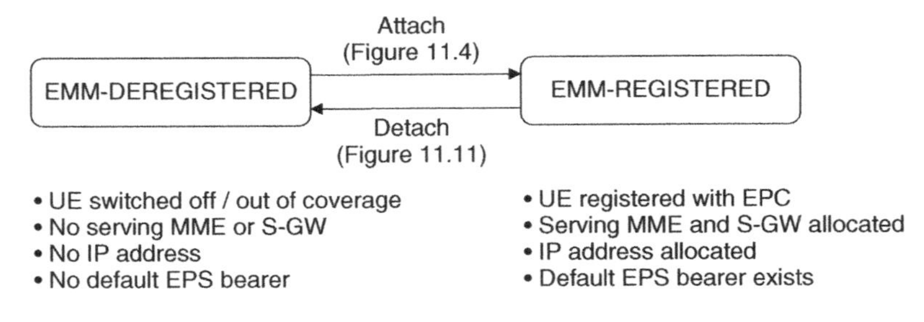

Figure 2.23 EPS mobility management (EMM) state diagram

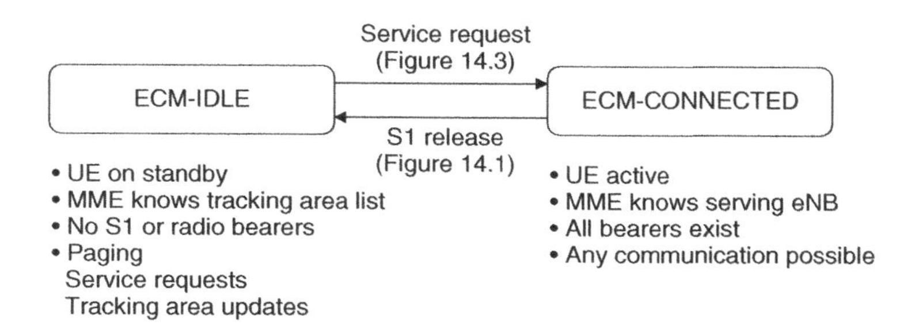

Figure 2.24 EPS connection management (ECM) state diagram

messages with the MME through a logical connection that is known as a *signalling connection* and can freely exchange data with the serving gateway.

When on standby, a mobile is in ECM-IDLE. In this state, it would be inappropriate to keep all the bearers in place, because the network would have to re-route them whenever the mobile moved from one cell to another, even though they would not be carrying any information. To avoid the resulting signalling overhead, the network tears down a mobile's S1 bearers and radio bearers whenever the mobile enters ECM-IDLE. The mobile can then freely move from one cell to another, without the need to re-route the bearers every time. However, the EPS bearers remain in place, so that a mobile retains its logical connections with the outside world. The S5/S8 bearers also remain in place, as the mobile changes its serving gateway only occasionally.

Furthermore, the MME does not know exactly where an idle mobile is located: instead, it just knows which tracking area the mobile is in. This allows the mobile to move from one cell to another without notifying the MME; instead, it only does so if it crosses a tracking area boundary. The MME can also create mobile-specific groups of tracking areas known as *tracking area lists*, and can tell a mobile to send a notification only if it moves outside a tracking area list. Tracking area lists can be useful for mobiles that are moving back and forth across tracking area boundaries, and can help a network operator to place those boundaries in populated areas without worrying about the impact on signalling.

In the state of ECM-IDLE, some limited communication is still possible. If the MME wishes to contact an idle mobile, then it can do so by sending an S1-AP *Paging* message to all the base stations in the mobile's tracking area list. The base stations react by transmitting an RRC Paging message, in the manner described below. If the mobile wishes to contact the network or reply to a paging message, then it sends the MME an EMM message called a *Service Request* and the MME reacts by moving the mobile into ECM-CONNECTED. Finally, the mobile can send an EMM *Tracking Area Update Request* to the MME, if it notices that it has moved into a tracking area in which it is not currently registered.

2.8.3 Radio Resource Control

The final state diagram (Figure 2.25) is for radio resource control (RRC). As the name implies, these states are managed by the RRC protocol in the mobile and the serving eNB.

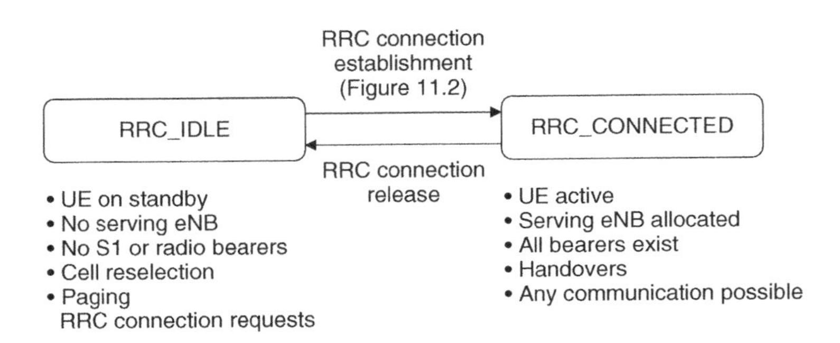

Figure 2.25 Radio resource control (RRC) state diagram

The mobile's RRC state depends on whether it is active or idle, from the viewpoint of the access stratum protocols and the E-UTRAN. An active mobile is in RRC_CONNECTED state. In this state, the mobile is assigned to a serving eNB, and can freely communicate with it using signalling messages on SRB 1.

When on standby, a mobile is in RRC_IDLE. In this state, the radio access network knows nothing about the mobile, so no serving eNB is assigned and SRB 1 is torn down. As before, however, some limited communication is still possible. If the radio access network wishes to contact the mobile, typically because it has received a paging request from the evolved packet core, then it can do so using an RRC *Paging* message. If the mobile wishes to contact the radio access network or reply to a paging message, then it can do so by initiating the RRC connection establishment procedure that we introduced earlier. In turn, the base station reacts by moving the mobile into RRC_CONNECTED.

The two RRC states handle moving devices in different ways. A mobile in RRC_CONNECTED state can be transmitting and receiving at a high data rate, so it is important for the radio access network to control which cell the mobile is communicating with. It does this using a procedure known as *handover*, in which the network switches the mobile's communication path from one cell to another. If the old and new cells are controlled by different base stations, then the network also re-routes the mobile's S1-U and S1-MME interfaces, so that they run directly between the new base station and the evolved packet core: the old base station drops out of the mobile's communication path altogether. In addition, the network will change the mobile's serving gateway and S5/S8 interface(s) if it moves into a new S-GW service area and will change the mobile's serving MME if it moves into a new MME pool area.

In RRC_IDLE state, the main motivations are to reduce signalling and maximize the mobile's battery life. To achieve this, the mobile decides which cell it will listen to, using a procedure known as *cell reselection*. The radio access network remains completely unaware of its location, while the EPC is only informed if a tracking area update is required. In turn, the tracking area update may lead to a change of serving gateway or serving MME in the manner described above.

Except in certain transient situations, the ECM and RRC state diagrams are always used together. An active mobile is always in ECM-CONNECTED and RRC_CONNECTED, while a mobile on standby is always in ECM-IDLE and RRC_IDLE.

2.9 Spectrum Allocation

The 3GPP specifications allow mobiles and base stations to use a large number of frequency bands [60, 61]. These are defined in response to decisions from the ITU and national regulators about the allocation of radio spectrum to mobile telecommunications. Table 2.3 lists the bands that supported frequency division duplex (FDD) mode in January 2014, mid-way through the specification process for Release 12, while Table 2.4 lists the bands that supported time division duplex (TDD).

The tables show the 3GPP release in which each frequency band was introduced. Unlike most features of the specifications, the LTE frequency bands are release independent; for example,

Table 2.3 FDD frequency bands

Band	Release	Uplink band (MHz)	Downlink band (MHz)	Main regions	Common name	Notes
1	R99	1920–1980	2110–2170	1, 3	2100	WCDMA
2	R99	1850–1910	1930–1990	2	1900	PCS
3	R5	1710–1785	1805–1880	1, 3	1800	GSM 1800
4	R6	1710–1755	2110–2155	2	1700/2100	AWS
5	R6	824–849	869–894	2, 3	850	GSM 850
6	–	–	–			Not used by LTE
7	R7	2500–2570	2620–2690	1, 2, 3	2600	
8	R7	880–915	925–960	1, 3	900	GSM 900
9	R7	1749.9–1784.9	1844.9–1879.9	Japan		
10	R7	1710–1770	2110–2170	2		AWS extension
11	R8	1427.9–1447.9	1475.9–1495.9	Japan	1500	
12	R8	699–716	729–746	USA	700	Lower band A, B, C
13	R8	777–787	746–756	USA	700	Upper band C
14	R8	788–798	758–768	USA	700	Upper D, public safety
15	–	–	–			Not used by 3GPP
16	–	–	–			Not used by 3GPP
17	R8	704–716	734–746	USA	700	Lower band B, C
18	R9	815–830	860–875	Japan		
19	R9	830–845	875–890	Japan		
20	R9	832–862	791–821	Europe	800	Digital dividend
21	R9	1447.9–1462.9	1495.9–1510.9	Japan	1500	
22	R10	3410–3490	3510–3590	1, 2, 3		
23	R10	2000–2020	2180–2200	USA		S band
24	R10	1626.5–1660.5	1525–1559	USA		L band
25	R10	1850–1915	1930–1995	2	1900	PCS extension
26	R11	814–849	859–894	2, 3		Bands 5, 18, 19
27	R11	807–824	852–869	2		
28	R11	703–748	758–803	3	700	Digital dividend
29	R11	–	717–728	USA		Carrier aggregation
30	R12	2305–2315	2350–2360	USA		WCS
31	R12	452.5–457.5	462.5–467.5	1, 2, 3	450	

Source: TS 36.101 and TS 36.104. Reproduced by permission of ETSI.

Table 2.4 TDD frequency bands

Band	Release	Frequency band (MHz)	Main regions	Common name	Notes
33	R99	1900–1920	1, 3		
34	R99	2010–2025	3		
35	R99	1850–1910	2		PCS
36	R99	1930–1990	2		PCS
37	R99	1910–1930	2		PCS
38	R7	2570–2620	1, 2, 3	2600	
39	R8	1880–1920	China		
40	R8	2300–2400	3	2300	
41	R10	2496–2690	USA	2600	
42	R10	3400–3600	1, 2, 3		
43	R10	3600–3800	1, 2, 3		
44	R11	703–803	3		

Source: TS 36.101 and TS 36.104. Reproduced by permission of ETSI.

a mobile can support bands that were introduced in later releases of the 3GPP specifications even if it otherwise conforms to Release 8.

The tables also show the main ITU regions and individual countries that use each band. ITU region 1 covers Europe, Africa and North West Asia (including the Middle East and the former Soviet Union), region 2 covers the Americas, and region 3 covers South East Asia (including India and China) and Australasia. The LTE frequency bands are often denoted by their approximate carrier frequencies or by the names of technologies that have previously used them. In the last two columns, the tables include alternative names for some of the more commonly used bands.

Some of the bands in these tables are being newly released for use by mobile telecommunications. In 2008, for example, the US *Federal Communications Commission* (FCC) auctioned frequencies at 700 MHz (FDD bands 12, 13 and 17) that had previously been used for analogue television broadcasting. In Europe, similar auctions have been taking place at 800 and 2600 MHz (FDD bands 7 and 20, and TDD band 38). Network operators can also re-allocate frequencies that they have previously used for other mobile communication systems, as their users migrate to LTE. Examples include FDD bands 1, 3 and 8 in Europe, which were originally used by WCDMA, GSM 1800 and GSM 900, and FDD bands 2, 4 and 5 in the United States.

To illustrate the proliferation of carrier frequencies, Figure 2.26 shows the numbers of operational and planned LTE networks in the most common frequency bands in November 2013, using data from 4G Americas [62, 63]. The most common choices to date have been bands 3, 7 and 20 (1800, 2600 and 800 MHz) in ITU region 1, bands 4, 12, 13 and 17 (1700/2100 and 700 MHz) in ITU region 2, and bands 3 and 7 (1800 and 2600 MHz) in ITU region 3.

In response, device manufacturers have tended to support only some of LTE frequency bands, which depend on the other technologies that the device supports and the countries in which it is sold. Such support was very limited in the case of early LTE devices but was subsequently extended. As an example, the iPhone 5 was launched in 2012 with three models that each

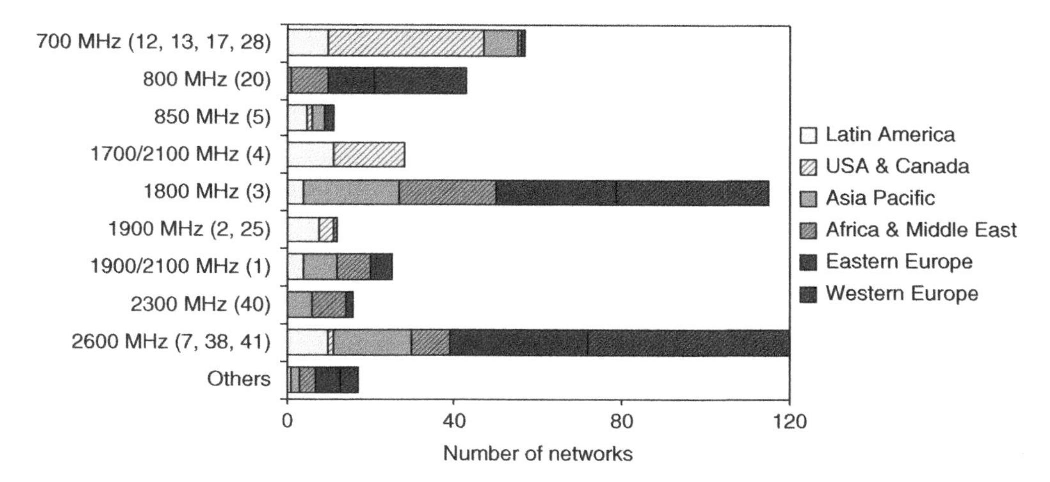

Figure 2.26 Frequency bands used by operational and planned LTE networks in November 2013. Source: http://www.4gamericas.org. Reproduced by permission of 4G Americas

supported between three and five FDD bands. The subsequent iPhone 5c and 5s models were launched in 2013 with four models each supporting between 7 and 13 bands and one model including support for TDD [64].

References

1. 3GPP TS 23.401 (2013) General Packet Radio Service (GPRS) Enhancements for Evolved Universal Terrestrial Radio Access Network (E-UTRAN) Access, Release 11, September 2013.
2. 3GPP TS 36.300 (2013) Evolved Universal Terrestrial Radio Access (E-UTRA) and Evolved Universal Terrestrial Radio Access Network (E-UTRAN); Overall Description; Stage 2, Release 11, September 2013.
3. 3GPP TS 23.002 (2013) Network Architecture, Release 11, June 2013.
4. 3GPP TS 36.401 (2013) Evolved Universal Terrestrial Radio Access Network (E-UTRAN); Architecture Description, Release 11, September 2013.
5. 3GPP TS 27.001 (2012) General on Terminal Adaptation Functions (TAF) for Mobile Stations (MS), Release 11, Section 4, September 2012.
6. 3GPP TS 31.102 (2013) Characteristics of the Universal Subscriber Identity Module (USIM) Application, Release 11, September 2013.
7. 3GPP TS 36.306 (2013) Evolved Universal Terrestrial Radio Access (E-UTRA); User Equipment (UE) Radio Access Capabilities, Release 11, September 2013.
8. 3GPP TS 36.331 (2013) Evolved Universal Terrestrial Radio Access (E-UTRA); Radio Resource Control (RRC); Protocol Specification, Release 11, Annexes B, C, September 2013.
9. 3GPP TS 36.401 (2013) Evolved Universal Terrestrial Radio Access Network (E-UTRAN); Architecture Description, Release 11, Section 6, September 2013.
10. 3GPP TS 36.300 (2013) Evolved Universal Terrestrial Radio Access (E-UTRA) and Evolved Universal Terrestrial Radio Access Network (E-UTRAN); Overall Description; Stage 2, Release 11, Section 4.6, September 2013.
11. 3GPP TS 23.401 (2013) General Packet Radio Service (GPRS) Enhancements for Evolved Universal Terrestrial Radio Access Network (E-UTRAN) Access, Release 11, Sections 4.2.1, 4.4, September 2013.
12. 3GPP TS 23.002 (2013) Network Architecture, Release 11, Section 4.1.4, June 2013.
13. 3GPP TS 23.003 (2013) Numbering, Addressing and Identification, Release 11, Section 9, September 2013.
14. 3GPP TS 22.168 (2012) Earthquake and Tsunami Warning System (ETWS) Requirements; Stage 1, Release 8, December 2012.

15. 3GPP TS 23.401 (2013) General Packet Radio Service (GPRS) Enhancements for Evolved Universal Terrestrial Radio Access Network (E-UTRAN) Access, Release 11, Section 4.2.2, September 2013.
16. GSM Association IR.34 (2013) Guidelines for IPX Provider Networks, Version 9.1, May 2013.
17. 3GPP TS 23.401 (2013) General Packet Radio Service (GPRS) Enhancements for Evolved Universal Terrestrial Radio Access Network (E-UTRAN) Access, Release 11, Section 4.3.8.1, September 2013.
18. 3GPP TS 23.401 (2013) General Packet Radio Service (GPRS) Enhancements for Evolved Universal Terrestrial Radio Access Network (E-UTRAN) Access, Release 11, Section 3.1, September 2013.
19. 3GPP TS 23.003 (2013) Numbering, Addressing and Identification, Release 11, Sections 2, 6, 19, September 2013.
20. 3GPP TS 36.401 (2013) Evolved Universal Terrestrial Radio Access Network (E-UTRAN); Architecture Description, Release 11, Section 5, September 2013.
21. 3GPP TS 36.321 (2013) Evolved Universal Terrestrial Radio Access (E-UTRA); Medium Access Control (MAC) Protocol Specification, Release 11, July 2013.
22. 3GPP TS 36.322 (2012) Evolved Universal Terrestrial Radio Access (E-UTRA); Radio Link Control (RLC) Protocol Specification, Release 11, September 2012.
23. 3GPP TS 36.323 (2013) Evolved Universal Terrestrial Radio Access (E-UTRA); Packet Data Convergence Protocol (PDCP) Specification, Release 11, March 2013.
24. IETF RFC 3031 (2001) Multiprotocol Label Switching Architecture, January 2001.
25. IETF RFC 3032 (2001) MPLS Label Stack Encoding, January 2001.
26. IETF RFC 791 (1981) Internet Protocol, September 1981.
27. IETF RFC 2460 (1998), Internet Protocol, Version 6 (IPv6) Specification, December 1998.
28. 3GPP TS 29.281 (2013) General Packet Radio System (GPRS) Tunnelling Protocol User Plane (GTPv1-U), Release 11, Section 4.4.1, March 2013.
29. 3GPP TS 36.414 (2012) Evolved Universal Terrestrial Radio Access Network (E-UTRAN); S1 Data Transport, Release 11, Section 5.3, September 2012.
30. 3GPP TS 36.424 (2012) Evolved Universal Terrestrial Radio Access Network (E-UTRAN); X2 Data Transport, Release 11, Section 5.3, September 2012.
31. IETF RFC 768 (1980) User Datagram Protocol, August 1980.
32. IETF RFC 793 (1981) Transmission Control Protocol, September 1981.
33. IETF RFC 4960 (2007) Stream Control Transmission Protocol, September 2007.
34. 3GPP TS 29.281 (2013) General Packet Radio System (GPRS) Tunnelling Protocol User Plane (GTPv1-U), Release 11, March 2013.
35. IETF RFC 2784 (2000) Generic Routing Encapsulation (GRE), March 2000.
36. 3GPP TS 36.331 (2013) Evolved Universal Terrestrial Radio Access (E-UTRA); Radio Resource Control (RRC); Protocol Specification, Release 11, September 2013.
37. 3GPP TS 36.413 (2013) Evolved Universal Terrestrial Radio Access Network (E-UTRAN); S1 Application Protocol (S1AP), Release 11, September 2013.
38. 3GPP TS 36.423 (2013) Evolved Universal Terrestrial Radio Access Network (E-UTRAN); X2 Application Protocol (X2AP), Release 11, September 2013.
39. 3GPP TS 24.301 (2013) Non-Access-Stratum (NAS) Protocol for Evolved Packet System (EPS); Stage 3, Release 11, September 2013.
40. IETF RFC 3588 (2003) Diameter Base Protocol, September 2003.
41. IETF RFC 2865 (2000) Remote Authentication Dial-In User Service (RADIUS), June 2000.
42. IETF RFC 2866 (2000) RADIUS Accounting, June 2000.
43. 3GPP TS 29.272 (2013) Evolved Packet System (EPS); Mobility Management Entity (MME) and Serving GPRS Support Node (SGSN) Related Interfaces Based on Diameter Protocol, Release 11, September 2013.
44. 3GPP TS 29.274 (2013) 3GPP Evolved Packet System (EPS); Evolved General Packet Radio Service (GPRS) Tunnelling Protocol for Control Plane (GTPv2-C); Stage 3, Release 11, September 2013.
45. 3GPP TS 29.275 (2013) Proxy Mobile IPv6 (PMIPv6) Based Mobility and Tunnelling Protocols; Stage 3, Release 11, June 2013.
46. IETF RFC 5213 (2008) Proxy Mobile IPv6, August 2008.
47. 3GPP TS 23.402 (2013) Architecture Enhancements for Non-3GPP Accesses, Release 11, June 2013.
48. 3GPP TS 36.331 (2013) Radio Resource Control (RRC); Protocol Specification, Release 11, Section 5.6.3, September 2013.
49. 3GPP TS 24.301 (2013) Non-Access-Stratum (NAS) Protocol for Evolved Packet System (EPS); Stage 3, Release 11, Section 5.4.1, September 2013.

50. 3GPP TS 36.413 (2013) Evolved Universal Terrestrial Radio Access Network (E-UTRAN); S1 Application Protocol (S1AP), Release 11, Section 8.6.2, September 2013.
51. 3GPP TS 36.331 (2013) Radio Resource Control (RRC); Protocol Specification, Release 11, Sections 5.6.1, 5.6.2, September 2013.
52. 3GPP TS 36.300 (2013) Evolved Universal Terrestrial Radio Access (E-UTRA) and Evolved Universal Terrestrial Radio Access Network (E-UTRAN); Overall Description; Stage 2, Release 11, Section 13, September 2013.
53. 3GPP TS 23.401 (2013) General Packet Radio Service (GPRS) Enhancements for Evolved Universal Terrestrial Radio Access Network (E-UTRAN) Access, Release 11, Section 4.7, September 2013.
54. 3GPP TS 24.007 (2012) Mobile Radio Interface Signalling Layer 3; General Aspects, Release 11, Section 11.2.3.1.5, June 2012.
55. 3GPP TS 23.402 (2013) Architecture Enhancements for Non-3GPP Accesses, Release 11, Section 4.10, June 2013.
56. 3GPP TS 36.331 (2013) Radio Resource Control (RRC); Protocol Specification, Release 11, Section 4.2.2, September 2013.
57. 3GPP TS 23.401 (2013) General Packet Radio Service (GPRS) Enhancements for Evolved Universal Terrestrial Radio Access Network (E-UTRAN) Access, Release 11, Section 4.6, September 2013.
58. 3GPP TS 24.301 (2013) Non-Access-Stratum (NAS) Protocol for Evolved Packet System (EPS); Stage 3, Release 11, Section 3.1, September 2013.
59. 3GPP TS 36.331 (2013) Radio Resource Control (RRC); Protocol Specification, Release 11, Section 4.2.8, September 2013.
60. 3GPP TS 36.101 (2013) Evolved Universal Terrestrial Radio Access (E-UTRA); User Equipment (UE) Radio Transmission and Reception, Release 11, Section 5.5, September 2013.
61. 3GPP TS 36.104 (2013) Evolved Universal Terrestrial Radio Access (E-UTRA); Base Station (BS) Radio Transmission and Reception, Release 11, Section 5.5, September 2013.
62. 4G Americas (2013) 3G/4G Deployment Status, http://www.4gamericas.org/index.cfm?fuseaction=page&pageid=939 (accessed 18 November 2013).
63. LteMaps (2013) Mapping LTE Deployments, http://ltemaps.org (accessed 15 October 2013).
64. Apple (2013) iPhone 5, http://www.apple.com/iphone/LTE/ (accessed 15 October 2013).

3

Digital Wireless Communications

The next three chapters describe the principles of radio transmission and reception in LTE. Here, we begin by reviewing the radio transmission techniques that LTE has inherited from 2G and 3G communication systems. The chapter covers the principles of modulation and demodulation, describes how these principles are applied to a mobile cellular network and shows how the received signal can be degraded by noise, fading and inter-symbol interference. It then discusses the techniques that are used to minimize the number of errors in the received signal, notably forward error correction, re-transmissions and hybrid automatic repeat request. For some detailed accounts of the material covered in this chapter, see for example References [1–6].

These three chapters contain more mathematics than the others in this book, but it has been kept reasonably lightweight to ensure that the material is accessible to those without a mathematical background. Some of the more detailed aspects have been ring-fenced into individual sections, which readers can skip without detracting from their overall appreciation of the subject.

3.1 Radio Transmission and Reception

3.1.1 Carrier Signal

A key part of any radio communication system is the creation and transmission of a radio wave, also known as a *carrier signal*. Mathematically, we can express the carrier signal as follows:

$$I(t) = a\cos(2\pi ft + \phi) \tag{3.1}$$

Here, a is the amplitude of the radio wave, f is its frequency and ϕ is its initial phase angle. The angles are measured in radians, with π radians equal to 180°.

We can visualize the radio wave by means of an object moving in an anti-clockwise circle around the origin, as shown on the left hand side of Figure 3.1. $I(t)$ is the object's position along the horizontal axis at time t. It is often known as the radio wave's *in-phase component*, and is shown at the bottom right of the figure. We can also define a quadrature component $Q(t)$,

An Introduction to LTE: LTE, LTE-Advanced, SAE, VoLTE and 4G Mobile Communications, Second Edition.
Christopher Cox.

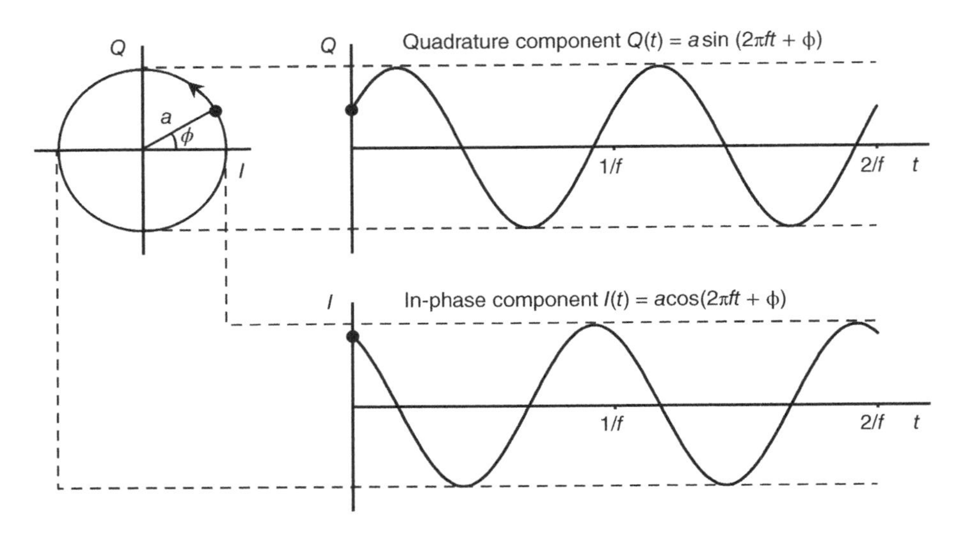

Figure 3.1 Generation of the in-phase and quadrature components of a carrier signal

which is the object's position along the vertical axis and is shown at the top right:

$$Q(t) = a\sin(2\pi ft + \phi) \tag{3.2}$$

The quadrature component will not be transmitted, but we will use it for internal bookkeeping purposes later on.

3.1.2 Modulation Techniques

A *modulator* encodes a sequence of bits onto the carrier signal by adjusting the parameters that describe it. In LTE, we choose to adjust the signal's amplitude a and/or its initial phase ϕ. Figure 3.2 shows an example modulation scheme, known as *quadrature phase shift keying* (QPSK). A QPSK modulator takes the incoming bits two at a time and transmits them using a radio wave that can have four different states, which are known as *symbols*. Each symbol

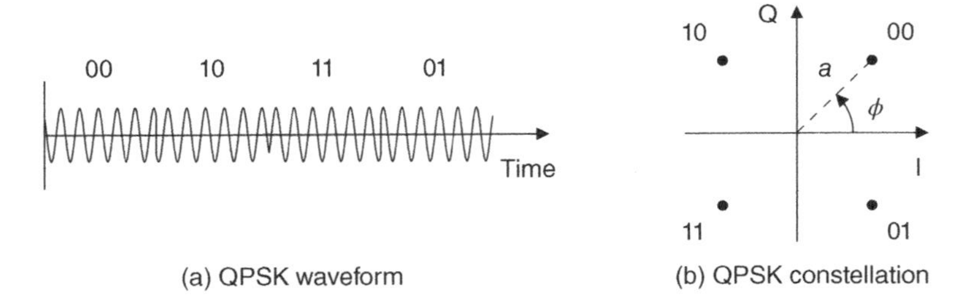

Figure 3.2 Quadrature phase shift keying. (a) Example QPSK waveform. (b) QPSK constellation diagram

is described using two numbers. These can either be the amplitude and initial phase of the resulting radio wave or the initial values of the wave's in-phase and quadrature components:

$$I_0 = a\cos\phi$$
$$Q_0 = a\sin\phi \tag{3.3}$$

In QPSK, the symbols have the same amplitude and have initial phases of 45°, 135°, 225° and 315° (Figure 3.2a), which correspond to bit combinations of 00, 10, 11 and 01 respectively. We can also represent the four QPSK symbols using the *constellation diagram* shown in Figure 3.2(b). In this diagram, the distance of each symbol from the origin represents the amplitude of the transmitted wave, while the angle (measured anti-clockwise from the x-axis) represents its initial phase.

As shown in Figure 3.3, LTE uses four modulation schemes altogether. *Binary phase shift keying* (BPSK) sends bits one at a time, using two symbols that can be interpreted either as initial phases of 0° and 180° or as signal amplitudes of +1 and −1. LTE uses this scheme for a limited number of control streams, but does not use it for normal data transmissions. *16 quadrature amplitude modulation* (16-QAM) sends bits four at a time using 16 symbols that have different amplitudes and phases. Similarly, 64-QAM sends bits six at a time using 64 different symbols, so it has a data rate six times greater than that of BPSK.

3.1.3 The Modulation Process

Figure 3.4 shows the most important components of a modulator, using the example of a QPSK signal. In outline, the transmitter accepts a stream of bits from the higher layer protocols, calculates the resulting symbols, and modulates the carrier signal by mixing the symbols and the carrier together. A real modulator also has other components, such as filters that smooth the abrupt phase transitions in the output signal, but we will ignore these so as to focus on the most important parts of the process.

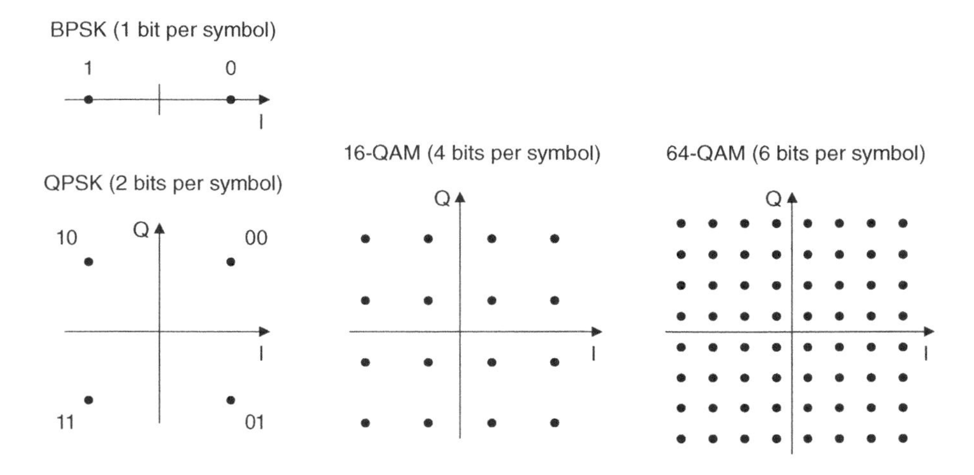

Figure 3.3 Modulation schemes used by LTE

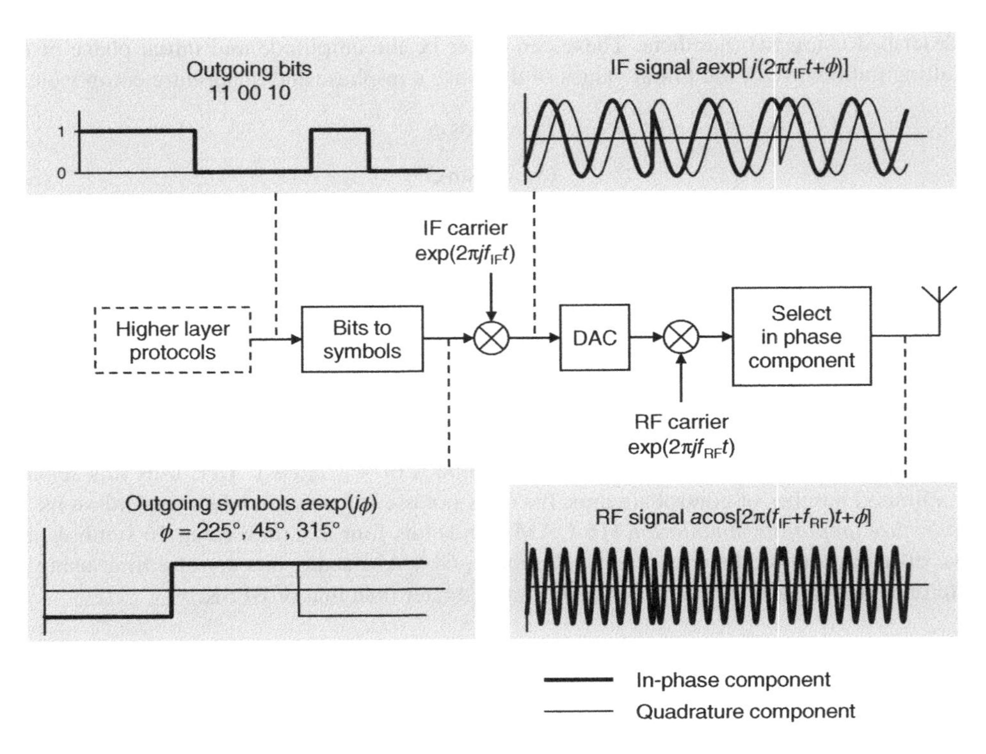

Figure 3.4 Block diagram of the modulator in a wireless communication system

In Figure 3.4, we have chosen to do the modulation process in two stages, by first mixing the symbols with an intermediate frequency (IF) carrier at a frequency f_{IF}, and then mixing the result with a much larger radio frequency (RF) carrier f_{RF}. The low frequency calculations are carried out digitally, so there is a digital to analogue converter (DAC) between the two stages. As a result of this process, the radio wave is eventually transmitted with a frequency of $f_{\mathrm{IF}} + f_{\mathrm{RF}}$. There are two justifications for this choice: practical modulators often use this technique, and the concept will be useful when we discuss orthogonal frequency division multiple access in Chapter 4.

To manipulate the symbols in the correct way, we have to keep track of two quantities throughout the modulation process. These quantities can be the amplitude and phase of the modulated signal, or they can be its in-phase and quadrature components. Mathematically, the easiest way to do this task is by interpreting the in-phase and quadrature components as the real and imaginary parts of the following complex number:

$$\begin{aligned} z(t) &= I(t) + jQ(t) \\ &= a\cos(2\pi ft + \phi) + ja\sin(2\pi ft + \phi) \end{aligned} \tag{3.4}$$

where $z(t)$ is the complex representation of the signal at time t and j is the square root of -1. The use of complex numbers allows us to exploit the following identity, which makes the calculations easier than they would otherwise be:

$$a\cos\theta + ja\sin\theta = a\exp(j\theta) = ae^{j\theta} \tag{3.5}$$

Here, e is approximately 2.718, and the two expressions on the right hand side simply show two different notations for the same quantity.

We can now manipulate the signal by keeping track of its complex representation. There are two rules to follow. Firstly, we mix two signals by multiplying their complex representations together, subject to the condition that $j^2 = -1$. Secondly, we only transmit the real part of the end result, on the grounds that the real world does not understand imaginary numbers. Following these rules, the transmitted signal is as follows:

$$I(t) = \mathrm{Re}[a\exp(j\phi)\exp(2\pi jf_{\mathrm{IF}}t)\exp(2\pi jf_{\mathrm{RF}}t)] \tag{3.6}$$

where the three terms on the right represent the transmitted symbols and the IF and RF carriers, and Re[] is the real part of a complex number. We then reach the following result:

$$\begin{aligned} I(t) &= \mathrm{Re}\{a\exp[2\pi j(f_{\mathrm{IF}} + f_{\mathrm{RF}})t + j\phi]\} \\ &= a\cos[2\pi(f_{\mathrm{IF}} + f_{\mathrm{RF}})t + \phi] \end{aligned} \tag{3.7}$$

which is the signal that we require. We could also reach this result without using complex numbers, but the calculations would be harder. The illustrations of modulation and upconversion in the LTE specifications [7] are equivalent to the last two steps shown in Figure 3.4, in which we mix the signal with the RF carrier and select the real part for transmission.

In Equation 3.7, the signal is transmitted at a carrier frequency that lies at a small offset f_{IF} above a radio frequency signal f_{RF}. The intermediate frequency f_{IF} could also be negative, in which case the carrier frequency would lie below f_{RF} instead of above. Although the idea of a negative frequency may be unfamiliar, we can easily visualize it by moving our object around the origin shown in Figure 3.1 in a clockwise instead of anti-clockwise direction. We cannot distinguish positive and negative frequencies using their in-phase components alone, so negative frequencies do not appear in the real world. However, we can distinguish them using the combination of their in-phase and quadrature components, so we can use negative frequencies for the signal manipulations that we have just described.

3.1.4 The Demodulation Process

The receiver accepts the incoming radio wave and recovers the bits by means of a *demodulator*. Figure 3.5 shows the most important parts of the demodulation process. In outline, the demodulator mixes the incoming signal with negative frequency replicas of the radio and intermediate frequencies. (Mathematically, each replica is created by a process known as *complex conjugation*, which changes the sign of its imaginary part.) By doing so, the demodulator mixes the incoming signal down to zero frequency, so as to recover the transmitted symbols and hence the transmitted bits.

There are three complications. Firstly, as shown in the diagram and discussed in more detail below, the incoming signal may be distorted by thermal noise and by interference from other transmitters. To overcome this problem, the integration stage adds up the in-phase and

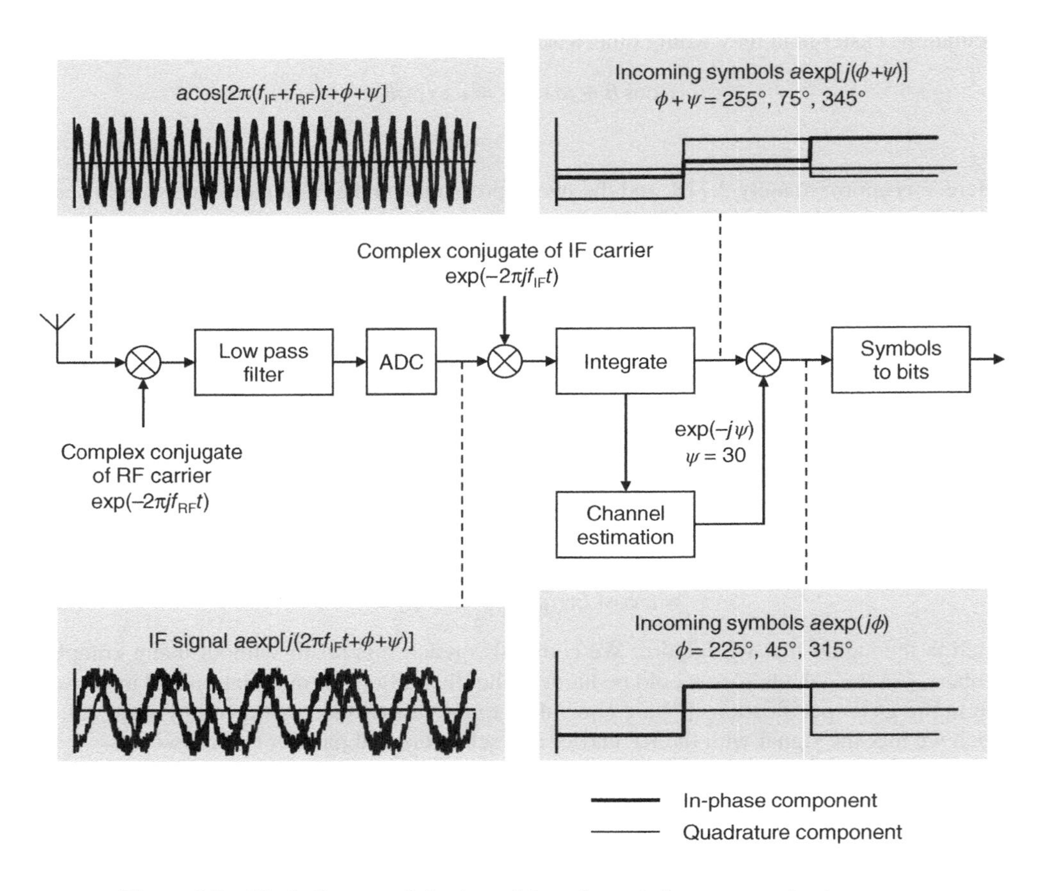

Figure 3.5 Block diagram of the demodulator in a wireless communication system

quadrature components of the incoming symbols, I_0 and Q_0, for the duration of each individual symbol. During this time, the contributions from the signal itself accumulate while the contributions from noise and interference average out, so the received signal-to-interference plus noise ratio (SINR) improves.

The second complication is the least important for us but is worth noting for completeness. It arises because the incoming signal is a real quantity, which we can write as follows:

$$I(t) = a\cos[2\pi(f_{\mathrm{IF}}+f_{\mathrm{RF}})t+\phi+\psi]$$
$$= \frac{a\exp\{j[2\pi(f_{\mathrm{IF}}+f_{\mathrm{RF}})t+\phi+\psi]\}+a\exp\{-j[2\pi(f_{\mathrm{IF}}+f_{\mathrm{RF}})t+\phi+\psi]\}}{2} \tag{3.8}$$

This equation shows that the incoming signal actually has two complex components, one with a positive frequency of $f_{\mathrm{IF}}+f_{\mathrm{RF}}$ and the other with a negative frequency of $-(f_{\mathrm{IF}}+f_{\mathrm{RF}})$. After mixing the signal with the replica of the radio frequency, the first component moves to the desired intermediate frequency f_{IF}. However, the second component moves to a frequency of $-(f_{\mathrm{IF}}+2f_{\mathrm{RF}})$, and causes rapid, unwanted fluctuations in the incoming signal. The low-pass filter removes those fluctuations, leaving only the signal that we require.

3.1.5 Channel Estimation

There is one more complication: the phase of the incoming signal depends not only on the phase of the transmitted signal but also on the receiver's exact position. If the receiver moves through half a wavelength of the carrier signal (a distance of 10 cm at a carrier frequency of 1500 MHz, for example), then the phase of the received signal changes by 180°. When using QPSK, this phase change turns bit pairs of 00 into 11 and vice versa, and completely destroys the received information. We can express this issue by including an arbitrary phase shift ψ in the received signal. In Figure 3.5, the phase shift is 30°.

To deal with this problem, the transmitter inserts occasional *reference symbols* into the data stream, which have a transmission time, amplitude and phase that are defined in the relevant specifications. In the receiver, a *channel estimation* function measures the incoming reference symbols, compares them with the ones that the specifications defined, and estimates the phase shift ψ that the air interface introduced. It can then remove this phase shift from the incoming symbols by multiplying them by the complex number $\exp(-j\psi)$. The phase shift does not change much from one symbol to the next, so the reference symbols only need to take up a small part of the transmitted data stream. The resulting overhead in LTE is about 10%.

3.1.6 Bandwidth of the Modulated Signal

There is one last point to make. The power of a modulated signal is not just confined to a single frequency; instead, it is spread over a range of frequencies known as the bandwidth. Roughly speaking, the bandwidth B and symbol duration T are related as follows:

$$B \approx \frac{1}{T} \tag{3.9}$$

Figure 3.6 shows the result. The carrier signal is transmitted at a single frequency, say f_C, so its power is confined to that frequency alone. If we modulate the carrier, then the power of the resulting transmission is spread over a bandwidth B. If, for example, the duration of each symbol is 1 μs, then the symbol rate is 1 Msps, and we can expect the transmitted signal to occupy a bandwidth of around 1 MHz.

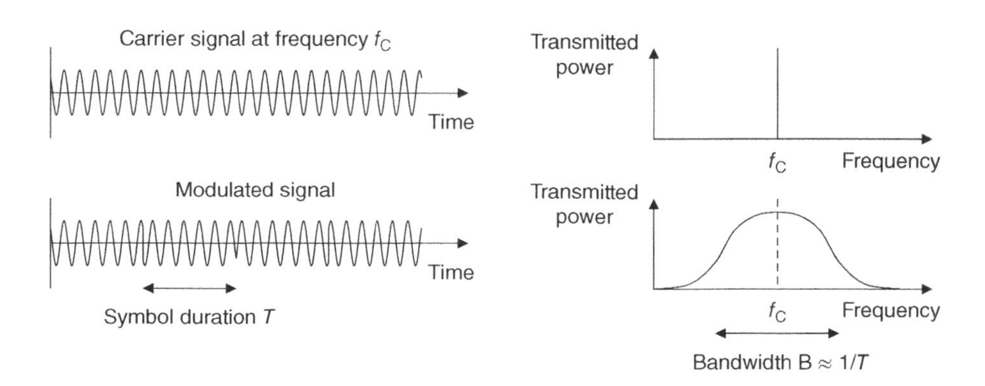

Figure 3.6 Relationship between the bandwidth and symbol duration of a modulated signal

3.2 Radio Transmission in a Mobile Cellular Network

3.2.1 Multiple Access Techniques

The techniques described so far work well for one-to-one communications. In a cellular network, however, a base station has to transmit to many different mobiles at once. It does this by sharing the resources of the air interface, in a technique known as *multiple access*.

Mobile communication systems use a few different multiple access techniques, two of which are shown in Figure 3.7. *Frequency division multiple access* (FDMA) was used by the first generation analogue systems. In this technique, each mobile receives on its own carrier frequency, which it distinguishes from the others by the use of analogue filters. The carriers are separated by unused *guard bands*, which minimizes the interference between them. In *time division multiple access* (TDMA), mobiles receive information on the same carrier frequency but at different times.

GSM uses a mix of frequency and time division multiple access, in which every cell has several carrier frequencies that are each shared amongst eight different mobiles. LTE uses another mixed technique known as *orthogonal frequency division multiple access* (OFDMA), which we will cover in Chapter 4.

Third generation communication systems used a different technique altogether, known as *code division multiple access* (CDMA). In this technique, mobiles receive on the same carrier frequency and at the same time, but the signals are labelled by the use of codes, which allow a mobile to separate its own signal from those of the others. LTE uses a few of the concepts from CDMA for some of its control signals, but does not implement the technique otherwise.

Multiple access is actually a generalization of a simpler technique known as *multiplexing*. The difference between the two is that a multiplexing system carries different data streams to or from a single device, while a multiple access system supports multiple devices.

3.2.2 FDD and TDD Modes

By using the multiple access techniques described above, a base station can distinguish the transmissions to and from the individual mobiles in the cell. However, we still need a way to distinguish the mobiles' transmissions from those of the base stations themselves.

To do this, a mobile communication system can operate in the transmission modes that we introduced in Chapter 1 (Figure 3.8). When using frequency division duplex (FDD), the base

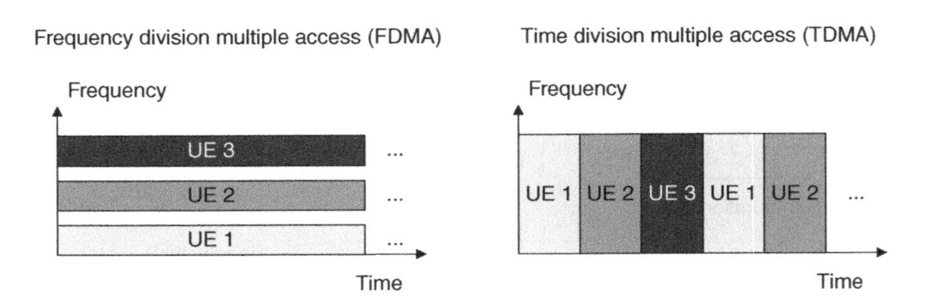

Figure 3.7 Example multiple access techniques

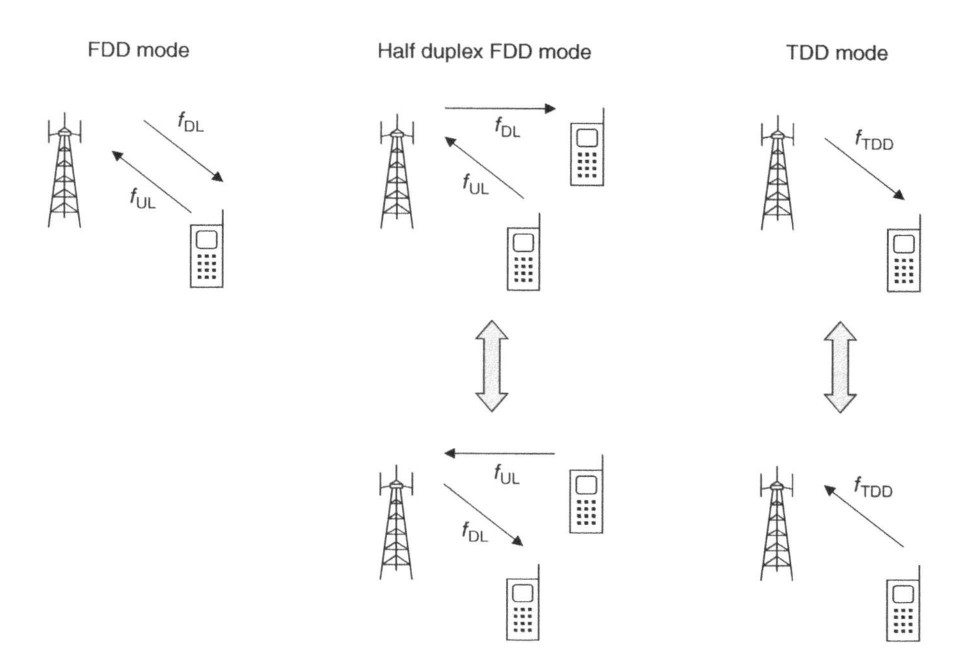

Figure 3.8 Operation of FDD and TDD modes

station and mobile transmit and receive at the same time, but using different carrier frequencies. Using time division duplex (TDD), they transmit and receive on the same carrier frequency but at different times.

FDD and TDD modes have different advantages and disadvantages. In FDD mode, the bandwidths of the uplink and downlink are fixed and are usually the same. This makes it suitable for voice communications, in which the uplink and downlink data rates are very similar. In TDD mode, the system can adjust how much time is allocated to the uplink and downlink. This makes it suitable for applications such as web browsing, in which the downlink data rate can be much greater than the rate on the uplink.

TDD mode can be badly affected by interference if, for example, one base station is transmitting while a nearby base station is receiving. To avoid this, nearby base stations must be carefully time synchronized and must use the same allocations for the uplink and downlink, so that they all transmit and receive at the same time. This makes TDD suitable for networks that are made from isolated hotspots, because each hotspot can have a different timing and resource allocation. In contrast, FDD is often preferred for wide-area networks that have no isolated regions.

When operating in FDD mode, the mobile usually has to contain a high-attenuation duplex filter that isolates the uplink transmitter from the downlink receiver. In a variation known as *half duplex FDD mode*, a base station can still transmit and receive at the same time, but a mobile can only do one or the other. This means that the mobile does not have to isolate the transmitter and receiver to the same extent, which eases the design of its radio hardware.

LTE supports each of the modes described above. A cell can use either FDD or TDD mode. A mobile can support any combination of full duplex FDD, half duplex FDD and TDD, although it will only use one of these at a time.

3.3 Impairments to the Received Signal

3.3.1 Propagation Loss

In a wireless communication system, the signal spreads out as it travels from the transmitter to the receiver, so the received power P_R is less than the transmitted power P_T. The *propagation loss* or *path loss, PL*, is the ratio of the two.

$$PL = \frac{P_T}{P_R} \tag{3.10}$$

If the signal is travelling through empty space, then at a distance r from the transmitter, it occupies a spherical surface with an area of $4\pi r^2$. The propagation loss is therefore proportional to r^2. In a cellular network, the signal can also be absorbed and reflected by obstacles such as buildings and the ground, which in turn affects the propagation loss. Experimentally, we find that the propagation loss in a cellular network is roughly proportional to r^m, where m typically lies between 3.5 and 4.

3.3.2 Noise and Interference

By itself, propagation loss would not be a problem. However, as we saw in Figure 3.5, the received signal is also distorted by thermal noise and interference from other transmitters. These effects mean that the receiver cannot make a completely accurate estimate of the transmitted amplitude and phase.

To help it deal with the problem, the receiver often reconstructs the received bits in two stages. In the first stage, the receiver uses its estimate of the incoming symbol's amplitude and phase to compute a *soft decision*. The soft decision expresses not only whether the incoming bit appears to be a 1 or a 0 but also the receiver's confidence in that result. In the second stage, the receiver makes a final decision between a 1 and a 0 by computing a *hard decision*. We will see an important use of soft decisions towards the end of the chapter.

If the noise and interference are large enough, then a bit of 1 can be misinterpreted as a bit of 0 and vice versa, leading to bit errors in the receiver. The error rate depends on the signal-to-interference plus noise ratio (SINR) at the receiver. In a fast modulation scheme such as 64-QAM, the signal can be transmitted in many different ways, using states in the constellation diagram that are packed closely together. As a result, 64-QAM is vulnerable to errors and can only be used if the SINR is high. In contrast, QPSK only has a few states, so is less vulnerable to errors and can be successfully used at a lower SINR. LTE exploits this by switching dynamically between different modulation schemes: it uses 64-QAM at high SINR to give a high data rate, but falls back to 16-QAM or QPSK at lower SINR to reduce the number of errors.

3.3.3 Multipath and Fading

Propagation loss and noise are not the only problem. As a result of reflections, rays can take several different paths from the transmitter to the receiver. This phenomenon is known as *multipath*.

At the receiver, the incoming rays can add together in different ways, which are shown in Figure 3.9. If the peaks of the incoming rays coincide then they reinforce each other, a situation known as *constructive interference*. If, however, the peaks of one ray coincide with the troughs of another, then the result is *destructive interference*, in which the rays cancel. Destructive interference can make the received signal power drop to a very low level, a situation known as *fading*. The resulting increase in the error rate makes fading a serious problem for any mobile communication system.

If the mobile moves from one place to another, then the ray geometry changes, so the interference pattern changes between constructive and destructive. Fading is therefore a function of time, as shown in Figure 3.10a. The amplitude and phase of the received signal vary over a timescale called the *coherence time*, T_c, which can be estimated as follows:

$$T_c \approx \frac{1}{f_D} \tag{3.11}$$

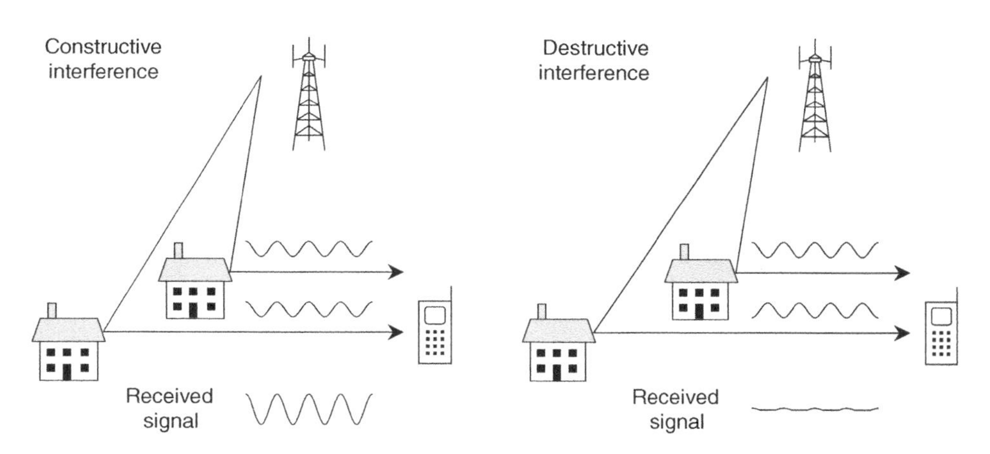

Figure 3.9 Generation of constructive interference, destructive interference and fading in a multipath environment

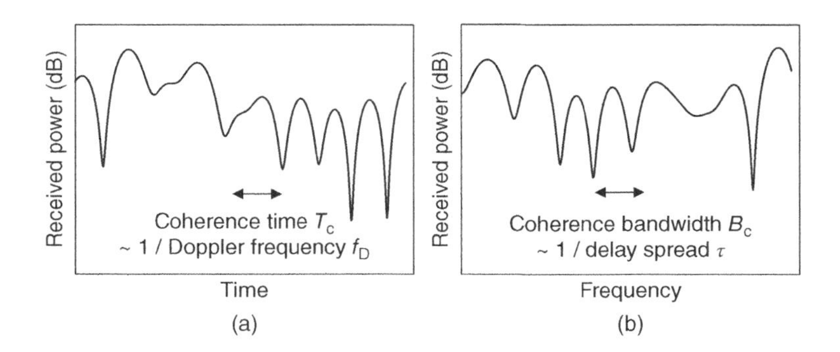

Figure 3.10 Fading as a function of (a) time and (b) frequency

Here f_D is the mobile's Doppler frequency:

$$f_D = \frac{v}{c} f_C \tag{3.12}$$

where f_C is the carrier frequency, v is the speed of the mobile and c is the speed of light ($3 \times 10^8\ \mathrm{ms}^{-1}$). For example, a pedestrian might walk with a speed of $1\ \mathrm{ms}^{-1}$ ($3.6\ \mathrm{km\,h}^{-1}$). At a carrier frequency of 1500 MHz, the resulting Doppler shift is 5 Hz, giving a coherence time of about 200 ms. Faster mobiles move through the interference pattern more quickly, so their coherence time is correspondingly less.

If the carrier frequency changes, then the wavelength of the radio signal changes. This also makes the interference pattern change between constructive and destructive, so fading is a function of frequency as well (Figure 3.10b). The amplitude and phase of the received signal vary over a frequency scale called the *coherence bandwidth,* B_c, which can be estimated as follows:

$$B_c \approx \frac{1}{\tau} \tag{3.13}$$

Here, τ is the *delay spread* of the radio channel, which is the difference between the arrival times of the earliest and latest rays. It can be calculated as follows:

$$\tau = \frac{\Delta L}{c} \tag{3.14}$$

where ΔL is the difference between the path lengths of the longest and shortest rays. In a macrocell, a typical path difference might be 100 m, giving a delay spread of 0.33 μs and a coherence bandwidth of around 3 MHz. Larger delay spreads can arise in larger cells. If, for example, the path difference increases to 1000 m, then the delay spread increases to 3.33 μs and the coherence bandwidth falls to around 300 kHz.

3.3.4 Inter-symbol Interference

If the path lengths of the longest and shortest rays are different, then symbols travelling on those rays will reach the receiver at different times. In particular, the receiver can start to receive one symbol on a short direct ray, while it is still receiving the previous symbol on a longer reflected ray. The two symbols therefore overlap at the receiver (Figure 3.11), causing another problem known as *inter-symbol interference* (ISI).

Let us continue the last example from the previous section, in which the delay spread τ was 3.33 μs. If the symbol rate is 120 ksps, then the symbol duration is 8.33 μs, so the symbols on the longest and shortest rays overlap by 40%. This causes a large amount of inter-symbol interference, which will greatly increase the error rate in the receiver. As the data rate increases, so the symbol duration falls and the problem becomes progressively worse. This makes inter-symbol interference a problem for any high-data rate communication system.

In these discussions, we have seen that frequency-dependent fading and inter-symbol interference are both important if the delay spread is large. In fact they are different ways of looking at the same underlying phenomenon: a large delay spread causes both frequency-dependent fading and ISI. 2G and 3G communication systems often combat the two effects using a device known as an *equalizer*, which passes the received signal through a filter that tries to model the time delays and undo their effect. Unfortunately equalizers are complex devices and are far

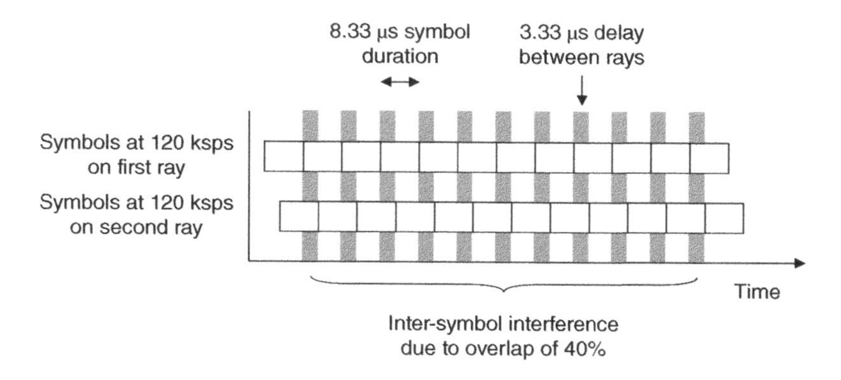

Figure 3.11 Generation of inter-symbol interference in a multipath environment

from perfect. In Chapter 4, we will see how the multiple access technique of OFDMA can deal with these issues in a far more direct way.

3.4 Error Management

3.4.1 Forward Error Correction

In the earlier sections, we saw that noise and interference lead to errors in a wireless communication receiver. These are bad enough during voice calls, but are even more damaging to important information such as web pages and emails. Fortunately, there are several ways to solve the problem.

The most important technique is *forward error correction*. In this technique, the transmitted information is represented using a *codeword* that typically contains two or three times as many bits. The extra bits supply additional, redundant data that allow the receiver to recover the original information sequence. For example, a transmitter might represent the information sequence 101 using the codeword 110010111. After an error in the second bit, the receiver might recover the codeword 100010111. If the coding scheme has been well designed, then the receiver can conclude that this is not a valid codeword, and that the most likely transmitted codeword was 110010111. The receiver has therefore corrected the bit error and can recover the original information. The effect is very like written English, which contains redundant letters that allow the reader to understand the underlying information, even in the presence of spelling mistakes.

The *coding rate* is the number of information bits divided by the number of transmitted bits ($^1/_3$ in the example above). Usually, forward error correction algorithms operate with a fixed coding rate. Despite this, a wireless transmitter can still adjust the coding rate using the two-stage process shown in Figure 3.12. In the first stage, the information bits are passed through a fixed-rate coder. The main algorithm used by LTE is known as *turbo coding* and has a fixed coding rate of $^1/_3$. In the second stage, called *rate matching*, some of the coded bits are selected for transmission, while the others are discarded in a process known as *puncturing*. The receiver has a copy of the puncturing algorithm, so it can insert dummy bits at the points

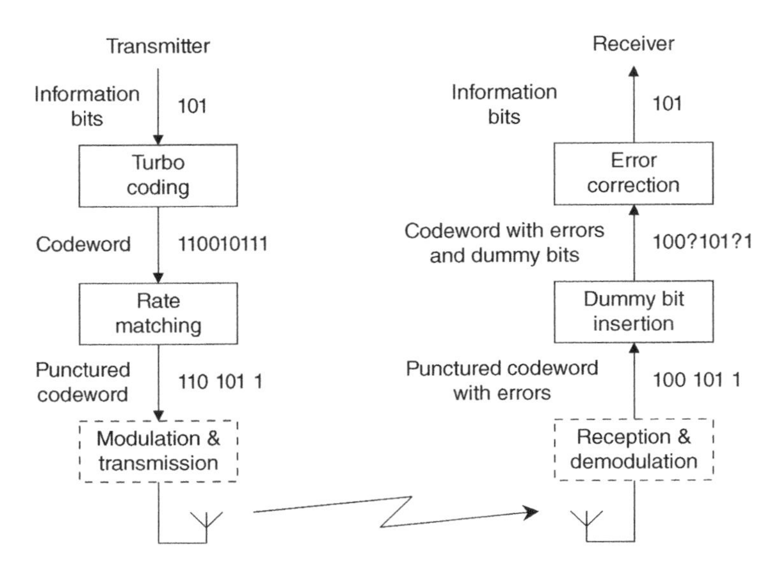

Figure 3.12 Block diagram of a transmitter and receiver using forward error correction and rate matching

where information was discarded. It can then pass the result through a turbo decoder for error correction.

Changes in the coding rate have a similar effect to changes in the modulation scheme. If the coding rate is low, then the transmitted data contain many redundant bits. This allows the receiver to correct a large number of errors and to operate successfully at a low SINR, but at the expense of a low information rate. If the coding rate is close to 1, then the information rate is higher but the system is more vulnerable to errors. LTE exploits this with a similar trade-off to the one we saw earlier, by transmitting with a high coding rate if the received SINR is high and vice versa.

3.4.2 Automatic Repeat Request

Automatic repeat request (ARQ) is another error management technique, which is shown in Figure 3.13. Here, the transmitter takes a block of information bits and uses them to compute some extra bits that are known as a *cyclic redundancy check* (CRC). It appends these to the information block and then transmits the two sets of data in the usual way.

The receiver separates the two fields and uses the information bits to compute the expected CRC bits. If the observed and expected CRC bits are the same, then it concludes that the information has been received correctly and sends a positive acknowledgement back to the transmitter. If the CRC bits are different, it concludes that an error has occurred and sends a negative acknowledgement to request a re-transmission. Positive and negative acknowledgements are often abbreviated to ACK and NACK respectively.

A wireless communication system often combines the two error management techniques that we have been describing. Such a system corrects most of the bit errors by the use of forward

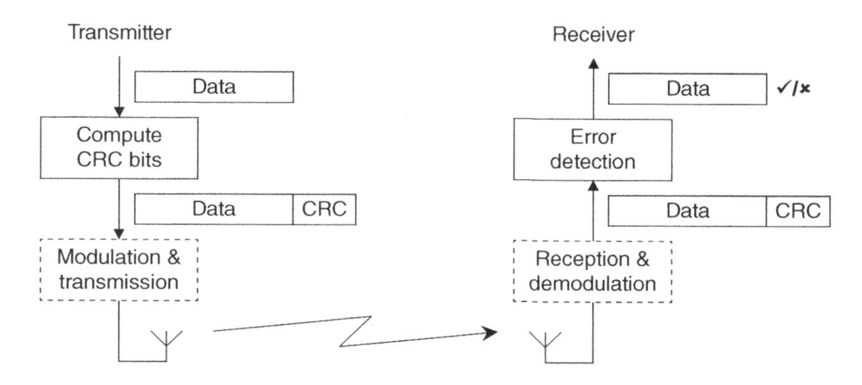

Figure 3.13 Block diagram of a transmitter and receiver using automatic repeat request

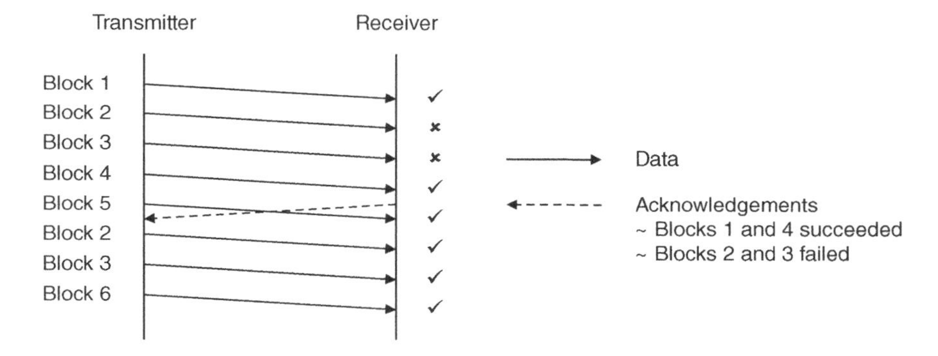

Figure 3.14 Operation of a selective re-transmission ARQ scheme

error correction and then uses automatic repeat requests to handle the remaining errors that leak through.

Normally, ARQ uses a technique called *selective re-transmission* (Figure 3.14), in which the receiver waits for several blocks of data to arrive before acknowledging them all. This allows the transmitter to continue sending data without waiting for an acknowledgement, but it means that any re-transmitted data can take a long time to arrive. Consequently, this technique is only suitable for non-real-time streams such as web pages and emails.

3.4.3 Hybrid ARQ

The ARQ technique from the previous section works well, but has one shortcoming. If a block of data fails the cyclic redundancy check, then the receiver throws it away, despite the fact that it contains some useful signal energy. If we could find a way to use that signal energy, then we might be able to design a more powerful receiver.

This idea is implemented in a technique known as *hybrid ARQ* (HARQ) which is shown in Figure 3.15. Here, the transmitter sends the data as before. The receiver demodulates the

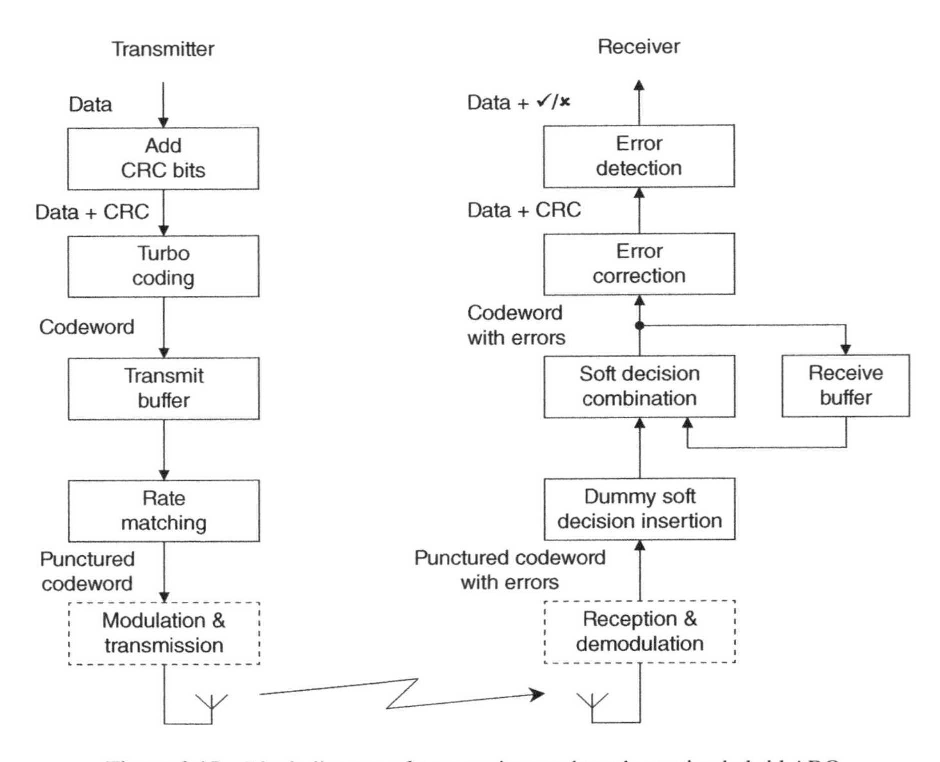

Figure 3.15 Block diagram of a transmitter and receiver using hybrid ARQ

incoming data, but this time it passes the soft decisions up to the next stage, instead of the hard decisions. It inserts zero soft decisions to account for any bits that the transmitter removed and stores the resulting codeword in a buffer. It then passes the codeword through the stages of error correction and error detection, and sends an acknowledgement back to the transmitter.

If the cyclic redundancy check fails, then the transmitter sends the data again. This time, however, the receiver combines the data from the first transmission and the re-transmission, by adding the soft decisions. This increases the signal energy at the receiver, so it increases the likelihood of a CRC pass. As a result, this scheme performs better than the basic ARQ technique, in which the first transmission was discarded.

Normally, hybrid ARQ uses a re-transmission technique called *stop-and-wait*, in which the transmitter waits for an acknowledgment before sending new data or a re-transmission. This simplifies the design and reduces the time delays in the system, which can make hybrid ARQ acceptable even for real-time streams such as voice. However, it also means that the transmitter has to pause while waiting for the acknowledgement to arrive. To prevent the throughput from falling, the system shares the data amongst several *hybrid ARQ processes*, which are multiple copies of Figure 3.15. One process can then transmit while the others are waiting for acknowledgements, in the manner shown in Figure 3.16.

The use of multiple hybrid ARQ processes means that the receiver decodes data blocks in a different order from the one in which they were transmitted. In Figure 3.16, for example, block 3 is transmitted four times and is only decoded some time after block 4. To deal with

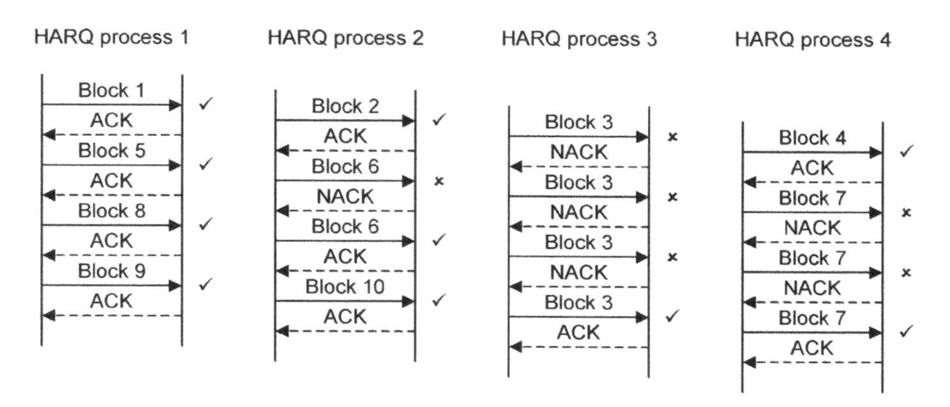

Figure 3.16 Operation of multiple hybrid ARQ processes in conjunction with a stop-and-wait re-transmission scheme

that problem, the receiver includes a re-ordering function that accepts the decoded blocks and returns them to their initial order.

There is one final problem. If an initial transmission is badly corrupted by interference, then it may take several re-transmissions before the interference is overcome. To limit the resulting time delays, a hybrid ARQ process is usually configured so that it gives up after a few unsuccessful attempts to transfer a block of data. At a higher level, a basic ARQ receiver can detect the problem, and can instruct the transmitter to send the block again from the beginning. LTE implements this technique, by the use of hybrid ARQ in the physical layer, backed up by a basic ARQ scheme in the radio link control protocol.

References

1. Goldsmith, A. (2005) *Wireless Communications*, Cambridge University Press.
2. Molisch, A.F. (2010) *Wireless Communications*, 2nd edn, John Wiley & Sons, Ltd, Chichester.
3. Rappaport, T.S. (2001) *Wireless Communications: Principles and Practice*, 2nd edn, Prentice Hall.
4. Tse, D. and Viswanath, P. (2005) *Fundamentals of Wireless Communication*, Cambridge University Press.
5. Parsons, J.D. (2000) *The Mobile Radio Propagation Channel*, 2nd edn, John Wiley & Sons, Ltd, Chichester.
6. Saunders, S. and Aragón-Zavala, A. (2007) *Antennas and Propagation for Wireless Communication Systems*, 2nd edn, John Wiley & Sons, Ltd, Chichester.
7. 3GPP TS 36.211 (2013) Physical Channels and Modulation, Release 11, Sections 5.8, 6.13, February 2013.

4

Orthogonal Frequency Division Multiple Access

The technique used for radio transmission and reception in LTE is known as orthogonal frequency division multiple access (OFDMA). OFDMA carries out the same functions as any other multiple access technique, by allowing the base station to communicate with several different mobiles at the same time. However, it is also a powerful way to improve the system's spectral efficiency and to minimize the problems of fading and inter-symbol interference that we introduced in Chapter 3. In this chapter, we will describe the basic principles of OFDMA and show the benefits that arise when it is used in a mobile cellular network. We will also cover a modified radio transmission technique, known as single carrier frequency division multiple access (SC-FDMA), which is used for the LTE uplink.

OFDMA is also used by several other radio communication systems, such as wireless local area networks (IEEE 802.11 versions a, g and n) and WiMAX (IEEE 802.16), as well as in digital television and radio broadcasting. However, LTE is the first system to have made use of SC-FDMA.

4.1 Principles of OFDMA

4.1.1 Sub-carriers

In Chapter 3, we saw how a traditional communication system transmits data by modulating a carrier signal. LTE uses a modified version of this technique known as *orthogonal frequency division multiplexing* (OFDM). As shown in Figure 4.1, an OFDM transmitter takes a block of symbols from the outgoing information stream and transmits each symbol on a different radio frequency that is known as a *sub-carrier*. The bandwidth of each individual sub-carrier is small, so it can only support a low symbol rate. Collectively, however, the sub-carriers occupy the same bandwidth as a traditional single carrier system. If other issues remain unchanged, their collective symbol rate is the same.

An Introduction to LTE: LTE, LTE-Advanced, SAE, VoLTE and 4G Mobile Communications, Second Edition.
Christopher Cox.

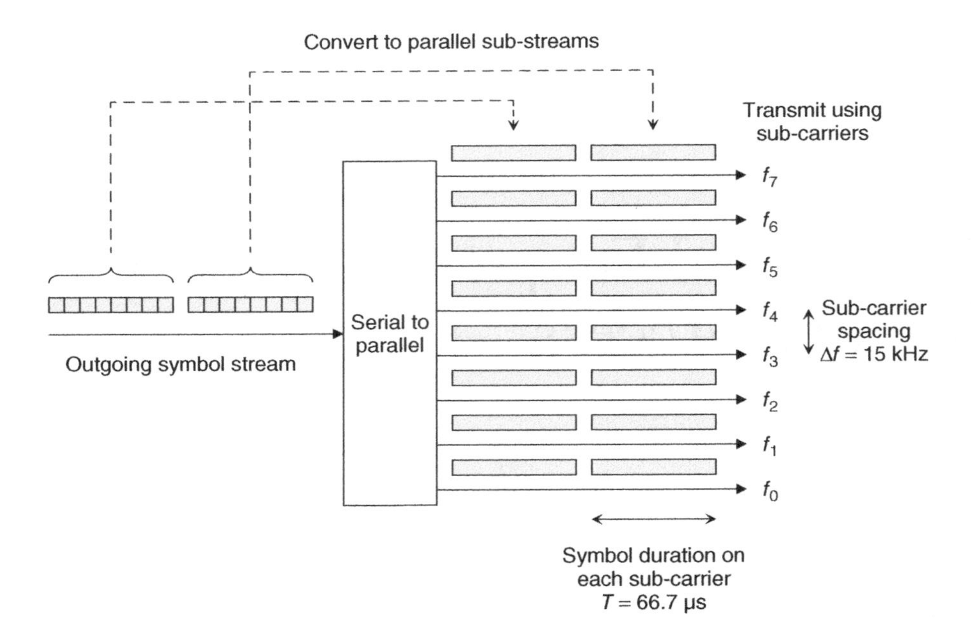

Figure 4.1 Division of the frequency band into sub-carriers using OFDM

In OFDM, the sub-carrier spacing Δf is related to the symbol duration on each individual sub-carrier, T, as follows:

$$\Delta f = \frac{1}{T} \tag{4.1}$$

For the moment this is just an arbitrary choice: the reasons will become clear later on, but it is consistent with the idea that the sub-carrier's bandwidth and symbol rate should be roughly the same. In LTE, the symbol duration is 66.7 μs, so the sub-carrier spacing is 15 kHz. Each cell can support a maximum of 1200 sub-carriers, which occupy the central 18 MHz of a 20 MHz allocation. If the cell's bandwidth is smaller, then we simply reduce the number of sub-carriers proportionately.

On the basis of these figures, we would expect each sub-carrier to have a symbol rate of 15 ksps. In practice, the symbol rate is slightly less, normally 14 ksps, because consecutive symbols are separated by a small gap known as the *cyclic prefix* (CP). We will see the reasons for introducing the cyclic prefix in due course.

4.1.2 The OFDM Transmitter

Figure 4.2 shows the most important components of an OFDM transmitter. The transmitter accepts a stream of bits from higher layer protocols and converts them to symbols using the chosen modulation scheme; for example, quadrature phase shift keying (QPSK).

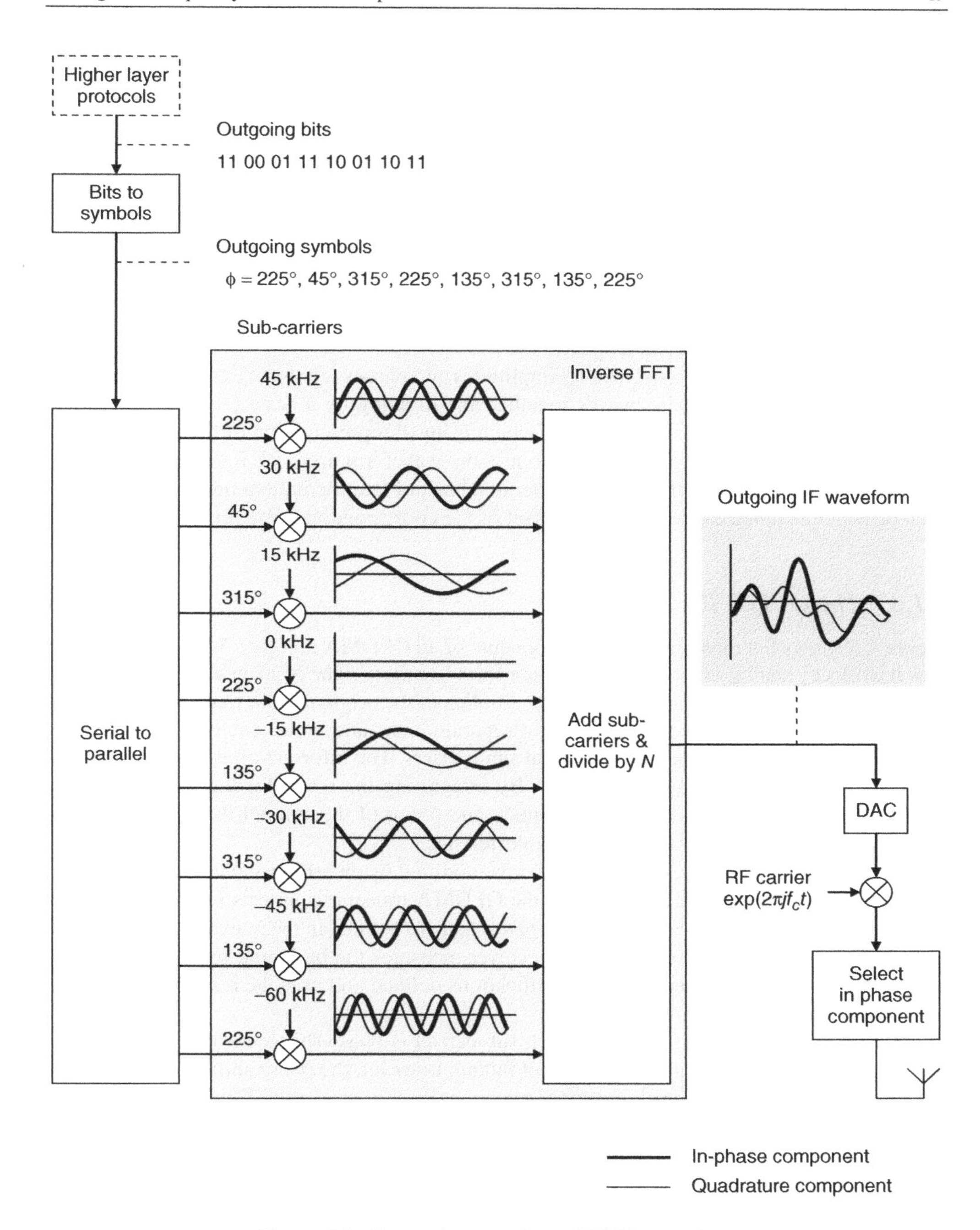

Figure 4.2 Processing steps in an OFDM transmitter

The serial-to-parallel converter then takes a block of *N* symbols, eight in this example, and directs them onto *N* parallel sub-streams.

The transmitter mixes each sub-stream with a sub-carrier, whose frequency is an integer multiple of 15 kHz. By analogy with the modulator from Chapter 3, we can interpret each sub-carrier as an intermediate frequency (IF), whose modulated signal will eventually be transmitted at a slightly different frequency from the radio frequency (RF) carrier, f_C. For consistency with the LTE specifications, it is convenient to place the sub-carriers at frequencies of -60, -45, ..., +45 kHz so that the RF carrier is roughly in the centre of the transmission band. The symbol duration is the reciprocal of the sub-carrier spacing, so the 15 kHz sub-carrier goes through one cycle during each 66.7 μs symbol, while the sub-carriers at 30 and 45 kHz go through two and three cycles respectively.

We now have eight sine waves, whose amplitudes and phases represent the eight transmitted symbols. By adding these sine waves together and dividing by a scale factor of *N*, we can generate a single time-domain waveform, which is an IF representation of the signal that we need to send. The only remaining task is to mix the waveform up to RF for transmission. We can therefore interpret the OFDM transmitter as a bank of intermediate frequency modulators, each of which is tuned to the frequency offset of the corresponding sub-carrier.

4.1.3 The OFDM Receiver

Figure 4.3 shows the most important components of an OFDMA receiver. We can understand how it works by analogy with the demodulator from Chapter 3. The receiver accepts the incoming signal, mixes it with a complex conjugate replica of the original radio frequency and filters the result. It then directs the signal to eight separate paths and mixes each one with a complex conjugate replica of one of the original sub-carriers. The information that arrived on that sub-carrier is now at a frequency of zero. By integrating the result for the duration of one symbol, the receiver can extract the amplitude and phase of the symbol that arrived on that sub-carrier, while rejecting any noise and interference.

As we noted in Chapter 3, each sub-carrier is modified by an arbitrary phase shift ψ before it reaches the receiver. To deal with this, the OFDMA transmitter injects reference symbols into the data stream with an amplitude and phase that are defined in the relevant specifications. In the process of channel estimation, the receiver measures the incoming reference symbols, compares them with the ones that the specifications defined and uses the result to remove the phase shift from the incoming signal.

In Figure 4.3, we have assumed that each sub-carrier is phase shifted by the same amount, 30°. In the presence of frequency-dependent fading, however, the phase shift is a slowly varying function of frequency as well as time. To account for this issue, the LTE reference symbols are scattered across the time and frequency domains in the manner that will be described in Chapter 7. The receiver can then measure the phase shift ψ as a function of frequency and can apply a different phase correction to each individual sub-carrier.

After the phase removal stage, the receiver returns the symbols to their original order by means of a parallel-to-serial converter, and recovers the transmitted bits. We can therefore interpret the OFDM receiver as a bank of intermediate frequency demodulators, each of which is tuned to the frequency offset of the corresponding sub-carrier.

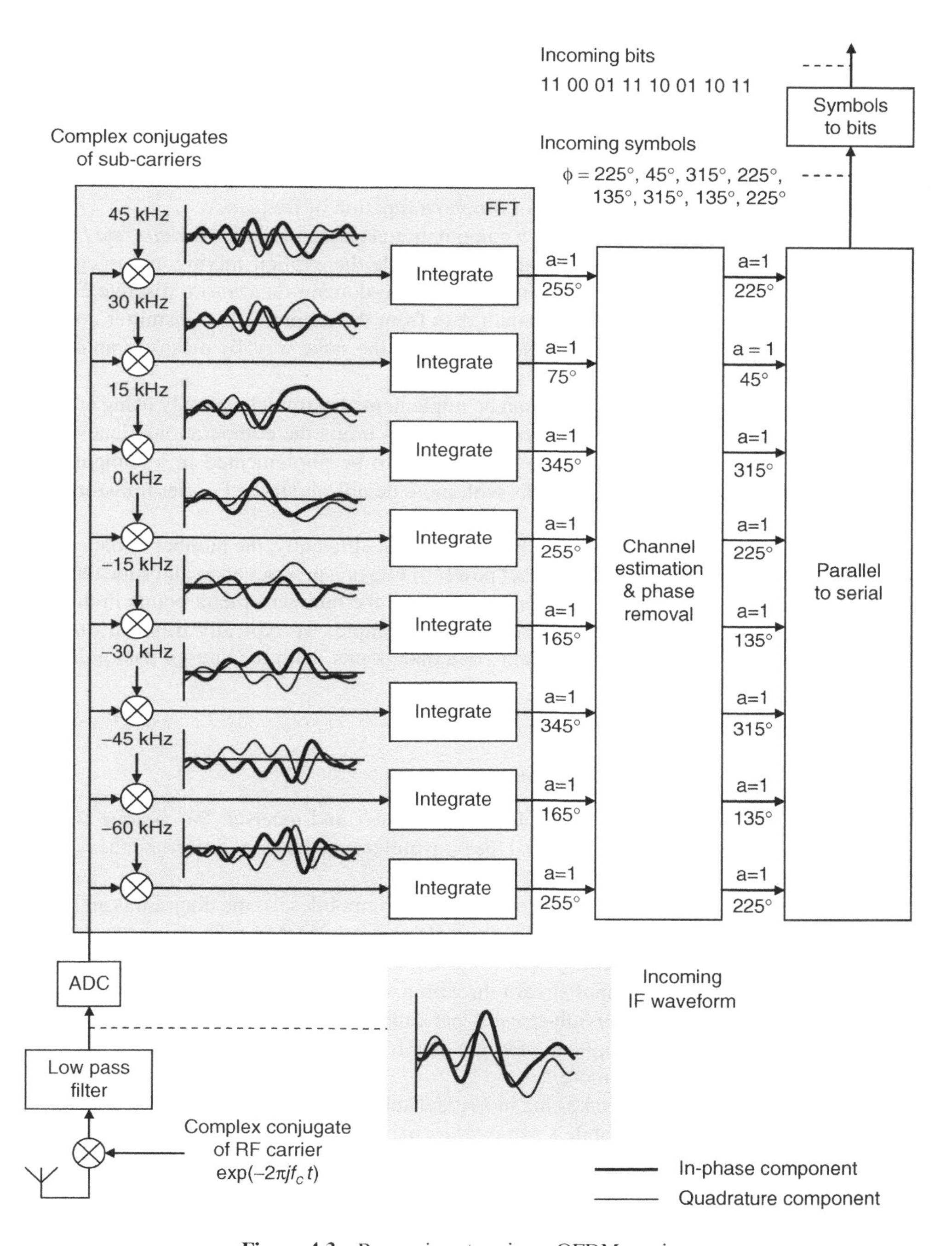

Figure 4.3 Processing steps in an OFDM receiver

4.1.4 The Fast Fourier Transform

Now let us look at two important points in the receiver's processing chain. Immediately after the low-pass filter, the data are the in-phase and quadrature components of the incoming signal as a function of time. After the integration stage later on, the data are the amplitude and phase of each sub-carrier, as a function of frequency. We can see that the mixing and integration steps have converted the data from a function of time to a function of frequency.

This conversion is actually a well-known computational technique called the *discrete Fourier transform* (DFT). By using this technique, we can hide the explicit mixing and integration steps in Figure 4.3. Instead, we can just pass the time-domain data into a discrete Fourier transform and pick up the frequency-domain data from the output. The transmitter converts frequency-domain data to time-domain data in exactly the same way, by means of an *inverse discrete Fourier transform.*

In turn, the discrete Fourier transform can be implemented extremely quickly using an algorithm known as the *fast Fourier transform* (FFT). This limits the computational load on the transmitter and receiver, and allows the two devices to be implemented in a computationally efficient way. There are several books with more details about the Fourier transform, for example References [1, 2].

There is one important restriction: for the FFT to work efficiently, the number of data points in each calculation should be either an exact power of two or a product of small prime numbers alone. We typically handle this restriction by rounding the number of data points in the FFT up to the next highest power of two. In LTE, for example, we typically transmit on 1200 sub-carriers by means of FFTs that contain 2048 data points, with the unused data points set to zero.

4.1.5 Block Diagram of OFDMA

Figure 4.4 is a block diagram of an OFDMA transmitter and receiver. We assume that the system is operating on the downlink, so that the transmitter is in the base station and the receiver is in the mobile.

The base station sends streams of bits to three different mobiles, so the diagram is an implementation of *orthogonal frequency division multiple access* (OFDMA). The base station modulates each bit stream independently, possibly using a different modulation scheme for each mobile. It then passes each symbol stream through a serial-to-parallel converter, to divide it into sub-streams. The number of sub-streams per mobile depends on the mobile's required data rate; for example, a voice application might only require a few sub-streams, while a video application might require many more.

The *resource element mapper* takes the individual sub-streams and chooses the sub-carriers on which to transmit them. A mobile's sub-carriers may lie in one contiguous block (as in the case of mobiles 1 and 3) or they may be divided (as for mobile 2). The resulting information is the amplitude and phase of each sub-carrier as a function of frequency. By passing it through an inverse FFT, we can compute the in-phase and quadrature components of the corresponding time-domain waveform, and can place them in the correct order by means of a parallel-to-serial converter (a stage that was hidden within Figure 4.2, but present nonetheless). After inserting the cyclic prefix that we noted earlier, the resulting signal can be mixed up to radio frequency and converted to analogue form for transmission.

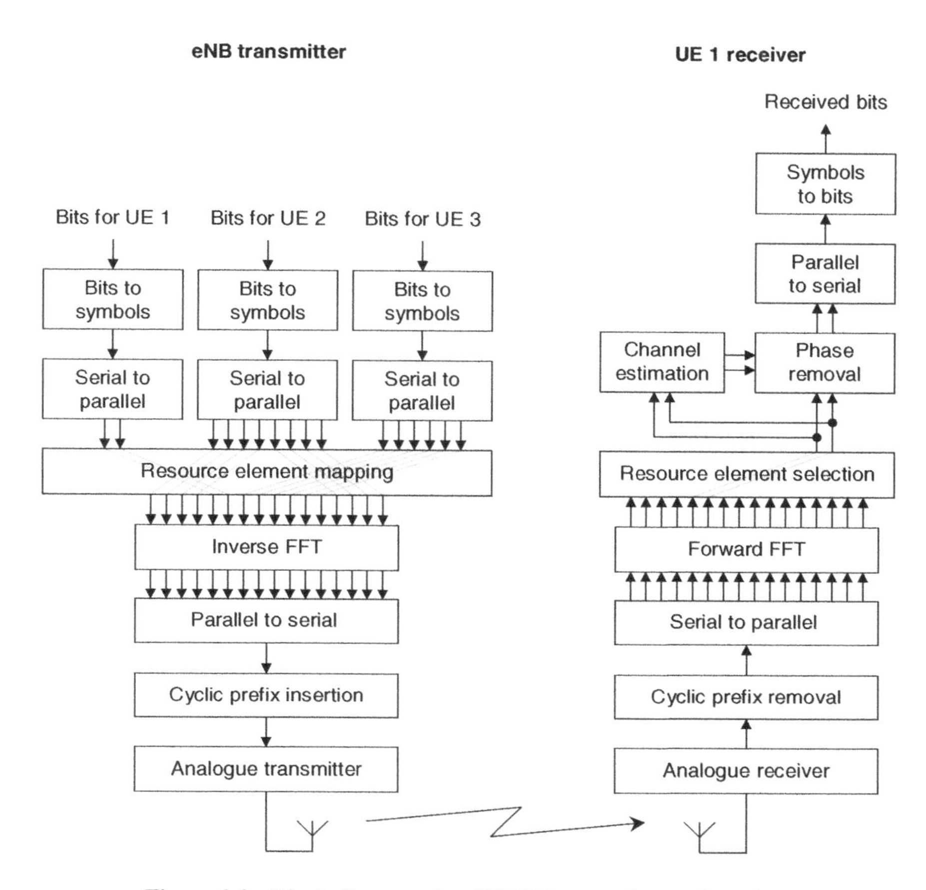

Figure 4.4 Block diagram of an OFDMA transmitter and receiver

The mobile reverses the process. It starts by sampling the incoming signal, converting it down to baseband and filtering it. It then removes the cyclic prefix and passes the data through a forward FFT, to recover the amplitude and phase of each sub-carrier. We now assume that the base station has already told the mobile which sub-carriers to use, through scheduling techniques that we will cover in Chapter 8. Using this knowledge, the mobile selects the required sub-carriers and recovers the transmitted information, while discarding the remainder. After the steps of channel estimation and phase removal, the mobile can recover the transmitted bits and can pass them to the higher layer protocols.

4.1.6 Details of the Fourier Transform

For the mathematically inclined, we now have enough information to write down the equation for the discrete Fourier transform. (Other readers may prefer to skip this section and move on.) Referring back to Figure 4.3, we can express the receiver's mixing and integration stages as follows:

$$Z(f_n) = \sum_{k=0}^{N-1} z(t_k) \exp(-2\pi j f_n t_k) \tag{4.2}$$

In this equation, $Z(f_n)$ is a complex number, which represents the amplitude and phase of the nth outgoing sub-carrier, at a frequency f_n, measured at the output from the integration stage. $z(t_k)$ is another complex number, which represents the in-phase and quadrature components of the kth incoming sample, at a time t_k, measured at the output from the low-pass filter. The term $\exp(-2\pi j f_n t_k)$ describes the receiver's complex conjugate replica of the original sub-carrier, while the summation describes the action of the integration stage. N is the number of incoming samples per symbol.

We now have to write down the values of t_k and f_n. If the incoming data samples are uniformly spaced with an interval Δt, then we can write the sample times as follows:

$$\begin{aligned} t_k &= k\Delta t \qquad k = 0, 1, 2, \ldots N-1 \\ &= \frac{kT}{N} \end{aligned} \tag{4.3}$$

where Δt is the sampling interval and T is the symbol duration. It is reasonable, and in fact correct, to assume that the number of sub-carriers should equal the number of samples per symbol, so that the output from Equation 4.2 carries the same amount of information as the input. We can then write the sub-carrier frequencies as follows, subject to one issue that we will revisit below:

$$\begin{aligned} f_n &= n\Delta f \qquad n = 0, 1, 2, \ldots N-1 \\ &= \frac{n}{T} \end{aligned} \tag{4.4}$$

where Δf is the sub-carrier spacing. Substituting these results into Equation 4.2 gives the equation for the discrete Fourier transform:

$$Z_n = \sum_{k=0}^{N-1} z_k \exp\left(-\frac{2\pi jnk}{N}\right) \tag{4.5}$$

where $z_k = z(t_k)$ and $Z_n = Z(f_n)$. In the same way, we can write the transmitter's mixing, addition and scaling stages from Figure 4.2 as follows:

$$z(t_k) = \frac{1}{N} \sum_{n=0}^{N-1} Z(f_n) \exp(2\pi j f_n t_k) \tag{4.6}$$

If we make the same substitutions as before, then we arrive at the equation for the inverse discrete Fourier transform:

$$z_k = \frac{1}{N} \sum_{n=0}^{N-1} Z_n \exp\left(\frac{2\pi jnk}{N}\right) \tag{4.7}$$

Earlier, we noted an issue with the choice of sub-carrier frequencies. In Equation 4.4, we placed the sub-carriers at frequencies of $n\Delta f$, where n ran from 0 to $N-1$. That choice implies that the eight sub-carriers from our earlier examples are at offsets of 0, 15, 30, ..., 105 kHz from the RF carrier. This sounds different from the choice we made in Figures 4.2 and 4.3, but is in fact exactly the same. When using eight sub-carriers, the sampling interval from

Equation 4.3 is 8.33 μs. During that time, the 60 kHz sub-carrier goes through exactly half a cycle, and is indistinguishable from a sub-carrier at −60 kHz that also goes through half a cycle. By the same argument, the sub-carriers at 75, 90 and 105 kHz are indistinguishable from sub-carriers at −45, −30 and −15 kHz respectively. So our choice of sub-carriers is in fact exactly the same as the one we made earlier: we have just labelled them in a different way.

4.2 Benefits and Additional Features of OFDMA

4.2.1 Orthogonal Sub-carriers

So far, OFDMA has just been a different way of sending data from a transmitter to a receiver. It does, however, bring several benefits. In this section we will run through the benefits of OFDMA and describe some of its additional features. To begin the discussion, Figure 4.5 adds some detail to our treatment of the OFDMA receiver, using an example in which information is arriving on the 15 kHz sub-carrier alone. When the incoming signal is mixed with a complex conjugate replica of the 15 kHz sub-carrier, the result is at a frequency of zero. The integration process then extracts the amplitude and phase of that sub-carrier, as required.

In the branch of the receiver that lies one level below, the incoming signal is mixed with a complex conjugate replica of the 0 kHz sub-carrier, so the result remains at a frequency of 15 kHz. During the integration period, the resulting signal goes through exactly one cycle, so its in-phase and quadrature components both sum to zero. We therefore conclude that no information has arrived on the 0 kHz sub-carrier, which is what we expect. One level further below, the incoming signal is mixed with a complex conjugate replica of the −15 kHz sub-carrier, so the result moves to a frequency of 30 kHz. During the integration period, the resulting signal goes through exactly two cycles, so its in-phase and quadrature components both sum to zero in exactly the same way.

We can draw the following conclusion. If the transmitter sends a signal on one sub-carrier, then the receiver detects a signal on that one sub-carrier alone and does not pick up interference on any of the others. Sub-carriers with this property are said to be *orthogonal*.

Because the sub-carriers are orthogonal, an OFDMA transmitter can pack them very closely together, without any risk of interference between them. This implies that OFDMA uses its frequency band in a very efficient way, and is one of the reasons why the spectral efficiency of LTE is so much greater than that of previous mobile telecommunication systems. Orthogonality relies on the fact that the symbol duration T is the reciprocal of the sub-carrier spacing Δf so that, in the example above, the inputs to the 0 kHz and −15 kHz integrators go through exactly one and two cycles respectively. It therefore justifies the choice of symbol duration that we made at the start of the chapter.

4.2.2 Choice of Sub-carrier Spacing

The argument in the previous section works fine if the mobile is stationary. If the mobile is moving, then the incoming signal is Doppler shifted to a higher or lower frequency. In Chapter 3, for example, we estimated that a moving vehicle has a Doppler shift of about 150 Hz, so the 15 kHz sub-carrier actually arrives at a frequency of 15.150 kHz. When the signal is mixed with a complex conjugate replica of the 0 kHz sub-carrier, the result remains at 15.150 kHz.

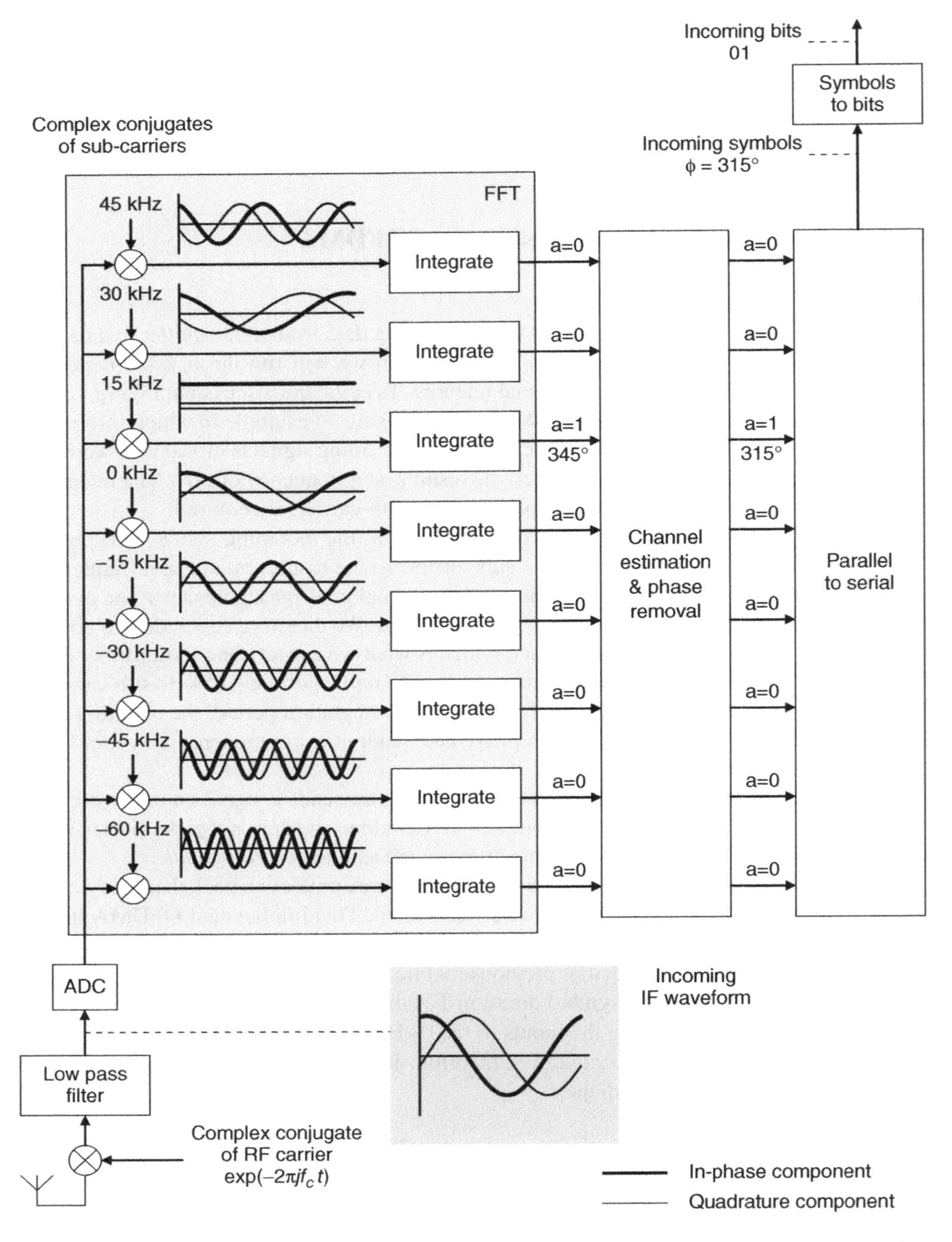

Figure 4.5 Processing steps in an OFDM receiver, in which the information arrives on the 15 kHz sub-carrier alone

During the integration period, that signal goes through slightly more than one cycle, so its in-phase and quadrature components sum to a result that is no longer zero. We therefore receive interference on the 0 kHz sub-carrier and, by extension, all of the others, so we have lost orthogonality.

We might think that the receiver could avoid the problem by applying the same Doppler shift to its replicas of the original sub-carriers. Unfortunately, in a multipath environment, a mobile can be moving towards some rays, which are shifted to higher frequencies, but away from others, whose frequencies are lower. As a result, the incoming signals are not simply shifted; instead, they are blurred across a range of frequencies, which makes the idea unusable.

The amount of interference will still be acceptable, however, if the Doppler shift is much less than the sub-carrier spacing. We therefore need to choose the sub-carrier spacing Δf as follows:

$$\Delta f >> f_D \tag{4.8}$$

where f_D is the Doppler shift from Equation 3.12. LTE is designed to operate with a maximum mobile speed of 350 km h^{-1} and a maximum carrier frequency of about 3.5 GHz, which gives a maximum Doppler shift of about 1.1 kHz. This is 7% of the sub-carrier spacing, so it satisfies the constraint above.

4.2.3 *Frequency-Specific Scheduling*

Earlier, we saw that a base station can transmit to several mobiles at the same time by allocating them different groups of sub-carriers. The base station's exact choice of sub-carriers is influenced by the existence of frequency-dependent fading, in the manner shown in Figure 4.6.

In Chapter 3, we noted that the mobiles' received signal power can be a function of frequency. When using OFDMA, the frequency band is divided into sub-carriers, so the mobile can measure the received signal power on each one by means of the reference signals that we introduced above. The mobile can then group the measurements from nearby sub-carriers and can send the results to the base station using a feedback signal known as the *channel quality indicator* (CQI).

The base station uses the channel quality indicator in two ways. Firstly, the base station can transmit to the mobile using sub-carriers on which the received signal power is the strongest.

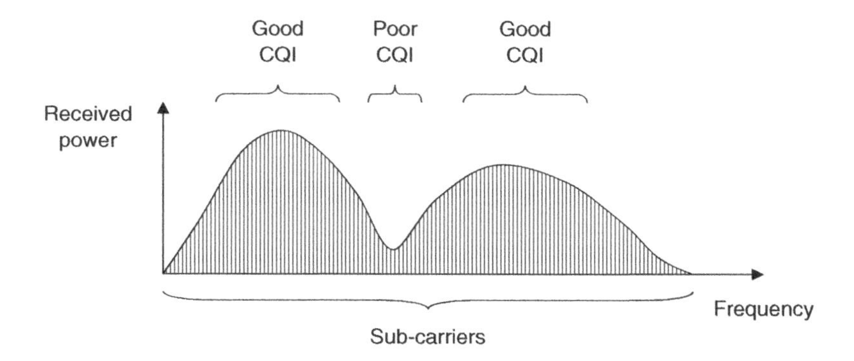

Figure 4.6 Frequency specific scheduling using OFDMA

The unwanted sub-carriers can be used by other mobiles because their fading patterns are different, while the allocation of sub-carriers can be regularly changed, once every millisecond in LTE. Secondly, the base station can use the result to determine the fastest modulation scheme and coding rate that the mobile can handle.

The technique does have limitations, as it is unlikely to help in scenarios with a low delay spread and a high coherence bandwidth. In other scenarios, however, an OFDMA transmitter can use this technique to minimize the impact of time- and frequency-dependent fading. We will give more details about the feedback process in Chapter 8.

4.2.4 *Reduction of Inter-symbol Interference*

In Chapter 3, we saw how high data rate transmission in a multipath environment leads to inter-symbol interference (ISI). In Figure 3.11, for example, the delay spread was 3.33 μs and the symbol rate was 120 ksps, so the symbols overlapped at the receiver by 40%. That led to interference and bit errors at the receiver.

If we share the data amongst several different sub-carriers by the use of OFDMA, then the symbol rate on each individual sub-carrier is several times lower than before, so the symbol duration is several times longer. This greatly reduces the levels of inter-symbol interference. Figure 4.7 shows a simple example. Here, we have divided the original data stream amongst eight sub-carriers, so the symbol rate on each sub-carrier is now 15 ksps and the symbol duration is 66.7 μs. If the delay spread remains at 3.33 μs, then the symbols overlap by only 5%. This reduces the amount of ISI to one eighth of what it was before and reduces the number of errors in the receiver. If the number of sub-carriers is larger, then the reduction of ISI is proportionally greater.

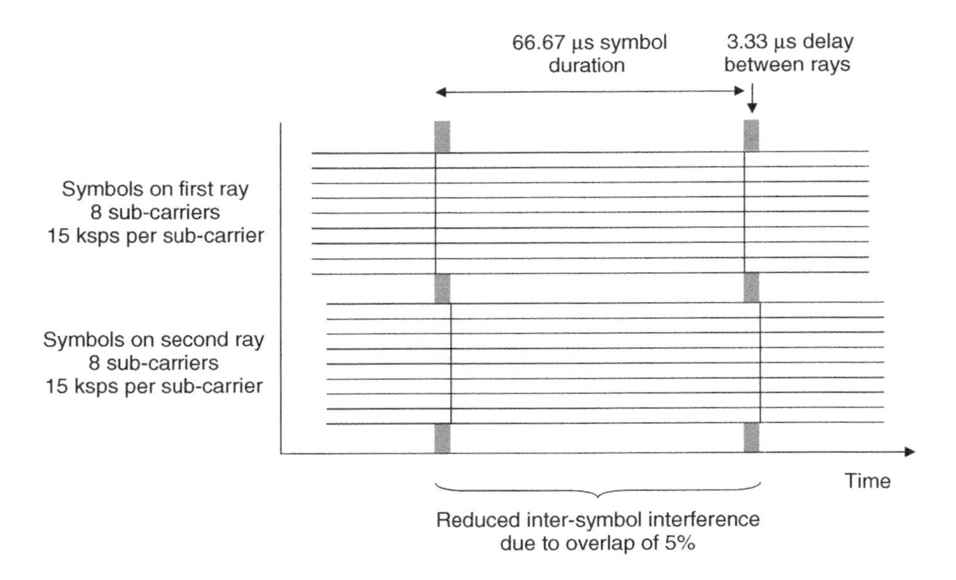

Figure 4.7 Reduction of inter-symbol interference by transmission on multiple sub-carriers

4.2.5 Cyclic Prefix Insertion

A further technique allows us to get rid of ISI altogether. The basic idea is to insert a *guard period* (GP) before each symbol, in which nothing is transmitted. If the guard period is longer than the delay spread, then the receiver can be confident of reading information from just one symbol at a time, without any overlap with the symbols that precede or follow. Naturally, the symbol reaches the receiver at different times on different rays, and some extra processing is required to tidy up the confusion. However, the extra processing is relatively straightforward.

LTE uses a slightly more complex technique known as *cyclic prefix insertion* (Figure 4.8). Here, the transmitter starts by inserting a guard period before each symbol, as before. However, it then copies data from the end of the symbol following, so as to fill up the guard period. If the cyclic prefix is longer than the delay spread, then the receiver can still be confident of reading information from just one symbol at a time.

We can see how cyclic prefix insertion works by looking at one sub-carrier (Figure 4.9). The transmitted signal is a sine wave, whose amplitude and phase change from one symbol to the next. As noted earlier, each symbol contains an exact number of cycles of the sine wave, so the amplitude and phase at the start of each symbol equal the amplitude and phase at the end. Because of this, the transmitted signal changes smoothly as we move from each cyclic prefix to the symbol following.

In a multipath environment, the receiver picks up multiple copies of the transmitted signal with multiple arrival times. These add together at the receive antenna, giving a sine wave with the same frequency but a different amplitude and phase. The received signal still changes smoothly at the transition from a cyclic prefix to the symbol that follows. There are a few glitches, but these are only at the start of the cyclic prefix and the end of the symbol, where the preceding and following symbols start to interfere.

The receiver processes the received signal within a window whose length equals the symbol duration, and discards the remainder. If the window is correctly placed, then the received signal is exactly what was transmitted, without any glitches, and subject only to an amplitude

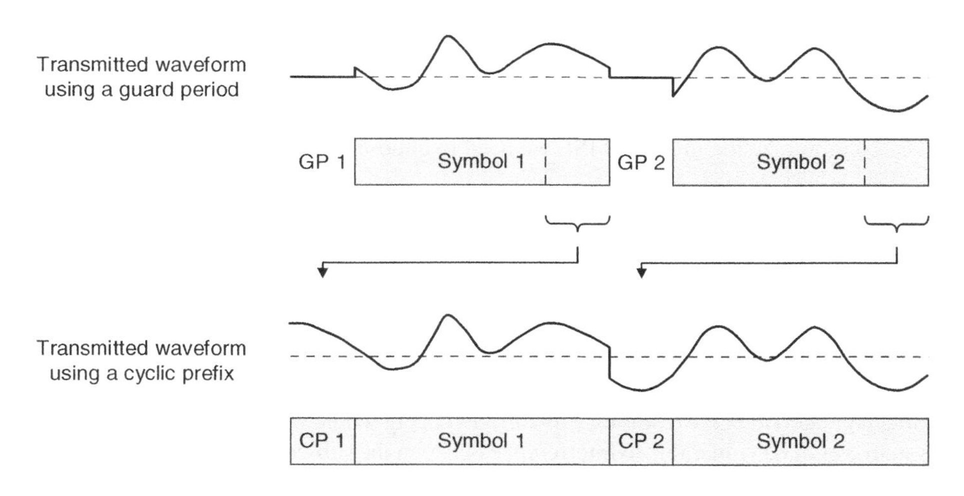

Figure 4.8 Operation of cyclic prefix insertion

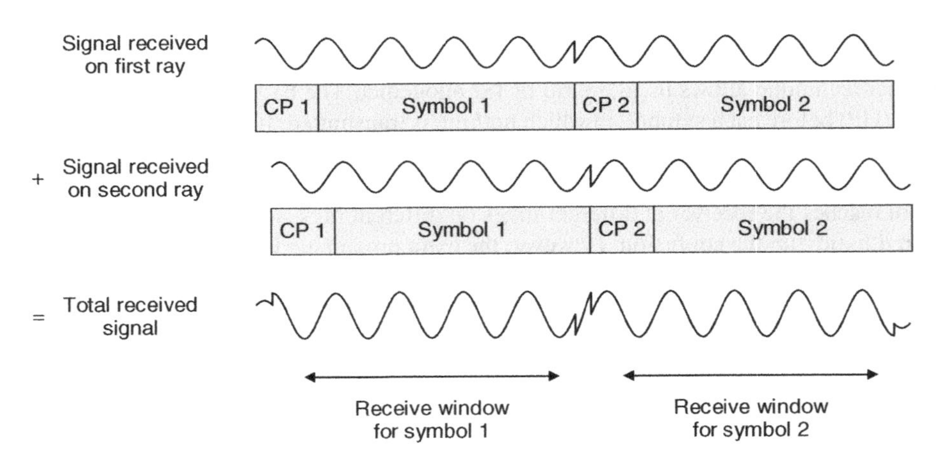

Figure 4.9 Operation of the cyclic prefix on a single sub-carrier

change and a phase shift. But the receiver can compensate for these using the channel estimation techniques described above. It can therefore handle the cyclic prefix without any extra processing at all.

Admittedly, the system uses multiple sub-carriers, not just one. However, we have already seen that the sub-carriers do not interfere with each other and can be treated independently, so the existence of multiple sub-carriers does not affect this argument at all.

Normally, LTE uses a cyclic prefix of about 4.7 μs. This corresponds to a maximum path difference of about 1.4 km between the lengths of the longest and shortest rays, which is enough for all but the very largest and most cluttered cells. The cyclic prefix reduces the symbol rate of each sub-carrier to the figure of 14 ksps that we quoted earlier, but this is a small price to pay for the removal of ISI.

4.2.6 *Choice of Symbol Duration*

The existence of inter-symbol interference places another constraint on the parameters used by LTE. To minimize the impact of ISI, we need to choose the symbol duration T as follows:

$$T >> \tau \tag{4.9}$$

where τ is the delay spread from Equation 3.14. As we noted earlier, LTE normally works with a maximum delay spread of about 4.7 μs. This is 7% of the 66.7 μs symbol duration, so it satisfies this second constraint.

We can draw the following conclusion. If the symbol duration were much less than 66.7 μs, then the system would be vulnerable to inter-symbol interference in large, cluttered cells. If it were much greater, then the resulting sub-carrier spacing would be much less than 15 kHz, and the system would be vulnerable to interference between the sub-carriers at high mobile speeds. The chosen symbol duration and sub-carrier spacing are the result of a trade-off between these two extremes.

4.2.7 Fractional Frequency Re-use

Using the techniques described above, one base station can send information to a large number of mobiles. However, a mobile communication system also has a large number of base stations, so every mobile has to receive a signal from one base station in the presence of interference from the others. We need a way to minimize the interference, so that the mobile can receive the information successfully.

Previous systems have used two different techniques. In GSM, nearby cells have different carrier frequencies. Typically, each cell might use a quarter of the total bandwidth, with a *re-use factor* of 25%. This technique reduces the interference between nearby cells, but it means that the frequency band is used inefficiently. In UMTS, each cell has the same carrier frequency, with a re-use factor of 100%. This technique uses the frequency band more efficiently than before, at the expense of increasing the interference in the system.

In an LTE network, every base station can use the same set of sub-carriers but can allocate them in a flexible way using a technique known as *fractional frequency re-use*. Figure 4.10 shows a simple static example for the LTE uplink, in which each base station controls one cell. The uplink sub-carriers are divided into three blocks, denoted as A, B and C. Each base station schedules transmissions from distant mobiles using one block of sub-carriers (block C in the case of the central base station) and schedules transmissions from nearby mobiles using the remaining two.

A distant mobile has to transmit with a high power to ensure that its signal can be successfully received. It therefore causes interference to neighbouring base stations, which are listening to mobiles of their own upon the same sub-carriers. By coordinating the use of sub-carriers as shown in Figure 4.10, we can ensure that the neighbouring base stations only use those sub-carriers for mobiles that are nearby. Those mobiles are only transmitting with a low power, so they can boost their power so as to overcome the interference and are unlikely to be badly affected. As a result, any interference problems can be significantly reduced.

LTE also supports dynamic techniques for fractional frequency re-use, in which nearby base stations coordinate their use of sub-carriers by means of signalling messages across the X2 interface. We will cover those signalling messages in Chapter 17.

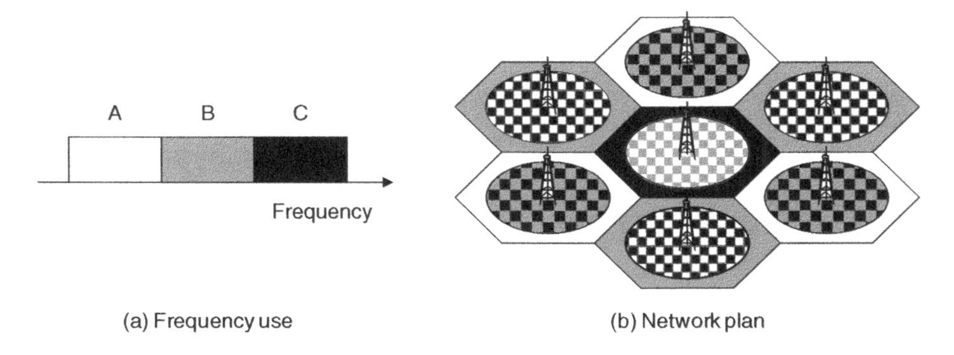

Figure 4.10 Example implementation of fractional frequency re-use when using OFDMA. (a) Use of the frequency domain. (b) Resulting network plan

4.3 Single Carrier Frequency Division Multiple Access

4.3.1 Power Variations From OFDMA

OFDMA works well on the LTE downlink. However, it has one disadvantage: the power of the transmitted signal is subject to rather large variations. To illustrate this, Figure 4.11a shows a set of sub-carriers that have been modulated using QPSK, and which therefore have constant power. The amplitude of the resulting signal (Figure 4.11b) varies widely, with maxima where the peaks of the sub-carriers coincide and zeros where they cancel. In turn, these variations are reflected in the power of the transmitted signal (Figure 4.11c), which is said to have a high *peak to average power ratio* (PAPR).

These power variations can cause problems for the transmitter's power amplifier. If the amplifier is linear, then the output power is proportional to the input, so the output waveform is exactly the shape that we require. If the amplifier is non-linear, then the output power is no longer proportional to the input, so the output waveform is distorted. Any distortion of the time-domain waveform will distort the frequency-domain power spectrum as well, so the signal will leak into adjacent frequency bands and will cause interference to other receivers.

In the downlink, the base station transmitters are large, expensive devices, so they can avoid the problem by using expensive power amplifiers that are very close to linear. In the uplink, a mobile transmitter has to be cheap, so does not have this option. This makes OFDMA unsuitable for the LTE uplink.

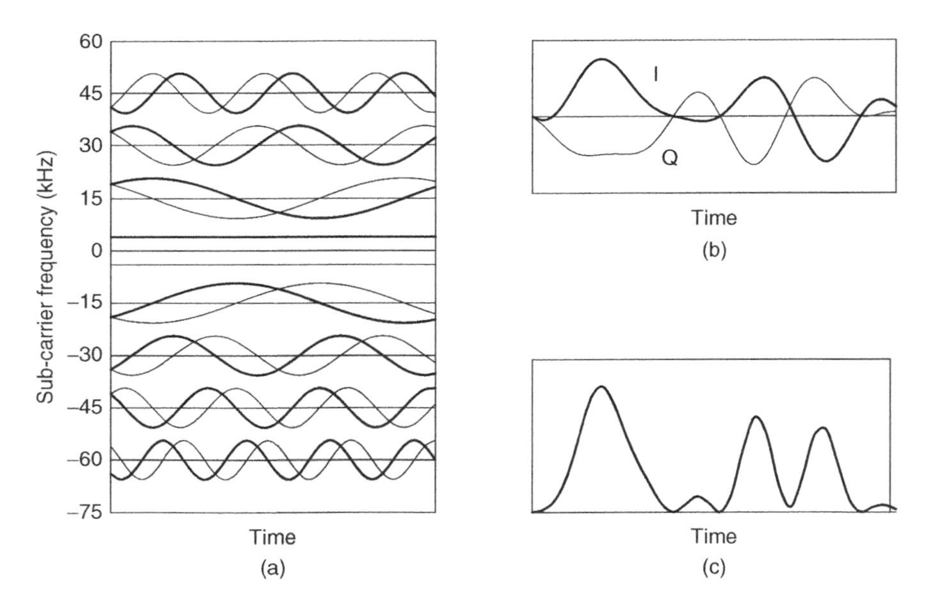

Figure 4.11 Example OFDMA waveform. (a) Amplitudes of the individual sub-carriers. (b) Amplitude of the resulting OFDMA waveform. (c) Power of the OFDMA waveform

4.3.2 *Block Diagram of SC-FDMA*

The power variations described above arise because there is a one-to-one mapping between symbols and sub-carriers. If we mixed the symbols together before placing them on the sub-carriers, then we might be able to adjust the transmitted signal and reduce its power variations. For example, when transmitting two symbols x_1 and x_2 on two sub-carriers, we might send their sum $x_1 + x_2$ on one sub-carrier, and their difference $x_1 - x_2$ on the other. We can use any mixing operation at all, as the receiver can reverse it: we just need to find one that minimizes the power variations in the transmitted signal.

It turns out that a suitable mixing operation is another FFT, this time a forward FFT. By including this operation, we arrive at a technique known as *single carrier frequency division multiple access* (SC-FDMA), which is illustrated in Figure 4.12.

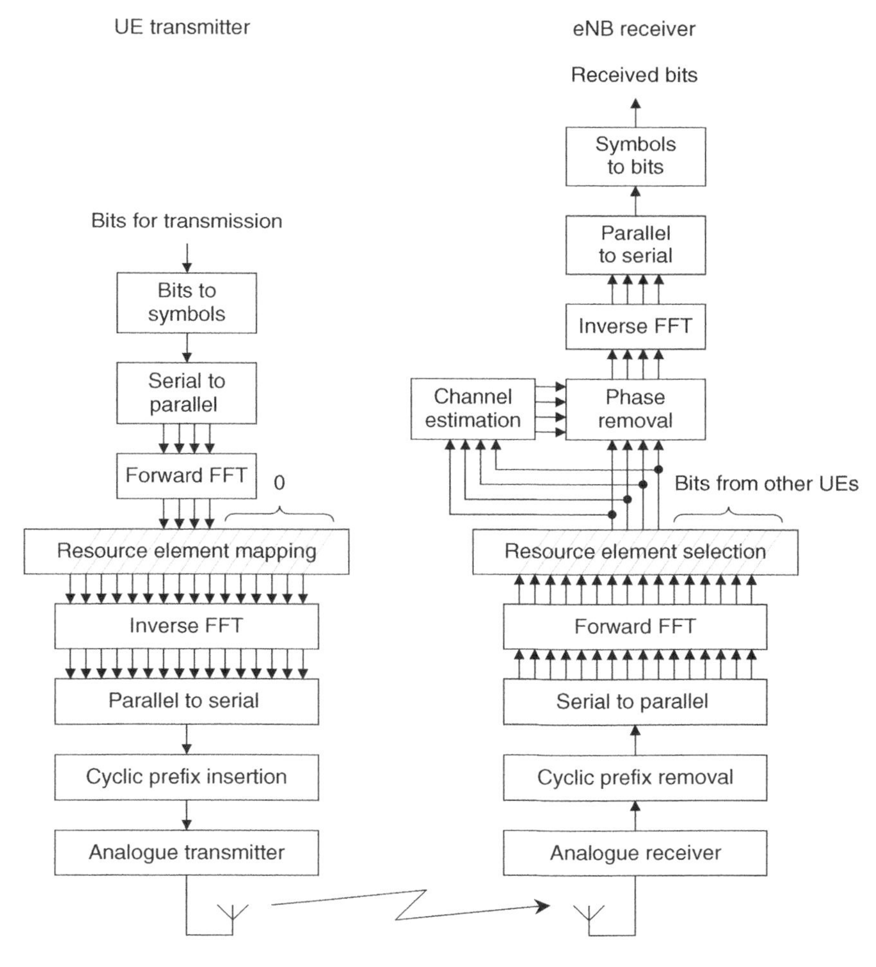

Figure 4.12 Block diagram of an SC-FDMA transmitter and receiver

In this diagram, there are three differences from OFDMA. The main difference is that the SC-FDMA transmitter includes an extra forward FFT, between the steps of serial-to-parallel conversion and resource element mapping. This mixes the symbols together in the manner required to minimize the power variations and is reversed by an inverse FFT in the receiver.

The second difference arises because the technique is used on the uplink. Because of this, the mobile transmitter only uses some of the sub-carriers: the others are set to zero, and are available for the other mobiles in the cell. Finally, each mobile transmits using a single, contiguous block of sub-carriers, without any internal gaps. This is implied by the name SC-FDMA and is necessary to keep the power variations to the lowest possible level.

We can understand how SC-FDMA works by looking at three key transmission steps: the forward FFT, the resource element mapper and the inverse FFT. The input to the forward FFT

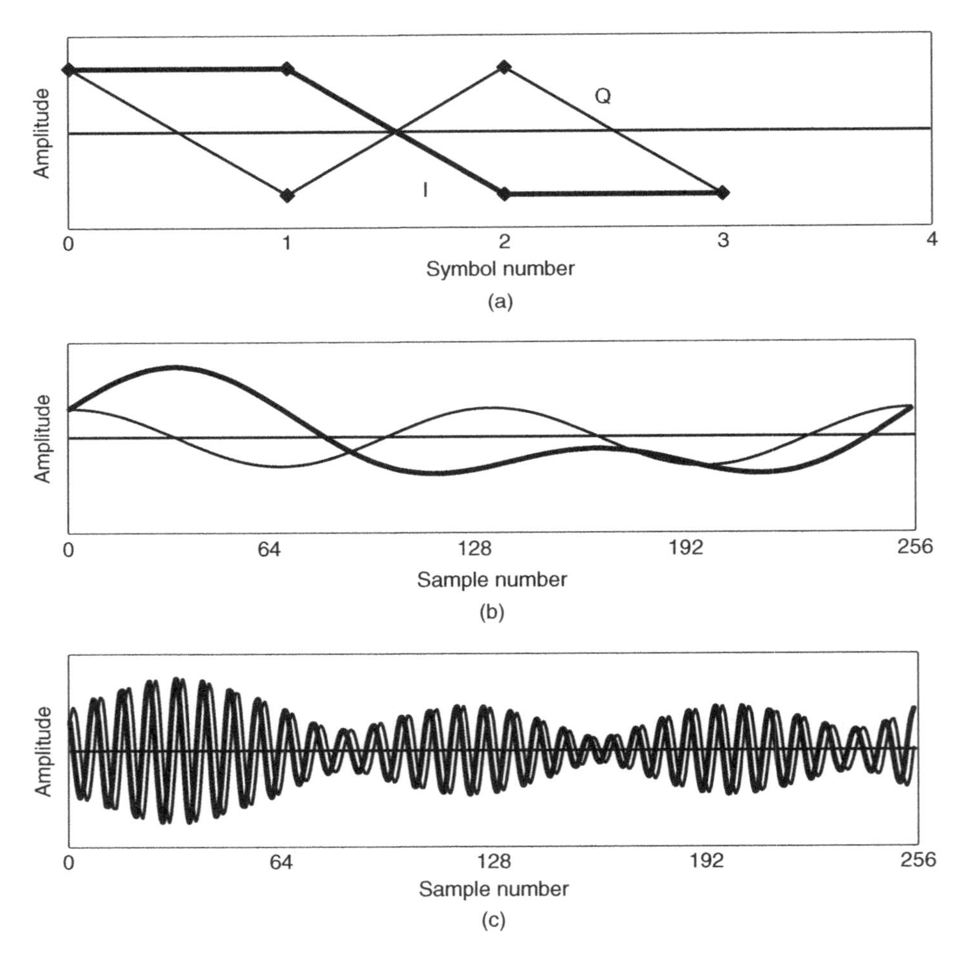

Figure 4.13 Example SC-FDMA waveform. (a) Transmitted symbols. (b) Resulting SC-FDMA waveform, if the data are transmitted on the central four sub-carriers out of 256. (c) SC-FDMA waveform, if the data are shifted by 32 sub-carriers

is a sequence of symbols in the time domain. The forward FFT converts these symbols to the frequency domain, the resource element mapper shifts them to the desired centre frequency and the inverse FFT converts them back to the time domain. Looking at these steps as a whole, we can see that the transmitted signal should be much the same as the original modulated waveform, except for a shift to another centre frequency. But the power of a QPSK signal is constant (at least in the absence of additional filtering), and it hardly varies at all in the cases of 16-QAM and 64-QAM. We have therefore achieved the result we require, of transmitting a signal with a roughly constant power.

Figure 4.13 shows the resulting waveforms, from an example in which the mobile is using four sub-carriers from a total of 256. The input (Figure 4.13a) is a sequence of four QPSK symbols, with [I, Q] values of [1, 1], [1, −1], [−1, 1] and [−1, −1]. If the data are transmitted on the central four sub-carriers, then the result (Figure 4.13b) looks very like the original QPSK waveform. The only difference is a smooth interpolation between the 256 samples in the time domain, which wraps round the ends of the data sequence due to the cyclic nature of the FFT. If we instead shift the data by 32 sub-carriers, then the only change (Figure 4.13c) is the introduction of some extra phase rotation into the resulting waveform.

We don't use SC-FDMA in the downlink, because the base station has to transmit to several mobiles, not just one. We could add one forward FFT per mobile to Figure 4.4, but that would destroy the single carrier nature of the transmission, and would allow the high power variations to return. Alternatively, we could add a single forward FFT across the whole of the downlink band. Unfortunately that would spread every mobile's data across the whole of the frequency domain, and would remove our ability to carry out frequency-dependent scheduling. Either way, SC-FDMA is unsuitable for the LTE downlink.

References

1. Smith, S.W. (1998) *The Scientist and Engineer's Guide to Digital Signal Processing*, California Technical Publishing, San Diego, CA.
2. Lyons, R.G. (2010) *Understanding Digital Signal Processing*, 3rd edn, Prentice Hall.

5

Multiple Antenna Techniques

From the beginning, LTE was designed so that the base station and mobile could both use multiple antennas for radio transmission and reception. This chapter covers the three main multiple antenna techniques, which have different objectives and which are implemented in different ways.

The most familiar is diversity processing, which increases the received signal power and reduces the amount of fading by using multiple antennas at the transmitter, the receiver or both. Diversity processing has been used since the early days of mobile communications, so we will only review it briefly.

In spatial multiplexing, the transmitter and receiver both use multiple antennas so as to increase the data rate. Spatial multiplexing is a relatively new technique that has only recently been introduced into mobile communications, so we will cover it in more detail than the others. Finally, beamforming uses multiple antennas at the base station in order to increase the coverage of the cell.

Spatial multiplexing is often described as the use of *multiple input multiple output* (MIMO) antennas. This name is derived from the inputs and outputs to the air interface, so that 'multiple input' refers to the transmitter and 'multiple output' to the receiver. Unfortunately, the name is a little ambiguous, as it can either refer to spatial multiplexing alone or include the use of transmit and receive diversity as well. For this reason, we will generally use the term 'spatial multiplexing' instead. For some reviews of multiple antenna techniques and their use in LTE, see References [1–4].

5.1 Diversity Processing

5.1.1 Receive Diversity

Receive diversity is most often used in the uplink, in the manner shown in Figure 5.1. Here, the base station uses two antennas to pick up two copies of the received signal. The signals reach the receive antennas with different phase shifts, but these can be removed by antenna-specific channel estimation. The base station can then add the signals together in-phase, without any risk of destructive interference between them.

An Introduction to LTE: LTE, LTE-Advanced, SAE, VoLTE and 4G Mobile Communications, Second Edition.
Christopher Cox.

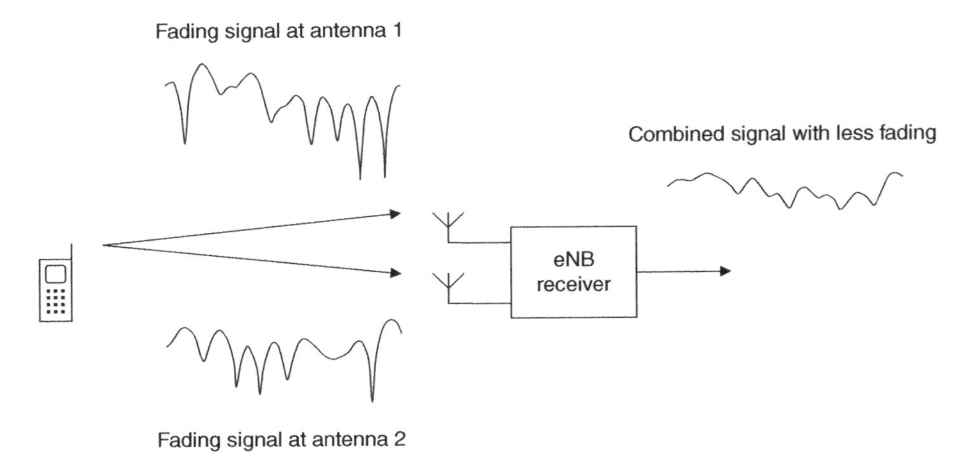

Figure 5.1 Reduction in fading by the use of a diversity receiver

The signals are both made up from several smaller rays, so they are both subject to fading. If the two individual signals undergo fades at the same time, then the power of the combined signal will be low. But if the antennas are far enough apart (a few wavelengths of the carrier frequency), then the two sets of fading geometries will be very different, so the signals will be far more likely to undergo fades at completely different times. We have therefore reduced the amount of fading in the combined signal, which in turn reduces the error rate.

Base stations usually have more than one receive antenna. In LTE, the mobile's test specifications assume that the mobile is using two receive antennas [5], so LTE systems are expected to use receive diversity on the downlink as well as the uplink. A mobile's antennas are closer together than a base station's, which reduces the benefit of receive diversity, but the situation can often be improved using antennas that measure two independent polarizations of the incoming signal.

5.1.2 *Closed Loop Transmit Diversity*

Transmit diversity reduces the amount of fading by using two or more antennas at the transmitter. It is superficially similar to receive diversity, but with a crucial problem: the signals add together at the single receive antenna, which brings a risk of destructive interference. There are two ways to solve the problem, the first of which is *closed loop transmit diversity* (Figure 5.2).

Here, the transmitter sends two copies of the signal in the expected way, but it also applies a phase shift to one or both signals before transmission. By doing this, it can ensure that the two signals reach the receiver in-phase, without any risk of destructive interference. The phase shift is determined by a *pre-coding matrix indicator* (PMI), which is calculated by the receiver and fed back to the transmitter. A simple PMI might indicate two options: either transmit both signals without any phase shifts or transmit the second with a phase shift of 180°. If the first option leads to destructive interference, then the second will automatically work. Once again, the amplitude of the combined signal is only low in the unlikely event that the two received signals undergo fades at the same time.

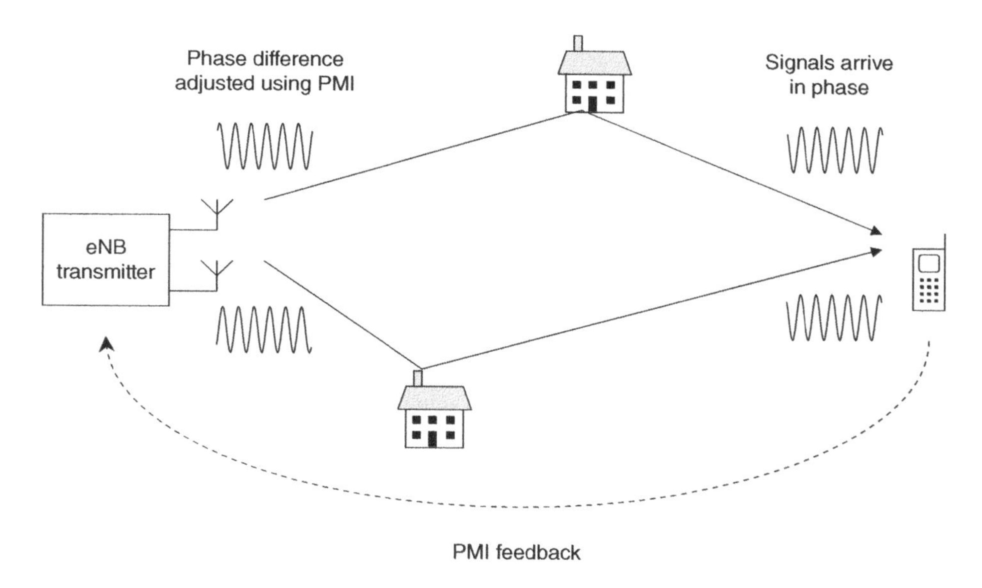

Figure 5.2 Operation of closed loop transmit diversity

The phase shifts introduced by the radio channel depend on the wavelength of the carrier signal and hence on its frequency. This implies that the best choice of PMI is a function of frequency as well. In LTE, the mobile can feed different PMI values back to the base station for different sets of downlink sub-carriers. However, the base station's downlink transmissions are usually over a much narrower bandwidth and only use a single PMI.

The best choice of PMI also depends on the position of the mobile, so a fast moving mobile will have a PMI that frequently changes. Unfortunately the feedback loop introduces time delays into the system, so in the case of fast moving mobiles, the PMI may be out of date by the time it is used. For this reason, closed loop transmit diversity is only suitable for mobiles that are moving sufficiently slowly. For fast moving mobiles, it is better to use the open loop technique described in the next section.

5.1.3 Open Loop Transmit Diversity

Figure 5.3 shows an implementation of *open loop transmit diversity* that is known as *Alamouti's technique* [6]. Here, the transmitter uses two antennas to send two symbols, denoted as s_1 and s_2, in two successive time steps. In the first step, the transmitter sends s_1 from the first antenna and s_2 from the second, while in the second step, it sends $-s_2^*$ from the first antenna and s_1^* from the second. (The symbol * indicates that the transmitter should change the sign of the quadrature component, in the process of complex conjugation.)

The receiver can now make two successive measurements of the received signal, which correspond to two different combinations of s_1 and s_2. It can then solve the resulting equations, so as to recover the two transmitted symbols. There are only two requirements: the fading patterns must stay roughly the same between the first time step and the second, and the two signals must not undergo fades at the same time. Both requirements are usually met.

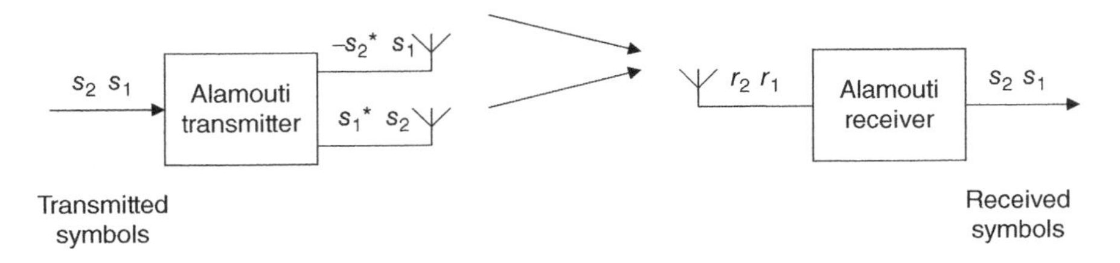

Figure 5.3 Operation of Alamouti's technique for open loop transmit diversity

There is no equivalent to Alamouti's technique for systems with more than two antennas. Despite this, some extra diversity gain can still be achieved in four antenna systems, by swapping back and forth between the two constituent antenna pairs. This technique is used for four antenna open loop diversity in LTE.

We can combine open and closed loop transmit diversity with the receive diversity techniques from earlier, giving a system that carries out diversity processing using multiple antennas at both the transmitter and the receiver. The technique is different from the spatial multiplexing techniques that we will describe next, although, as we will see, a spatial multiplexing system can fall back to diversity transmission and reception if the conditions require.

5.2 Spatial Multiplexing

5.2.1 Principles of Operation

Spatial multiplexing has a different purpose from diversity processing. If the transmitter and receiver both have multiple antennas, then we can set up multiple parallel data streams between them, so as to increase the data rate. In a system with N_T transmit and N_R receive antennas, often known as an $N_T \times N_R$ spatial multiplexing system, the peak data rate is proportional to $\min(N_T, N_R)$.

Figure 5.4 shows a basic spatial multiplexing system, in which the transmitter and receiver both have two antennas. In the transmitter, the antenna mapper takes symbols from the modulator two at a time, and sends one symbol to each antenna. The antennas transmit the two symbols simultaneously, so as to double the transmitted data rate.

The symbols travel to the receive antennas by way of four separate radio paths, so the received signals can be written as follows:

$$\begin{aligned} y_1 &= H_{11}x_1 + H_{12}x_2 + n_1 \\ y_2 &= H_{21}x_1 + H_{22}x_2 + n_2 \end{aligned} \tag{5.1}$$

Here, x_1 and x_2 are the signals sent from the two transmit antennas, y_1 and y_2 are the signals that arrive at the two receive antennas, and n_1 and n_2 represent the received noise and interference. H_{ij} expresses the way in which the transmitted symbols are attenuated and phase-shifted, as they travel to receive antenna i from transmit antenna j. (The subscripts i and j may look the wrong way round, but this is for consistency with the usual mathematical notation for matrices.)

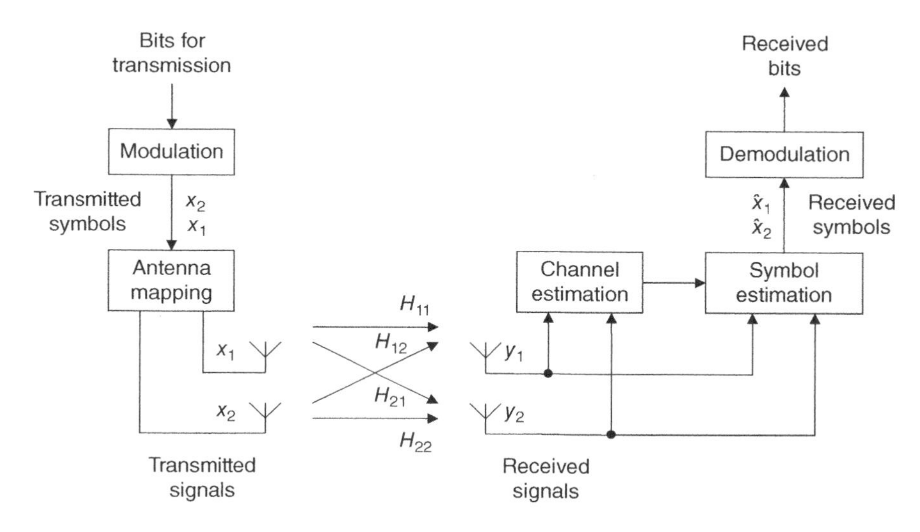

Figure 5.4 Basic principles of a 2 × 2 spatial multiplexing system

In general, all the terms in the equation above are complex. In the transmitted and received symbols x_j and y_i and the noise terms n_i, the real and imaginary parts are the amplitudes of the in-phase and quadrature components. Similarly, in each of the channel elements H_{ij}, the magnitude represents the attenuation of the radio signal, while the phase represents the phase shift. However, the use of complex numbers would make the examples unnecessarily complicated without adding much extra information, so we will simplify the examples by using real numbers alone. To do this, we will assume that the transmitter is modulating the bits using binary phase shift keying, so that the in-phase components are +1 and −1, and the quadrature components are zero. We will also assume that the radio channel can attenuate or invert the signal, but does not introduce any other phase shifts.

Consistent with these assumptions, let us consider the following example:

$$\begin{aligned} H_{11} &= 0.8 \quad H_{12} = 0.6 \quad x_1 = +1 \quad n_1 = +0.02 \\ H_{21} &= 0.2 \quad H_{22} = 0.4 \quad x_2 = -1 \quad n_2 = -0.02 \end{aligned} \tag{5.2}$$

Substituting these numbers into Equation 5.1 shows that the received signals are as follows:

$$\begin{aligned} y_1 &= +0.22 \\ y_2 &= -0.22 \end{aligned} \tag{5.3}$$

The receiver's first task is to estimate the four channel elements H_{ij}. To help it do this, the transmitter broadcasts reference symbols that follow the basic technique described in Chapter 3, but with one extra feature: when one antenna transmits a reference symbol, the other antenna keeps quiet and sends nothing at all. The receiver can then estimate the channel elements H_{11} and H_{21}, by measuring the two received signals at the times when transmit antenna 1 is sending a reference symbol. It can then wait until transmit antenna 2 sends a reference symbol, before estimating the channel elements H_{12} and H_{22}.

The receiver now has enough information to estimate the transmitted symbols x_1 and x_2. There are several ways for it to do this, but the simplest is a *zero-forcing detector*, which operates as follows. If we ignore the noise and interference, then Equation 5.1 is a pair of simultaneous equations for two unknown quantities, x_1 and x_2. These equations can be inverted as follows:

$$\hat{x}_1 = \frac{\hat{H}_{22}y_1 - \hat{H}_{12}y_2}{\hat{H}_{11}\hat{H}_{22} - \hat{H}_{21}\hat{H}_{12}}$$
$$\hat{x}_2 = \frac{\hat{H}_{11}y_2 - \hat{H}_{21}y_1}{\hat{H}_{11}\hat{H}_{22} - \hat{H}_{21}\hat{H}_{12}} \tag{5.4}$$

Here, $\hat{H}_{ij}$ is the receiver's estimate of the channel element H_{ij}. (This quantity may be different from H_{ij} because of noise and other errors in the channel estimation process.) Similarly, $\hat{x}_1$ and $\hat{x}_2$ are the receiver's estimates of the transmitted symbols x_1 and x_2. Substituting the numbers from Equations 5.2 and 5.3 gives the following result:

$$\hat{x}_1 = +1.1$$
$$\hat{x}_2 = -1.1 \tag{5.5}$$

This is consistent with transmitted symbols of +1 and −1. We have therefore transferred two symbols at the same time using the same sub-carriers, and have doubled the data rate.

5.2.2 *Open Loop Spatial Multiplexing*

There is a problem with the technique described above. To illustrate this, let us change one of the channel elements, H_{11}, to give the following example:

$$H_{11} = 0.3 \quad H_{12} = 0.6$$
$$H_{21} = 0.2 \quad H_{22} = 0.4 \tag{5.6}$$

If we try to estimate the transmitted symbols using Equation 5.4, we find that $H_{11}H_{22} - H_{21}H_{12}$ is zero. We therefore end up dividing by zero, which is nonsense. So, for this choice of channel elements, the technique has failed. We can see what has gone wrong by substituting the channel elements into Equation 5.1 and writing the received signals as follows:

$$y_1 = 0.3(x_1 + 2x_2) + n_1$$
$$y_2 = 0.2(x_1 + 2x_2) + n_2 \tag{5.7}$$

By measuring the received signals y_1 and y_2, we were expecting to measure two different pieces of information, from which we could recover the transmitted data. This time, however, we have measured the same piece of information, namely, $x_1 + 2x_2$, twice. As a result, we do not have enough information to recover x_1 and x_2 independently. Furthermore, this is not just an isolated special case. If $H_{11}H_{22} - H_{21}H_{12}$ is small but non-zero, then our estimates of x_1 and x_2 turn out to be badly corrupted by noise and are completely unusable.

The solution comes from the knowledge that we can still send one symbol at a time, by the use of diversity processing. We therefore require an adaptive system, which can use spatial multiplexing to send two symbols at a time if the channel elements are well behaved and can fall back to diversity processing otherwise. Such a system is shown in Figure 5.5. Here, the receiver measures the channel elements and works out a *rank indication* (RI), which indicates the number of symbols that it can successfully receive. It then feeds the rank indication back to the transmitter.

If the rank indication is two, then the system operates in the same way that we described earlier. The transmitter's *layer mapper* grabs two symbols, s_1 and s_2, from the transmit buffer, so as to create two independent data streams that are known as *layers*. The *antenna mapper* then sends one symbol to each antenna, by a straightforward mapping operation:

$$
\begin{aligned}
x_1 &= s_1 \\
x_2 &= s_2
\end{aligned}
\tag{5.8}
$$

The receiver measures the incoming signals and recovers the transmitted symbols as before.

If the rank indication is one, then the layer mapper only grabs one symbol, s_1, which the antenna mapper sends to both transmit antennas as follows:

$$
\begin{aligned}
x_1 &= s_1 \\
x_2 &= s_1
\end{aligned}
\tag{5.9}
$$

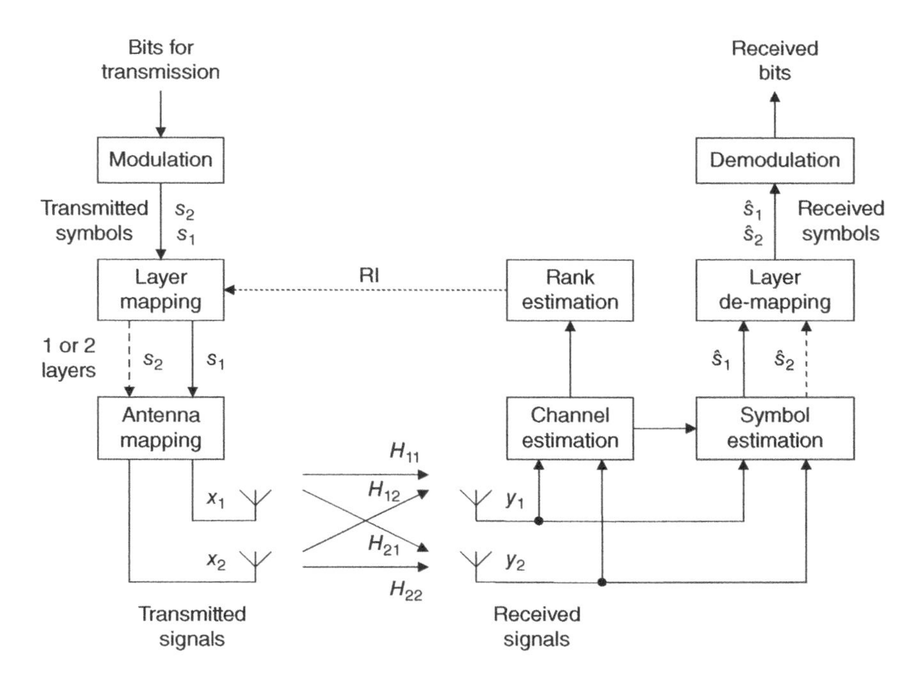

Figure 5.5 Operation of a 2 × 2 open loop spatial multiplexing system

Under these assumptions, Equation 5.7 becomes the following:

$$
\begin{aligned}
y_1 &= 0.9s_1 + n_1 \\
y_2 &= 0.6s_1 + n_2
\end{aligned}
\tag{5.10}
$$

The receiver now has two measurements of the transmitted symbol s_1, and can combine these in a diversity receiver so as to recover the transmitted data.

The effect is as follows. If the channel elements are well behaved, then the transmitter sends two symbols at a time and the receiver recovers them using a spatial multiplexing receiver. Sometimes this is not possible, in which case the transmitter falls back to sending one symbol at a time and the receiver falls back to diversity reception. The transmitter indicates its choice for each block of transmitted data by means of low level control information, to ensure that the receiver can process the data correctly. This technique is implemented in LTE and, for reasons that will become clear in the next section, is known as *open loop spatial multiplexing*.

5.2.3 Closed Loop Spatial Multiplexing

There is one remaining problem. To illustrate this, let us change two more of the channel elements, so that:

$$
\begin{aligned}
H_{11} &= 0.3 \quad H_{12} = -0.3 \\
H_{21} &= 0.2 \quad H_{22} = -0.2
\end{aligned}
\tag{5.11}
$$

These channel elements are badly behaved, in that $H_{11}H_{22} - H_{21}H_{12}$ is zero. But if we try to handle the situation in the manner described above, by sending the same symbol from both transmit antennas, then the received signals are as follows:

$$
\begin{aligned}
y_1 &= 0.3s_1 - 0.3s_1 + n_1 \\
y_2 &= 0.2s_1 - 0.2s_1 + n_2
\end{aligned}
\tag{5.12}
$$

So the transmitted signals cancel out at both receive antennas and we are left with measurements of the incoming noise and interference. We therefore have insufficient information even to recover s_1.

To see the way out, consider what happens if we send one symbol at a time as before, but invert the signal that is sent from the second antenna:

$$
\begin{aligned}
x_1 &= s_1 \\
x_2 &= -s_1
\end{aligned}
\tag{5.13}
$$

The received signal can now be written as follows:

$$
\begin{aligned}
y_1 &= 0.3s_1 + 0.3s_1 + n_1 \\
y_2 &= 0.2s_1 + 0.2s_1 + n_2
\end{aligned}
\tag{5.14}
$$

This time, we can recover the transmitted symbol s_1.

So we now require two levels of adaptation. If the rank indication is two, then the transmitter sends two symbols at a time using the antenna mapping of Equation 5.8. If the rank indication is one, then the transmitter falls back to diversity processing and sends one symbol at a time. In doing so, it chooses an antenna mapping such as Equations 5.9 or 5.13, which depends on the exact nature of the channel elements and which guarantees a strong signal at the receiver.

Such a system is shown in Figure 5.6. Here, the receiver measures the channel elements as before and uses them to feed back two quantities, namely, the rank indication and a pre-coding matrix indicator (PMI). The PMI controls a *pre-coding* step in the transmitter, which implements an adaptive antenna mapping using (for example) Equations 5.8, 5.9 and 5.13, to ensure that the signals reach the receiver without cancellation. (In fact the PMI has exactly the same role that we saw earlier when discussing closed loop transmit diversity, which is why its name is the same.) In the receiver, the *post-coding* step reverses the effect of pre-coding and also includes the soft decision estimation step from earlier. As before, the transmitter indicates its choice for each block of transmitted data by means of low level control information, to ensure that the receiver can process the data correctly.

This technique is also implemented in LTE, and is known as *closed loop spatial multiplexing*. In this expression, the term 'closed loop' refers specifically to the loop that is created by feeding back the PMI. The technique from Section 5.2.2 is known as 'open loop spatial multiplexing', even though the receiver is still feeding back a rank indication.

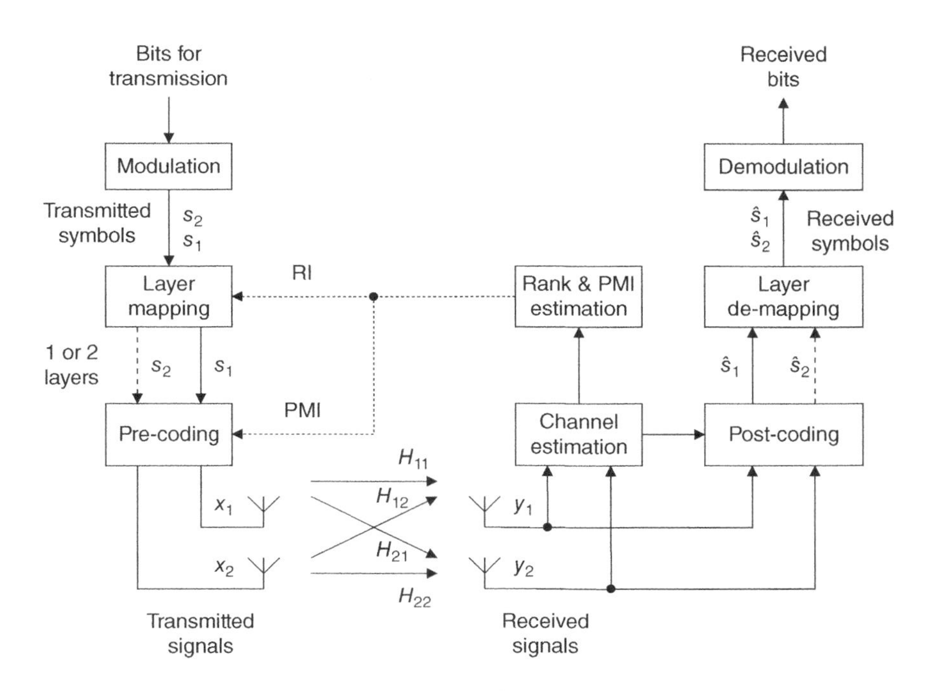

Figure 5.6 Operation of a 2 × 2 closed loop spatial multiplexing system

5.2.4 Matrix Representation

We have now covered the basic principles of spatial multiplexing. To go further, we need a more mathematical description in terms of matrices. Readers who are unfamiliar with matrices may prefer to skip this section and to resume the discussion in Section 5.2.5 below.

In matrix notation, we can write the received signal (Equation 5.1) as follows:

$$\boldsymbol{y} = \boldsymbol{H}.\boldsymbol{x} + \boldsymbol{n} \tag{5.15}$$

Here, $\boldsymbol{x}$ is a column vector that contains the signals that are sent from the N_T transmit antennas. Similarly, $\boldsymbol{n}$ and $\boldsymbol{y}$ are column vectors containing the noise and the resulting signals at the N_R receive antennas. The *channel matrix* $\boldsymbol{H}$ expresses the amplitude changes and phase shifts that the air interface introduced. The matrix has N_R rows and N_T columns, so it is known as an $N_R \times N_T$ matrix. In the examples we considered earlier, the system had two transmit and two receive antennas, so the matrix equation above could be written as follows:

$$\begin{bmatrix} y_1 \\ y_2 \end{bmatrix} = \begin{bmatrix} H_{11} & H_{12} \\ H_{21} & H_{22} \end{bmatrix} \cdot \begin{bmatrix} x_1 \\ x_2 \end{bmatrix} + \begin{bmatrix} n_1 \\ n_2 \end{bmatrix} \tag{5.16}$$

Now let us assume that the numbers of transmit and receive antennas are equal, so that $N_R = N_T = N$, and let us ignore the noise and interference as before. We can then invert the channel matrix and derive the following estimate of the transmitted symbols:

$$\widehat{\boldsymbol{x}} = \widehat{\boldsymbol{H}}^{-1}.\boldsymbol{y} \tag{5.17}$$

Here, $\widehat{H}^{-1}$ is the receiver's estimate of the inverse of the channel matrix, while $\widehat{x}$ is its estimate of the transmitted signal. This is the zero-forcing detector from earlier. The detector runs into problems if the noise and interference are too great, but, in these circumstances, a *minimum mean square error* (MMSE) detector gives a more accurate answer.

If the channel matrix is well behaved, then we can measure the signals that arrive at the N receive antennas and use a suitable detector to estimate the symbols that were transmitted. As a result, we can increase the data rate by a factor N. The channel matrix may, however, be *singular* (as in Equations 5.6 and 5.11), in which case its inverse does not exist. Alternatively, the matrix may be *ill conditioned*, in which case its inverse is corrupted by noise. Either way, we need to find another solution.

The solution comes from a technique known as *singular value decomposition* (SVD) [7], in which we write the channel matrix $\boldsymbol{H}$ as follows:

$$\boldsymbol{H} = \boldsymbol{U}.\Sigma.\boldsymbol{V}^{\dagger} \tag{5.18}$$

Here, $\boldsymbol{U}$ is an $N_R \times N_R$ matrix and $\boldsymbol{V}$ is an $N_T \times N_T$ matrix. $\boldsymbol{V}^{\dagger}$ is the *Hermitian conjugate* of $\boldsymbol{V}$, in other words the complex conjugate of its transpose. Both $\boldsymbol{U}$ and $\boldsymbol{V}$ are *unitary* matrices, so that their inverses equal their Hermitian conjugates. Σ is an $N_R \times N_T$ diagonal matrix, whose elements σ_i are either positive or zero and are known as the *singular values* of $\boldsymbol{H}$. In the two antenna example, the diagonal matrix is:

$$\Sigma = \begin{bmatrix} \sigma_1 & 0 \\ 0 & \sigma_2 \end{bmatrix} \tag{5.19}$$

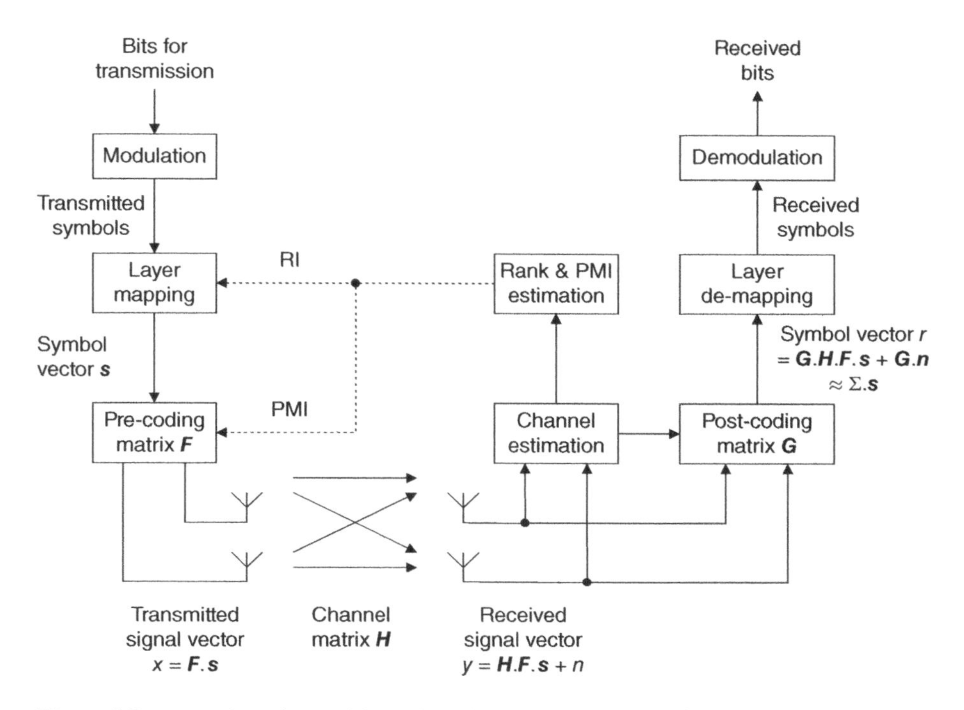

Figure 5.7 Operation of a spatial multiplexing system with an arbitrary number of antennas

The technique is loosely related to the more familiar technique of eigenvalue decomposition, as the eigenvalues of $\boldsymbol{H}.\boldsymbol{H}^{\dagger}$ are the squares of the singular values of $\boldsymbol{H}$.

Now let us transmit the symbols in the manner shown in Figure 5.7. At the output from the post-coding stage, the received symbol vector is:

$$\boldsymbol{r} = \boldsymbol{G}.\boldsymbol{H}.\boldsymbol{F}.\boldsymbol{s} + \boldsymbol{G}.\boldsymbol{n} \tag{5.20}$$

where $\boldsymbol{s}$ contains the transmitted symbols at the input to the pre-coding stage, $\boldsymbol{F}$ is the pre-coding matrix, $\boldsymbol{H}$ is the usual channel matrix and $\boldsymbol{G}$ is the post-coding matrix. If we now choose the pre- and post-coding matrices so that they are good approximations to the unitary matrices from earlier:

$$\begin{aligned} \boldsymbol{F} &\approx \boldsymbol{V} \\ \boldsymbol{G} &\approx \boldsymbol{U}^{\dagger} \end{aligned} \tag{5.21}$$

then the received symbol vector becomes the following:

$$\begin{aligned} \boldsymbol{r} &\approx \boldsymbol{U}^{\dagger}.\boldsymbol{H}.\boldsymbol{V}.\boldsymbol{s} + \boldsymbol{V}.\boldsymbol{n} \\ &\approx \Sigma.\boldsymbol{s} + \boldsymbol{V}.\boldsymbol{n} \end{aligned} \tag{5.22}$$

Ignoring the noise, we can now write the received symbols in a two antenna spatial multiplexing system as follows:

$$\begin{bmatrix} r_1 \\ r_2 \end{bmatrix} \approx \begin{bmatrix} \sigma_1 & 0 \\ 0 & \sigma_2 \end{bmatrix} \cdot \begin{bmatrix} s_1 \\ s_2 \end{bmatrix} \tag{5.23}$$

We therefore have two independent data streams, without any coupling between them. It is now trivial for the receiver to recover the transmitted symbols, as follows:

$$\hat{s}_i = \frac{r_i}{\sigma_i} \tag{5.24}$$

So, by a suitable choice of pre- and post-coding matrices, $\boldsymbol{F}$ and $\boldsymbol{G}$, we can greatly simplify the design of the receiver.

If the channel matrix $\boldsymbol{H}$ is singular, then some of its singular values σ_i are zero. If it is ill-conditioned, then some of the singular values are very small, so the reconstructed symbols are badly corrupted by noise. The condition number of $\boldsymbol{H}$ equals the largest singular value divided by the smallest, and measures just how ill-conditioned the matrix is. The *rank* of $\boldsymbol{H}$ is the number of usable singular values, and the rank indication from Section 5.2.2 equals the rank of $\boldsymbol{H}$. In a two antenna system with a rank of 1, for example, the received symbol vector is as follows:

$$\begin{bmatrix} r_1 \\ r_2 \end{bmatrix} \approx \begin{bmatrix} \sigma_1 & 0 \\ 0 & 0 \end{bmatrix} \cdot \begin{bmatrix} s_1 \\ s_2 \end{bmatrix} \tag{5.25}$$

The system can exploit this behaviour in the following way. The receiver estimates the channel matrix and feeds back the rank indication along with the pre-coding matrix $\boldsymbol{F}$. If the rank indication is two, then the transmitter sends two symbols, s_1 and s_2, and the receiver reconstructs them from Equation 5.23. If the rank indication is one, then the transmitter just sends one symbol, s_1, and doesn't bother with s_2 at all. The receiver can then reconstruct the transmitted symbol from Equation 5.25.

In practice, the receiver does not pass a full description of $\boldsymbol{F}$ back to the transmitter, as that would require too much feedback. Instead, it selects the closest approximation to $\boldsymbol{V}$ from a *codebook* and indicates its choice using the pre-coding matrix indicator, PMI.

Inspection of Equations 5.22 and 5.23 shows that the received symbols r_1 and r_2 can have different signal-to-noise ratios, which depend on the corresponding singular values σ_1 and σ_2. In LTE, the transmitter can exploit this by sending the two symbols with different modulation schemes and coding rates, and also with different transmit powers.

We can use the technique in exactly the same way if the numbers of transmit and receive antennas are different. With four transmit and two receive antennas, for example, Equation 5.23 becomes the following:

$$\begin{bmatrix} r_1 \\ r_2 \end{bmatrix} \approx \begin{bmatrix} \sigma_1 & 0 & 0 & 0 \\ 0 & \sigma_2 & 0 & 0 \end{bmatrix} \cdot \begin{bmatrix} s_1 \\ s_2 \\ s_3 \\ s_4 \end{bmatrix} \tag{5.26}$$

The channel matrix has a rank of two, so we can successfully reconstruct only two transmitted symbols, s_1 and s_2, not four. In the general case, the maximum data rate is proportional to $\min(N_T, N_R)$, with any extra antennas providing additional transmit or receive diversity.

5.2.5 *Implementation Issues*

Spatial multiplexing is implemented in the downlink of LTE Release 8, using a maximum of four transmit antennas on the base station and four receive antennas on the mobile. There are similar implementation issues to diversity processing. Firstly, the antennas at the base station and mobile should be reasonably far apart, ideally a few wavelengths of the carrier frequency, or should handle different polarizations. If the antennas are too close together, then the channel elements H_{ij} will be very similar. This can easily take us into the situation from Section 5.2.2, where spatial multiplexing was unusable and we had to fall back to diversity processing.

A similar situation can easily arise in the case of line-of-sight transmission and reception. This leads us to an unexpected conclusion: spatial multiplexing actually works best in conditions with no direct line-of-sight and significant multipath because, in these conditions, the channel elements H_{ij} are uncorrelated with each other. In line-of-sight conditions, we often have to fall back to diversity processing.

As in the case of closed loop transmit diversity, the PMI depends on the carrier frequency and the position of the mobile. For fast moving mobiles, delays in the feedback loop can make the PMI unreliable by the time the transmitter comes to use it, so open loop spatial multiplexing is often preferred.

5.2.6 *Multiple User MIMO*

Figure 5.8 shows a slightly different technique. Here, two transmit and two receive antennas are sharing the same transmission times and frequencies, in the same way as before. This time,

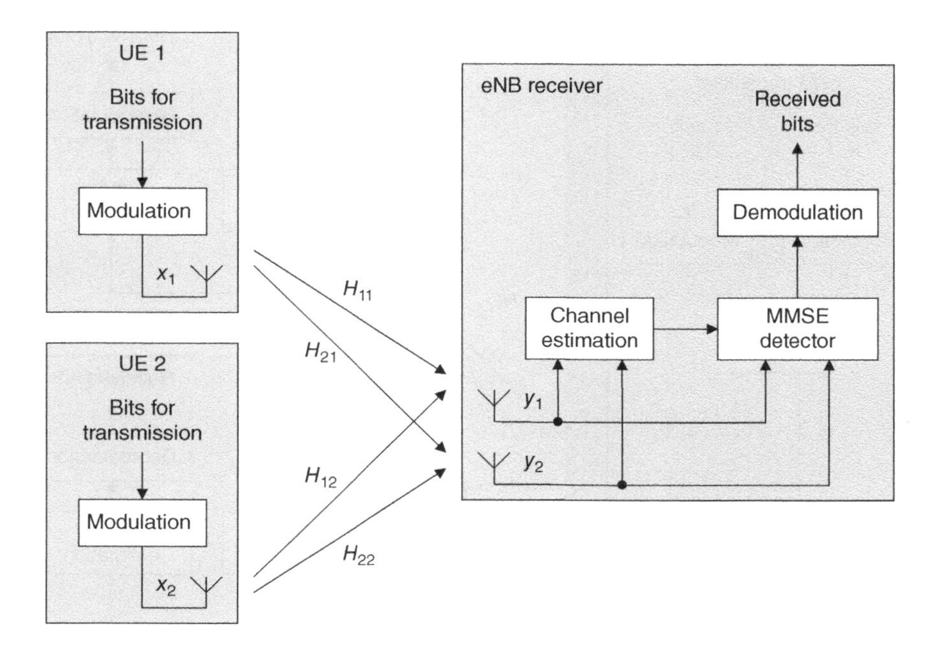

Figure 5.8 Uplink multiple user MIMO

however, the mobile antennas are on two different mobiles instead of one. This technique is known as *multiple user MIMO* (MU-MIMO), in contrast with the earlier spatial multiplexing techniques, which are sometimes known as *single user MIMO* (SU-MIMO).

Figure 5.8 specifically shows the implementation of multiple user MIMO on the uplink, which is the more common situation. Here, the mobiles transmit at the same time and on the same carrier frequency, but without using any pre-coding and without even knowing that they are part of a spatial multiplexing system. The base station receives their transmissions and separates them using (for example) the minimum mean square error detector that we noted earlier.

This technique only works if the channel matrix is well behaved, but we can usually guarantee this for two reasons. Firstly, the mobiles are likely to be far apart, so their ray paths are likely to be very different. Secondly, the base station can freely choose the mobiles that are taking part, so it can freely choose mobiles that lead to a well-behaved channel matrix.

Uplink multiple user MIMO does not increase the peak data rate of an individual mobile, but it is still beneficial because of the increase in cell throughput. It can also be implemented using inexpensive mobiles that just have one power amplifier and one transmit antenna, not two. For these reasons, multiple user MIMO is the standard technique in the uplink of LTE Release 8: single user MIMO is not introduced into the uplink until Release 10.

We can also apply multiple user MIMO to the downlink, as shown in Figure 5.9. This time, however, there is a problem. Mobile 1 can measure its received signal y_1 and the channel elements H_{11} and H_{12}, in the same way as before. However, it has no knowledge of the

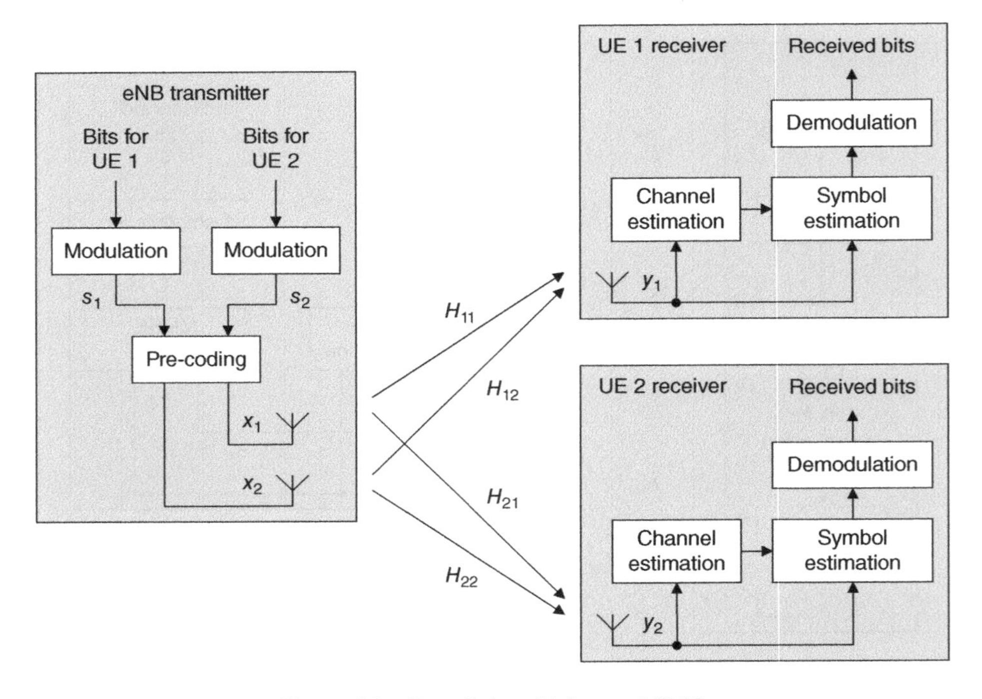

Figure 5.9 Downlink multiple user MIMO

other received signal y_2, or of the other channel elements H_{21} and H_{22}. The opposite situation applies for mobile 2. Neither mobile has complete knowledge of the channel elements or of the received signals, which invalidates the techniques we have been using.

The solution is to implement downlink multiple user MIMO by adapting another multiple antenna technique, known as beamforming. We will cover beamforming in the next section and then return to downlink multiple user MIMO at the end of the chapter.

5.3 Beamforming

5.3.1 Principles of Operation

Beamforming is similar to diversity transmission and reception, but has a different geometry and uses different techniques. The base station still has multiple antennas, but their separation is less than before, typically half a wavelength of the carrier frequency. We also assume that there is little multipath propagation between the base station and the mobile. With that geometry, the mobile receives highly correlated signals from the base station's transmit antennas, so the spatial multiplexing techniques from the previous section are inappropriate.

Instead, the multiple antennas increase the coverage of the cell following the principles shown in Figure 5.10. Here, mobile 1 is a long way from the base station, on a line-of-sight that is at right angles to the antenna array. The signals from each antenna reach mobile 1 in phase, so they interfere constructively, and the received signal power is high. On the other hand, mobile 2 is at an oblique angle and receives signals from alternate antennas that are 180° out of phase. These signals interfere destructively, so the received signal power is low. We have therefore created a synthetic antenna beam, which has a main beam pointing towards mobile 1 and a null pointing towards mobile 2. The beamwidth is narrower than one from a

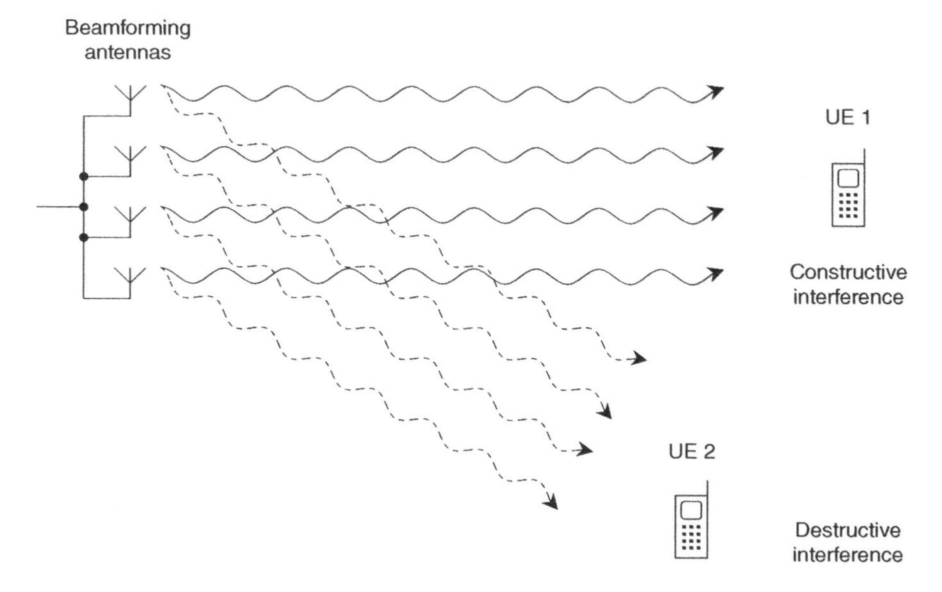

Figure 5.10 Basic principles of beamforming

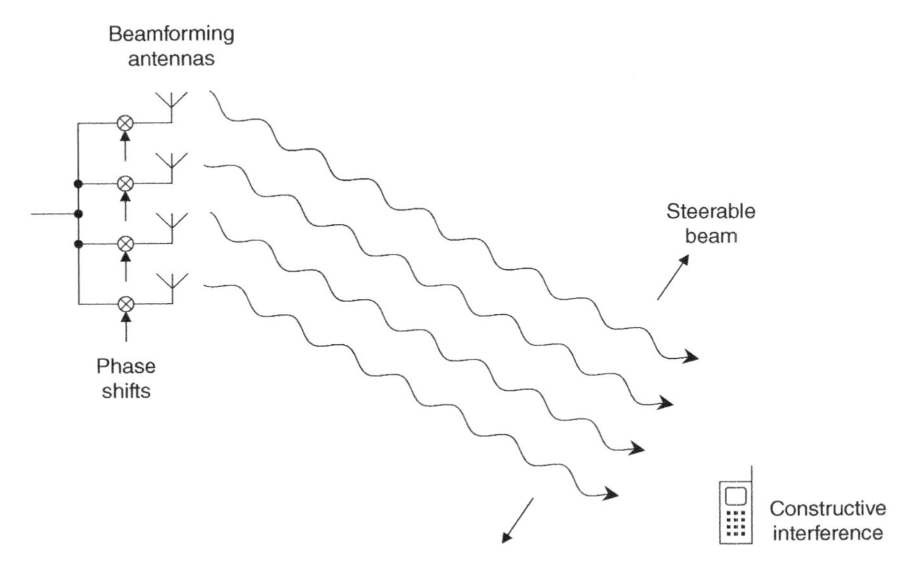

Figure 5.11 Beam steering using a set of phase shifts

single antenna, so the transmitted power is focussed towards mobile 1. As a result, the range of the base station in the direction of mobile 1 is greater than before.

As shown in Figure 5.11, we can go a step further. By applying a phase ramp to the transmitted signals, we can change the direction at which constructive interference arises, so we can direct the beam towards any direction we choose. More generally, we can adjust the amplitudes and phases of the transmitted signals, by applying a suitable set of antenna weights. In a system with N antennas, this allows us to adjust the direction of the main beam and up to $N-2$ nulls or sidelobes.

We can use the same technique to construct a synthetic reception beam for the uplink. By applying a suitable set of antenna weights at the base station receiver, we can ensure that the received signals add together in phase and interfere constructively. As a result, we can increase the range in the uplink as well.

5.3.2 Beam Steering

Mathematically, the antenna weights in a beamforming system are the same as the pre-coding weights that we use for transmit diversity and spatial multiplexing. However, in the case of beamforming, the different ray paths are highly correlated with each other, so the antenna weights remain stable for long periods of time. This often makes it feasible for the base station to estimate the weights without any feedback from the mobile.

For the reception beams on the uplink, there are two main techniques [8, 9]. Using the *reference signal technique*, the base station adjusts the antenna weights so as to reconstruct the mobile's reference symbols with the correct signal phase and the greatest possible signal-to-interference plus noise ratio (SINR). An alternative is the *direction of arrival technique*, in which the base station measures the signals that are received by each antenna and estimates the direction of the target mobile. From this quantity, it can estimate the antenna weights that are needed for satisfactory reception.

For the transmission beams on the downlink, the answer depends on the base station's mode of operation. In TDD mode, the uplink and the downlink use the same carrier frequency, so the base station can use the same antenna weights on the downlink that it calculated for the uplink. In FDD mode, the carrier frequencies are different, so the downlink antenna weights are different and are harder to estimate. For this reason, beamforming is more common in systems that are using TDD rather than FDD.

Despite this restriction, the lack of feedback brings a further benefit: we can simplify the system design by making the beamforming process transparent to the mobile. To achieve this, the base station applies its antenna weights to everything that it transmits, including not only the data but also the reference signals, which are now specific to the target mobile. As a result, the mobile receives all its downlink transmissions with the correct amplitude and phase, so the base station does not have to signal any information about its choice of weights at all.

Even though the reference signals are mobile specific, an OFDMA base station can still transmit to more than one mobile at the same time. To achieve this, it simply has to process different sets of sub-carriers using different sets of antenna weights, so as to create antenna beams that point towards different mobiles. Furthermore, the base station may also be able to transmit to multiple mobiles on the same set of sub-carriers by implementing the downlink multiple user MIMO techniques that we discuss below.

5.3.3 Downlink Multiple User MIMO Revisited

At the end of Section 5.2.6, we tried to implement downlink multiple user MIMO using the same techniques that we had previously used for spatial multiplexing. We discovered that the mobiles did not have enough information to recover the transmitted symbols, so that the previous techniques were inappropriate.

We can, however, get round the problem in the manner shown in Figure 5.12. Here, the base station sends two different data streams into its antenna array, instead of just one. It then processes the data using two different sets of antenna weights and adds the results together before transmission. In doing so, it has created two separate antenna beams, which share the same sub-carriers but carry two different sets of information. The base station can then adjust the antenna weights so as to steer the beams and nulls towards two different mobiles, so that the first mobile receives constructive interference from beam 1 and destructive interference from beam 2, and vice-versa. By doing this, the base station can double the capacity of the cell, in exactly the manner that downlink multiple user MIMO requires. We conclude that downlink MU-MIMO is best treated as a variety of beamforming and is best implemented using base station antennas that are close together rather than far apart.

For the technique to work effectively, we have to steer each antenna beam so that it points towards the target mobile and steer the corresponding nulls so that they point towards the other mobiles that are participating. This is hard to achieve, so LTE does not offer full support for downlink multiple user MIMO until Release 10. In that release, the specifications introduce accurate PMI feedback that is designed for multiple user MIMO and helps the base station steer the beams and nulls with the accuracy required.

There is, however, limited support for downlink multiple user MIMO in Release 8. There are two possible techniques. The main technique uses the same PMI feedback that single user MIMO does, so it only works effectively if the codebook happens to contain a pre-coding

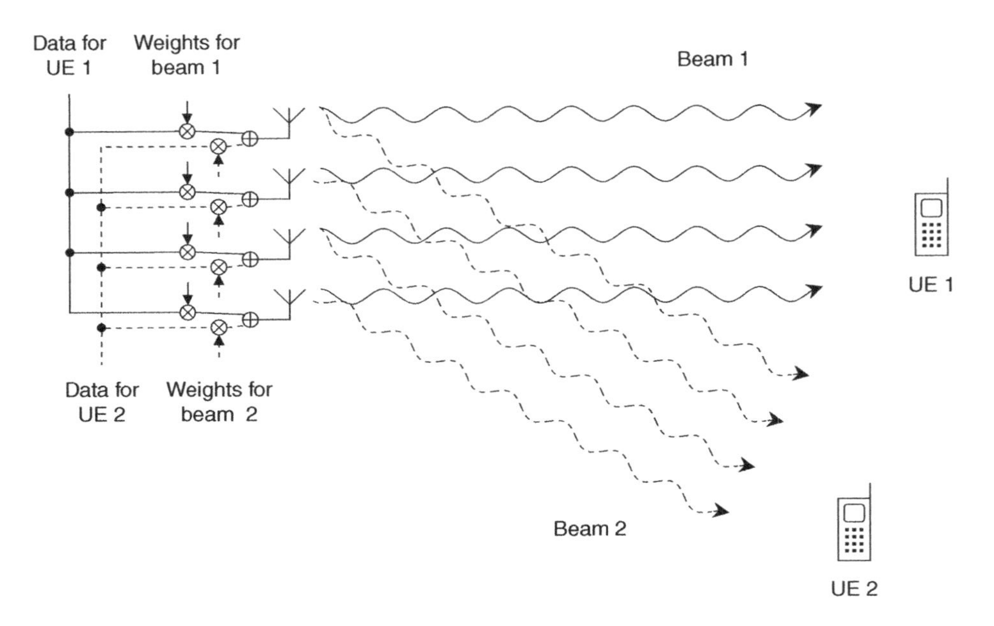

Figure 5.12 Beamforming using two parallel sets of antenna weights

matrix that is suitable for multiple user MIMO. Often it does not, so the performance of this technique in Release 8 is comparatively poor. Alternatively, the base station can estimate the antenna weights without any feedback at all, using the techniques that we described earlier. This is only really effective in TDD mode and is generally inappropriate in the case of FDD.

References

1. Biglieri, E., Calderbank, R., Constantinides, A. *et al.* (2010) *MIMO Wireless Communications*, Cambridge University Press.
2. 4G Americas (2009) MIMO Transmission Schemes for LTE and HSPA Networks, June 2009.
3. 4G Americas (2010) MIMO and Smart Antennas for 3G and 4G Wireless Systems: Practical Aspects and Deployment Considerations, May 2010.
4. Lee, J., Han, J.K. and Zhang, J. (2009) MIMO technologies in 3GPP LTE and LTE-advanced. *EURASIP Journal on Wireless Communications and Networking*, **2009**, 1–10, article ID 302092.
5. 3GPP TS 36.101 (2013) User Equipment (UE) Radio Transmission and Reception, Release 11, Section 7.2, September 2013.
6. Alamouti, S. (1998) Space block coding: a simple transmitter diversity technique for wireless communications. *IEEE Journal on Selected Areas in Communications*, **16**, 1451–1458.
7. Press, W.H., Teukolsky, S.A., Vetterling, W.T. and Flannery, B.P. (2007) *Numerical Recipes*, 3rd edn, Section 2.6, Cambridge University Press.
8. Godara, L.C. (1997) Applications of antenna arrays to mobile communications, part I: performance improvement, feasibility, and system considerations. *Proceedings of the IEEE*, **85**, 1031–1060.
9. Godara, L.C. (1997) Application of antenna arrays to mobile communications, part II: beam-forming and direction-of-arrival considerations. *Proceedings of the IEEE*, **85**, 1195–1245.

6

Architecture of the LTE Air Interface

Now that we have covered the principles of the air interface, we can explain how those principles are actually implemented in LTE. This task is the focus of the next five chapters.

In this chapter, we will cover the air interface's high-level architecture. We begin by reviewing the air interface protocol stack, and by listing the channels and signals that carry information between the different protocols. We then describe how the OFDMA and SC-FDMA air interfaces are organized as a function of time and frequency in a resource grid and discuss how LTE implements transmissions from multiple antennas using multiple copies of the grid. Finally, we bring the preceding material together by illustrating how the channels and signals are mapped onto the resource grids that are used in the uplink and downlink.

6.1 Air Interface Protocol Stack

Figure 6.1 reviews the protocols that are used in the air interface, from the viewpoint of the mobile. As well as the information presented in Chapter 2, the figure adds some detail to the physical layer and shows the information flows between the different levels of the protocol stack.

Let us consider the transmitter. In the user plane, the application creates data packets that are processed by protocols such as TCP, UDP and IP, while in the control plane, the radio resource control (RRC) protocol [1] writes the signalling messages that are exchanged between the base station and the mobile. In both cases, the information is processed by the packet data convergence protocol (PDCP) [2], the radio link control (RLC) protocol [3] and the medium access control (MAC) protocol [4], before being passed to the physical layer for transmission.

The physical layer has three parts. The *transport channel processor* [5] applies the error management procedures that we covered in Section 3.4, while the *physical channel processor* [6] applies the techniques of OFDMA, SC-FDMA and multiple antenna transmission from Chapters 4 and 5. Finally, the *analogue processor* [7, 8] converts the information to analogue

An Introduction to LTE: LTE, LTE-Advanced, SAE, VoLTE and 4G Mobile Communications, Second Edition.
Christopher Cox.

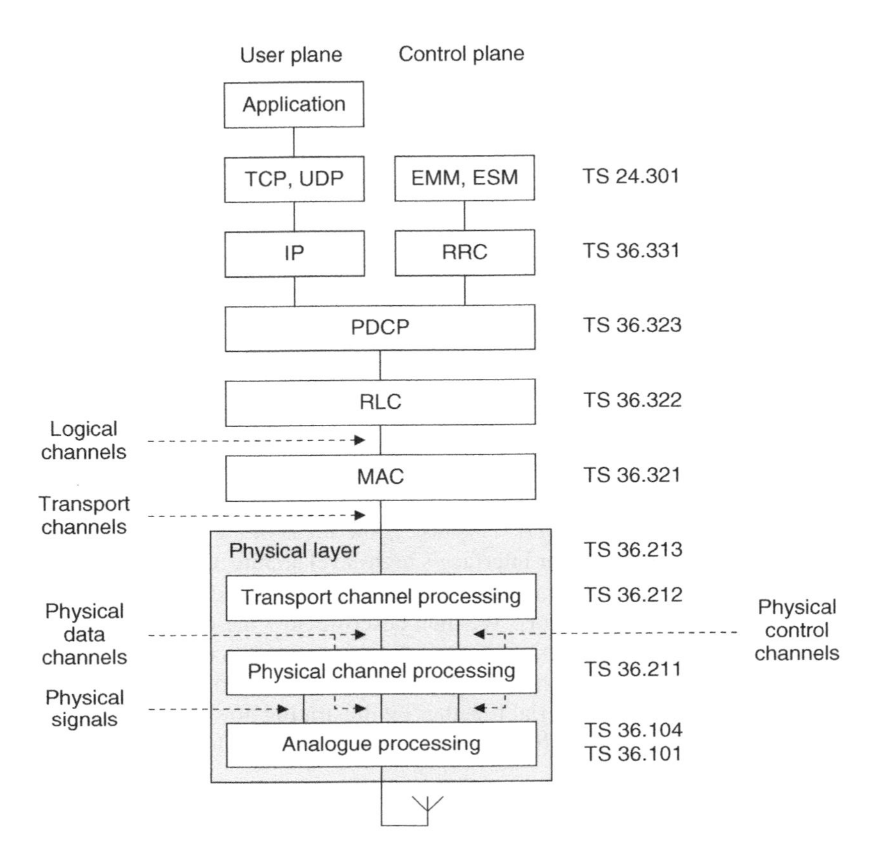

Figure 6.1 Architecture of the air interface protocol stack

form, filters it and mixes it up to radio frequency for transmission. A separate specification [9] describes the procedures that straddle the individual parts of the physical layer.

The information flows between the different protocols are known as channels and signals. Data and signalling messages are carried on *logical channels* between the RLC and MAC protocols, *transport channels* between the MAC and the physical layer, and *physical data channels* between the different levels of the physical layer. LTE uses several different types of logical, transport and physical channel, which are distinguished by the kind of information they carry and by the way in which the information is processed.

In the transmitter, the transport channel processor also creates *control information* that supports the low-level operation of the physical layer and sends this information to the physical channel processor in the form of *physical control channels*. The information travels as far as the transport channel processor in the receiver, but is completely invisible to higher layers. Similarly, the physical channel processor creates *physical signals*, which support the lowest-level aspects of the system. These travel as far as the physical channel processor in the receiver, but once again are invisible to higher layers.

6.2 Logical, Transport and Physical Channels

6.2.1 Logical Channels

Table 6.1 lists the logical channels that are used by LTE [10]. They are distinguished by the information they carry and can be classified in two ways. Firstly, logical traffic channels carry data in the user plane, while logical control channels carry signalling messages in the control plane. Secondly, dedicated logical channels are allocated to a specific mobile, while common logical channels can be used by more than one.

The most important logical channels are the *dedicated traffic channel* (DTCH), which carries data to or from a single mobile, and the *dedicated control channel* (DCCH), which carries the large majority of signalling messages. To be exact, the dedicated control channel carries all the mobile-specific signalling messages on signalling radio bearers 1 and 2, for mobiles that are in RRC_CONNECTED state.

The *broadcast control channel* (BCCH) carries RRC system information messages, which the base station broadcasts across the whole of the cell to tell the mobiles about how the cell is configured. These messages are divided into two unequal groups, which are handled differently by lower layers. The *master information block* (MIB) carries a few important parameters such as the downlink bandwidth, while several *system information blocks* (SIBs) carry the remainder.

The *paging control channel* (PCCH) carries *paging messages*, which the base station transmits if it wishes to contact mobiles that are in RRC_IDLE. The *common control channel* (CCCH) carries messages on signalling radio bearer 0, for mobiles that are moving from RRC_IDLE to RRC_CONNECTED in the procedure of RRC connection establishment.

Like the other tables in this chapter, Table 6.1 lists the channels that were introduced in every release of LTE. The *multicast traffic channel* (MTCH) and *multicast control channel* (MCCH) first appeared in LTE Release 9, to handle a service known as the *multimedia broadcast/multicast service* (MBMS). We will discuss these channels in Chapter 18.

6.2.2 Transport Channels

The transport channels [11] are listed in Table 6.2. They are distinguished by the ways in which the transport channel processor manipulates them.

Table 6.1 Logical channels

Channel	Release	Name	Information carried	Direction
DTCH	R8	Dedicated traffic channel	User plane data	UL, DL
DCCH	R8	Dedicated control channel	Signalling on SRB 1 and 2	
CCCH	R8	Common control channel	Signalling on SRB 0	
PCCH	R8	Paging control channel	Paging messages	DL
BCCH	R8	Broadcast control channel	System information	
MCCH	R9	Multicast control channel	MBMS signalling	
MTCH	R9	Multicast traffic channel	MBMS data	

Table 6.2 Transport channels

Channel	Release	Name	Information carried	Direction
UL-SCH	R8	Uplink shared channel	Uplink data and signalling	UL
RACH	R8	Random access channel	Random access requests	
DL-SCH	R8	Downlink shared channel	Downlink data and signalling	DL
PCH	R8	Paging channel	Paging messages	
BCH	R8	Broadcast channel	Master information block	
MCH	R8/R9	Multicast channel	MBMS	

The most important transport channels are the *uplink shared channel* (UL-SCH) and the *downlink shared channel* (DL-SCH), which carry the large majority of data and signalling messages across the air interface. The *paging channel* (PCH) carries paging messages that originated from the paging control channel. The *broadcast channel* (BCH) carries the broadcast control channel's master information block: the remaining system information messages are handled by the downlink shared channel, as if they were normal downlink data. The *multicast channel* (MCH) was fully specified in Release 8, to carry data from the multimedia broadcast/multicast service. However, it was not actually usable until the introduction of the actual service in Release 9.

The base station usually schedules the transmissions that a mobile makes, by granting it resources for uplink transmission at specific times and on specific sub-carriers. The *random access channel* (RACH) is a special channel through which the mobile can contact the network without any prior scheduling. Random access transmissions are composed by the mobile's MAC protocol and travel as far as the MAC protocol in the base station, but are completely invisible to higher layers.

The main differences between the transport channels lie in their approaches to error management. In particular, the uplink and downlink shared channels are the only transport channels that use the techniques of automatic repeat request and hybrid ARQ, and are the only channels that can adapt their coding rate to changes in the received signal to interference plus noise ratio (SINR). The other transport channels use forward error correction alone and have a fixed coding rate. The same restrictions apply to the control information that we will discuss below.

6.2.3 Physical Data Channels

Table 6.3 lists the physical data channels [12]. They are distinguished by the ways in which the physical channel processor manipulates them, and by the ways in which they are mapped onto the symbols and sub-carriers used by OFDMA.

The most important physical channels are the *physical downlink shared channel* (PDSCH) and the *physical uplink shared channel* (PUSCH). The PDSCH carries data and signalling messages from the downlink shared channel, as well as paging messages from the paging channel. The PUSCH carries data and signalling messages from the uplink shared channel and can sometimes carry the uplink control information (UCI) that is described below.

The *physical broadcast channel* (PBCH) carries the master information block from the broadcast channel, while the *physical random access channel* (PRACH) carries random access

Table 6.3 Physical data channels

Channel	Release	Name	Information carried	Direction
PUSCH	R8	Physical uplink shared channel	UL-SCH and/or UCI	UL
PRACH	R8	Physical random access channel	RACH	
PDSCH	R8	Physical downlink shared channel	DL-SCH and PCH	DL
PBCH	R8	Physical broadcast channel	BCH	
PMCH	R8/R9	Physical multicast channel	MCH	

transmissions from the random access channel. The *physical multicast channel* (PMCH) was fully specified in Release 8, to carry data from the multicast channel, but is not usable until Release 9.

The PDSCH and PUSCH are the only physical channels that can adapt their modulation schemes in response to changes in the received SINR. The other physical channels all use a fixed modulation scheme, usually QPSK. At least in LTE Release 8, the PDSCH is the only physical channel that uses the techniques of spatial multiplexing and beamforming from Sections 5.2 and 5.3, or the technique of closed loop transmit diversity from Section 5.1.2. The other channels are sent from a single antenna, or can use open loop transmit diversity in the case of the downlink. Once again, the same restrictions apply to the physical control channels that we will list below.

6.2.4 Control Information

The transport channel processor composes several types of control information, to support the low-level operation of the physical layer. These are listed in Table 6.4.

The *uplink control information* contains several fields. *Hybrid ARQ acknowledgements* are the mobile's acknowledgements of the base station's transmissions on the DL-SCH. The channel quality indicator (CQI) was introduced in Chapter 4 and describes the received SINR as

Table 6.4 Control information

Field	Release	Name	Information carried	Direction
UCI	R8	Uplink control information	Hybrid ARQ acknowledgements Channel quality indicators (CQI) Pre-coding matrix indicators (PMI) Rank indications (RI) Scheduling requests (SR)	UL
DCI	R8	Downlink control information	Downlink scheduling commands Uplink scheduling grants Uplink power control commands	DL
CFI	R8	Control format indicator	Size of downlink control region	DL
HI	R8	Hybrid ARQ indicator	Hybrid ARQ acknowledgements	

a function of frequency in support of frequency-dependent scheduling, while the precoding matrix indicator (PMI) and rank indication (RI) were introduced in Chapter 5 and support the use of spatial multiplexing. Collectively, the channel quality indicator, precoding matrix indicator and rank indication are sometimes known as *channel state information* (CSI), although this term does not actually appear in the specifications until Release 10. Finally, the mobile sends a *scheduling request* (SR) if it wishes to transmit uplink data on the PUSCH, but does not have the resources to do so.

The *downlink control information* (DCI) contains most of the downlink control fields. Using *scheduling commands* and *scheduling grants*, the base station can alert the mobile to forthcoming transmissions on the downlink shared channel and grant it resources for transmissions on the uplink shared channel. It can also adjust the power with which the mobiles are transmitting, by the use of *power control commands*.

The other sets of control information are less important. *Control format indicators* (CFIs) tell the mobiles about the organization of data and control information on the downlink, while *hybrid ARQ indicators* (HIs) are the base station's acknowledgements of the mobiles' uplink transmissions on the UL-SCH.

6.2.5 Physical Control Channels

The physical control channels are listed in Table 6.5.

In the downlink, there is a one-to-one mapping between the physical control channels and the control information listed above. As such, the *physical downlink control channel* (PDCCH), *physical control format indicator channel* (PCFICH) and *physical hybrid ARQ indicator channel* (PHICH) carry the downlink control information, control format indicators and hybrid ARQ indicators respectively. The *relay physical downlink control channel* (R-PDCCH) and *enhanced physical downlink control channel* (EPDCCH) are variants of the PDCCH that were introduced in Releases 10 and 11 respectively.

The uplink control information is sent on the PUSCH if the mobile is transmitting uplink data at the same time and on the *physical uplink control channel* (PUCCH) otherwise. The PUSCH and PUCCH are transmitted on different sets of sub-carriers, so this arrangement preserves the single carrier nature of the uplink transmission, in accordance with the requirements of SC-FDMA.

Table 6.5 Physical control channels

Channel	Release	Name	Information carried	Direction
PUCCH	R8	Physical uplink control channel	UCI	UL
PCFICH	R8	Physical control format indicator channel	CFI	DL
PHICH	R8	Physical hybrid ARQ indicator channel	HI	
PDCCH	R8	Physical downlink control channel	DCI	
R-PDCCH	R10	Relay physical downlink control channel	DCI	
EPDCCH	R11	Enhanced physical downlink control channel	DCI	

Table 6.6 Physical signals

Signal	Release	Name	Use	Direction
DRS	R8	Demodulation reference signal	Channel estimation	UL
SRS	R8	Sounding reference signal	Scheduling	
PSS	R8	Primary synchronization signal	Acquisition	DL
SSS	R8	Secondary synchronization signal	Acquisition	
RS	R8	Cell specific reference signal	Channel estimation and scheduling	DL
	R8	UE specific reference signal	Channel estimation	
	R8/R9	MBMS reference signal	Channel estimation	
	R9	Positioning reference signal	Location services	
	R10	CSI reference signal	Scheduling	

6.2.6 *Physical Signals*

The final information streams are the physical signals, which support the lowest-level operation of the physical layer. These are listed in Table 6.6.

In the uplink, the mobile transmits the *demodulation reference signal* (DRS) at the same time as the PUSCH and PUCCH, as a phase reference for use in channel estimation. It can also transmit the *sounding reference signal* (SRS) at times configured by the base station, as a power reference in support of frequency-dependent scheduling.

The downlink usually combines these two roles in the form of the *cell specific reference signal* (RS). *UE specific reference signals* are less important, and are sent to mobiles that are using beamforming in support of channel estimation. The specifications introduce other downlink reference signals as part of Releases 9 and 10. The base station also transmits two other physical signals, which help the mobile acquire the base station after it first switches on. These are known as the *primary synchronization signal* (PSS) and the *secondary synchronization signal* (SSS).

6.2.7 *Information Flows*

Tables 6.1 to 6.6 contain a large number of channels, but LTE uses them in just a few types of information flow. Figure 6.2 shows the information flows that are used in the uplink, with the arrows drawn from the viewpoint of the base station, so that uplink channels have arrows pointing upwards, and vice versa. Figure 6.3 shows the corresponding situation in the downlink.

6.3 The Resource Grid

6.3.1 *Slot Structure*

LTE maps the physical channels and physical signals onto the OFDMA symbols and sub-carriers that we introduced in Chapter 4. To understand how it does this, we first need to understand how LTE organizes its symbols and sub-carriers in the time and frequency domains [13].

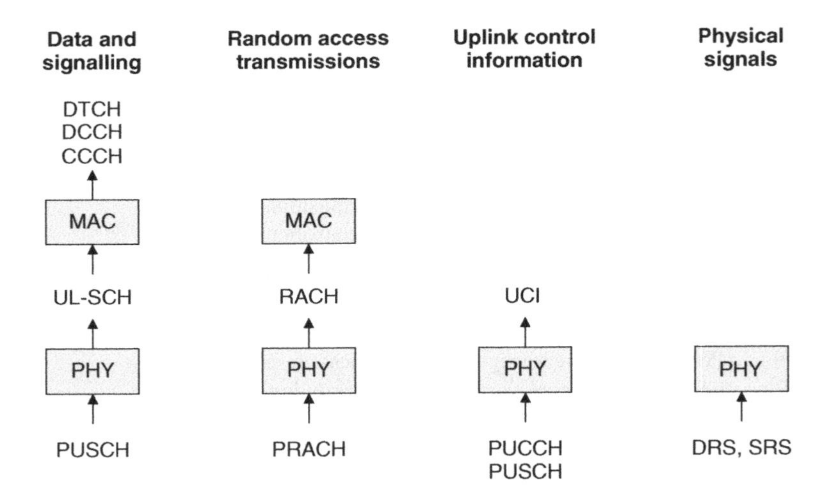

Figure 6.2 Uplink information flows used by LTE

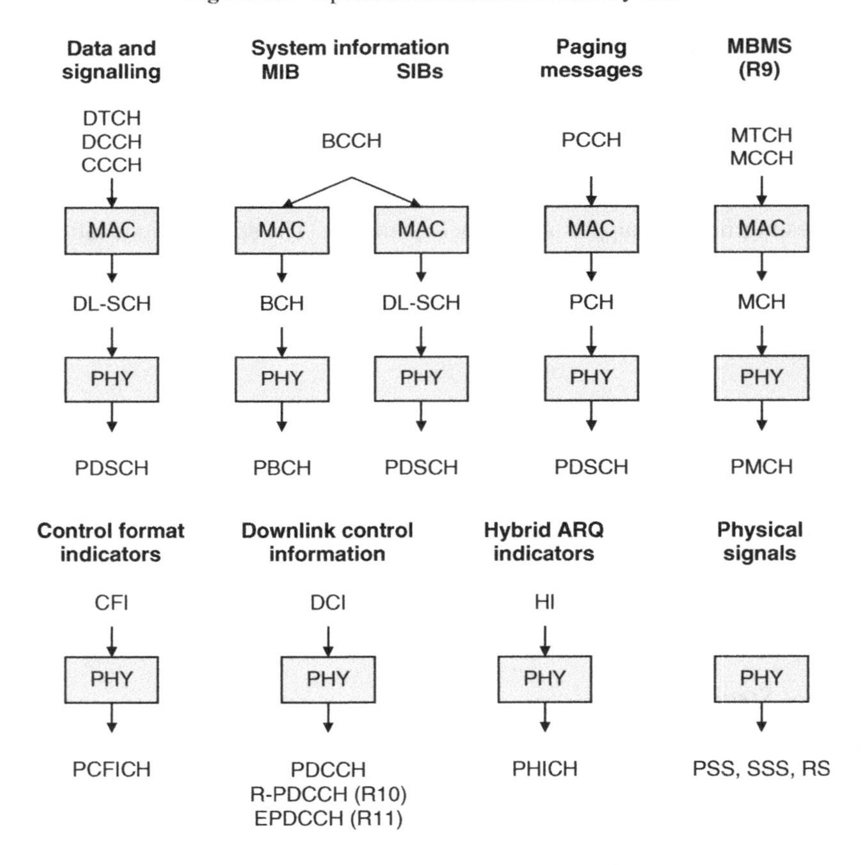

Figure 6.3 Downlink information flows used by LTE

First consider the time domain. The timing of the LTE transmissions is based on a time unit T_s, which is defined as follows:

$$T_s = \frac{1}{2048 \times 15\,000} \text{seconds} \approx 32.6\,\text{ns} \tag{6.1}$$

T_s is the shortest time interval that is of interest to the physical channel processor. (To be exact, T_s is the sampling interval Δt from Equation 4.3 if the system uses a fast Fourier transform that contains 2048 points, which is the largest value ever likely to be used.) The 66.7 μs symbol duration is then equal to 2048 T_s.

The symbols are grouped into *slots*, whose duration is 0.5 ms (15 360 T_s). This can be done in two ways, as shown in Figure 6.4. With the *normal cyclic prefix*, each symbol is preceded by a cyclic prefix that is usually 144 T_s (4.7 μs) long. The first cyclic prefix has a longer duration of 160 T_s (5.2 μs), to tidy up the unevenness that results from fitting seven symbols into a slot.

Using the normal cyclic prefix, the receiver can remove inter-symbol interference with a delay spread of 4.7 μs, corresponding to a path difference of 1.4 km between the lengths of the longest and shortest rays. This is normally plenty, but may not be enough if the cell is unusually large or cluttered. To deal with this possibility, LTE also supports an *extended cyclic prefix*, in which the number of symbols per slot is reduced to six. This allows the cyclic prefix to be extended to 512 T_s (16.7 μs), to support a maximum path difference of 5 km.

With one exception, related to the multimedia broadcast/multicast service in Release 9, the base station sticks with either the normal or extended cyclic prefix in the downlink and does not change between the two. Mobiles generally use the same cyclic prefix duration in the uplink, but the base station can force a different choice by the use of its system information. The normal cyclic prefix is far more common, so we will use it almost exclusively.

6.3.2 Frame Structure

At a higher level, the slots are grouped into subframes and frames [14]. In FDD mode, this is done using *frame structure type 1*, which is shown in Figure 6.5.

Two slots make one *subframe*, which is 1 ms long (30 720 T_s). Subframes are used for scheduling. When a base station transmits to a mobile on the downlink, it schedules its PDSCH

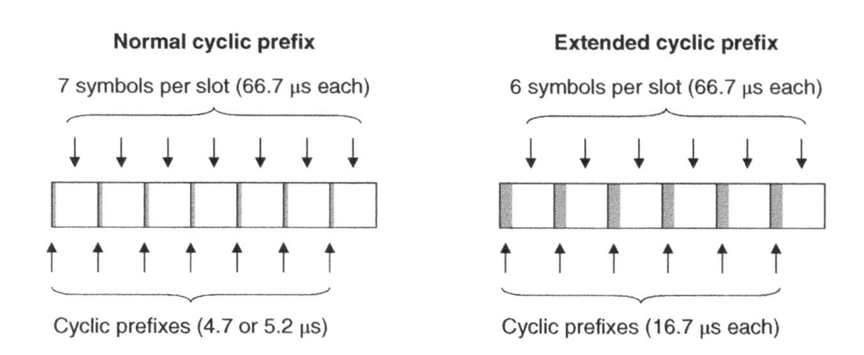

Figure 6.4 Organization of symbols into slots using the normal and extended cyclic prefix

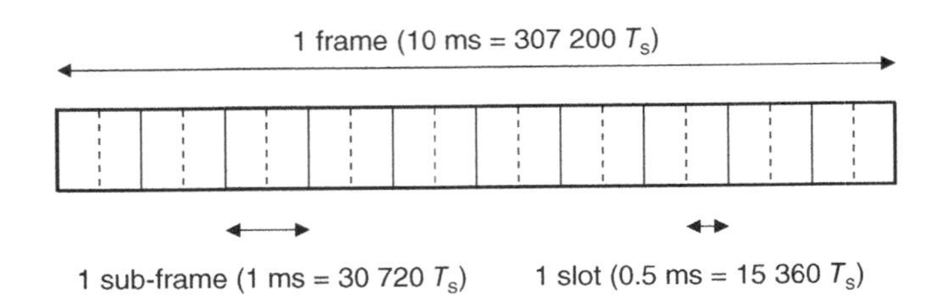

Figure 6.5 Frame structure type 1, used in FDD mode. Source: TS 36.211. Reproduced by permission of ETSI

transmissions one subframe at a time, and maps each block of data onto a set of sub-carriers within that subframe. A similar process happens on the uplink.

In turn, 10 subframes make one *frame*, which is 10 ms long (307 200 T_s). Each frame is numbered using a *system frame number* (SFN), which runs repeatedly from 0 to 1023. Frames help to schedule a number of slowly changing processes, such as the transmission of system information and reference signals.

TDD mode uses *frame structure type 2*. In this structure, the slots, subframes and frames have the same duration as before, but each subframe can be allocated to either the uplink or downlink using one of the *TDD configurations* shown in Figure 6.6.

Different cells can have different TDD configurations, which are advertised as part of the cells' system information. Configuration 1 might be suitable if the data rates are similar on the uplink and downlink, for example, while configuration 5 might be used in cells that are dominated by downlink transmissions. Nearby cells should generally use the same TDD configuration, to minimize the interference between the uplink and downlink.

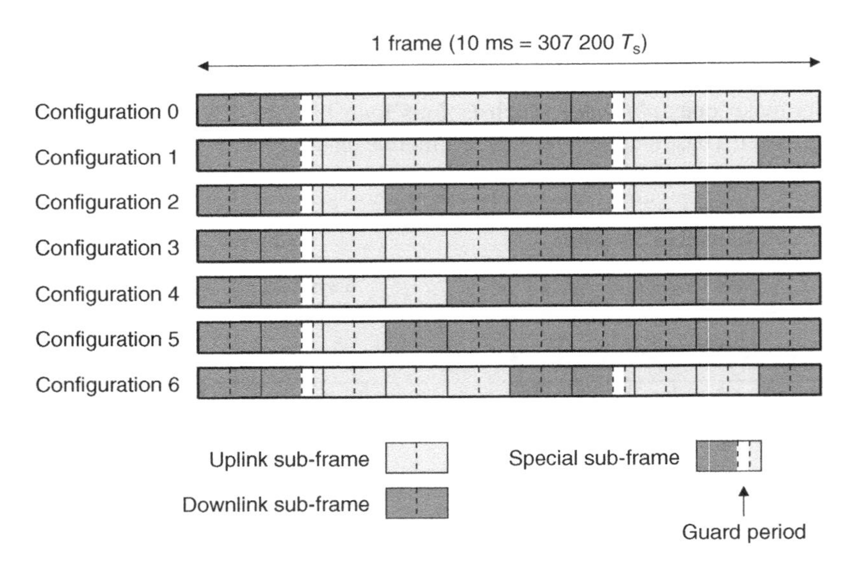

Figure 6.6 TDD configurations using frame structure type 2

Special subframes are used at the transitions from downlink to uplink transmission. They contain three regions. The *special downlink region* occupies 3 to 12 symbols and is used in the same way as any other downlink region. The *special uplink region* occupies either one or two symbols, and is only used by the random access channel and the sounding reference signal. The two regions are separated by a *guard period* of 1 to 10 symbols that supports the timing advance procedure described below. The cell can adjust the size of each region using a *special subframe configuration*, which again is advertised in the system information.

6.3.3 Uplink Timing Advance

In LTE, a mobile starts transmitting its uplink frames at a time TA before the arrival of the corresponding frames on the downlink [15] (Figure 6.7). TA is known as the *timing advance* and is used for the following reason. Even travelling at the speed of light, a mobile's transmissions take time (typically a few microseconds) to reach the base station. However, the signals from different mobiles have to reach the base station at roughly the same time, with a spread less than the cyclic prefix duration, to prevent any risk of inter-symbol interference between them. To enforce this requirement, distant mobiles have to start transmitting slightly earlier than they otherwise would.

Because the uplink transmission time is based on the downlink arrival time, the timing advance has to compensate for the round-trip travel time between the base station and the mobile:

$$\mathrm{TA} \approx \frac{2L}{c} \tag{6.2}$$

Here, L is the distance between the mobile and the base station, and c is the speed of light. The timing advance does not have to be completely accurate, as the cyclic prefix can handle any remaining errors.

The specifications define the timing advance as follows:

$$\mathrm{TA} = (N_{\mathrm{TA}} + N_{\mathrm{TAoffset}})T_{\mathrm{s}} \tag{6.3}$$

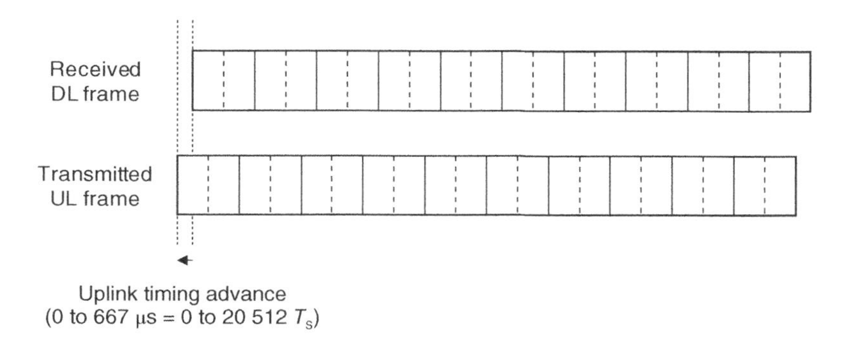

Figure 6.7 Timing relationship between the uplink and downlink in FDD mode. Source: TS 36.211. Reproduced by permission of ETSI

Here, N_{TA} lies between 0 and 20 512. This gives a maximum timing advance of about 667 μs (two-thirds of a subframe), which supports a maximum cell size of 100 km. N_{TA} is initialized by the random access procedure described in Chapter 9, and updated by the timing advance procedure from Chapter 10.

$N_{TAoffset}$ is zero in FDD mode, but 624 in TDD mode. This creates a small gap at the transition from uplink to downlink transmissions, which gives the base station time to switch from one to the other. The guard period in each special subframe creates a longer gap at the transition from downlink to uplink, which allows the mobile to advance its uplink frames without them colliding with the frames received on the downlink.

6.3.4 Resource Grid Structure

In LTE, information is organized as a function of frequency as well as time, using a *resource grid* [16]. Figure 6.8 shows the resource grid for the case of a normal cyclic prefix. (There is a similar grid for the extended cyclic prefix, which uses six symbols per slot rather than seven.)

The basic unit is a *resource element* (RE), which spans one symbol by one sub-carrier. Each resource element usually carries two, four or six physical channel bits, depending on whether the modulation scheme is QPSK, 16-QAM or 64-QAM. Resource elements are grouped into *resource blocks* (RBs), each of which spans 0.5 ms (one slot) by 180 kHz (12 sub-carriers). The

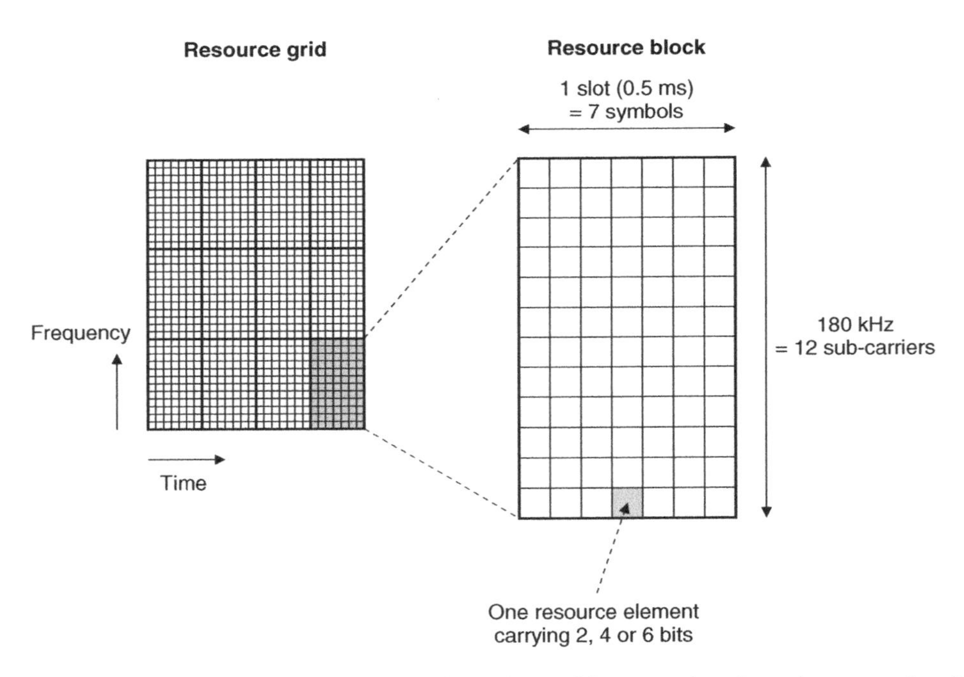

Figure 6.8 Structure of the LTE resource grid in the time and frequency domains, using a normal cyclic prefix

base station uses resource blocks for frequency-dependent scheduling, by allocating the symbols and sub-carriers within each subframe in units of resource blocks. The term *resource block pair* is sometimes used for the two consecutive resource blocks that use the same sub-carriers within a particular subframe.

One subtle point is that the downlink resource grid does not use the 0 kHz sub-carrier; instead, the resource blocks on either side use sub-carriers from +15 to +180 kHz and from −15 to −180 kHz respectively. The reason is that the 0 kHz sub-carrier reaches the mobile's OFDMA receiver at a frequency of zero, where it can suffer from high levels of noise and interference. We still use that sub-carrier on the uplink, because the SC-FDMA symbols are spread across sub-carriers so are less vulnerable to interference, and because its omission would increase the power variations of the SC-FDMA waveform.

6.3.5 Bandwidth Options

A cell can be configured with several different bandwidths [17], which are listed in Table 6.7. In a 5 MHz band, for example, the base station transmits using 25 resource blocks (300 sub-carriers), giving a transmission bandwidth of 4.5 MHz. That arrangement leaves room for guard bands at the upper and lower edges of the frequency band, which minimize the amount of interference with the next band along. The two guard bands are usually the same width, but the network operator can adjust them if necessary by shifting the centre frequency in units of 100 kHz.

The existence of all these bandwidth options makes it easy for network operators to deploy LTE in a variety of spectrum management regimes. For example, 1.4 MHz is close to the bandwidths previously used by cdma2000 and TD-SCDMA, 5 MHz is the same bandwidth used by WCDMA, while 20 MHz allows an LTE base station to operate at its highest possible data rate. In FDD mode, the uplink and downlink bandwidths are usually the same. If they are different, then the base station signals the uplink bandwidth as part of its system information.

In Chapter 4, we noted that the fast Fourier transform operates most efficiently if the number of data points is an exact power of 2. This is easy to achieve, because the transmitter can simply round up the number of sub-carriers to the next highest power of 2, and can fill the extreme ones with zeros. In a 20 MHz bandwidth, for example, it will generally process the data using a 2048 point FFT, which is consistent with the value of T_s that we introduced earlier.

Table 6.7 Cell bandwidths supported by LTE

Total bandwidth	Number of resource blocks	Number of sub-carriers	Occupied bandwidth	Usual guard bands
1.4 MHz	6	72	1.08 MHz	2×0.16 MHz
3 MHz	15	180	2.7 MHz	2×0.15 MHz
5 MHz	25	300	4.5 MHz	2×0.25 MHz
10 MHz	50	600	9 MHz	2×0.5 MHz
15 MHz	75	900	13.5 MHz	2×0.75 MHz
20 MHz	100	1200	18 MHz	2×1 MHz

6.4 Multiple Antenna Transmission

6.4.1 Downlink Antenna Ports

In the downlink, multiple antenna transmissions are organized using *antenna ports*, each of which has its own copy of the resource grid that we introduced above. Table 6.8 lists the base station antenna ports that LTE uses. Ports 0 to 3 are used for single antenna transmission, transmit diversity and spatial multiplexing, while port 5 is reserved for beamforming. The remaining antenna ports are introduced in Releases 9 to 11 and will be covered towards the end of the book.

It is worth noting that an antenna port is not necessarily the same as a physical antenna; instead, it is an output from the base station transmitter that can drive one or more physical antennas. In particular, as shown in Figure 6.9, port 5 will always drive several physical

Table 6.8 Antenna ports used by the LTE downlink

Antenna port	Release	Application
0	R8	Single antenna transmission 2 and 4 antenna transmit diversity and spatial multiplexing
1	R8	2 and 4 antenna transmit diversity and spatial multiplexing
2	R8	4 antenna transmit diversity and spatial multiplexing
3	R8	4 antenna transmit diversity and spatial multiplexing
4	R8 / R9	MBMS
5	R8	Beamforming
6	R9	Positioning reference signals
7–8	R9	Dual layer beamforming 8 antenna spatial multiplexing
9–14	R10	8 antenna spatial multiplexing
15–22	R10	CSI reference signals
107–110	R11	EPDCCH

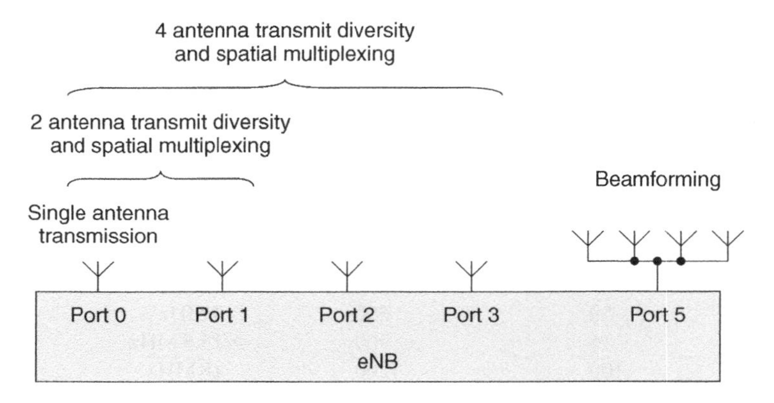

Figure 6.9 Antenna ports used by a Release 8 base station

Table 6.9 Downlink transmission modes

Mode	Release	Purpose	Uplink feedback required		
			CQI	RI	PMI
1	R8	Single antenna transmission	✓		
2	R8	Open loop transmit diversity	✓		
3	R8	Open loop spatial multiplexing	✓	✓	
4	R8	Closed loop spatial multiplexing	✓	✓	✓
5	R8	Multiple user MIMO	✓		✓
6	R8	Closed loop transmit diversity	✓		✓
7	R8	Beamforming	✓		
8	R9	Dual layer beamforming	✓	Configurable	
9	R10	Eight layer spatial multiplexing	✓	Configurable	
10	R11	Coordinated multipoint transmission	✓	Configurable	

antennas that are used for beamforming. The base station constructs the beam by applying different weights to each of those antennas, but the calculations are an internal matter for the base station, so the 3GPP specifications do not have to refer to the individual antennas at all.

6.4.2 Downlink Transmission Modes

To support the use of multiple antennas, the base station can optionally configure the mobile into one of the *downlink transmission modes* that are listed in Table 6.9.

The transmission mode defines the type of multiple antenna processing that the base station will use for its transmissions on the PDSCH, and hence the type of processing that the mobile should use for PDSCH reception. It also defines the feedback that the base station will expect from the mobile, in the manner listed in the table.

If the base station does not configure the mobile in this way, then it transmits the PDSCH using either a single antenna or open loop transmit diversity, depending on the total number of antenna ports that it has.

6.5 Resource Element Mapping

6.5.1 Downlink Resource Element Mapping

The LTE physical layer transmits the physical channels and physical signals by mapping them onto the resource elements that we introduced above. The exact mapping depends on the exact configuration of the base station and mobile, so we will cover it one channel at a time as part of Chapters 7 to 9. However, it is instructive to show some example mappings for the uplink and downlink for a typical system configuration.

Figure 6.10 shows an example resource element mapping for the downlink. The figure assumes the use of FDD mode, the normal cyclic prefix and a bandwidth of 3 MHz. Time is plotted horizontally and spans the 10 subframes (20 slots) that make up one frame. Frequency is plotted vertically and spans the 15 resource blocks that make up the transmission band.

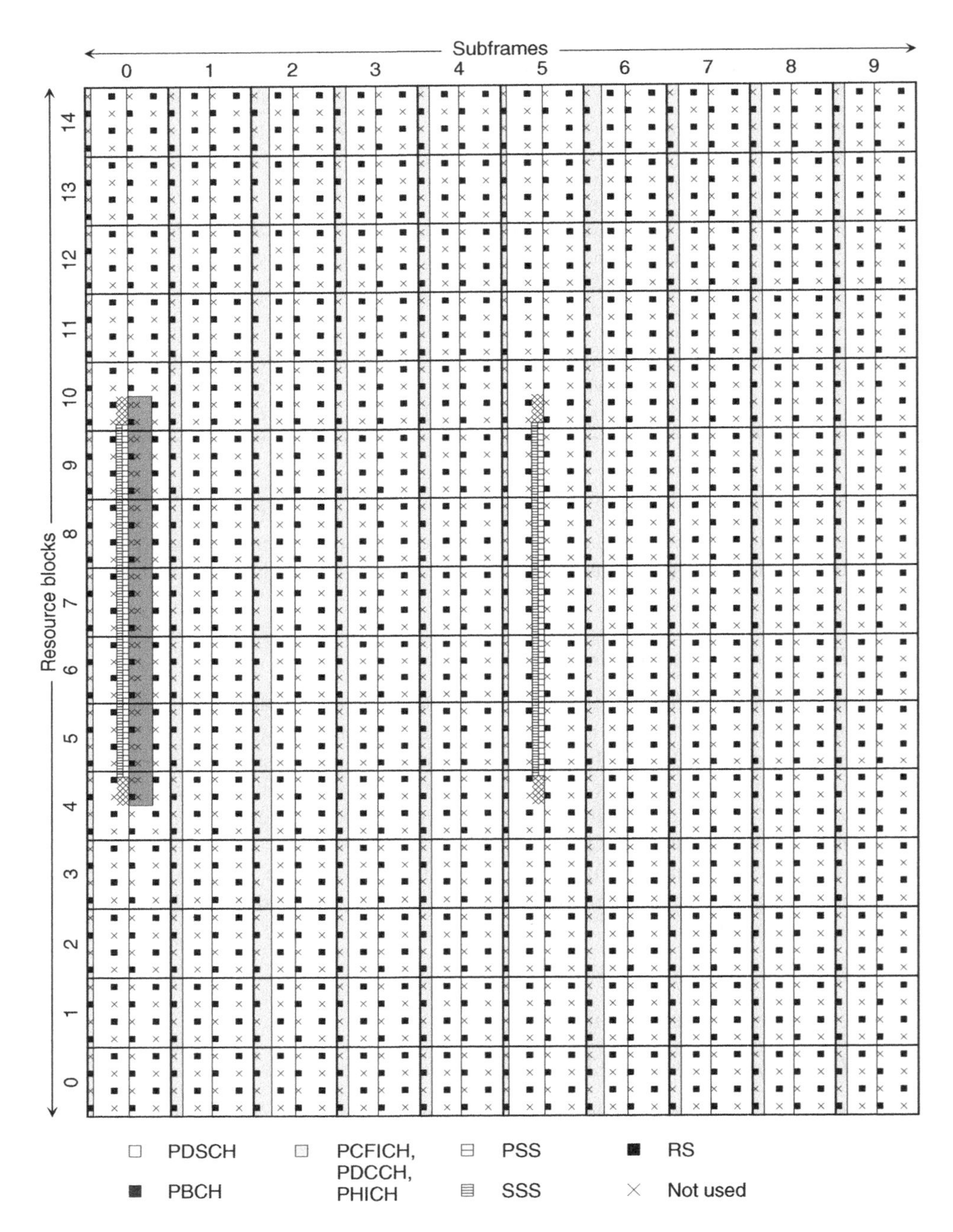

Figure 6.10 Example mapping of physical channels to resource elements in the downlink, using FDD mode, a normal cyclic prefix, a 3 MHz bandwidth, the first antenna port of two and a physical cell ID of 1

The cell specific reference signals are scattered across the time and frequency domains. While one antenna port is sending a reference signal, the others keep quiet, so that the mobile can measure the received reference signal from one antenna port at a time. The diagram assumes the use of two antenna ports and shows the reference signals that are sent from port 0. The exact mapping depends on the physical cell identity from Chapter 2: the one shown is suitable for a physical cell identity of 1, 7, 13, ...

Within each frame, certain resource elements are reserved for the primary and secondary synchronization signals and for the physical broadcast channel, and are read during the acquisition procedure that is described in Chapter 7. This information is only sent on the central 72 sub-carriers (1.08 MHz), which is the narrowest bandwidth ever used by LTE. This allows the mobile to read it without prior knowledge of the downlink bandwidth.

At the start of each subframe, a few symbols are reserved for the control information that the base station transmits on the PCFICH, PDCCH and PHICH. The number of control symbols can vary from one subframe to the next, depending on how much control information the base station needs to send. The rest of the subframe is reserved for data transmissions on the PDSCH and is allocated to individual mobiles in units of resource blocks within each subframe.

6.5.2 Uplink Resource Element Mapping

Figure 6.11 shows the corresponding situation on the uplink. Once again, the figure assumes the use of FDD mode, the normal cyclic prefix and a bandwidth of 3 MHz.

The outermost parts of the band are reserved for uplink control information on the PUCCH and for the associated demodulation reference signals. The PUCCH is divided into two parts, which have various different formats that depend on the information that the mobile has to transmit. The outer part has a fixed bandwidth and is used for PUCCH formats known as 2, 2a and 2b, which have five control symbols per slot and two reference symbols. The inner part has a variable bandwidth and is used for PUCCH formats 1, 1a and 2b, which have four control symbols per slot and three reference symbols.

The rest of the band is mainly used by the PUSCH and is allocated to individual mobiles in units of resource blocks within each subframe. PUSCH transmissions contain six data symbols per slot and one reference symbol.

The base station also reserves certain resource blocks for random access transmissions on the PRACH. The PRACH has a bandwidth of six resource blocks and a duration from one to three subframes, while its locations in the resource grid are configured by the base station. In the example shown, the base station has reserved resource blocks 7–12 in subframe 1 by using a PRACH frequency offset of 7 and a PRACH configuration index of 3. Many other configurations are possible.

Furthermore, the base station can reserve the last symbol of certain subframes for the transmission of sounding reference signals. In the example shown, the base station has reserved the last symbol of subframes 2 and 7 by using an SRS subframe configuration index of 5. Within the reserved region, an individual mobile transmits on alternate sub-carriers, using a mobile-specific bandwidth and frequency offset. Once again, many other configurations are possible.

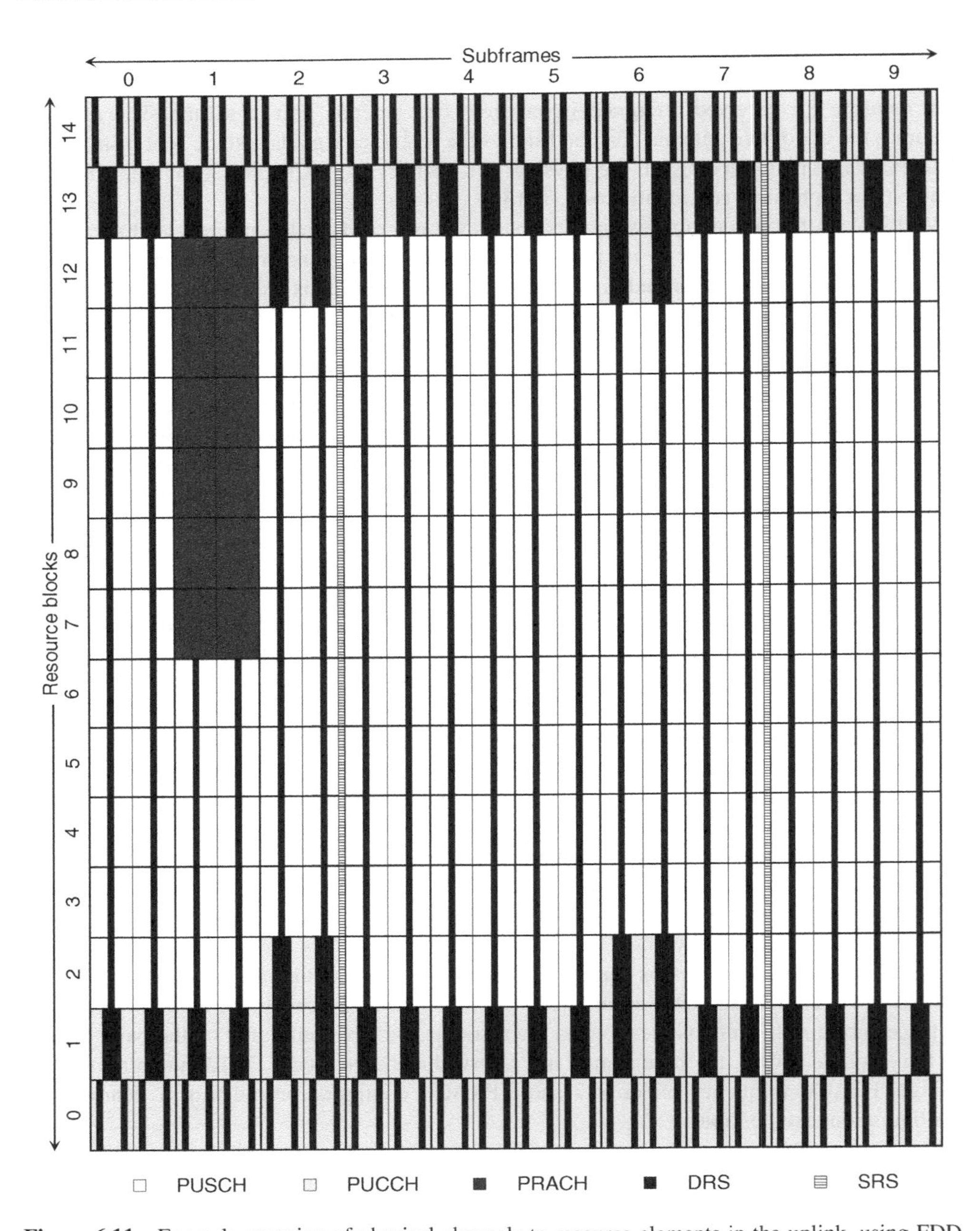

Figure 6.11 Example mapping of physical channels to resource elements in the uplink, using FDD mode, a normal cyclic prefix, a 3 MHz bandwidth, a PRACH configuration index of 3 and an SRS subframe configuration index of 5

References

1. 3GPP TS 36.331 (2013) Radio Resource Control (RRC); Protocol Specification, Release 11, September 2013.
2. 3GPP TS 36.323 (2013) Packet Data Convergence Protocol (PDCP) Specification, Release 11, March 2013.
3. 3GPP TS 36.322 (2012) Radio Link Control (RLC) Protocol Specification, Release 11, September 2012.
4. 3GPP TS 36.321 (2013) Medium Access Control (MAC) Protocol Specification, Release 11, July 2013.
5. 3GPP TS 36.212 (2013) Evolved Universal Terrestrial Radio Access (E-UTRA); Multiplexing and Channel Coding, Release 11, June 2013.
6. 3GPP TS 36.211 (2013) Evolved Universal Terrestrial Radio Access (E-UTRA); Physical Channels and Modulation, Release 11, September 2013.
7. 3GPP TS 36.101 (2013) User Equipment (UE) Radio Transmission and Reception, Release 11, September 2013.
8. 3GPP TS 36.104 (2013) Base Station (BS) Radio Transmission and Reception, Release 11, September 2013.
9. 3GPP TS 36.213 (2013) Evolved Universal Terrestrial Radio Access (E-UTRA); Physical Layer Procedures, Release 11, September 2013.
10. 3GPP TS 36.321 (2013) Medium Access Control (MAC) Protocol Specification, Release 11, Section 4.5, July 2013.
11. 3GPP TS 36.212 (2013) Multiplexing and Channel Coding, Release 11, Section 4, June 2013.
12. 3GPP TS 36.211 (2013) Physical Channels and Modulation, Release 11, Sections 5.1, 6.1, September 2013.
13. 3GPP TS 36.211 (2013) Physical Channels and Modulation, Release 11, Sections 5.6, 6.12, September 2013.
14. 3GPP TS 36.211 (2013) Physical Channels and Modulation, Release 11, Section 4, September 2013.
15. 3GPP TS 36.211 (2013) Physical Channels and Modulation, Release 11, Section 8, September 2013.
16. 3GPP TS 36.211 (2013) Physical Channels and Modulation, Release 11, Sections 5.2, 6.2, September 2013.
17. 3GPP TS 36.101 (2013) User Equipment (UE) Radio Transmission and Reception, Release 11, Section 5.6, September 2013.

7

Cell Acquisition

After a mobile switches on, it runs a low-level acquisition procedure so as to identify the nearby LTE cells and discover how they are configured. In doing so, it receives the primary and secondary synchronization signals, reads the master information block from the physical broadcast channel and reads the remaining system information blocks from the physical downlink shared channel. It also starts reception of the downlink reference signals and the physical control format indicator channel, which it will need throughout the process of data transmission and reception later on. In this chapter, we begin by summarizing the acquisition procedure and then move to a discussion of the individual steps.

The most important specification for this chapter is the one for the physical channel processor, TS 36.211 [1]. The system information blocks form part of the radio resource control protocol and are defined in TS 36.331 [2].

7.1 Acquisition Procedure

The acquisition procedure is summarized in Table 7.1. There are several steps. The mobile starts by receiving the synchronization signals from all the nearby cells. From the primary synchronization signal (PSS), it discovers the symbol timing and gets some incomplete information about the physical cell identity. From the secondary synchronization signal (SSS), it discovers the frame timing, the physical cell identity, the transmission mode (FDD or TDD) and the cyclic prefix duration (normal or extended).

At this point, the mobile starts reception of the cell-specific reference signals. These provide an amplitude and phase reference for the channel estimation process, so they are essential for everything that follows. The mobile then receives the physical broadcast channel and reads the master information block. By doing so, it discovers the number of transmit antennas at the base station, the downlink bandwidth, the system frame number and a quantity called the PHICH configuration that describes the physical hybrid ARQ indicator channel.

The mobile can now start reception of the physical control format indicator channel (PCFICH), so as to read the control format indicators. These indicate how many symbols are reserved at the start of each downlink subframe for the physical control channels and

An Introduction to LTE: LTE, LTE-Advanced, SAE, VoLTE and 4G Mobile Communications, Second Edition.
Christopher Cox.

Table 7.1 Steps in the cell acquisition procedure

Step	Task	Information obtained
1	Receive PSS	Symbol timing Cell identity within group
2	Receive SSS	Frame timing Physical cell identity Transmission mode Cyclic prefix duration
3	Start reception of RS	Amplitude and phase reference for demodulation Power reference for channel quality estimation
4	Read MIB from PBCH	Number of transmit antennas Downlink bandwidth System frame number PHICH configuration
5	Start reception of PCFICH	Number of control symbols per subframe
6	Read SIBs from PDSCH	System information

how many are available for data transmissions. Finally, the mobile can start reception of the physical downlink control channel (PDCCH). This allows the mobile to read the remaining system information blocks (SIBs), which are sent on the physical downlink shared channel (PDSCH). By doing this, it discovers all the remaining details about how the cell is configured, such as the identities of the networks that it belongs to.

7.2 Synchronization Signals

7.2.1 Physical Cell Identity

The physical cell identity is a number between 0 and 503, which is transmitted on the synchronization signals [3] and used in three ways. Firstly, it determines the exact set of resource elements that are used for the cell-specific reference signals and the PCFICH. Secondly, it influences a downlink transmission process known as scrambling, in a bid to minimize interference between nearby cells. Thirdly, it identifies individual cells during RRC procedures such as measurement reporting and handover. The physical cell identity is assigned during network planning or self-configuration. Nearby cells should always receive different physical cell identities, to ensure that each of these roles is properly fulfilled.

It would be hard for a mobile to find the physical cell identities in one step, so they are organized into *cell identity groups* as follows:

$$N_{\mathrm{ID}}^{\mathrm{cell}} = 3N_{\mathrm{ID}}^{(1)} + N_{\mathrm{ID}}^{(2)} \tag{7.1}$$

In this equation, $N_{\mathrm{ID}}^{\mathrm{cell}}$ is the physical cell identity. $N_{\mathrm{ID}}^{(1)}$ is the cell identity group, which runs from 0 to 167 and is signalled using the SSS. $N_{\mathrm{ID}}^{(2)}$ is the cell identity within the group, which runs from 0 to 2 and is signalled using the PSS. Using this arrangement, a network planner can give each nearby base station a different cell identity group, and can distinguish its sectors using the cell identity within the group.

7.2.2 Primary Synchronization Signal

Figure 7.1 shows the time domain mapping of the primary and secondary synchronization signals. The signals are both transmitted twice per frame. In FDD mode, the PSS is transmitted in the last symbol of slots 0 and 10, while the SSS is sent one symbol earlier. In TDD mode, the PSS is transmitted in the third symbol of slots 2 and 12, while the SSS is sent three symbols earlier.

In the frequency domain, the base station maps the synchronization signals onto the central 62 sub-carriers, and pads the resulting signal with zeros so that it occupies the central 72 sub-carriers (1.08 MHz). This second bandwidth is the smallest transmission band that LTE supports, which ensures that the mobile can receive both signals without prior knowledge of the downlink bandwidth.

The base station creates the actual signal using a *Zadoff-Chu sequence* [4, 5]. A Zadoff-Chu sequence is a complex-valued sequence with a length of N_{ZC} data points. For each sequence length, we can generate a maximum of N_{ZC} different *root sequences* using a sequence number that runs from 0 to $N_{ZC} - 1$ and is not an integer factor of N_{ZC}. We can then adjust each root sequence further by applying a maximum of N_{ZC} different *cyclic shifts*, in which the data points are shifted along the sequence and wrapped around the ends. Zadoff-Chu sequences are also used by the uplink reference signals (Chapter 8) and by the physical random access channel (Chapter 9).

The PSS indicates the three possible values of $N_{ID}^{(2)}$ using three root sequences. To process the PSS, the mobile receives the incoming signal for a time of at least 5 ms and compares it with those three root sequences. By doing so, it measures the times at which the primary synchronization signal arrives from each of the nearby cells and finds the cell identity within the group, $N_{ID}^{(2)}$.

Zadoff-Chu sequences are useful because they have good correlation properties. In practical terms, this means that there is little risk of the mobile making a mistake in its measurement

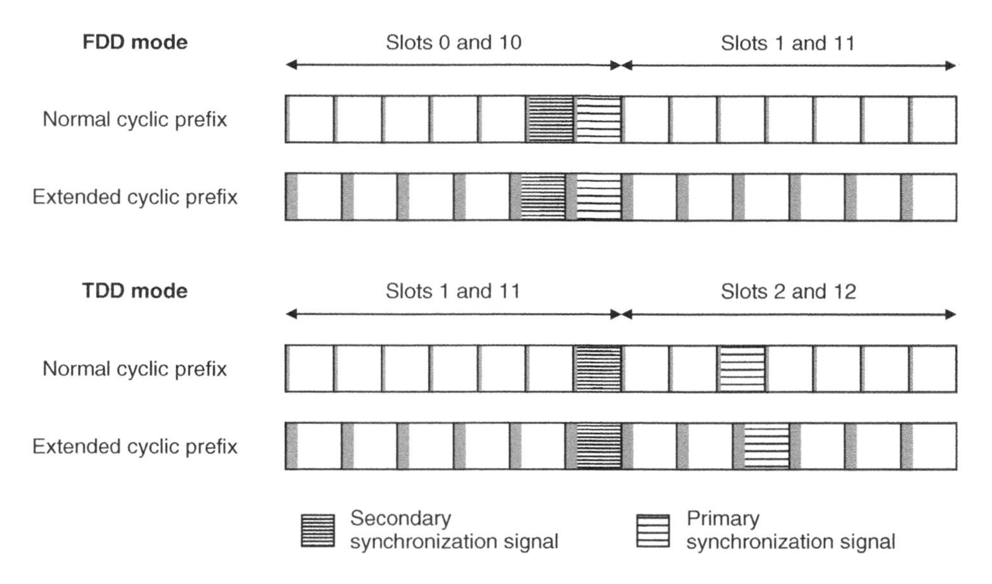

Figure 7.1 Time domain mapping of the primary and secondary synchronization signals

of $N_{\mathrm{ID}}^{(2)}$, even if the received signal-to-interference plus noise ratio (SINR) is low. Adding some more detail, different cyclic shifts of the same root sequence are orthogonal: if a mobile receives one cyclic shift and compares it with another, then the result is zero if the two cyclic shifts are different and non-zero only if they are the same. Different root sequences are not orthogonal but are still valuable: if a mobile receives one root sequence and compares it with another, then the result is small if the two sequences are different and large only if they are the same. Unlike the cyclic shifts, the root sequences maintain their good correlation properties if the base station and mobile are not properly time or frequency aligned with one another, so they are preferred in the case of the PSS.

7.2.3 Secondary Synchronization Signal

The base station transmits the SSS immediately before the PSS in FDD mode, or three symbols before it in TDD mode. The exact transmission time depends on the cyclic prefix duration (normal or extended), giving four possible transmission times altogether. The actual signal is created using pseudo-random sequences known as *Gold sequences* [6]. The exact sequence indicates the cell identity group, $N_{\mathrm{ID}}^{(1)}$, and is different for the first and second transmission of the signal within the frame.

To receive the secondary synchronization signal, the mobile inspects each of the four possible transmission times relative to the PSS and looks for each of the possible SSS sequences. By finding the time when the signal was transmitted, it can deduce the transmission mode that the cell is using and the cyclic prefix duration. By identifying the transmitted sequence, it can deduce the cell identity group and hence the physical cell identity. It can also deduce whether the sequence was from the first or the second transmission of the SSS within the frame, so it can find the time at which a 10 ms frame begins.

7.3 Downlink Reference Signals

The mobile now starts reception of the downlink reference signals [7]. These are used in two ways. Their immediate role is to give the mobile an amplitude and phase reference for use in channel estimation. Later on, the mobile will use them to measure the received signal power as a function of frequency and to calculate the channel quality indicators.

Release 8 uses two types of downlink reference signal, but the cell-specific reference signals are the most important. Figure 7.2 shows how these signals are mapped to resource elements, for the case of a normal cyclic prefix. As shown in the figure, the mapping depends on the number of antenna ports that the base station is using and on the antenna port number. While one antenna is transmitting a reference signal, all the others stay silent, in the manner required for spatial multiplexing.

This basic pattern is offset in the frequency domain, to minimize interference between the reference signals transmitted from nearby cells. The sub-carrier offset is:

$$v_{\mathrm{shift}} = N_{\mathrm{ID}}^{\mathrm{cell}} \bmod 6 \tag{7.2}$$

where $N_{\mathrm{ID}}^{\mathrm{cell}}$ is the physical cell identity that the mobile has already found. Previously, Figure 6.10 showed the resource element mapping in the case where $v_{\mathrm{shift}} = 1$, so that the physical cell identity could have values of 1, 7, 13 ...

The resource elements are filled with a Gold sequence that depends on the physical cell identity, again in a bid to minimize interference. By measuring the received reference symbols

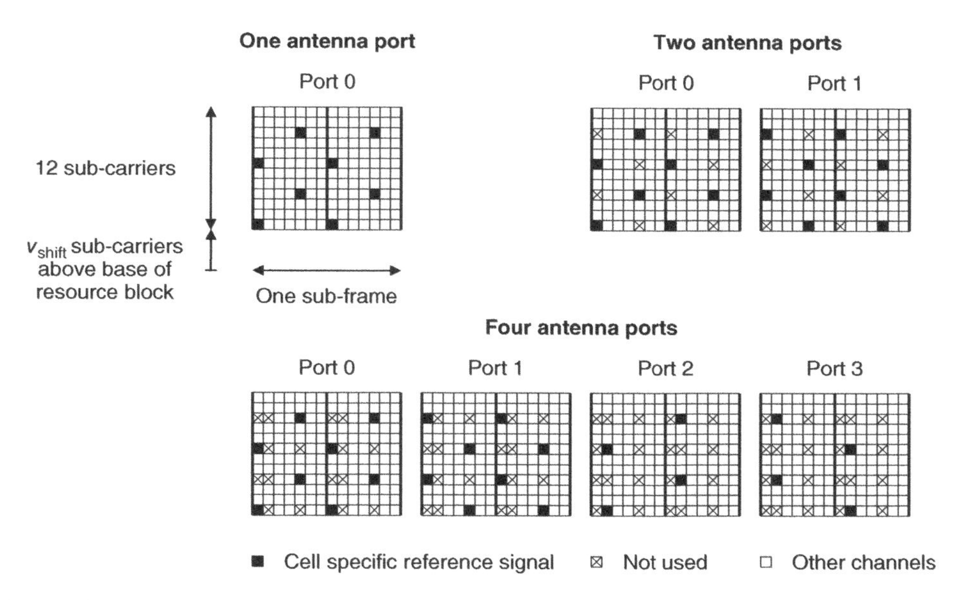

Figure 7.2 Resource element mapping for the cell-specific reference signals, using a normal cyclic prefix. Source: TS 36.211. Reproduced by permission of ETSI

and comparing them with the ones transmitted, the mobile can measure the amplitude changes and phase shifts that the air interface has introduced. It can then estimate those quantities at intervening resource elements by interpolation.

Antenna ports 0 and 1 use four reference symbols per resource block, while antenna ports 2 and 3 use only two. This is because a cell is only likely to use four antenna ports when it is dominated by slowly moving mobiles, for which the amplitude and phase of the received signal will only vary slowly with time.

A Release 8 base station can also transmit UE-specific reference signals from antenna port 5, as an amplitude and phase reference for mobiles that are using beamforming. These differ from the cell-specific reference signals in three ways. Firstly, the base station only transmits them in the resource blocks that it is using for beamforming transmissions on the PDSCH. Secondly, the Gold sequence depends on the identity of the target mobile, as well as that of the base station. Thirdly, the base station precodes the UE-specific reference signals using the same antenna weights that it applies to the PDSCH. This last step ensures that the reference signals are directed towards the target mobile and also ensures that the weighting process is completely transparent: by recovering the original reference signals during channel estimation, the mobile automatically removes all the phase shifts that the weighting process introduced.

7.4 Physical Broadcast Channel

The master information block [8] contains the downlink bandwidth and the eight most significant bits of the 10-bit system frame number. It also contains a quantity known as the *PHICH configuration*, which indicates the resource elements that the base station has reserved for the physical hybrid ARQ indicator channel.

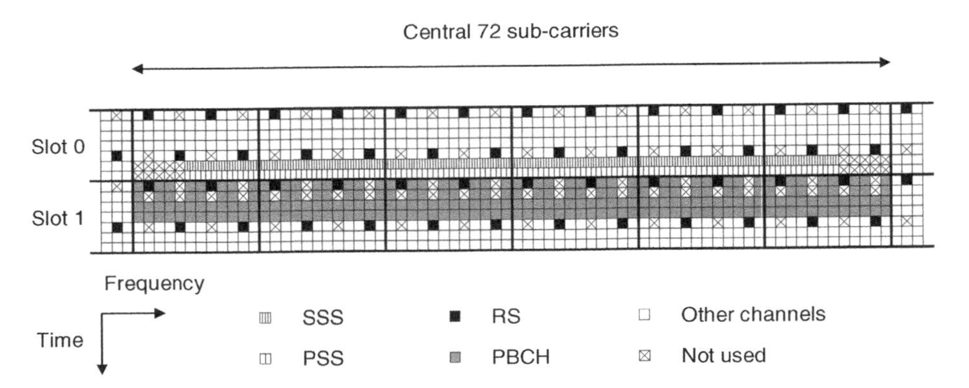

Figure 7.3 Resource element mapping for the PBCH, using FDD mode, a normal cyclic prefix, a 10 or 20 MHz bandwidth, the first antenna port of two and a physical cell ID of 1

The base station transmits the master information block on the broadcast channel and the physical broadcast channel [9, 10]. It processes the information in much the same way as any other downlink channel: in the next chapter, we will illustrate these steps using the PDSCH as an example. The only significant difference is that the base station manipulates the broadcast channel in a manner that indicates the two least significant bits of the system frame number and the number of antenna ports. As noted in Chapter 6, the broadcast channel uses a fixed coding rate and a fixed modulation scheme (QPSK), and is transmitted using either one antenna or open loop transmit diversity.

The base station maps the master information block across four successive frames, beginning in frames where the system frame number is a multiple of four. When using a normal cyclic prefix, it transmits the physical broadcast channel on the central 72 sub-carriers, using the first four symbols of slot 1. Figure 7.3 shows an example. As shown in the figure, the mapping skips the resource elements used for cell-specific reference signals from a base station with four antenna ports, irrespective of how many ports it actually has. (Note that, in a departure from our usual convention, this diagram shows time plotted vertically and frequency horizontally.)

The mobile processes the physical broadcast channel blindly using all the possible ways in which the base station might have manipulated the information, by choosing each combination of one, two and four antenna ports and each combination of the two least significant bits of the system frame number. Only the correct choice allows the cyclic redundancy check to pass. By reading the master information block, the mobile can then discover the downlink bandwidth, the remaining bits of the system frame number and the PHICH configuration.

7.5 Physical Control Format Indicator Channel

Every downlink subframe starts with a control region that contains the PCFICH, PHICH and PDCCH and continues with a data region that contains the PDSCH. The number of control symbols per subframe can be 2, 3 or 4 in a bandwidth of 1.4 MHz and 1, 2 or 3 otherwise, and can change from one subframe to the next. Every subframe, the base station indicates the number of control symbols using the control format indicator and transmits the information on the PCFICH [11, 12]. The mobile's next task is to start receiving this channel.

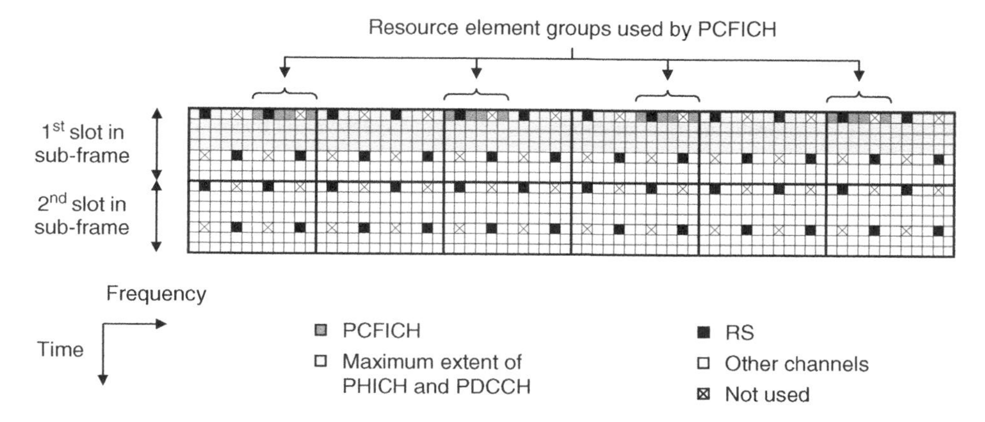

Figure 7.4 Resource element mapping for the PCFICH, using a normal cyclic prefix, a 1.4 MHz bandwidth, the first antenna port of two and a physical cell ID of 1

The PCFICH is mapped onto 16 resource elements in the first symbol of every subframe, with the precise mapping depending on the physical cell identity and the downlink bandwidth. Figure 7.4 shows an example, for a bandwidth of 1.4 MHz and a physical cell identity of 1.

As shown in the figure, the selected resource elements are organized into four *resource element groups* (REGs). Each group contains four resource elements, which are sent on nearby sub-carriers that are not required by the cell-specific reference signals. Resource element groups are used throughout the downlink control region, so they are also used by the PHICH and PDCCH.

At the start of every subframe, the mobile goes to the resource elements that are occupied by the PCFICH, reads the control format indicator and determines the size of the downlink control region. Using the PHICH configuration, it can work out which of the remaining resource element groups are used by the PHICH and which by the PDCCH. It can then go on to receive downlink control information on the PDCCH and downlink data on the PDSCH. In particular, it can read the rest of the cell's system information, in the manner described below.

7.6 System Information

7.6.1 Organization of the System Information

Every cell broadcasts RRC *system information* messages [13] that indicate how it has been configured. The base station transmits these messages on the PDSCH, in a way that is almost identical to any other data transmission. The mobile's final acquisition task is to read this information.

The system information is organized into the master information block that we discussed above and into several numbered system information blocks. These are listed in Table 7.2, along with examples of the information elements that we will use in the remaining chapters.

SIB 1 defines the way in which the other system information blocks will be scheduled. It also includes the parameters that the mobile will need for network and cell selection (Chapter 11), such as the tracking area code and a list of networks that the cell belongs to. This list can

Table 7.2 Organization of the system information

Block	Release	Information	Examples
MIB	R8	Master information block	Downlink bandwidth PHICH configuration System frame number/4
SIB 1	R8	Cell selection parameters Scheduling of other SIBs	PLMN identity list Tracking area code CSG identity TDD configuration Q_{rxlevmin} SIB mapping, period, window size
SIB 2	R8	Radio resource configuration	Downlink reference signal power Default DRX cycle length Time alignment timer
SIB 3	R8	Common cell reselection data Cell-independent intra-frequency data	$S_{\mathrm{IntraSearchP}}$, $S_{\mathrm{NonIntraSearchP}}$, Q_{hyst}
SIB 4	R8	Cell-specific intra-frequency data	$Q_{\mathrm{offset},\,s,\,n}$
SIB 5	R8	Inter-frequency reselection	Target carrier frequency $\mathrm{Thresh}_{x,\,\mathrm{LowP}}$, $\mathrm{Thresh}_{x,\,\mathrm{HighP}}$
SIB 6	R8	Reselection to UMTS	UMTS carrier frequencies
SIB 7	R8	Reselection to GSM	GSM carrier frequencies
SIB 8	R8	Reselection to cdma2000	cdma2000 carriers and cells
SIB 9	R8	Home eNB identifier	Name of home eNB
SIB 10	R8	ETWS primary notification	ETWS alert about natural disaster
SIB 11	R8	ETWS secondary notification	Supplementary ETWS information
SIB 12	R9	CMAS notification	CMAS emergency message
SIB 13	R9	MBMS information	Details of MBSFN areas
SIB 14	R11	Extended access barring	Barring parameters for MTC
SIB 15	R11	MBMS service area identities	Identities of this and other carriers
SIB 16	R11	Timing information	Universal, GPS and local time

identify up to six networks, which allows a base station to be easily shared amongst different network operators. SIB 2 contains parameters that describe the cell's radio resources and physical channels, such as the power that the base station is transmitting on the downlink reference signals.

SIBs 3 to 8 help to specify the cell reselection procedures that are used by mobiles in RRC_IDLE (Chapters 14 and 15). SIB 3 contains the parameters that a mobile will need for any type of cell reselection, as well as the cell-independent parameters that it will need for reselection on the same LTE carrier frequency. SIB 4 is optional and contains any cell-specific parameters that the base station might define for that process. SIB 5 covers reselection to a different LTE frequency, while SIBs 6 to 8 respectively cover reselection to UMTS (both WCDMA and TD-SCDMA), GSM and cdma2000.

The remaining SIBs are more specialized. If the base station belongs to a closed subscriber group, then SIB 9 identifies its name. The mobile can then indicate this to the user, in support of closed subscriber group selection. SIBs 10 and 11 contain notifications from the earthquake and tsunami warning system. In this system, the cell broadcast centre can receive an alert about a natural disaster and can distribute it to all the base stations in the network, which then broadcast the alert through their system information. SIB 10 contains a primary notification that has to be distributed in seconds, while SIB 11 contains a secondary notification that includes less urgent supplementary information. SIBs 12 and 13 are introduced in Release 9, while SIBs 14 to 16 are introduced in Release 11.

7.6.2 *Transmission and Reception of the System Information*

The base station can transmit its system information using two techniques. In the first technique, the base station broadcasts the system information across the whole of the cell, for use by mobiles in RRC_IDLE state and by mobiles in RRC_CONNECTED that have just handed over to a new cell. It does this in much the same way as any other downlink transmission (Chapter 8), but with a few differences. The system information transmissions do not support automatic repeat requests, which are unsuitable for one-to-many broadcast transmissions. The base station sends the system information using one antenna or open loop transmit diversity, depending on the number of antenna ports that it has, and the modulation scheme is fixed at QPSK.

There are a few rules about the timing of these system information broadcasts. The base station transmits SIB 1 in subframe 5 of frames with an even system frame number, with a full transmission taking eight frames altogether. It defines the choice of sub-carriers using its downlink scheduling command. The base station then collects the remaining SIBs into RRC *System Information* messages, and sends each message within a transmission window that has a duration of 1–40 ms and a period of 80–5120 ms. SIB 1 defines the mapping of SIBs onto messages, the period of each message and the window duration, while the downlink scheduling command defines the exact transmission time and the choice of sub-carriers.

If the base station wishes to update the system information that it is broadcasting, it first notifies the mobiles using the paging procedure (Chapter 8). It also increments a *value tag* in SIB 1, for use by mobiles that are returning from a region of poor coverage in which they may have missed a paging message. The base station then changes the system information on a pre-defined modification period boundary.

In the second technique, the base station can update the system information being used by a mobile in RRC_CONNECTED state, by sending it an explicit System Information message. It does this in the same way as any other downlink signalling transmission.

7.7 Procedures after Acquisition

After the mobile has completed the acquisition procedure, it has to run two higher-level procedures before it can exchange data with the network.

In the random access procedure (Chapter 9), the mobile acquires three pieces of information: an initial value for the uplink timing advance, an initial set of parameters for the transmission of uplink data on the physical uplink shared channel and a quantity known as the cell radio network temporary identifier (C-RNTI) that the base station will use to identify it. In the RRC connection establishment procedure (Chapter 11), the mobile acquires several other pieces of information, notably a set of parameters for the transmission of uplink control information on the physical uplink control channel and a set of protocol configurations for its data and signalling radio bearers.

Before looking at these, however, we need to cover the underlying procedure that LTE uses for data transmission and reception. That is the subject of the next chapter.

References

1. 3GPP TS 36.211 (2013) Physical Channels and Modulation, Release 11, September 2013.
2. 3GPP TS 36.331 (2013) Radio Resource Control (RRC); Protocol Specification, Release 11, September 2013.
3. 3GPP TS 36.211 (2013) Physical Channels and Modulation, Release 11, Section 6.11, September 2013.
4. Frank, R., Zadoff, S. and Heimiller, R. (1962) Phase shift pulse codes with good periodic correlation properties. *IEEE Transactions on Information Theory*, **8**, 381–382.
5. Chu, D. (1972) Polyphase codes with good periodic correlation properties. *IEEE Transactions on Information Theory*, **18**, 531–532.
6. Gold, R. (1967) Optimal binary sequences for spread spectrum multiplexing. *IEEE Transactions on Information Theory*, **13**, 619–621.
7. 3GPP TS 36.211 (2013) Physical Channels and Modulation, Release 11, Section 6.10, September 2013.
8. 3GPP TS 36.331 (2013) Radio Resource Control (RRC); Protocol Specification, Release 11, Section 6.2.2 (MasterInformationBlock), September 2013.
9. 3GPP TS 36.211 (2013) Physical Channels and Modulation, Release 11, Section 6.6, September 2013.
10. 3GPP TS 36.212 (2013) Multiplexing and Channel Coding, Release 11, Section 5.3.1, June 2013.
11. 3GPP TS 36.211 (2013) Physical Channels and Modulation, Release 11, Sections 6.2.4, 6.7, September 2013.
12. 3GPP TS 36.212 (2013) Multiplexing and Channel Coding, Release 11, Section 5.3.4, June 2013.
13. 3GPP TS 36.331 (2013) Radio Resource Control (RRC); Protocol Specification, Release 11, Sections 5.2, 6.2.2 (SystemInformation, SystemInformationBlockType1), 6.3.1, September 2013.

8

Data Transmission and Reception

Data transmission and reception is one of the more complex parts of LTE. In this chapter, we begin with an overview of the transmission and reception procedures that are used in the uplink and downlink. We then cover the three main stages of those procedures in turn, namely, the delivery of scheduling messages from the base station, the actual process of data transmission and the delivery of acknowledgements and any associated control information from the receiver. We also cover the transmission of uplink reference signals as well as two associated procedures, power control and discontinuous reception.

Several 3GPP specifications are relevant to this chapter. Data transmission and reception is defined by the physical layer specifications that we noted earlier, particularly TS 36.211 [1], TS 36.212 [2] and TS 36.213 [3], and is controlled by the MAC protocol in the manner defined by TS 36.321 [4]. In addition, the base station configures the mobile's physical layer and MAC protocols by means of RRC signalling [5].

8.1 Data Transmission Procedures

8.1.1 Downlink Transmission and Reception

Figure 8.1 shows the procedure that is used for downlink transmission and reception [6, 7]. The base station begins the procedure by sending the mobile a *scheduling command* (step 1), which is written using the downlink control information (DCI) and transmitted on the physical downlink control channel (PDCCH). The scheduling command alerts the mobile to a forthcoming data transmission and states how it will be sent, by specifying parameters such as the amount of data, the resource block allocation and the modulation scheme.

In step 2, the base station transmits the data on the downlink shared channel (DL-SCH) and the physical downlink shared channel (PDSCH). The data comprise either one or two *transport blocks*, whose duration is known as the *transmission time interval* (TTI), which equals the subframe duration of 1 ms. In response (step 3), the mobile composes a hybrid ARQ acknowledgement to indicate whether the data arrived correctly. It sends the acknowledgement on the physical uplink shared channel (PUSCH) if it is transmitting uplink data in the same subframe and on the physical uplink control channel (PUCCH) otherwise.

An Introduction to LTE: LTE, LTE-Advanced, SAE, VoLTE and 4G Mobile Communications, Second Edition.
Christopher Cox.

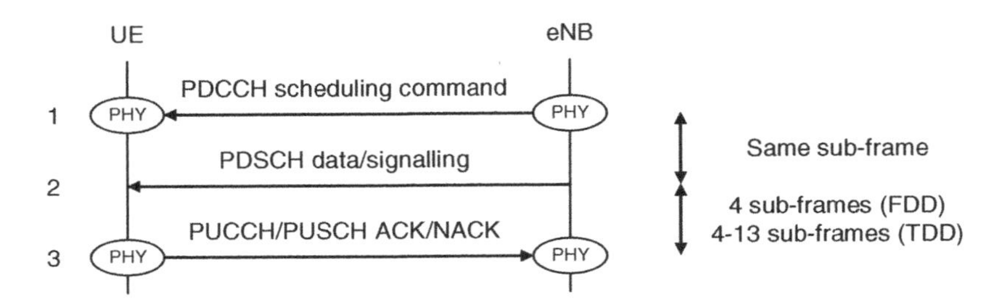

Figure 8.1 Downlink transmission and reception procedure

Usually, the base station moves to a new transport block after a positive acknowledgement and re-transmits the original one after a negative acknowledgement. If, however, the base station reaches a certain maximum number of re-transmissions without receiving a positive response, then it moves to a new transmission anyway, on the grounds that the mobile's receive buffer may have been corrupted by a burst of interference. The radio link control (RLC) protocol then picks up the problem, for example by sending the transport block again from the beginning.

The downlink transmission timing is as follows. The scheduling command lies in the control region at the beginning of a downlink subframe, while the transport block lies in the data region of the same subframe. In FDD mode, there is a fixed time delay of four subframes between the transport block and the corresponding acknowledgement, which helps the base station to match the two pieces of information together. In TDD mode, the delay is between four and 13 subframes, according to a mapping that depends on the TDD configuration. Figure 8.2 shows an example mapping, for the case of TDD configuration 1.

As described in Chapter 3, the downlink uses several parallel hybrid ARQ processes, each with its own copy of Figures 8.1 and 8.2. In FDD mode, the maximum number of hybrid ARQ processes is eight. In TDD mode, the maximum number depends on the TDD configuration, up to an absolute maximum of 15 for TDD configuration 5. The LTE downlink uses a technique known as *asynchronous hybrid ARQ*, in which the base station explicitly specifies the hybrid

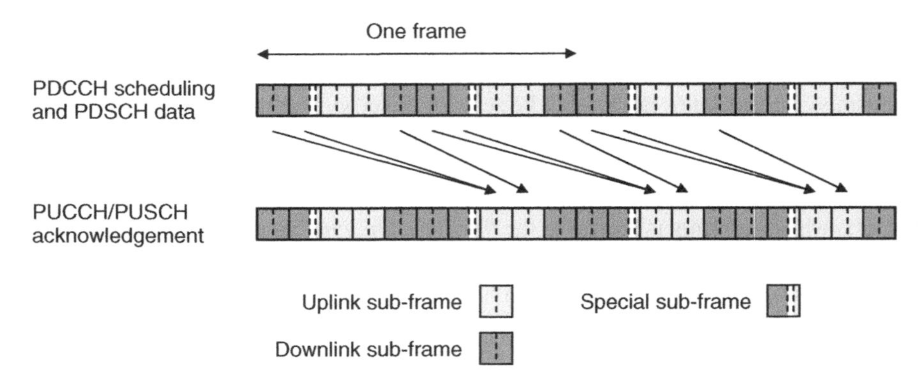

Figure 8.2 Relationship between the timing of downlink data and uplink acknowledgements, for TDD configuration 1

ARQ process number in every scheduling command. As a result, there is no need to define the timing delay between a negative acknowledgement and a re-transmission: instead, the base station schedules a re-transmission whenever it likes and simply states the hybrid ARQ process number that it is using.

8.1.2 Uplink Transmission and Reception

Figure 8.3 shows the corresponding procedure for the uplink [8, 9]. As in the downlink, the base station starts the procedure by sending the mobile a *scheduling grant* on the PDCCH (step 1). This grants permission for the mobile to transmit and states all the transmission parameters that it should use, for example the transport block size, the resource block allocation and the modulation scheme. In response, the mobile carries out an uplink data transmission on the uplink shared channel (UL-SCH) and the PUSCH (step 2).

If the base station fails to receive the data correctly then there are two ways for it to respond. In one technique, the base station can trigger a *non adaptive re-transmission* by sending the mobile a negative acknowledgement on the PHICH. The mobile then re-transmits the data with the same parameters that it used first time around. Alternatively, the base station can trigger an *adaptive re-transmission* by explicitly sending the mobile another scheduling grant on the PDCCH. It can do this to change the parameters that the mobile uses for the re-transmission, such as the resource block allocation or the uplink modulation scheme.

If the base station does receive the data correctly then it can respond in two similar ways, either by sending a positive acknowledgement on the PHICH so as to end the procedure or by sending a new scheduling grant on the PDCCH so as to request a new transmission. If the mobile receives a PHICH acknowledgement and a PDCCH scheduling grant in the same subframe, then the scheduling grant takes priority.

In the diagram, steps 3 to 5 assume that the base station fails to decode the mobile's first transmission, but succeeds with the second. If the mobile reaches a maximum number of re-transmissions without receiving a positive reply, then it moves to a new transmission anyway and leaves the RLC protocol to solve the problem.

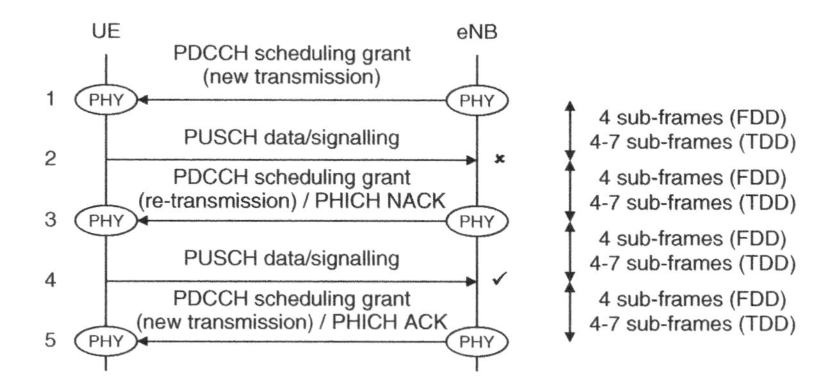

Figure 8.3 Uplink transmission and reception procedure

Once again, the uplink uses several hybrid ARQ processes, each with its own copy of Figure 8.3. In FDD mode, the maximum number of hybrid ARQ processes is eight. In TDD mode, the absolute maximum is seven in TDD configuration 0.

The uplink uses a technique known as *synchronous hybrid ARQ*, in which the hybrid ARQ process number is not signalled explicitly, but is instead defined using the transmission timing. In FDD mode, there is a delay of four subframes between the scheduling grant and the corresponding uplink transmission, and another four subframe delay before any re-transmission request on the same hybrid ARQ process. This gives all the information that the devices need to match up the scheduling grants, transmissions, acknowledgements and re-transmissions.

As in the downlink, TDD mode uses a variable set of delays, according to a mapping that depends on the TDD configuration. Figure 8.4 shows an example mapping, for the case of TDD configuration 1.

The mobile can trigger the procedure in three ways. If the mobile is in RRC_IDLE then it can alert the base station using the random access procedure (Chapter 9). If the mobile is in RRC_CONNECTED but is not yet transmitting on the PUSCH then it can send the base station a scheduling request on the PUCCH (Section 8.5.6). Finally, if the mobile is transmitting on the PUSCH, then it can keep the base station informed about its transmit buffer occupancy using control elements known as buffer status reports (Chapter 10).

In applications such as voice over IP, the mobile's uplink and downlink data rates are the same, but the mobile's maximum distance from the base station is usually limited by the uplink. We can improve a mobile's uplink coverage in these situations using a technique known as *TTI bundling*. In this technique, the base station sends the mobile a single scheduling grant in the usual way, but the mobile responds by transmitting the same data in four consecutive subframes so as to boost the received signal energy. TTI bundling is often used in conjunction with the semi persistent scheduling techniques that we describe below.

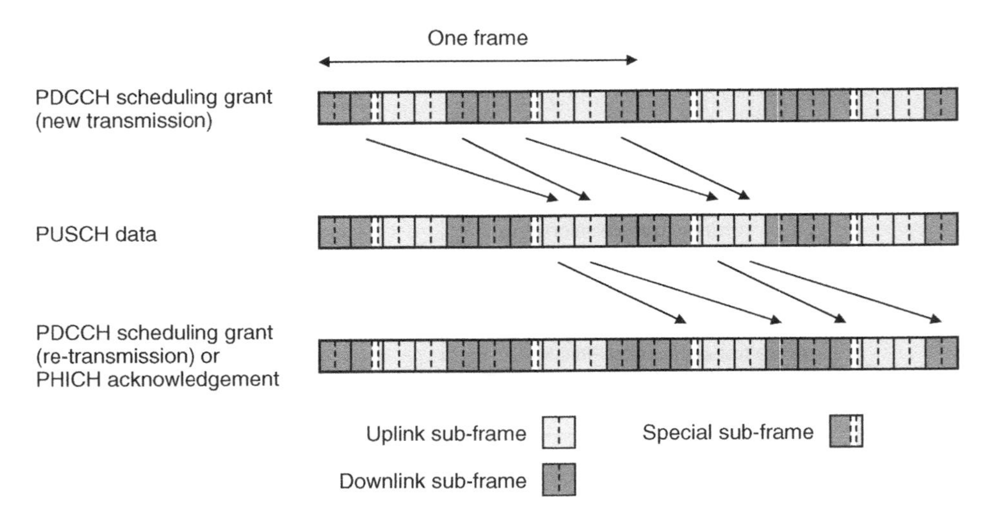

Figure 8.4 Relationship between the timing of scheduling grants, uplink data and downlink acknowledgements, for TDD configuration 1

8.1.3 Semi Persistent Scheduling

When using *semi persistent scheduling* (SPS) [10–12], the base station can schedule several transmissions spanning several subframes, by sending the mobile a single scheduling message that contains just one resource allocation. Semi persistent scheduling is designed for services such as voice over IP. For these services, the data rate is low, so the overhead of the scheduling message can be high. However the data rate is also constant, so the base station can confidently use the same resource allocation from one transmission to the next.

The base station configures a mobile for semi persistent scheduling by means of a mobile-specific RRC signalling message. As part of the message, it specifies the interval between transmissions, which lies between 10 and 640 subframes. Later on, the base station can activate semi persistent scheduling by sending the mobile a specially formatted scheduling command or scheduling grant.

In the downlink, the base station sends new transmissions on the PDSCH at the interval defined by the mobile's SPS configuration, in the manner indicated by the original scheduling command. The mobile cycles the hybrid ARQ process number for every new transmission, because the base station has no opportunity to specify it. However the base station continues to schedule all its re-transmissions explicitly, in the manner shown in Figure 8.1. It can therefore specify different transmission parameters for them such as different resource block allocations or different modulation schemes. A similar situation applies in the uplink: the mobile sends new transmissions on the PUSCH at the interval defined by its SPS configuration, but the base station continues to schedule any re-transmissions in the manner shown in Figure 8.2.

Eventually, the base station can release the SPS assignment by sending the mobile another specially formatted scheduling message. In addition, the mobile can implicitly release an uplink SPS assignment if it has reached a maximum number of transmission opportunities without having any data to send.

8.2 Transmission of Scheduling Messages on the PDCCH

8.2.1 Downlink Control Information

In looking at the details of the transmission and reception procedures, we will begin with the transmission of downlink control information on the PDCCH. The base station uses its downlink control information to send downlink scheduling commands, uplink scheduling grants and uplink power control commands to the mobile. The DCI can be written using several different formats, which are listed in Table 8.1 [13]. Each format contains a specific set of information and has a specific purpose.

DCI format 0 contains scheduling grants for the mobile's uplink transmissions. The scheduling commands for downlink transmissions are more complicated, and are handled in Release 8 by DCI formats 1 to 1D and 2 to 2A.

DCI format 1 schedules data that the base station will transmit using one antenna, open loop diversity or beamforming, for mobiles that already have been configured into one of the downlink transmission modes 1, 2 or 7. When using this format, the base station can allocate the downlink resource blocks in a flexible way, by means of two resource allocation schemes known as type 0 and type 1 that we will describe below.

Table 8.1 List of DCI formats and their applications

DCI format	Release	Purpose		Resource allocation	DL mode
0	R8	UL scheduling grants	1 antenna	–	–
1	R8	DL scheduling commands	1 antenna, open loop diversity, beamforming	Type 0, 1	1, 2, 7
1A	R8		1 antenna, open loop diversity	Type 2	Any
1B	R8		Closed loop diversity	Type 2	6
1C	R8		System information, paging, random access responses	Type 2	Any
1D	R8		Multiple user MIMO	Type 2	5
2	R8	DL scheduling commands	Closed loop MIMO	Type 0, 1	4
2A	R8		Open loop MIMO	Type 0, 1	3
2B	R9		Dual layer beamforming	Type 0, 1	8
2C	R10		8 layer MIMO	Type 0, 1	9
2D	R11		CoMP	Type 0, 1	10
3	R8	UL power control	2 bit power adjustments	–	–
3A	R8		1 bit power adjustments	–	–
4	R10	UL scheduling grants	Closed loop MIMO	–	–

Format 1A is similar, but the base station uses a compact form of resource allocation known as type 2. Format 1A can also be used in any downlink transmission mode. If the mobile has previously been configured into one of transmission modes 3 to 7, then it receives the data by falling back to single antenna reception if the base station has one antenna port, or open loop transmit diversity otherwise.

Skipping a line, format 1C uses a very compact format that only specifies the resource allocation and the amount of data that the base station will send. In the ensuing data transmission, the modulation scheme is fixed at QPSK, and hybrid ARQ is not used. Format 1C is only used to schedule system information messages, paging messages and random access responses, for which this very compact format is appropriate.

Formats 1B, 1D, 2 and 2A are respectively used for closed loop transmit diversity, the Release 8 implementation of multiple user MIMO, and closed and open loop spatial multiplexing. They include extra fields to signal information such as the precoding matrix that the base station will apply to the PDSCH and the number of layers that the base station will transmit.

Unlike the others, DCI formats 3 and 3A do not schedule any transmissions: instead, they control the power that the mobile transmits on the uplink by means of embedded power control commands. We will cover this procedure later in the chapter. Formats 2B, 2C, 2D and 4 are introduced in Releases 9 to 11, and are covered towards the end of the book.

8.2.2 Resource Allocation

The base station has various ways of allocating the resource blocks to individual mobiles in the uplink and downlink [14, 15]. In the downlink, as noted above, it can use two flexible resource allocation formats that are known as types 0 and 1, and a compact format known as type 2.

When using downlink resource allocation type 0, the base station collects the resource blocks into *resource block groups* (RBGs), which it assigns individually using a bitmap. With resource allocation type 1, it can assign individual resource blocks within a group, but has less flexibility over the assignment of the groups themselves. Allocation type 1 might be suitable in environments with severe frequency-dependent fading, in which the frequency resolution of type 0 might be too coarse.

When using resource allocation type 2, the base station gives the mobile a contiguous allocation of *virtual resource blocks* (VRBs). In the downlink, these come in two varieties: localized and distributed. Localized virtual resource blocks are identical to the *physical resource blocks* (PRBs) that we have been considering elsewhere, so, when using these, the mobile simply receives a contiguous resource block allocation. Distributed virtual resource blocks are related to physical resource blocks by a mapping operation, which is different in the first and second slots of a subframe. The use of distributed virtual resource blocks gives the mobile extra frequency diversity and is suitable in environments that are subject to frequency-dependent fading.

The mobile also receives a contiguous allocation of virtual resource blocks for its uplink transmissions. Their meaning depends on whether the base station has requested the use of *frequency hopping* in DCI format 0. If frequency hopping is disabled, then the uplink virtual resource blocks map directly onto physical resource blocks. If frequency hopping is enabled, then the virtual and physical resource blocks are related using a mapping that is either explicitly signalled (type 1 hopping) or follows a pseudo-random pattern (type 2 hopping). A mobile can also change transmission frequency in every subframe or in every slot, depending on a hopping mode that is configured using RRC signalling.

In the uplink, the number of resource blocks per mobile must be either 1, or a number whose prime factors are 2, 3 or 5. The reason lies in the extra Fourier transform used by SC-FDMA, which runs quickly if the number of sub-carriers is a power of 2 or a product of small prime numbers alone, but slowly if a large prime number is involved.

8.2.3 Example: DCI Format 1

To illustrate the DCI formats, Table 8.2 shows the contents of DCI format 1 in Release 8. The other formats are not that different: the details can be found in the specifications.

The base station indicates whether the mobile should use resource allocation type 0 or 1 using the resource allocation header, and carries out the allocation using the resource block assignment. In a bandwidth of 1.4 MHz, allocation type 0 is not supported, so the header field is omitted.

The modulation and coding scheme is a five-bit number, from which the mobile can look up the modulation scheme that the PDSCH will use (QPSK, 16-QAM or 64-QAM). By combining the modulation and coding scheme with the number of resource blocks in its allocation, the mobile can also look up the number of bits in the transport block. By comparing the transport block size with the number of resource elements in its allocation, the mobile can calculate the coding rate for the DL-SCH.

As noted earlier, the base station explicitly signals the hybrid ARQ process number in every downlink scheduling command. The base station also toggles the new data indicator for every new transmission, while leaving it unchanged for a re-transmission. The redundancy version

Table 8.2 Contents of DCI format 1 in 3GPP Release 8

Field	Number of bits
Resource allocation header	0 (1.4 MHz) or 1 (otherwise)
Resource block assignment	6 (1.4 MHz) to 25 (20 MHz)
Modulation and coding scheme	5
HARQ process number	3 (FDD) or 4 (TDD)
New data indicator	1
Redundancy version	2
TPC command for PUCCH	2
Downlink assignment index	2 (TDD only)
Padding	0 or 1

indicates which of the turbo coded bits will be transmitted after the rate matching stage and which will be punctured.

The base station uses the *transmit power control* (TPC) command for PUCCH to adjust the power that the mobile will use when sending uplink control information on the PUCCH. (This is an alternative technique to the adjustment of transmit power using DCI formats 3 and 3A.) In TDD mode, it uses the downlink assignment index to assist the mobile's transmission of uplink acknowledgements, in the manner that we will describe later on.

There is a significant omission from Table 8.2: there is no header field to indicate what the DCI format actually is. Although some of the other formats do contain such headers, the mobile usually distinguishes the different DCI formats by the fact that they contain different numbers of bits. The base station occasionally adds a padding bit to the end of the scheduling command, to ensure that format 1 contains a different number of bits from all the others.

8.2.4 Radio Network Temporary Identifiers

The base station transmits a PDCCH scheduling message by addressing it to a *radio network temporary identifier* (RNTI) [16]. In LTE, an RNTI defines two things: the identity of the mobile(s) that should read the scheduling message and the type of information that is being scheduled. Table 8.3 lists the RNTIs that are used by LTE, along with the hexadecimal values that they can use.

The *cell RNTI* (C-RNTI) is the most important. The base station assigns a unique C-RNTI to a mobile as part of the random access procedure. Later on, it can schedule a transmission that extends over one subframe, by addressing a scheduling message to the mobile's C-RNTI.

The *SPS C-RNTI* is used for semi persistent scheduling. The base station first assigns an SPS C-RNTI to a mobile using mobile-specific RRC signalling. Later on, it can schedule a transmission that extends over several subframes by addressing a specially formatted scheduling message to the SPS C-RNTI.

The *paging RNTI* (P-RNTI) and *system information RNTI* (SI-RNTI) are fixed values, which are used to schedule the transmission of paging and system information messages to all the mobiles in the cell. The *temporary C-RNTI* and *random access RNTI* (RA-RNTI) are temporary fields during the random access procedure (Chapter 9), while the *MBMS*

Table 8.3 List of radio network temporary identifiers and their applications

Type of RNTI	Release	Information scheduled	Hex value
RA-RNTI	R8	Random access response	0001 - 003C
Temporary C-RNTI	R8	Random access contention resolution	003D - FFF3
C-RNTI	R8	One UL or DL transmission	
SPS C-RNTI	R8	Several UL or DL transmissions	
TPC-PUCCH-RNTI	R8	Embedded PUCCH TPC command	
TPC-PUSCH-RNTI	R8	Embedded PUSCH TPC command	
M-RNTI	R9	MBMS change notification	FFFD
P-RNTI	R8	Paging message	FFFE
SI-RNTI	R8	System information message	FFFF

Source: TS 36.321. Reproduced by permission of ETSI.

RNTI (M-RNTI) is used by the multimedia broadcast/multicast service (Chapter 18). Finally, the TPC-PUCCH-RNTI and TPC-PUSCH-RNTI are used to send embedded uplink power control commands using DCI formats 3 and 3A.

8.2.5 Transmission and Reception of the PDCCH

We are now in a position to discuss how the PDCCH is transmitted and received, a process that is summarized in Figure 8.5. In its transport channel processor, the base station first manipulates the DCI by attachment of a cyclic redundancy check (CRC) and error correction coding [17], in a manner that depends on the RNTI of the target mobile. In the physical channel

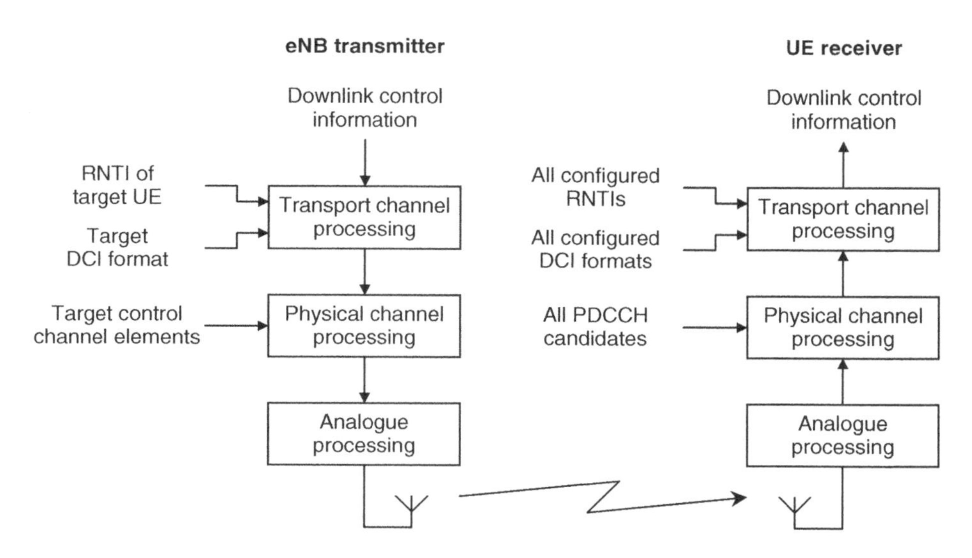

Figure 8.5 Transmission and reception of the PDCCH

processor, it then processes the PDCCH using QPSK modulation and either single antenna transmission or open loop transmit diversity, depending on the number of antenna ports that it has [18]. Finally, the base station maps the PDCCH onto the chosen resource elements.

The resource element mapping for the PDCCH is organized using *control channel elements* (CCEs) [19], each of which contains nine resource element groups that have not already been assigned to the physical control format indicator channel (PCFICH) or the PHICH. Depending on the length of the DCI message, the base station can transmit a PDCCH scheduling message by mapping it onto one, two, four or eight consecutive CCEs; in other words onto 36, 72, 144 or 288 resource elements.

In turn, the control channel elements are organized into *search spaces*. These come in two types. *Common search spaces* are available to all the mobiles in the cell and are located at fixed positions within the downlink control region. *UE-specific search spaces* are assigned to groups of mobiles and have locations that depend on the mobiles' RNTIs. Each search space contains up to 16 control channel elements, so it contains several locations where the base station might transmit downlink control information. The base station can therefore use these search spaces to send several PDCCH messages to several different mobiles at the same time.

A mobile then receives the PDCCH as follows. Every subframe, the mobile reads the control format indicator, and establishes the size of the downlink control region and the locations of the common and UE-specific search spaces. Within each search space, it identifies the possible *PDCCH candidates*, which are control channel elements where the base station might have transmitted downlink control information. The mobile then attempts to process each PDCCH candidate, using all the combinations of RNTI and DCI format that it has been configured to look for. If the observed CRC bits match the ones expected, then it concludes that the message was sent using the RNTI and DCI format that it was looking for. It then reads the downlink control information and acts upon it.

The cyclic redundancy check can fail for several reasons: the base station may not have sent a scheduling message in those control channel elements, or it may have sent a scheduling message using a different DCI format or a different RNTI, or the mobile may have failed to read the message due to an uncorrected bit error. Whichever situation applies, the mobile's response is the same: it moves on to the next combination of PDCCH candidate, RNTI and DCI format, and tries again.

8.3 Data Transmission on the PDSCH and PUSCH

8.3.1 Transport Channel Processing

After the base station has sent the mobile a scheduling command, it can transmit the DL-SCH in the way that the scheduling command defined. After reception of an uplink scheduling grant, the mobile can transmit the UL-SCH in a similar way. Figure 8.6 shows the steps that the transport channel processor uses to send the data [20].

At the top of the figure, the medium access control (MAC) protocol sends information to the physical layer in the form of transport blocks. The size of each transport block is defined by the downlink control information, while its duration is the 1 ms transmission time interval.

In the uplink, the mobile sends one transport block at a time. In the downlink, the base station usually sends one transport block to each mobile, but can send two when using spatial multiplexing (DCI formats 2 to 2D). The two transport blocks can have different modulation

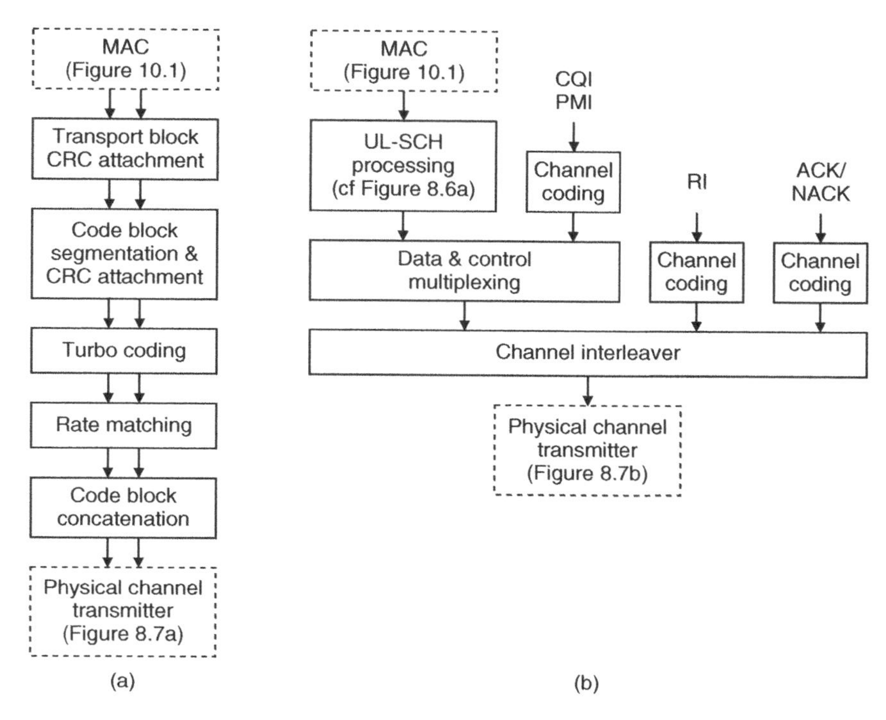

Figure 8.6 Transport channel processing in Releases 8 and 9, for the (a) DL-SCH (b) UL-SCH Source: TS 36.212. Reproduced by permission of ETSI

schemes and coding rates, are mapped to different layers and are separately acknowledged. This increases the amount of signalling, so it adds some overhead to the transmission. As noted in Chapter 5, however, different layers can reach the mobile with different values of the signal to interference plus noise ratio, so we can improve the performance of the air interface by transmitting a high SINR layer using a fast modulation scheme and coding rate, and vice versa. By limiting the maximum number of transport blocks to two rather than four, we reach a compromise between these two conflicting criteria.

In the downlink (Figure 8.6a), the base station adds a 24-bit CRC to each DL-SCH transport block, which the mobile will eventually use for error detection. If the resulting block is longer than 6144 bits, then the base station segments it into smaller code blocks and adds another CRC to each one. It then passes the data through a $^1/_3$ rate turbo coder. The rate matching stage stores the resulting bits in a circular buffer and then selects bits from the buffer for transmission. The number of transmitted bits is determined by the size of the resource allocation and the exact choice is determined by the redundancy version. Finally, the base station reassembles the coded transport blocks and sends them to the physical channel processor in the form of *codewords*.

The mobile processes the received data in the manner described in Chapter 3. The turbo decoding algorithm is an iterative one, which continues until the code block CRC is passed. The receiver then reassembles each transport block, examines the transport block CRC and determines whether the data have arrived correctly.

In the uplink (Figure 8.6b), the mobile transmits the UL-SCH by means of the same steps that the base station used on the downlink. If the mobile is sending uplink control information in the same subframe, then it processes the control bits using forward error correction and multiplexes them into the UL-SCH, in the manner indicated by the diagram.

8.3.2 Physical Channel Processing

The transport channel processor passes the outgoing codeword(s) to the physical channel processor, which transmits them in the manner shown in Figure 8.7 [21].

In the downlink (Figure 8.7a), the scrambling stage mixes each codeword with a pseudo-random sequence that depends on the physical cell ID and the target RNTI, to reduce the interference between transmissions from nearby cells. The modulation mapper takes the resulting bits in groups of two, four or six and maps them onto the in-phase and quadrature components using QPSK, 16-QAM or 64-QAM.

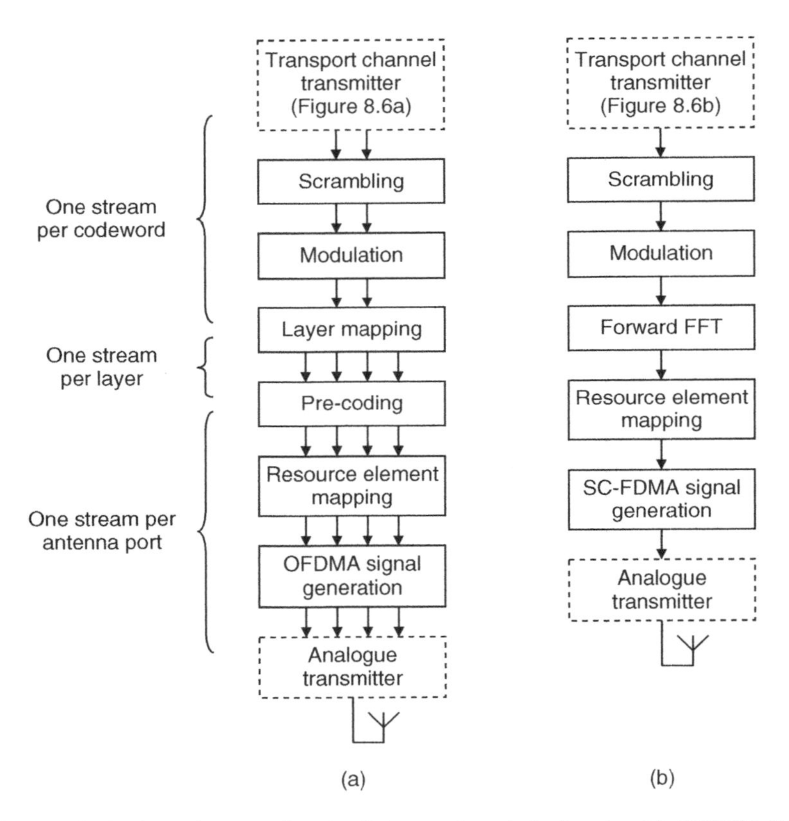

Figure 8.7 Physical channel processing in Releases 8 and 9, for the (a) PDSCH (b) PUSCH Source: TS 36.211. Reproduced by permission of ETSI

The next two stages implement the multiple antenna transmission techniques from Chapter 5. The layer mapping stage takes the codewords and maps them onto one to four independent layers, while the precoding stage applies the chosen precoding matrix and maps the layers onto the different antenna ports.

The resource element mapper carries out a serial-to-parallel conversion and maps the resulting sub-streams onto the chosen sub-carriers, along with the sub-streams resulting from all the other data transmissions, control channels and physical signals. The PDSCH occupies resource elements in the data region of each subframe that have not been assigned to other channels or signals, in the manner shown in Figure 6.10. Finally, the OFDMA signal generator applies an inverse fast Fourier transform and a parallel-to-serial conversion and inserts the cyclic prefix. The result is a digital representation of the time-domain data that will be transmitted from each antenna port.

There are just a few differences in the uplink (Figure 8.7b). Firstly, the process includes the forward FFT that is the distinguishing feature of SC-FDMA. Secondly, there is no layer mapping or precoding because the uplink does not use single user MIMO in LTE Release 8. Thirdly, the PUSCH occupies a contiguous set of resource blocks towards the centre of the uplink band, with the edges reserved for the PUCCH. Each subframe contains six PUSCH symbols and one demodulation reference symbol, in the manner illustrated in Figure 8.8.

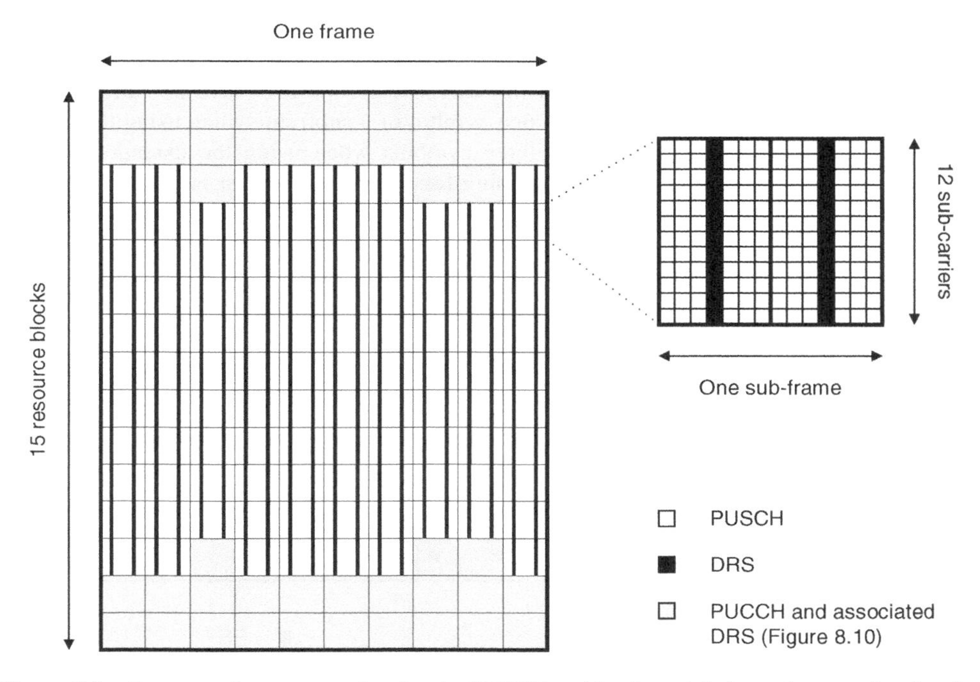

Figure 8.8 Resource element mapping for the PUSCH and its demodulation reference signal, using FDD mode, a normal cyclic prefix, a 3 MHz bandwidth and an example allocation for the PUCCH

8.4 Transmission of Hybrid ARQ Indicators on the PHICH

8.4.1 Introduction

We can now start to discuss the feedback that the receiver sends back to the transmitter. The base station's feedback is easier to understand than the mobile's and is a better place to begin.

During the procedure for uplink transmission and reception, the base station sends acknowledgements to the mobiles in the form of hybrid ARQ indicators and transmits them on the physical hybrid ARQ indicator channel [22–25]. The exact transmission technique depends on the cell's PHICH configuration, which contains two parameters: the *PHICH duration* (normal or extended) and a parameter N_g that can take values of $^1/_6$, $^1/_2$, 1 or 2. The transmission technique also depends on the cyclic prefix duration.

In the discussion that follows, we will generally assume that the base station is using the normal PHICH duration and the normal cyclic prefix. The details of the other techniques are rather different, but the underlying principles remain the same.

8.4.2 Resource Element Mapping of the PHICH

The base station transmits each hybrid ARQ indicator in the downlink control region, using a set of three resource element groups (12 resource elements) that is known as a *PHICH group*. The number of PHICH groups depends on the cell bandwidth and the value of N_g. It is constant in FDD mode, but can vary from one subframe to the next in TDD mode because the base station has to send more acknowledgements in some TDD subframes than in others.

Each PHICH group is mapped to resource element groups that have not already been assigned to the PCFICH. These lie in the first symbol of a subframe when using the normal PHICH duration, but can cover two or three symbols when using the extended PHICH duration. Figure 8.9 shows an example mapping for a base station that is using the normal PHICH duration and two PHICH groups.

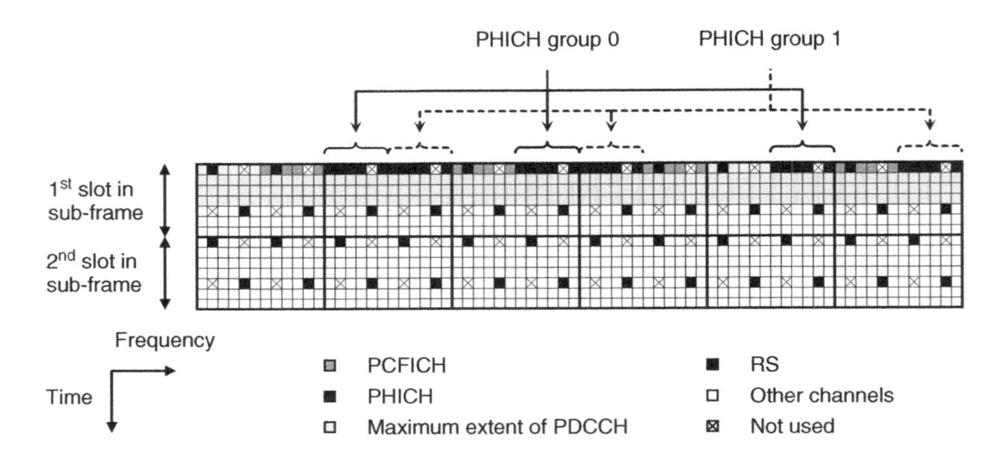

Figure 8.9 Resource element mapping for the PHICH, using a normal PHICH duration, a normal cyclic prefix, a 1.4 MHz bandwidth, the first antenna port of two, a physical cell ID of 1 and two PHICH groups

A PHICH group is not dedicated to a single mobile; instead, it is shared amongst eight mobiles, by assigning each mobile a different *orthogonal sequence index*. A mobile determines the PHICH group number and orthogonal sequence index that it should inspect using two parameters from its original scheduling grant, namely the first physical resource block that it used for the uplink transmission and a parameter called the cyclic shift that we will see in Section 8.6. Together, the PHICH group number and orthogonal sequence index are known as a *PHICH resource*.

8.4.3 Physical Channel Processing of the PHICH

To transmit a hybrid ARQ indicator, the base station modulates it by means of BPSK, using symbols of +1 and −1 for positive and negative acknowledgements respectively. It then spreads each indicator across the four symbols in a resource element group, by multiplying it by the chosen orthogonal sequence. There are four basic sequences available to the base station, namely [+1 +1 +1 +1], [+1 −1 +1 −1], [+1 +1 −1 −1] and [+1 −1 −1 +1], but each can be applied to the in-phase and quadrature components of the signal, making a total of eight orthogonal sequences in all. The base station can send simultaneous acknowledgements to the eight mobiles in a PHICH group, by assigning them different orthogonal sequence indexes and adding the resulting symbols. This technique will be familiar to those with experience of code division multiple access and is one of the few uses that LTE makes of CDMA.

The PHICH symbols are then repeated across three resource element groups to increase the received symbol energy and are transmitted in a similar manner to the other downlink physical channels.

8.5 Uplink Control Information

8.5.1 Hybrid ARQ Acknowledgements

The mobile sends three types of uplink control information to the base station [26–28]: hybrid ARQ acknowledgements of the base station's downlink transmissions, uplink scheduling requests and channel state information. In turn, the channel state information comprises the channel quality indicator (CQI), the precoding matrix indicator (PMI) and the rank indication (RI).

First, let us consider the hybrid ARQ acknowledgements. In FDD mode, the mobile computes one or two acknowledgements per subframe, depending on the number of transport blocks that it received. It then transmits them four subframes later.

In TDD mode, things are more complicated. If the mobile is acknowledging one downlink subframe at a time, then it does so in the same way as in FDD mode. There are two ways for it to acknowledge multiple subframes. Using *ACK/NACK bundling*, the mobile sends a maximum of two acknowledgements, one for each parallel stream of transport blocks. Each acknowledgement is positive if it successfully received the corresponding transport block in all the downlink subframes and negative otherwise. Using *ACK/NACK multiplexing*, the mobile computes one acknowledgement for each downlink subframe. Each acknowledgement is positive if it successfully received both transport blocks in that subframe and negative otherwise. When using ACK/NACK multiplexing, the specifications only require the mobile to transmit a maximum of four acknowledgements at the same time, for the data received in four downlink subframes. To achieve this, the technique is not supported in TDD configuration 5.

In TDD mode, the scheduling command included a quantity known as the downlink assignment index. This indicates the total number of downlink transmissions that the mobile should be acknowledging at the same time as the scheduled data. It reduces the risk of a wrongly formatted acknowledgment if the mobile missed an earlier scheduling command, so reduces the overall error rate on the air interface.

8.5.2 *Channel Quality Indicator*

The channel quality indicator is a four-bit quantity, which indicates the maximum data rate that the mobile can handle with a block error ratio of 10% or below. The CQI mainly depends on the received signal to interference plus noise ratio, because a high data rate can only be received successfully at a high SINR. However, it also depends on the implementation of the mobile receiver, because an advanced receiver can successfully process the incoming data at a lower SINR than a more basic one.

Table 8.4 shows how the CQI is interpreted in terms of the downlink modulation scheme and coding rate. The last column shows the number of information bits per symbol and is calculated by multiplying the coding rate by 2, 4 or 6.

Because of frequency-dependent fading, the channel quality can often vary across the downlink band. To reflect this, the base station can configure the mobile to report the CQI in three different ways. *Wideband reporting* covers the whole of the downlink band. For *higher layer*

Table 8.4 Interpretation of the channel quality indicator, in terms of the modulation scheme and coding rate that a mobile can successfully receive

CQI	Modulation scheme	Coding rate (units of 1/1024)	Information bits per symbol
0	n/a	0	0.00
1	QPSK	78	0.15
2	QPSK	120	0.23
3	QPSK	193	0.38
4	QPSK	308	0.60
5	QPSK	449	0.88
6	QPSK	602	1.18
7	16-QAM	378	1.48
8	16-QAM	490	1.91
9	16-QAM	616	2.41
10	64-QAM	466	2.73
11	64-QAM	567	3.32
12	64-QAM	666	3.90
13	64-QAM	772	4.52
14	64-QAM	873	5.12
15	64-QAM	948	5.55

Source: TS 36.213. Reproduced by permission of ETS.

configured sub-band reporting, the base station divides the downlink band into sub-bands, and the mobile reports one CQI value for each. For *UE selected sub-band reporting*, the mobile selects the sub-bands that have the best channel quality and reports their locations, together with one CQI that spans them and a separate wideband CQI. If the mobile is receiving more than one transport block, then it can also report a different CQI value for each, to reflect the fact that different layers can reach the mobile with different values of the SINR.

The base station uses the received CQI in its calculation of the modulation scheme and coding rate, and in support of frequency-dependent scheduling. Despite the frequency dependence of the CQI, however, the base station only uses one frequency-independent modulation scheme and coding rate per transport block when it comes to transmit the downlink data.

8.5.3 Rank Indication

The mobile reports a rank indication when it is configured for spatial multiplexing in transmission mode 3 or 4. The rank indication lies between 1 and the number of base station antenna ports and indicates the maximum number of layers that the mobile can successfully receive.

The mobile reports a single rank indication, which applies across the whole of the downlink band. The rank indication can be calculated jointly with the PMI, by choosing the combination that maximizes the expected downlink data rate.

8.5.4 Precoding Matrix Indicator

The mobile reports a precoding matrix indicator when it is configured for closed loop spatial multiplexing, multiple user MIMO or closed loop transmit diversity, in transmission modes 4, 5 or 6. The PMI indicates the precoding matrix that the base station should apply before transmitting the signal.

The PMI can vary across the downlink band, in a similar way to the CQI. To reflect this, there are two options for PMI reporting. The mobile can report a *single PMI* spanning the whole downlink band or spanning all of the UE-selected sub-bands. When using *multiple PMIs*, it either reports both of these quantities, or reports one PMI for each higher layer configured sub-band.

The base station uses the received PMI to calculate the precoding matrix that it should apply to its next downlink transmission. Once again, the base station actually transmits the data using one frequency-independent precoding matrix, despite the frequency dependence of the PMI.

8.5.5 Channel State Reporting Mechanisms

The mobile can return channel state information to the base station in two ways. *Periodic reporting* is carried out at regular intervals, which lie between 2 and 160 ms for the CQI and PMI and are up to 32 times greater for the RI. The information is usually carried by the PUCCH, but is transferred to the PUSCH if the mobile is sending uplink data in the same subframe. The maximum number of bits in each periodic report is 11, to reflect the low data rate that is available on the PUCCH.

Aperiodic reporting is carried out at the same time as a PUSCH data transmission and is requested using a field in the mobile's scheduling grant. If both types of report are scheduled in the same subframe, then the aperiodic report takes priority.

For both techniques, the base station can configure the mobile into a *channel quality reporting mode* using RRC signalling. The reporting mode defines the type of channel quality information that the base station requires, in the manner defined by Tables 8.5 and 8.6. In each mode, the first number describes the type of CQI feedback that the base station requires, while the second describes the type of PMI feedback. The precise definitions of each reporting mode are covered in the specifications, and are different for periodic and aperiodic reporting because of the need to limit the amount of data transferred on the PUCCH. In particular, periodic mode 2-0 is defined differently from aperiodic mode 2-0.

8.5.6 Scheduling Requests

If a mobile is in RRC_CONNECTED and has data waiting for transmission on the PUSCH, then it can request a scheduling grant by composing a one-bit scheduling request for transmission on the PUCCH. The mobile does not send the request right away, because it has to share the PUCCH with other mobiles. Instead, it transmits the scheduling request in a subframe that is configured by RRC signalling, which recurs with a period between 5 and 80 ms. The mobile never sends channel state information at the same time as a scheduling request; instead, the scheduling request takes priority.

A well behaved base station should reply to a scheduling request by giving the mobile a scheduling grant. However, it is not obliged to do so. If the mobile reaches a maximum number of scheduling requests without receiving a reply, then it triggers the random access procedure

Table 8.5 Channel quality reporting modes for periodic reporting on the PUCCH or PUSCH

		CQI feedback type	
PMI feedback type	Downlink transmission modes	Wideband	UE selected sub-bands
None	1, 2, 3, 7	Mode 1-0	Mode 2-0
Single	4, 5, 6	Mode 1-1	Mode 2-1

Source: TS 36.213. Reproduced by permission of ETSI.

Table 8.6 Channel quality reporting modes for aperiodic reporting on the PUSCH

		CQI feedback type		
PMI feedback type	Downlink transmission modes	Wideband	UE selected sub-bands	Higher layer configured sub-bands
None	1, 2, 3, 7	–	Mode 2-0	Mode 3-0
Single	4, 5, 6	–	–	Mode 3-1
Multiple	4, 6	Mode 1-2	Mode 2-2	–

Source: TS 36.213. Reproduced by permission of ETSI.

that is covered in Chapter 9. The base station is obliged to give the mobile a scheduling grant as part of that procedure, which solves the problem.

A mobile in RRC_IDLE state cannot transmit on the PUCCH, so it cannot send a scheduling request at all. Instead, it uses the random access procedure right away.

8.6 Transmission of Uplink Control Information on the PUCCH

8.6.1 PUCCH Formats

If the mobile wishes to send uplink control information and is not carrying out a PUSCH transmission in the same subframe, then it transmits the information on the physical uplink control channel [29–32]. The PUCCH can be transmitted using several different formats. Table 8.7 shows how these formats are used for the case of a normal cyclic prefix.

When using PUCCH formats 2, 2a and 2b, the transport channel processor applies error correction coding to the channel state information, which increases the number of CSI bits to 20. However, it sends the scheduling request and acknowledgement bits directly down to the physical layer, without any coding at all.

The mobile transmits the PUCCH at the edges of the uplink band (Figure 8.10), to keep it separate from the PUSCH. The base station reserves resource blocks at the extreme edges of the band for PUCCH formats 2, 2a and 2b, with the exact number of blocks advertised in SIB 2. Formats 1, 1a and 1b use resource blocks further in, with the number of blocks varying dynamically from one subframe to the next depending on the number of acknowledgements that the base station is expecting. The base station can also share an intermediate pair of resource blocks amongst all the PUCCH formats, which can be useful if the bandwidth is small. When using the normal cyclic prefix, formats 1, 1a and 1b use four PUCCH symbols per slot and three demodulation reference symbols, while formats 2, 2a and 2b use five PUCCH symbols per slot and two demodulation reference symbols.

An individual mobile transmits the PUCCH using two resource blocks, which are in the first and second slots of a subframe and at opposite sides of the frequency band. However a mobile does not have these resource blocks to itself. In PUCCH formats 2, 2a and 2b, each pair of resource blocks is shared amongst 12 mobiles, using a mobile-specific parameter known as

Table 8.7 List of PUCCH formats and their applications in the case of a normal cyclic prefix

PUCCH format	Release	Application	Number of UCI bits	Number of PUCCH bits
1	R8	SR	1	1
1a	R8	1 bit HARQ-ACK and optional SR	1 or 2	1 or 2
1b	R8	2 bit HARQ-ACK and optional SR	2 or 3	2 or 3
2	R8	CQI, PMI, RI	≤11	20
2a	R8	CQI, PMI, RI and 1 bit HARQ-ACK	≤12	21
2b	R8	CQI, PMI, RI and 2 bit HARQ-ACK	≤13	22
3	R10	20 bit HARQ-ACK and optional SR	≤21	48

Source: TS 36.211. Reproduced by permission of ETSI.

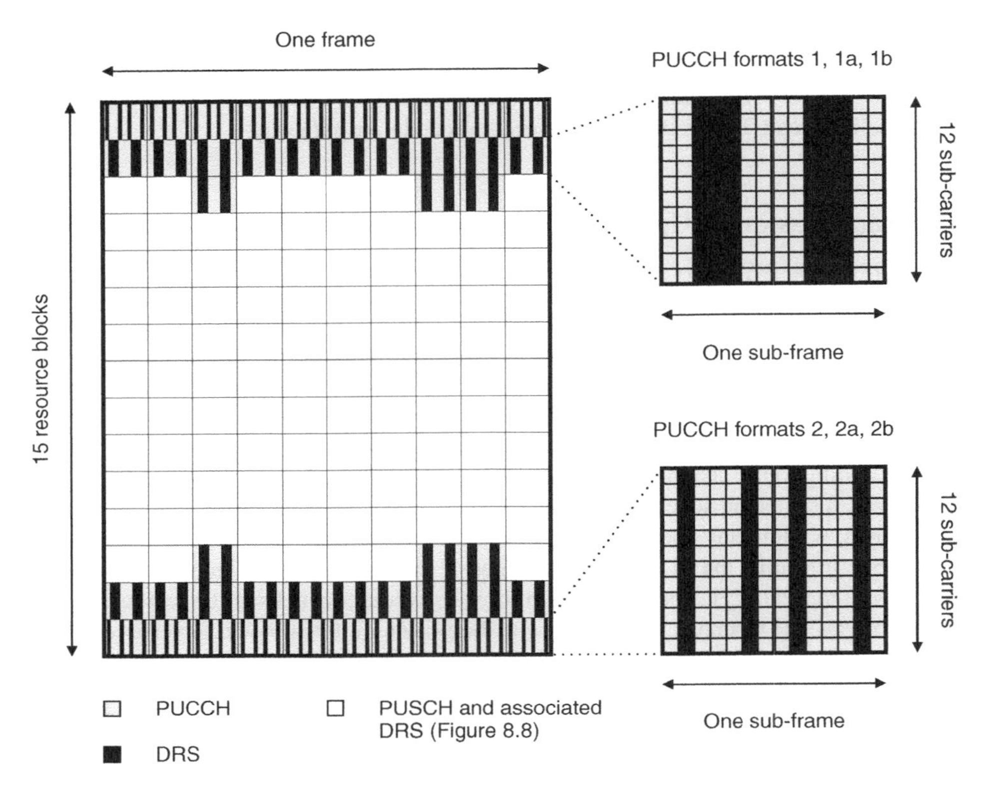

Figure 8.10 Resource element mapping for the PUCCH and its demodulation reference signal in Releases 8 and 9, using FDD mode, a normal cyclic prefix, a 3 MHz bandwidth, one pair of resource blocks for PUCCH formats 2, 2a and 2b, and an example allocation for PUCCH formats 1, 1a and 1b

the *cyclic shift* that runs from 0 to 11. In PUCCH formats 1, 1a and 1b, the resource blocks are shared amongst 36 mobiles, using the cyclic shift and another mobile-specific parameter, the *orthogonal sequence index*, which runs from 0 to 2.

8.6.2 PUCCH Resources

A *PUCCH resource* is a number that determines three things: the resource blocks on which the mobile should transmit the PUCCH, and the orthogonal sequence index and cyclic shift that it should use. The base station can assign three types of PUCCH resource to each mobile.

The first PUCCH resource, denoted $n^{(1)}_{\text{PUCCH}}$, is used for stand-alone hybrid ARQ acknowledgements in formats 1a and 1b. The mobile calculates $n^{(1)}_{\text{PUCCH}}$ dynamically, using the index of the first control channel element that the base station used for its downlink scheduling command.

The second PUCCH resource, denoted $n^{(1)}_{\text{PUCCH, SRI}}$, is used for scheduling requests in format 1. The third, denoted $n^{(2)}_{\text{PUCCH}}$, is used for channel state information and optional acknowledgements in formats 2, 2a and 2b. The mobile receives both of these resources by means of mobile-specific RRC signalling messages, during the procedures for RRC connection establishment or reconfiguration.

If a mobile wishes to send hybrid ARQ acknowledgements at the same time as a scheduling request, then it processes the acknowledgements in the usual way, but transmits them using $n^{(1)}_{\text{PUCCH, SRI}}$. The base station is already expecting the acknowledgements, so it knows how to process them, while it recognizes the scheduling request from the mobile's use of $n^{(1)}_{\text{PUCCH, SRI}}$.

If the mobile is using ACK/NACK multiplexing in TDD mode, then it may have to send up to four acknowledgements in one uplink subframe. The mobile usually does this by transmitting two bits within one of four PUCCH resources, denoted by $n^{(1)}_{\text{PUCCH, 0}}$ to $n^{(1)}_{\text{PUCCH, 3}}$, which are calculated from the first CCE in a similar way to $n^{(1)}_{\text{PUCCH}}$. A lookup table determines the mapping from the acknowledgement bits to the transmitted bits and the choice of PUCCH resource. If, however, the mobile wishes to send a scheduling request or channel state information at the same time, then it compresses the hybrid ARQ acknowledgements down to two bits by means of a different lookup table and sends them on $n^{(1)}_{\text{PUCCH, SRI}}$ or $n^{(2)}_{\text{PUCCH}}$ in the usual way.

8.6.3 Physical Channel Processing of the PUCCH

We now have enough information to describe the physical channel processing for the PUCCH.

When using PUCCH formats 1, 1a and 1b, the mobile modulates the bits onto one symbol, using on-off modulation for a scheduling request, BPSK for a one-bit acknowledgement and QPSK for a two-bit acknowledgement. It then spreads the information in the time domain using the orthogonal sequence index, usually across four symbols, but across three symbols in slots which support a sounding reference signal that is taking priority over these PUCCH formats (Section 8.7.2). The spreading process follows a technique similar to the one that the base station used for the PHICH and allows the symbols to be shared amongst three different mobiles.

The mobile then spreads the information across 12 sub-carriers in the frequency domain using the cyclic shift. This technique is implemented differently from the one above but has the same objective, namely sharing the sub-carriers amongst 12 different mobiles. Finally, the mobile repeats its transmission in the first and second slots of the subframe.

When using PUCCH format 2, the mobile modulates the channel state information bits onto 10 symbols using QPSK and spreads the information in the frequency domain using the cyclic shift. It can also send simultaneous acknowledgements in formats 2a and 2b, by modulating the second reference symbol in each subframe using BPSK or QPSK.

8.7 Uplink Reference Signals

8.7.1 Demodulation Reference Signal

The mobile transmits the demodulation reference signal [33] along with the PUSCH and PUCCH, to help the base station carry out channel estimation. As shown in Figures 8.8 and 8.10, the signal occupies three symbols per slot when the mobile is using PUCCH formats 1, 1a and 1b, two when using PUCCH formats 2, 2a and 2b and one when using the PUSCH.

The demodulation reference signal can contain 12, 24, 36, ... data points, corresponding to transmission bandwidths of 1, 2, 3, ... resource blocks. To generate the signal, each cell is assigned to one of 30 *sequence groups*. With one exception, described at the end of this section, each sequence group contains one *base sequence* of every possible length, which is generated either from a Zadoff-Chu sequence or, in the case of the very shortest sequences, from a look-up table. The base sequence is then modified by one of 12 cyclic shifts, to generate the reference signal itself.

There are two ways to assign the sequence groups. In *sequence group planning*, each cell is permanently assigned to one of the sequence groups during radio network planning. Nearby cells should lie in different sequence groups so as to minimize the interference between them. In *sequence group hopping*, the sequence group changes from one slot to the next according to one of 510 pseudo-random hopping patterns. The hopping pattern depends on the physical cell identity and can be calculated without the need for any further planning.

When sending PUSCH reference signals, the mobile calculates the cyclic shift from a field that the base station supplied in its scheduling grant. In the case of uplink multiple user MIMO, the base station can distinguish different mobiles that are sharing the same resource blocks by giving them different cyclic shifts. The remaining cyclic shifts can be used to distinguish nearby cells that share the same sequence group.

When sending PUCCH reference signals, the mobile applies the same cyclic shift that it used for the PUCCH transmission itself, and modifies the demodulation reference signal further in the case of formats 1, 1a and 1b using the orthogonal sequence index. This process allows the base station to distinguish the reference signals from all the mobiles that are sharing each pair of resource blocks.

There are two other complications. Firstly, each sequence group actually contains two base sequences for every transmission bandwidth of six resource blocks or more. In *sequence hopping*, a mobile can be configured to switch between the two sequences according to a pseudo-random pattern. Secondly, *cyclic shift hopping* makes the cyclic shift change in a pseudo-random manner from one slot to the next. Both techniques reduce the interference between nearby cells that share the same sequence group.

8.7.2 Sounding Reference Signal

The mobile transmits the sounding reference signal (SRS) [34–36] to help the base station measure the received signal power across a wide transmission bandwidth. The base station then uses the information for frequency dependent scheduling.

The base station controls the timing of sounding reference signals in two ways. Firstly, it tells the mobiles which subframes support sounding, using a parameter in SIB 2 called the *SRS subframe configuration*. Secondly, it configures each mobile with a sounding period of 2 to 320 subframes and an offset within that period, using a mobile-specific parameter called the *SRS configuration index*. A mobile transmits the sounding reference signal whenever the resulting transmission times coincide with a subframe that supports sounding.

The mobile usually sends the sounding reference signal in the last symbol of the subframe, as shown in Figure 8.11. In TDD mode, it can also send the signal in the uplink region of a special subframe. The mobile generates the signal in a similar way to the demodulation reference signal described above. The main difference is that the sounding reference signal uses eight cyclic shifts rather than 12, so that eight mobiles can share the same set of resource elements.

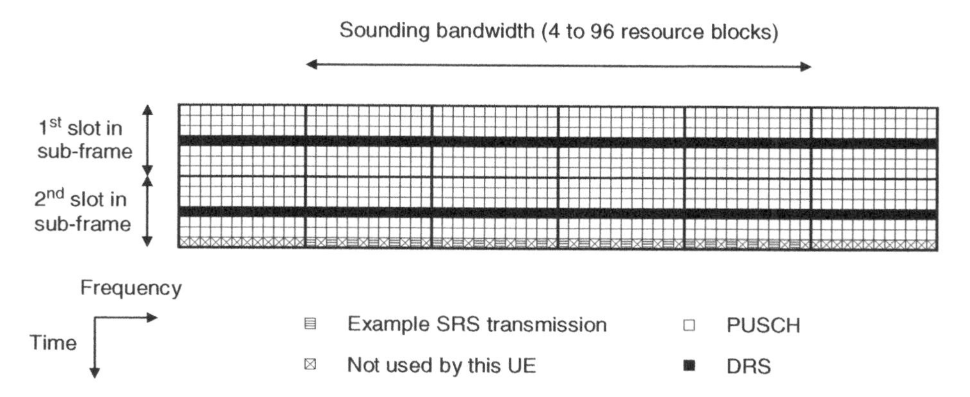

Figure 8.11 Example resource element mapping for the sounding reference signal, using a normal cyclic prefix

In the frequency domain, the base station controls the starting position and transmission bandwidth using cell- and mobile-specific parameters called the SRS bandwidth configuration, SRS bandwidth, frequency domain position and SRS hopping bandwidth. As shown in the figure, an individual mobile transmits on alternate sub-carriers, as configured by a transmission comb.

There are various ways to prevent clashes between the sounding reference signal and the mobile's other transmissions. A mobile does not transmit the PUSCH in the last symbol of subframes that support sounding, so it can always send the PUSCH and SRS in the same subframe. PUCCH formats 2, 2a and 2b take priority over the sounding reference signal, as they have reserved frequencies at the edge of the transmission band that are of no interest to the sounding procedure. The base station can configure PUCCH formats 1, 1a and 1b to use either technique by means of RRC signalling.

8.8 Power Control

8.8.1 Uplink Power Calculation

The uplink power control procedure [37, 38] sets the mobile's transmit power to the smallest value that is consistent with satisfactory reception of the signal. This reduces the interference between mobiles that are transmitting on the same resource elements in nearby cells and increases the mobile's battery life. In LTE, the mobile estimates its transmit power as well as it can and the base station adjusts this estimate using power control commands. The mobile uses slightly different calculations for the PUSCH, PUCCH and SRS, so, to illustrate the principles, we will just look at the PUSCH.

The PUSCH transmit power is calculated as follows:

$$P_{\mathrm{PUSCH}}(i) = \min(P(i), P_{\mathrm{CMAX}}) \tag{8.1}$$

In this equation, $P_{\mathrm{PUSCH}}(i)$ is the power transmitted on the PUSCH in subframe i, measured in decibels relative to 1 mW (dBm). P_{CMAX} is the mobile's maximum transmit power, while

$P(i)$ is calculated as follows:

$$P(i) = P_{O\_PUSCH} + 10\log_{10}(M_{PUSCH}(i)) + \Delta_{TF}(i) + \alpha.PL + f(i) \tag{8.2}$$

Here, P_{O_PUSCH} is the power that the base station expects to receive over a bandwidth of one resource block. It has two components, a cell-specific baseline $P_{O\_NOMINAL\_PUSCH}$ and a mobile-specific adjustment $P_{O\_UE\_PUSCH}$, which are sent to the mobile using RRC signalling. $M_{PUSCH}(i)$ is the number of resource blocks on which the mobile is transmitting in subframe i. $\Delta_{TF}(i)$ is an optional adjustment for the data rate in subframe i, which ensures that the mobile uses a higher transmit power for a larger coding rate or a faster modulation scheme such as 64-QAM.

PL is the downlink path loss. The base station advertises the power transmitted on the downlink reference signals as part of SIB 2, so the mobile can estimate PL by reading this quantity and subtracting the power received. α is a weighting factor that reduces the impact of changes in the path loss, in a technique known as *fractional power control*. By setting α to a value between zero and one, the base station can ensure that mobiles at the cell edge transmit a weaker signal than would otherwise be expected. This reduces the interference that they send into nearby cells and can increase the capacity of the system.

Using the parameters covered so far, the mobile can make its own estimate of the PUSCH transmit power. However this estimate may not be accurate, particularly in FDD mode, where the fading patterns are likely to be different on the uplink and downlink. The base station therefore adjusts the mobile's power using power control commands, which are handled by the last parameter, $f(i)$.

8.8.2 Uplink Power Control Commands

The base station can send power control commands for the PUSCH in two ways. Firstly, it can send stand-alone power control commands to groups of mobiles using DCI formats 3 and 3A. When using these formats, the base station addresses the PDCCH message to a radio network identity known as the TPC-PUSCH-RNTI, which is shared amongst all the mobiles in the group. The message contains a power control command for each of the group's mobiles, which is found using an offset that has previously been configured by means of RRC signalling.

The mobile then accumulates its power control commands in the following way:

$$f(i) = f(i-1) + \delta_{PUSCH}(i - K_{PUSCH}) \tag{8.3}$$

Here, the mobile receives a power adjustment of δ_{PUSCH} in subframe $i - K_{PUSCH}$ and applies it in subframe i. K_{PUSCH} is four in FDD mode, while in TDD mode, it can lie between four and seven in the usual way. When using DCI format 3, the power control command contains two bits and causes power adjustments of −1, 0, 1 and 3 dB. When using DCI format 3A, the command only contains one bit and causes power adjustments of −1 and 1 dB.

The base station can also send two-bit power control commands to one mobile as part of an uplink scheduling grant. Usually, the mobile interprets them in the manner described above. However, the base station can also disable the accumulation of power control commands using RRC signalling, in which case the mobile interprets them as follows:

$$f(i) = \delta_{PUSCH}(i - K_{PUSCH}) \tag{8.4}$$

In this case, the power adjustment δ_{PUSCH} can take values of −4, −1, 1 and 4 dB.

8.8.3 Downlink Power Control

Downlink power control is more straightforward. The downlink transmit power is quantified using the *energy per resource element* (EPRE) of an individual channel or signal. The base station can use a different EPRE for the downlink reference signals and for the PDSCH transmissions to individual mobiles, and can tell the mobiles about the chosen values by means of RRC signalling messages. However each value is independent of frequency and is only occasionally changed; instead, the base station adapts to changes in a mobile's downlink propagation loss by adjusting the modulation scheme and coding rate. This is consistent with the idea that downlink transmit power is a shared resource, and prevents the base station from allocating too much power to distant mobiles that cannot use it efficiently.

8.9 Discontinuous Reception

8.9.1 Discontinuous Reception and Paging in RRC_IDLE

When a mobile is in a state of *discontinuous reception* (DRX), the base station only sends it downlink control information on the PDCCH in certain subframes. Between those subframes, the mobile can stop monitoring the PDCCH and can enter a low-power state known as *sleep mode*, so as to maximize its battery life. Discontinuous reception is implemented using two different mechanisms, which support paging in RRC_IDLE and low data rate transmission in RRC_CONNECTED.

In RRC_IDLE state, discontinuous reception is defined using a *DRX cycle* [39, 40], which lies between 32 and 256 frames (0.32 and 2.56 s). The base station specifies a default DRX cycle length in SIB 2, but the mobile can request a different cycle length during an attach request or a tracking area update (Chapters 11 and 14).

As shown in Figure 8.12, the mobile wakes up once every DRX cycle frames, in a *paging frame* whose system frame number depends on the mobile's international mobile subscriber identity. Within that frame, the mobile inspects a subframe known as a *paging occasion*, which also depends on the IMSI. If the mobile finds downlink control information addressed to the P-RNTI at the beginning of the subframe, then it goes on to receive an RRC *Paging* message on the PDSCH in the remainder of the subframe. The network knows the mobile's IMSI, so it can send the control information and the paging message in the correct subframe.

Several mobiles can share the same paging occasion. To resolve this conflict, the Paging message contains the identity of the target mobile, using the S-TMSI (if available) or the IMSI (otherwise). If the mobile detects a match, then it responds to the paging message using an EPS mobility management procedure known as a service request, in the manner described in Chapter 14.

8.9.2 Discontinuous Reception in RRC_CONNECTED

In RRC_CONNECTED state, the base station configures a mobile's discontinuous reception parameters by means of mobile-specific RRC signalling. During discontinuous reception (Figure 8.13), the mobile wakes up every *DRX cycle* subframes, in a subframe defined by a *DRX start offset*. It monitors the PDCCH continuously for a duration known as the *active time* and then goes back to sleep [41, 42].

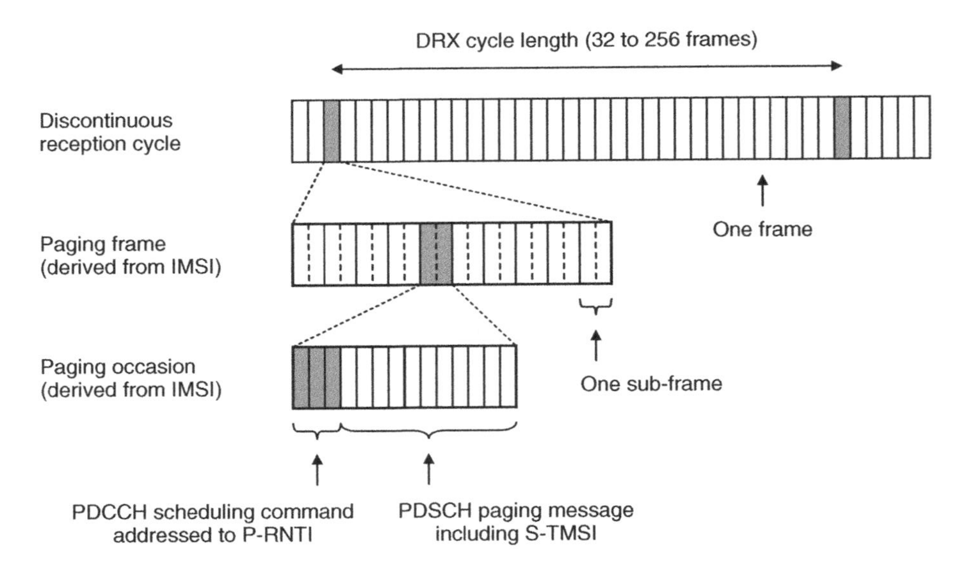

Figure 8.12 Operation of discontinuous reception and paging in RRC_IDLE

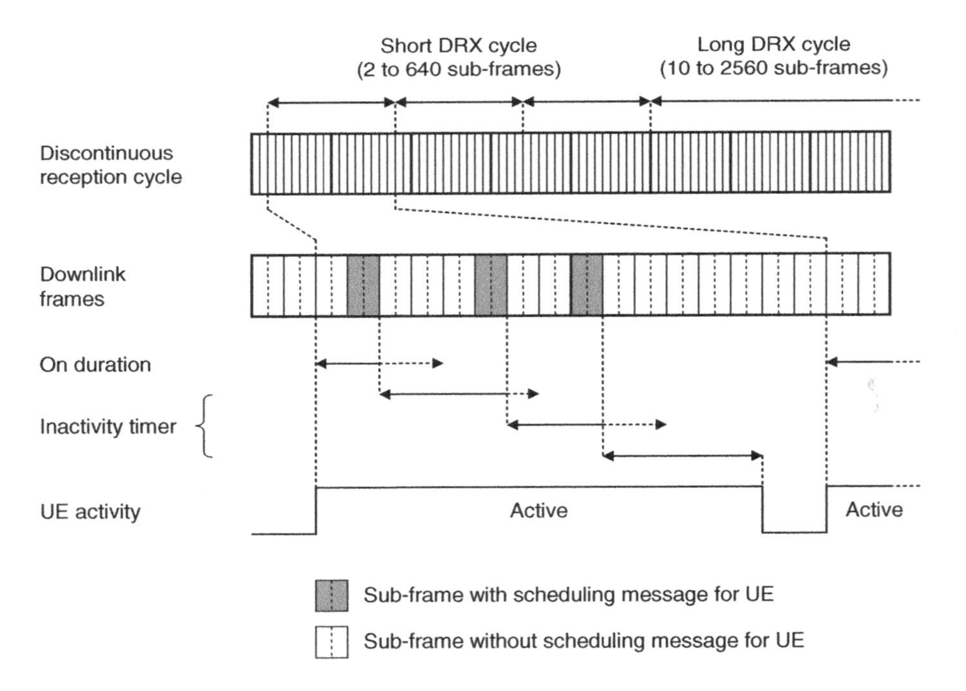

Figure 8.13 Operation of discontinuous reception in RRC_CONNECTED

Several timers contribute to the active time. Initially, the mobile stays awake for a time of *on duration* (1 to 200 subframes), waiting for a scheduling message on the PDCCH. If one arrives, then the mobile stays awake for a time of *DRX inactivity timer* (1 to 2560 subframes) after every PDCCH command. Other timers ensure that the mobile stays awake while waiting for information such as hybrid ARQ re-transmissions, but, if all the timers expire, then the mobile goes back to sleep. The base station can also send the mobile to sleep explicitly, by sending it a MAC control element (Chapter 10) known as a *DRX command*.

There are actually two discontinuous reception cycles, the *long DRX cycle* (10 to 2560 subframes) and the optional *short DRX cycle* (2 to 640 subframes). If both are configured, then the mobile starts by using the short cycle, but moves to the long cycle if it goes for *DRX short cycle timer* (1 to 16) cycles without receiving a PDCCH command.

References

1. 3GPP TS 36.211 (2013) Physical Channels and Modulation, Release 11, September 2013.
2. 3GPP TS 36.212 (2013) Multiplexing and Channel Coding, Release 11, June 2013.
3. 3GPP TS 36.213 (2013) Physical Layer Procedures, Release 11, September 2013.
4. 3GPP TS 36.321 (2013) Medium Access Control (MAC) Protocol Specification, Release 11, July 2013.
5. 3GPP TS 36.331 (2013) Radio Resource Control (RRC); Protocol Specification, Release 11, September 2013.
6. 3GPP TS 36.321 (2013) Medium Access Control (MAC) Protocol Specification, Release 11, Section 5.3, July 2013.
7. 3GPP TS 36.213 (2013) Physical Layer Procedures, Release 11, Section 7, September 2013.
8. 3GPP TS 36.321 (2013) Medium Access Control (MAC) Protocol Specification, Release 11, Section 5.4, July 2013.
9. 3GPP TS 36.213 (2013) Physical Layer Procedures, Release 11, Section 8, September 2013.
10. 3GPP TS 36.321 (2013) Medium Access Control (MAC) Protocol Specification, Release 11, Section 5.10, July 2013.
11. 3GPP TS 36.213 (2013) Physical Layer Procedures, Release 11, Section 9.2, September 2013.
12. 3GPP TS 36.331 (2013) Radio Resource Control (RRC); Protocol Specification, Release 11, Section 6.3.2 (SPS-Config), September 2013.
13. 3GPP TS 36.212 (2013) Multiplexing and Channel Coding, Release 11, Section 5.3.3.1, June 2013.
14. 3GPP TS 36.213 (2013) Physical Layer Procedures, Release 11, Sections 7.1.6, 8.1, 8.4, September 2013.
15. 3GPP TS 36.211 (2013) Physical Channels and Modulation, Release 11, Section 5.2.3, 5.3.4, 6.2.3, September 2013.
16. 3GPP TS 36.321 (2013) Medium Access Control (MAC) Protocol Specification, Release 11, Section 7.1m, July 2013.
17. 3GPP TS 36.212 (2013) Multiplexing and Channel Coding, Release 11, Sections 5.3.3.2, 5.3.3.3, 5.3.3.4, June 2013.
18. 3GPP TS 36.211 (2013) Physical Channels and Modulation, Release 11, Section 6.8, September 2013.
19. 3GPP TS 36.213 (2013) Physical Layer Procedures, Release 11, Section 9.1.1, September 2013.
20. 3GPP TS 36.212 (2013) Multiplexing and Channel Coding, Release 11, Sections 5.1, 5.2.2, 5.2.4, 5.3.2, June 2013.
21. 3GPP TS 36.211 (2013) Physical Channels and Modulation, Release 11, Sections 5.3, 5.6, 5.8, 6.3, 6.4, 6.12, 6.13, September 2013.
22. 3GPP TS 36.211 (2013) Physical Channels and Modulation, Release 11, Section 6.9, September 2013.
23. 3GPP TS 36.212 (2013) Multiplexing and Channel Coding, Release 11, Section 5.3.5, June 2013.
24. 3GPP TS 36.213 (2013) Physical Layer Procedures, Release 11, Section 9.1.2, September 2013.
25. 3GPP TS 36.331 (2013) Radio Resource Control (RRC); Protocol Specification, Release 11, Section 6.3.2 (PHICH-Config), September 2013.
26. 3GPP TS 36.213 (2013) Physical Layer Procedures, Release 11, Sections 7.2, 7.3, 10.1, September 2013.
27. 3GPP TS 36.321 (2013) Medium Access Control (MAC) Protocol Specification, Release 11, Section 5.4.4, July 2013.

28. 3GPP TS 36.331 (2013) Radio Resource Control (RRC); Protocol Specification, Release 11, Section 6.3.2 (CQI-ReportConfig, SchedulingRequestConfig), September 2013.
29. 3GPP TS 36.211 (2013) Physical Channels and Modulation, Release 11, Section 5.4, September 2013.
30. 3GPP TS 36.212 (2013) Multiplexing and Channel Coding, Release 11, Section 5.2.3, June 2013.
31. 3GPP TS 36.213 (2013) Physical Layer Procedures, Release 11, Section 10.1, September 2013.
32. 3GPP TS 36.331 (2013) Radio Resource Control (RRC); Protocol Specification, Release 11, Section 6.3.2 (PUCCH-Config), September 2013.
33. 3GPP TS 36.211 (2013) Physical Channels and Modulation, Release 11, Sections 5.5.1, 5.5.2, September 2013.
34. 3GPP TS 36.211 (2013) Physical Channels and Modulation, Release 11, Section 5.5.3, September 2013.
35. 3GPP TS 36.213 (2013) Physical Layer Procedures, Release 11, Section 8.2, September 2013.
36. 3GPP TS 36.331 (2013) Radio Resource Control (RRC); Protocol Specification, Release 11, Section 6.3.2 (SoundingRS-UL-Config), September 2013.
37. 3GPP TS 36.213 (2013) Physical Layer Procedures, Release 11, Section 5, September 2013.
38. 3GPP TS 36.331 (2013) Radio Resource Control (RRC); Protocol Specification, Release 11, Section 6.3.2 (Uplink Power Control, TPC-PDCCH-Config), September 2013.
39. 3GPP TS 36.304 (2013) Evolved Universal Terrestrial Radio Access (E-UTRA); User Equipment (UE) Procedures in Idle Mode, Release 11, Section 7, September 2013.
40. 3GPP TS 36.331 (2013) Radio Resource Control (RRC); Protocol Specification, Release 11, Sections 5.3.2, 6.2.2 (Paging), September 2013.
41. 3GPP TS 36.321 (2013) Medium Access Control (MAC) Protocol Specification, Release 11, Section 5.7, July 2013.
42. 3GPP TS 36.331 (2013) Radio Resource Control (RRC); Protocol Specification, Release 11, Section 6.3.2 (MAC-MainConfig), September 2013.

9

Random Access

As described in the previous chapter, the base station explicitly schedules all the transmissions that the mobile carries out on the physical uplink shared channel. If the mobile wishes to transmit on the PUSCH but does not have the resources to do so, then it usually sends a scheduling request on the physical uplink control channel. If it does not have the resources to do that, then it initiates the random access procedure. This can happen in a few different situations, primarily during the establishment of an RRC connection, during a handover or if the mobile has lost timing synchronization with the base station. The base station can also trigger the random access procedure, if it wishes to transmit to the mobile after a loss of timing synchronization.

The procedure begins when the mobile transmits a random access preamble on the physical random access channel (PRACH). This initiates an exchange of messages between the mobile and the base station that has two main variants, non-contention-based and contention-based. As a result of the procedure, the mobile receives three quantities: resources for an uplink transmission on the PUSCH, an initial value for the uplink timing advance and, if it does not already have one, a C-RNTI.

The random access procedure is defined by the same specifications that were used for data transmission and reception. The most important are TS 36.211 [1], TS 36.213 [2], TS 36.321 [3] and TS 36.331 [4].

9.1 Transmission of Random Access Preambles on the PRACH

9.1.1 Resource Element Mapping

The best place to start discussing the PRACH [5–7] is with the resource element mapping. In the frequency domain, a PRACH transmission has a bandwidth of six resource blocks. In the time domain, the transmission is usually one subframe long, but it can be longer or shorter. Figure 9.1 shows an example, for the case of FDD mode and a bandwidth of 3 MHz.

Looking in more detail, the PRACH transmission comprises a cyclic prefix, a preamble sequence and a guard period. In turn, the preamble sequence contains one or two PRACH symbols, which are usually 800 μs long. The mobile transmits the PRACH without any timing advance, but the guard period prevents it from colliding at the base station with the symbols that follow.

An Introduction to LTE: LTE, LTE-Advanced, SAE, VoLTE and 4G Mobile Communications, Second Edition. Christopher Cox.

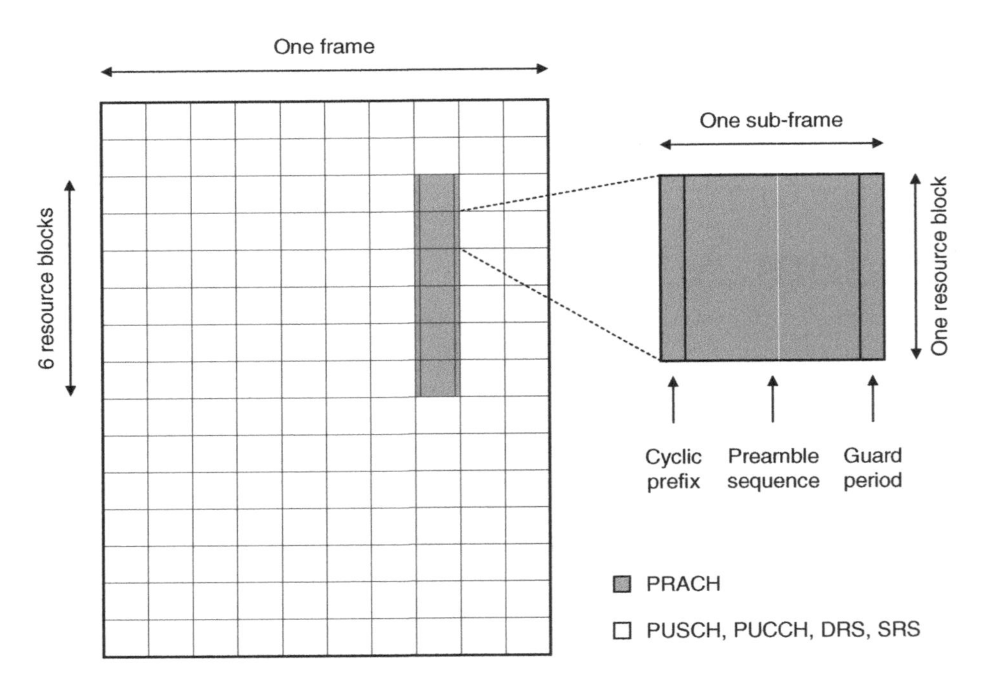

Figure 9.1 Resource element mapping for the physical random access channel, using FDD mode, a normal cyclic prefix, a 3 MHz bandwidth, a PRACH configuration index of 5 and a PRACH frequency offset of 7

The base station can specify the duration of each component using the preamble formats listed in Table 9.1. The most common is format 0, in which the transmission takes up one subframe. Formats 1 and 3 have long guard periods, so are useful in large cells, while formats 2 and 3 have two PRACH symbols, so are useful if the received signal is weak. Format 4 is only used by small TDD cells, and maps the PRACH onto the uplink part of a special subframe.

The base station reserves specific resource blocks for the PRACH using two parameters that it advertises in SIB 2, namely the *PRACH configuration index* and the *PRACH frequency offset*. These parameters have different meanings in FDD and TDD modes.

Table 9.1 Random access preamble formats

	Approximate duration (μs)				
Format	Cyclic prefix	Preamble	Guard period	Total	Application
0	103	800	97	1000	Normal cells
1	684	800	516	2000	Large cells
2	203	1600	197	2000	Weak signals
3	684	1600	716	3000	Large cells and weak signals
4	15	133	9	157	Small TDD cells

Source: TS 36.211. Reproduced by permission of ETSI.

In FDD mode, the PRACH configuration index specifies the preamble format and the subframes in which random access transmissions can begin, while the PRACH frequency offset specifies their location in the frequency domain. In Figure 9.1, for example, the PRACH configuration index is 5, which supports PRACH transmissions in subframe 7 of every frame using preamble format 0. The PRACH frequency offset is 7, so the transmissions occupy resource blocks 7 to 12 in the frequency domain.

In TDD mode, random access transmissions cannot take place during downlink subframes, so there are fewer opportunities for them. To compensate for this, mobiles can transmit the PRACH at a maximum of six locations in the frequency domain, instead of one. Together, the PRACH configuration index and frequency offset define a mapping into the resource grid, from which the mobile can discover the times and frequencies that support the PRACH.

9.1.2 Preamble Sequence Generation

Each cell supports 64 different preamble sequences. The base station can distinguish mobiles that are transmitting on the same set of resource blocks, provided that their preamble sequences are different.

The mobile generates its preamble sequences from the Zadoff-Chu sequences that we introduced in Chapter 7, using three parameters that the base station advertises in SIB 2. The first is the *zero correlation zone configuration*, which determines the number of cyclic shifts that the mobile can generate from a single Zadoff-Chu root sequence. In a large cell the mobile is only allowed to use a few widely spaced cyclic shifts, because there would otherwise be a risk of the base station confusing one cyclic shift from a nearby mobile with a slightly different cyclic shift from a distant one. In a small cell that risk is absent, so the mobiles use a large number of cyclic shifts to take advantage of their mutual orthogonality.

The *root sequence index* identifies the first Zadoff-Chu root sequence that the cell is actually using. The mobile generates as many cyclic shifts from this root sequence as the zero correlation zone configuration allows, moves to the root sequences that follow and continues until it has generated all 64 preambles. Nearby cells should use different root sequence indices so as to minimize the interference between their mobiles' random access transmissions. The indices can be assigned either during network planning or by the self-optimization functions that we discuss in Chapter 17.

The calculations are modified in cells that contain fast moving mobiles because there is a risk that the base station will confuse one cyclic shift from a slow moving mobile with another from a fast moving one. The base station can deal with the problem by setting a *high speed flag*, which restricts the number of cyclic shifts that are available in such cells.

The base station then reserves some of the 64 preamble sequences for the non-contention-based random access procedure that we discuss next, and assigns them to individual mobiles by means of RRC signalling. The remainder are available for the contention-based procedure and are chosen at random by the mobile.

9.1.3 Signal Transmission

To transmit the physical random access channel, the mobile simply generates the appropriate time domain preamble sequence and passes it to the forward Fourier transform in its physical layer. There are, however, some differences from the usual techniques for resource element mapping that we introduced in Chapter 6. In particular, the PRACH symbol duration is 800 μs in formats 0 to 3, instead of the usual value of 66.7 μs. That implies that the sub-carrier spacing is 1250 Hz, instead of the usual value of 15 kHz. In format 4, the symbol duration is 133 μs, so the sub-carrier spacing is 7500 Hz. The use of a smaller sub-carrier spacing means that the PRACH sub-carriers are not orthogonal to the sub-carriers used by the PUCCH and PUSCH. Because of this, the PRACH transmission band contains small guard bands at its upper and lower edges, to minimize the amount of interference that occurs.

Power control works differently on the random access channel from the other uplink channels. The mobile first transmits a random access preamble with the following power:

$$P_{\mathrm{PRACH}} = \min(P_{\mathrm{PREAMBLE,\ INITIAL}} + PL, P_{\mathrm{CMAX}}) \tag{9.1}$$

Here, P_{CMAX} is the mobile's maximum transmit power, PL is its estimate of the downlink path loss and $P_{\mathrm{PREAMBLE,\ INITIAL}}$ is a parameter that the base station advertises in SIB 2, which describes the power that it expects to receive.

The mobile then awaits a response from the base station, in a random access window whose duration lies between 2 and 10 subframes. If it does not receive a response within this time, then it assumes that the transmit power was too low for the base station to hear it, so it increases the transmit power by a value that lies between 0 and 6 decibels and repeats the transmission. This process continues until the mobile receives a response or until it reaches a maximum number of re-transmissions.

9.2 Non-Contention-Based Procedure

When the mobile sends a PRACH transmission to the base station, it initiates the random access procedure [8–10]. There are two variants of this procedure, namely non-contention-based and contention-based.

If the network can reserve a preamble sequence for a mobile, then it can guarantee that no other mobile will be using that sequence in the same set of resource blocks. This idea is the basis of the non-contention-based random access procedure, which is typically used as part of a handover as shown in Figure 9.2.

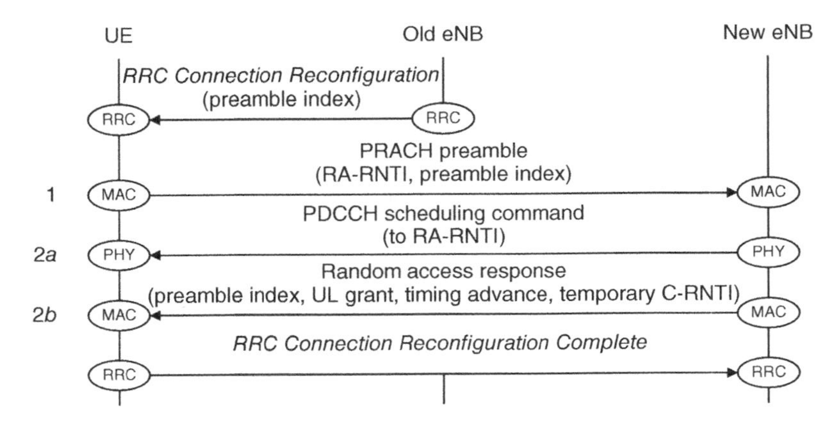

Figure 9.2 Non-contention-based random access procedure, as used during a handover

Before the procedure begins, the old base station sends the mobile an RRC message known as *RRC Connection Reconfiguration*. This tells the mobile how to reconfigure itself for communication with the new base station and identifies a preamble sequence that the new base station has reserved for it. The mobile reads the RRC message and reconfigures itself as instructed. However, it does not yet have timing synchronization, so it triggers the random access procedure.

The mobile reads the new cell's random access configuration from SIB 2, chooses the next available PRACH transmission time and sends a preamble using the requested sequence (step 1). The transmission frequency is fixed in FDD mode, while in TDD mode it is chosen at random. Together, the transmission time and frequency determine a mobile identity known as the random access RNTI (RA-RNTI). If necessary, the mobile repeats the transmission in the manner described above, until it receives a response.

Once the base station receives the preamble, it measures the arrival time and calculates the required timing advance. It replies first with a PDCCH scheduling command (step 2a), which it writes using DCI format 1A or 1C and addresses to the mobile's RA-RNTI. It follows this with a *random access response* (step 2b), which identifies the preamble sequence that the mobile used, and gives the mobile an uplink scheduling grant and an initial value for the uplink timing advance. (The base station also gives the mobile an identity known as the temporary C-RNTI, but the mobile does not actually use it in this version of the procedure.) The base station can identify several preamble sequences in one response, so it can simultaneously reply to all the mobiles that transmitted on the same resource blocks but with different preambles.

The mobile receives the base station's response and initializes its timing advance. It can then reply to the base station's signalling message, using an *RRC Connection Reconfiguration Complete*.

A base station can also initiate the non-contention-based random access procedure if it wishes to transmit to the mobile on the downlink, but has lost timing synchronization with it. To do this, it triggers the procedure using a variant of DCI format 1A known as a *PDCCH order* [11]. The procedure then continues in the manner described above.

9.3 Contention-Based Procedure

A mobile uses the contention-based random access procedure if it has not been allocated a preamble index. This typically happens as part of a procedure known as RRC connection establishment, in the manner shown in Figure 9.3.

In this example, the mobile wishes to send the base station an RRC message known as an *RRC Connection Request*, in which it asks to move from RRC_IDLE to RRC_CONNECTED. It has no PUSCH resources on which to send the message and no PUCCH resources on which to send a scheduling request, so it triggers the random access procedure.

The mobile reads the cell's random access configuration from SIB 2 and chooses a preamble sequence at random from the ones available for the contention-based procedure. Optionally, the base station can divide these into two further groups, namely group A, which is used either for small packets or for large packets in poor radio conditions, and group B, which is used for large packets in good radio conditions. A preamble in group A will eventually lead to a small scheduling grant that is suitable for a small transmission or for a buffer status report. A preamble in group B will lead to a larger scheduling grant, with which the mobile can start a larger uplink transmission and may even be able to complete it.

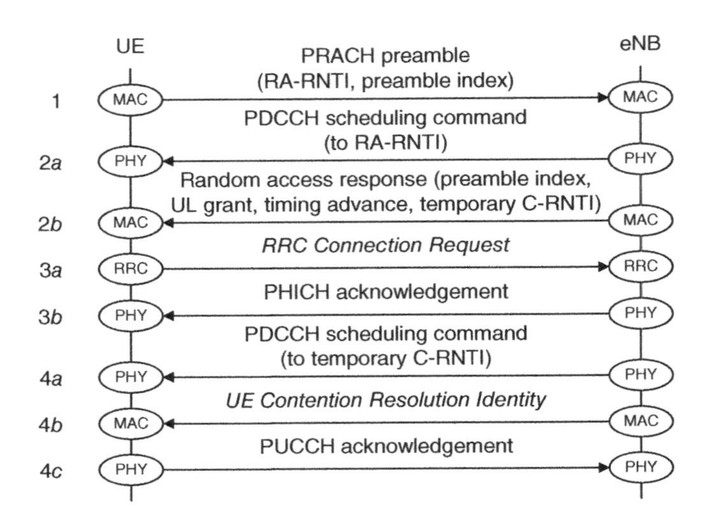

Figure 9.3 Contention-based random access procedure, as used during RRC connection establishment

The mobile then transmits the preamble in the usual way (step 1). There is a risk of contention, if two or mobiles transmit on the same resource blocks using the same preamble sequence. As before, the base station sends the mobile a scheduling command followed by a random access response (steps 2a and 2b).

Using the uplink grant, the mobile sends its RRC message in the usual way (step 3a). As part of the message, the mobile uniquely identifies itself using either its S-TMSI or a random number (Section 11.3.1). There is still a risk of contention between the mobiles that initiated the procedure, but if one of the transmissions is much stronger than the others, then the base station will be able to decode it. The other transmissions will only cause interference. The base station sends an acknowledgement using the PHICH resource that was indicated by the scheduling grant (step 3b).

The base station now sends the mobile another scheduling command (step 4a), which it addresses to the temporary C-RNTI that it allocated earlier. It follows the command with a MAC control element called the *UE contention resolution identity* (step 4b). This echoes back the RRC message that the mobile transmitted in step 3, so it includes the identity of the successful mobile.

If a mobile receives an echo of the message that it originally transmitted, then it sends an acknowledgement using the PUCCH resource indicated by the scheduling command (step 4c). It then promotes the temporary C-RNTI to a full C-RNTI and continues the RRC procedure. If the message does not match, then the mobile discards the temporary C-RNTI and tries the random access procedure again at a later time. As a result, the base station has selected one of the mobiles that were originally competing for its attention and has told the others to back off.

A mobile can also initiate the contention-based procedure in RRC_CONNECTED state, if it wishes to transmit to the base station but has lost timing synchronization, or if it has reached a maximum number of scheduling requests without receiving a reply. In this situation, however, the mobile already has a C-RNTI. In step 3 of the procedure, it replaces the RRC message with a C-RNTI MAC control element (Chapter 10) and the base station then uses the C-RNTI as the basis for contention resolution.

References

1. 3GPP TS 36.211 (2013) Physical Channels and Modulation, Release 11, September 2013.
2. 3GPP TS 36.213 (2013) Physical Layer Procedures, Release 11, September 2013.
3. 3GPP TS 36.321 (2013) Medium Access Control (MAC) Protocol Specification, Release 11, July 2013.
4. 3GPP TS 36.331 (2013) Radio Resource Control (RRC) Protocol Specification, Release 11, September 2013.
5. 3GPP TS 36.211 (2013) Physical Channels and Modulation, Release 11, Section 5.7, September 2013.
6. 3GPP TS 36.212 (2013) Multiplexing and Channel Coding, Release 11, Section 5.2.1, June 2013.
7. 3GPP TS 36.331 (2013) Radio Resource Control (RRC) Protocol Specification, Release 11, Section 6.3.2 (PRACH-Config, RACH-ConfigCommon, RACH-ConfigDedicated), September 2013.
8. 3GPP TS 36.300 (2013) Evolved Universal Terrestrial Radio Access (E-UTRA) and Evolved Universal Terrestrial Radio Access Network (E-UTRAN) Overall Description; Stage 2, Release 11, Section 10.1.5, September 2013.
9. 3GPP TS 36.321 (2013) Medium Access Control (MAC) Protocol Specification, Release 11, Sections 5.1, 6.1.3.2, 6.1.3.4, 6.1.5, 6.2.2, 6.2.3, July 2013.
10. 3GPP TS 36.213 (2013) Physical Layer Procedures, Release 11, Section 6, September 2013.
11. 3GPP TS 36.212 (2013) Multiplexing and Channel Coding, Release 11, Section 5.3.3.1.3, June 2013.

10

Air Interface Layer 2

We have now completed our survey of the air interface's physical layer. In this chapter, we round off our discussion of the LTE air interface by describing the three protocols in the data link layer, layer 2 of the OSI model. The medium access control protocol schedules all the transmissions that are made on the LTE air interface and controls the low-level operation of the physical layer. The radio link control protocol maintains the data link across the air interface, if necessary by re-transmitting packets that the physical layer has not delivered correctly. Finally, the packet data convergence protocol maintains the security of the air interface, compresses the headers of IP packets and ensures the reliable delivery of packets after a handover.

10.1 Medium Access Control Protocol

10.1.1 Protocol Architecture

The medium access control (MAC) protocol [1, 2] schedules the transmissions that are carried out on the air interface and controls the low-level operation of the physical layer. There is one MAC entity in the base station and one in the mobile. To illustrate its architecture, Figure 10.1 is a high-level block diagram of the mobile's MAC protocol.

In the transmitter, the *logical channel prioritization* function determines how much data the mobile should transmit from each incoming logical channel in every transmission time interval. In response, the mobile grabs the required data from transmit buffers in the RLC protocol, in the form of MAC service data units (SDUs). The *multiplexing* function combines the service data units together, attaches a header and sends the resulting data on a transport channel to the physical layer. It does this by way of the *hybrid ARQ transmission* function, which controls the operation of the physical layer's hybrid ARQ protocol. The output data packets are known as MAC protocol data units (PDUs) and are identical to the transport blocks that we saw in Chapter 8. The functions are reversed in the mobile's receiver.

The principles are much the same in the base station, with the downlink channels transmitted and the uplink channels received. However, there are two main differences. Firstly, the base station's MAC protocol has to carry out transmissions to different mobiles on the downlink and receive transmissions from different mobiles on the uplink. Secondly, the protocol includes a

An Introduction to LTE: LTE, LTE-Advanced, SAE, VoLTE and 4G Mobile Communications, Second Edition.
Christopher Cox.

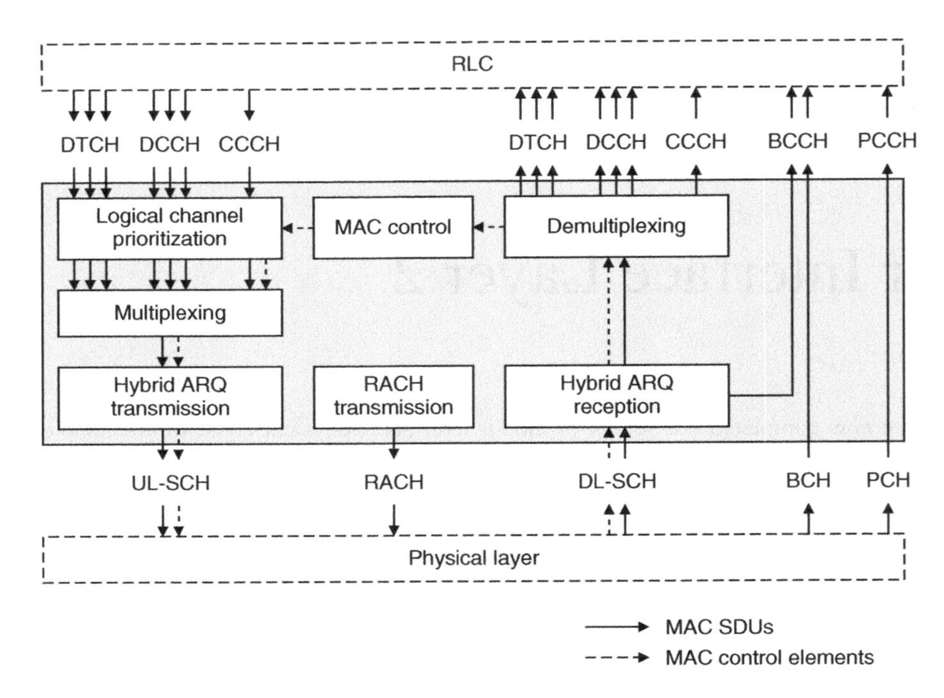

Figure 10.1 High-level architecture of the mobile's MAC protocol. Source: TS 36.321. Reproduced by permission of ETSI

scheduling function, which organizes the base station's transmissions on the downlink and the mobiles' transmissions on the uplink, and which ultimately determines the contents of the PDCCH scheduling commands and scheduling grants.

The MAC protocol also sends and receives a number of *MAC control elements*, which control the low-level operation of the physical layer. There are several types of control element, which are listed in Table 10.1. We have already seen three of them. During discontinuous reception in RRC_CONNECTED, the base station can send the mobile to sleep using a *DRX command*,

Table 10.1 List of MAC control elements

MAC control element	Release	Application	Direction
Buffer status report	R8	UE transmit buffer occupancy	UL
C-RNTI	R8	UE identification option during random access	
Power headroom	R8	UE transmit power headroom	
Extended power headroom	R10	Power headroom during carrier aggregation	
DRX command	R8	Sends UE to sleep during DRX	DL
Timing advance command	R8	Adjusts UE timing advance	
UE contention resolution identity	R8	Resolves contention during random access	
MCH scheduling information	R9	Informs UE about scheduling of MBMS	
Activation/deactivation	R10	Activates/deactivates secondary cells	

while the *UE contention resolution identity* and *C-RNTI* control elements are both used by the contention-based random access procedure. We will discuss the remaining Release 8 control elements below, before moving on to the remaining blocks in the diagram.

10.1.2 Timing Advance Commands

After initializing a mobile's timing advance using the random access procedure, the base station updates it using MAC control elements known as *Timing advance commands*. Each command adjusts the timing advance by an amount ranging from $-496T_s$ to $+512T_s$, with a resolution of $16T_s$ [3]. This corresponds to a change of −2.4 to +2.5 km in the distance between the mobile and the base station, with a resolution of 80 m.

The mobile expects to receive timing advance commands from the base station at regular intervals. The maximum permitted interval is a quantity known as *timeAlignmentTimer*, which can take a value from 500 to 10 240 subframes (0.5 to 10.24 s) or can be infinite if the cell size is small [4]. If the time elapsed since the previous timing advance command exceeds this value, then the mobile concludes that it has lost timing synchronization with the base station. In response, it releases all its PUSCH and PUCCH resources, notably the parameters $n^{(1)}_{\mathrm{PUCCH}}$, $n^{(1)}_{\mathrm{PUCCH,\,SRI}}$ and $n^{(2)}_{\mathrm{PUCCH}}$ from Chapter 8. Any subsequent attempt to transmit will trigger the random access procedure, through which the mobile can recover its timing synchronization.

10.1.3 Buffer Status Reporting

The mobile transmits *Buffer status report* (BSR) MAC control elements to tell the base station about how much data it has available for transmission. There are three types of buffer status report, of which the most important is the *regular BSR*. A mobile sends this in three situations: if data become ready for transmission when the transmit buffers were previously empty, or if data become ready for transmission on a logical channel with a higher priority than the buffers were previously storing, or if a timer expires while data are waiting for transmission. The mobile expects the base station to reply with a scheduling grant.

If the mobile wishes to send a regular BSR, but does not have the PUSCH resources on which to do so, then it instead sends the base station a scheduling request on the PUCCH. (In fact a scheduling request is always triggered in this way, by an inability to send a regular BSR.) If, however, the mobile is in RRC_IDLE or has lost timing synchronization with the base station, then it has no PUCCH resources either. In that situation, it runs the random access procedure instead.

There are two other types of buffer status report. The mobile transmits *periodic BSRs* at regular intervals during data transmission on the PUSCH and *padding BSRs* if it has enough spare room during a normal PUSCH transmission.

10.1.4 Power Headroom Reporting

A mobile's power headroom is the difference between its maximum transmit power and the power requested for its PUSCH transmission [5]. The power headroom is usually positive, but it can be negative if the requested power exceeds the power available.

The mobile reports its power headroom to the base station using a *Power headroom* MAC control element. It can do so in two situations: periodically, or if the downlink path loss has changed significantly since the last report. The base station uses the information to support its uplink scheduling procedure, typically by limiting the data rate at which it asks the mobile to transmit.

10.1.5 Multiplexing and De-multiplexing

We can now discuss the internal structure of a MAC protocol data unit (PDU) and the way in which it is assembled. In its most general form (Figure 10.2), a MAC protocol data unit contains several MAC service data units and several control elements, which collectively make up the MAC payload. Each service data unit contains data received from the RLC on a single logical channel, while each control element is one of the elements listed in Table 10.1. The payload can also contain padding, which rounds the protocol data unit up to one of the permitted transport block sizes.

Each item in the MAC payload is associated with a sub-header. The SDU sub-header identifies the size of the service data unit and the logical channel from which it originated, while the control element sub-header identifies the control element's size and type.

In the uplink, the mobile discovers the required PDU size from the base station's scheduling grant. Using the prioritization algorithm described below, the mobile decides how it will fill the available space in the protocol data unit, grabs service data units from the buffers in the overlying RLC protocol and grabs control elements from the MAC control unit. The multiplexing function then writes the corresponding sub-headers and assembles the PDU. The same technique is used in the downlink, except that, as we will see, the prioritization process is rather different.

10.1.6 Logical Channel Prioritization

We saw above that the base station tells the mobile about the size of every uplink MAC protocol data unit using its scheduling grant. However the scheduling grant says nothing about what the protocol data unit should contain. The mobile therefore runs a prioritization algorithm, to decide how to fill it.

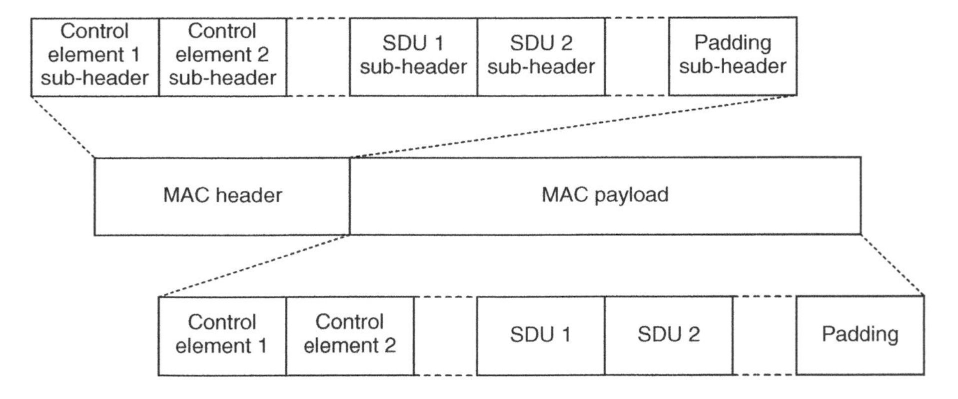

Figure 10.2 Structure of a MAC PDU. Source: TS 36.321. Reproduced by permission of ETSI

To support the algorithm, each logical channel is associated with a priority from 1 to 16, where a small number corresponds to a high priority. The logical channel is also associated with a *prioritized bit rate* (PBR), which runs from 0 to 256 kbps and is a target for the long-term average bit rate. The specifications also support an infinite prioritized bit rate, with the interpretation 'as fast as possible'. Ultimately, these parameters are all derived from the quality of service parameters that we will discuss in Chapter 13.

The algorithm is fully defined by the MAC protocol specification and the principles are as follows. First, the MAC runs through the logical channels in order of priority and grabs data from channels that have fallen behind their long-term average bit rates. It then runs through the channels in priority order once again and fills any space in the PDU that remains. The algorithm also prioritizes the control elements, and the resulting priority order is as follows: data on the common control channel together with any associated C-RNTI control elements, regular or periodic buffer status reports, power headroom reports, data on other logical channels and finally padding buffer status reports.

Prioritization in the downlink is rather different because the base station is free to fill up the PDUs in any way that it likes. In practice, the downlink prioritization algorithm will form part of the proprietary scheduling algorithm that we discuss below.

10.1.7 Scheduling of Transmissions on the Air Interface

The base station's scheduling algorithm has to decide the contents of every downlink scheduling command and uplink scheduling grant, on the basis of all the information available to it at the time. The specifications say nothing about how it should work, but to illustrate its operation, Figure 10.3 shows some of the main inputs and outputs.

The information available to the downlink scheduler includes the following. Each bearer is associated with a buffer occupancy, as well as information about its quality of service such as the priority and prioritized bit rate that we introduced earlier. To support the scheduling

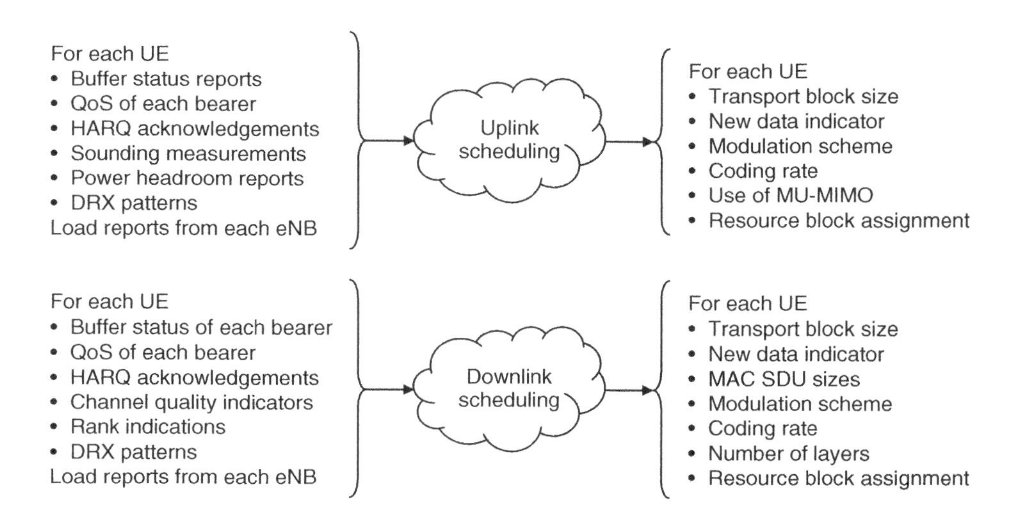

Figure 10.3 Inputs and outputs for the uplink and downlink scheduling algorithms

function, the mobile returns hybrid ARQ acknowledgements, channel quality indicators and rank indications. The base station also knows the discontinuous reception pattern for every mobile in the cell and can receive load information from nearby cells about their own use of the downlink sub-carriers.

Using this information, the scheduler has to decide how many information bits to send to each mobile, whether to send a new transmission or a re-transmission and how to divide new transmissions amongst the available bearers. It also has to decide the modulation schemes and coding rates to use, the number of layers in the case of spatial multiplexing, and the allocation of resource blocks to every mobile.

The uplink scheduler follows the same principles, although some of the inputs and outputs are different. For example, the base station does not have complete knowledge of the uplink buffer occupancy and does not tell the mobiles which logical channels they should use for their uplink transmissions. In addition, the base station derives its channel quality information from the sounding procedure instead of from the mobiles' channel quality indicators.

At a basic level, two extreme scheduling algorithms are available. A *maximum rate scheduler* allocates resources to the mobiles with the highest signal-to-noise ratios, which can transmit or receive at the highest data rates. This maximizes the throughput of the cell but is grossly unfair, as distant mobiles may not get a chance to transmit or receive at all. At the other extreme, a *round-robin scheduler* gives the same data rate to every mobile. This is appropriate for real-time, constant data rate services such as live video. However it is inefficient for non-real-time, variable data rate services such as web browsing, because distant mobiles with low signal-to-noise ratios will dominate the cell's use of resources. In practice, techniques such as *proportional fair scheduling* try to strike a balance between the two extremes. In a simple example, a base station might allocate the same number of resource blocks to every mobile that is using a particular service. Nearby mobiles would then have a higher data rate than distant ones through the use of a faster modulation scheme and a higher coding rate, but distant mobiles would still have some of the cell's resources.

To summarize, the scheduler is a complex piece of software and one of the harder parts of the system to implement effectively. It is likely to be an important differentiator between equipment manufacturers and network operators.

10.2 Radio Link Control Protocol

10.2.1 Protocol Architecture

The radio link control (RLC) protocol [6, 7] maintains the layer 2 data link between the mobile and the base station, for example by ensuring reliable delivery for data streams that have to reach the receiver correctly. Figure 10.4 shows the high-level architecture of the RLC. The transmitter receives service data units from higher layers in the form of modified IP packets or signalling messages and sends PDUs to the MAC protocol on the logical channels. The process is reversed in the receiver.

The RLC has three modes of operation, namely *transparent mode* (TM), *unacknowledged mode* (UM) and *acknowledged mode* (AM). These are set up on a channel-by-channel basis so that each logical channel is associated with an RLC object that is configured in one of these modes. The transparent and unacknowledged mode RLC objects are uni-directional, while the acknowledged mode object is bi-directional.

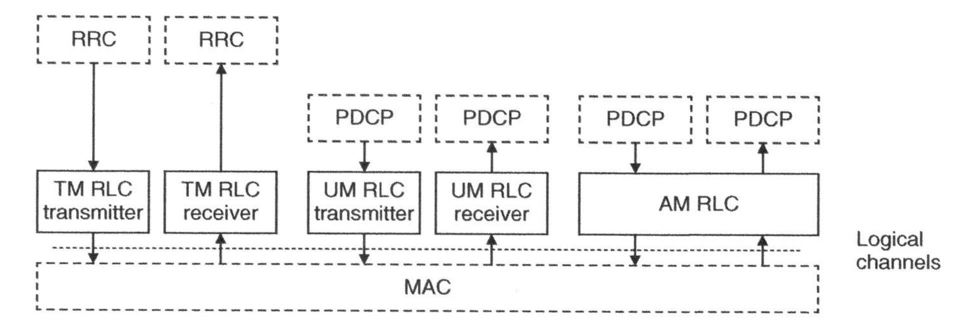

Figure 10.4 High-level architecture of the RLC protocol. Source: TS 36.322. Reproduced by permission of ETSI

10.2.2 Transparent Mode

Transparent mode handles three types of signalling message: system information messages on the broadcast control channel, paging messages on the paging control channel and RRC connection establishment messages on the common control channel. Its architecture (Figure 10.5) is very simple.

In the transmitter, the RLC receives signalling messages directly from the RRC protocol and stores them in a buffer. The MAC protocol grabs the messages from the buffer as RLC PDUs, without any modification. (The messages are short enough to fit into a single transport block, without segmenting them.) In the receiver, the RLC passes the received messages directly up to the RRC.

10.2.3 Unacknowledged Mode

Unacknowledged mode handles data streams on the dedicated traffic channel for which timely delivery is more important than reliability, such as voice over IP and streaming video. Its architecture is shown in Figure 10.6.

The RLC transmitter receives service data units from the PDCP in the form of modified IP packets and stores them in a buffer in the same way as before. This time, however, the MAC

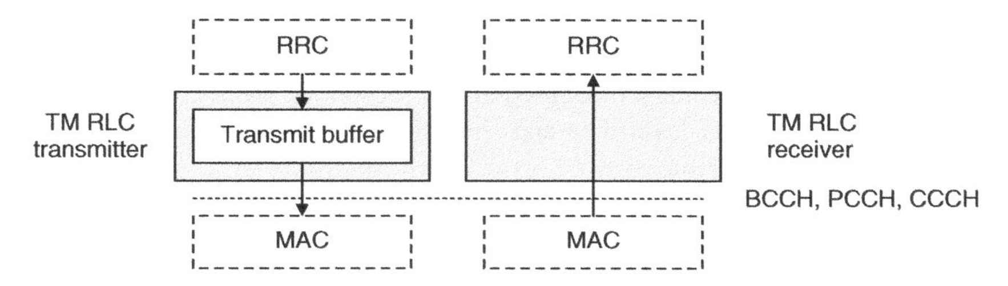

Figure 10.5 Internal architecture of the RLC protocol in transparent mode. Source: TS 36.322. Reproduced by permission of ETSI

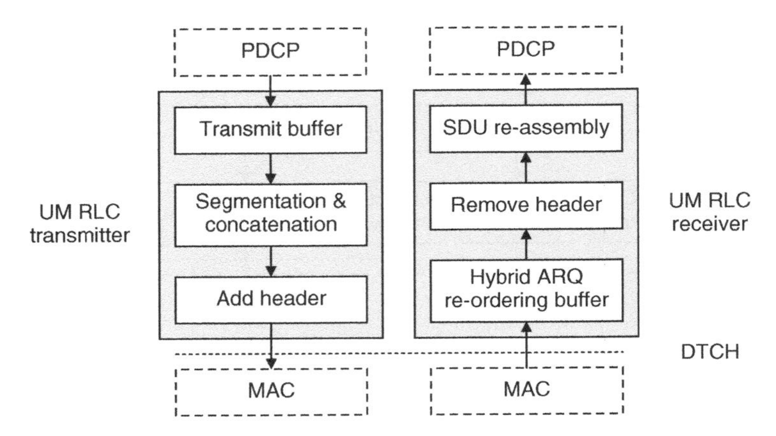

Figure 10.6 Internal architecture of the RLC protocol in unacknowledged mode. Source: TS 36.322. Reproduced by permission of ETSI

protocol tells the RLC to send a PDU with a specific size. In response, the segmentation and concatenation function cuts up the buffered IP packets and splices their ends together, so as to deliver a PDU with the correct size down to the MAC. As a result, the output PDU size does not bear any resemblance to the size of the input SDU. Finally, the RLC adds a header that contains two important pieces of information: a PDU sequence number and a description of any segmentation and concatenation that it has done.

The PDUs can reach the receiver's RLC protocol in a different order because of the underlying hybrid ARQ processes. To deal with this problem, the hybrid ARQ re-ordering function stores the received PDUs in a buffer and uses their sequence numbers to send them upwards in the correct order. The receiver can then remove the header from every PDU, use the header information to undo the segmentation and concatenation process and reconstruct the original packets.

10.2.4 Acknowledged Mode

Acknowledged mode handles two types of information: data streams on the dedicated traffic channel such as web pages and emails, for which reliability is more important than speed of delivery, and mobile-specific signalling messages on the dedicated control channel. It is similar to unacknowledged mode but also re-transmits any packets that have not reached the receiver correctly. The architecture (Figure 10.7) is bi-directional, in the sense that the same acknowledged mode object handles transmission and reception. There are also two types of PDU: data PDUs carry higher-layer data and signalling messages, and control PDUs carry RLC-specific control information.

The transmitter sends data packets in a similar way to the unacknowledged mode RLC. This time, however, it stores the transmitted PDUs in a re-transmission buffer, until it knows that they have reached the receiver correctly. At regular intervals, the transmitter also sets a *polling*

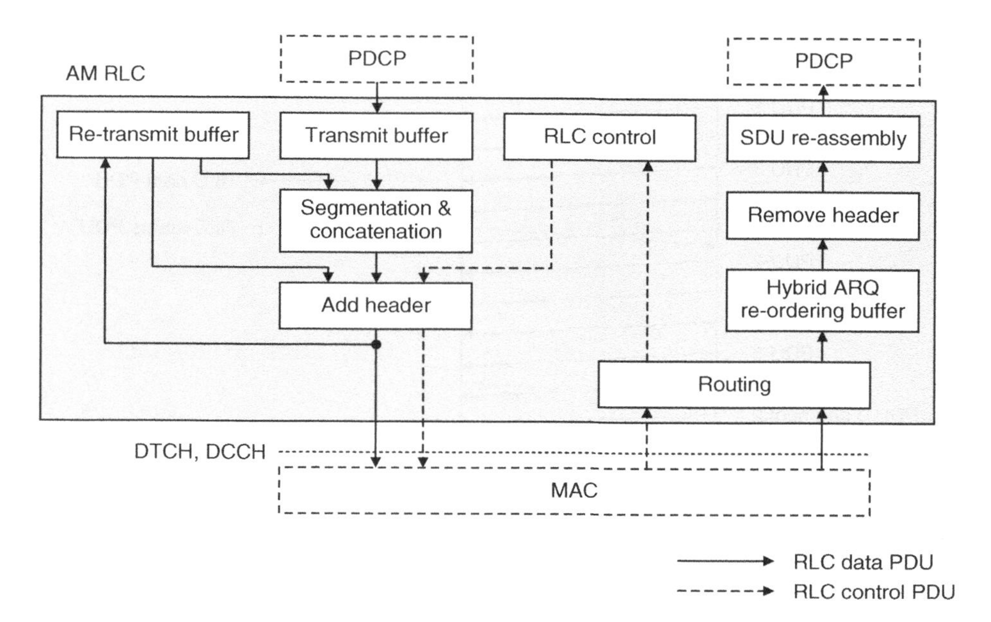

Figure 10.7 Internal architecture of the RLC protocol in acknowledged mode. Source: TS 36.322. Reproduced by permission of ETSI

bit in one of the data PDU headers. This tells the receiver to return a type of control PDU known as a *status PDU*, which lists the data PDUs it has received and the ones it has missed. In response, the transmitter discards the data PDUs that have arrived correctly and re-transmits the ones that have not.

There is one problem. If the SINR is falling, then the MAC protocol may request a smaller PDU size for the re-transmission than it did first time around. The PDUs in the re-transmission buffer will then be too large to send. To solve the problem, the RLC protocol can cut a previously transmitted PDU into smaller segments. To support this process, the data PDU header includes extra fields, which describe the position of a re-transmitted segment within a previously transmitted PDU. The receiver can acknowledge each segment individually, using similar fields in the status PDU.

Figure 10.8 shows an example. At the start of the sequence, the transmitter sends four data PDUs to the receiver and labels each one with a sequence number. PDUs 1 and 4 reach the receiver correctly, but PDUs 2 and 3 are lost. (To be exact, the hybrid ARQ transmitter reaches its maximum number of re-transmissions and moves on to the next PDU.)

The transmitter sets a polling bit in PDU 4 and the receiver replies by returning a status PDU. The transmitter can re-transmit PDU 2, but a fall in the SINR means that PDU 3 is now too large. In response, the transmitter cuts PDU 3 into two segments and re-transmits them individually. The first segment of PDU 3 arrives correctly, but the second is lost. In response to another status PDU, the transmitter can discard the first segment and can re-transmit the second.

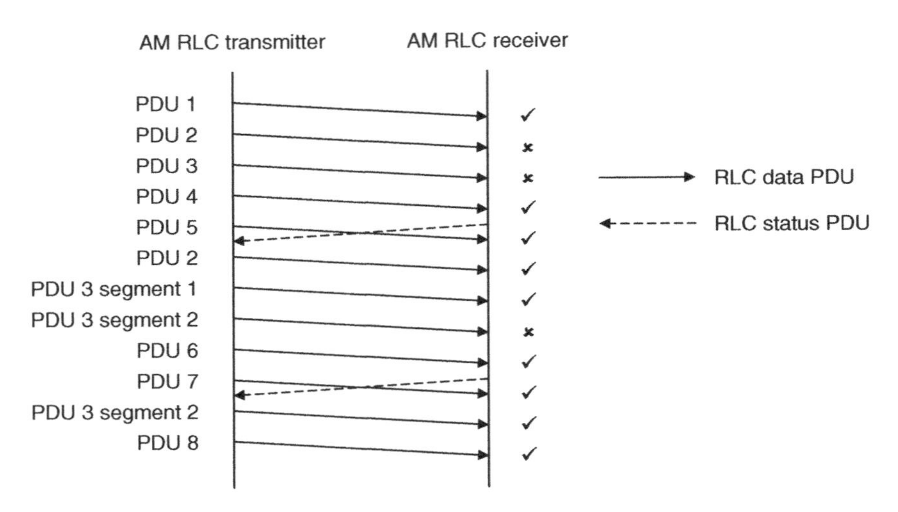

Figure 10.8 Operation of transmission, re-segmentation and re-transmission in RLC acknowledged mode

10.3 Packet Data Convergence Protocol

10.3.1 Protocol Architecture

The packet data convergence protocol (PDCP) [8, 9] supports data streams that are using RLC acknowledged mode, by ensuring that none of their packets are lost during a handover. It also manages three air interface functions, namely header compression, ciphering and integrity protection. We will cover header compression here, while leaving ciphering and integrity protection until our discussion of security in Chapter 12.

The PDCP is only used by the dedicated traffic and control channels, for which the underlying RLC protocol is operating in unacknowledged or acknowledged mode. As shown in Figure 10.9, there are different architectures for data and signalling.

In the user plane, the transmitter receives PDCP service data units in the form of IP packets, adds a PDCP sequence number, and stores any packets that are using RLC acknowledged mode in a re-transmission buffer. It then compresses the IP headers, ciphers the information, adds a PDCP header and outputs the resulting PDU. The receiver reverses the process, stores incoming packets in the receive buffer, and uses the sequence number to deliver them in the correct order to higher layers.

In the control plane, the signalling messages are protected by an extra security procedure, known as integrity protection. There is no re-transmission buffer, because the re-transmission function is only used for data and there are no IP headers to compress.

10.3.2 Header Compression

Headers can make up a large proportion of a slow packet data stream. In the case of voice over IP, for example, the narrowband *adaptive multi-rate* (AMR) *codec* has a bit rate of 4.75

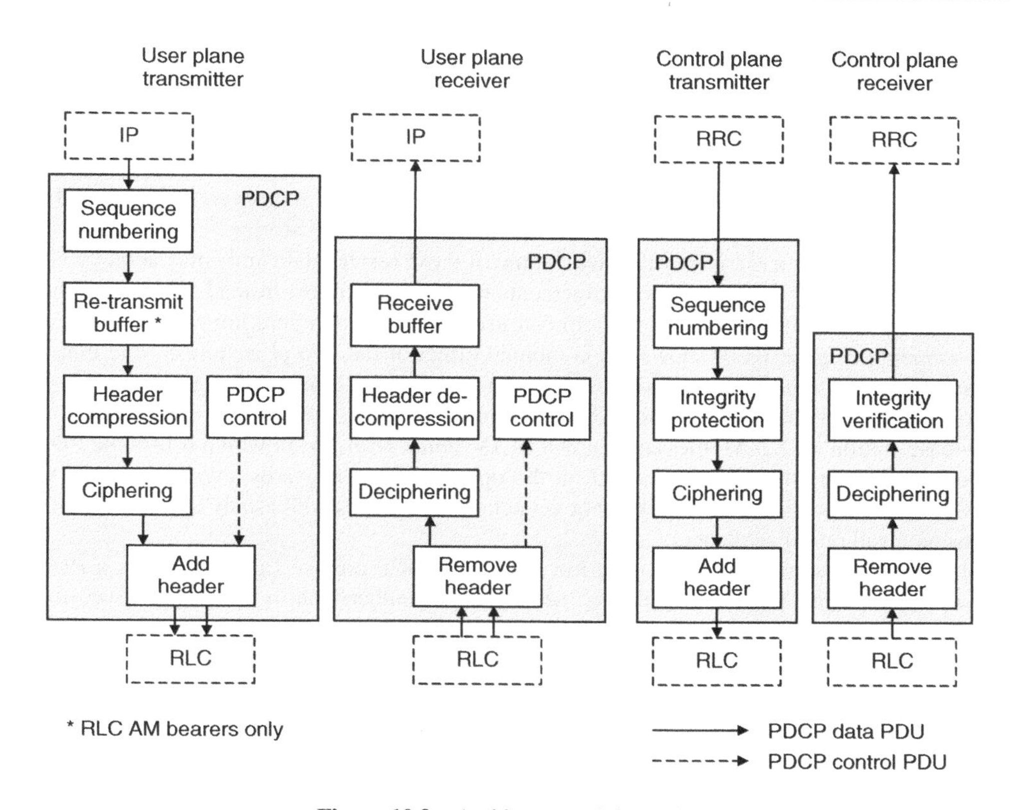

Figure 10.9 Architecture of the PDCP

to 12.2 kbps and a packet duration of 20 ms, giving a typical packet size of around 15 to 30 bytes [10]. However, the header typically contains 40 to 60 bytes, comprising 12 bytes from the *real-time protocol* (RTP), 8 bytes from UDP and either 20 bytes from IP version 4 or 40 bytes from IP version 6. The resulting overhead does not matter so much in the evolved packet core, which is likely to be dominated by fast data services. However it is inappropriate across the air interface, because the wireless link acts as a bottleneck and because an individual cell might occasionally be dominated by voice calls.

To solve the problem, the PDCP includes an IETF protocol known as *robust header compression* (ROHC) [11]. The principle is that the transmitter sends the full header in the first packet, but only sends differences in subsequent packets. Most of the header stays the same from one packet to the next, so the difference fields are considerably smaller. The protocol can compress the original 40 and 60 byte headers to as little as 1 and 3 bytes respectively, which greatly reduces the overhead.

Robust header compression has an advantage over other header compression protocols, in that it is designed to work well even if the underlying rate of packet loss is high. This makes it suitable for an unreliable data link such as the LTE air interface, particularly for real-time data streams, such as voice over IP, that are using RLC unacknowledged mode.

10.3.3 Prevention of Packet Loss during Handover

When transmitting data streams that are using RLC acknowledged mode, the PDCP stores each service data unit in a re-transmission buffer, until the RLC tells it that the SDU has been successfully received. During a handover, the processes of transmission and reception are briefly interrupted, so there is a risk of packet loss. On completion of the handover, the PDCP solves the problem by re-transmitting any service data units that it is still storing.

There is, however, a secondary problem: some of those service data units may actually have reached the receiver, but the acknowledgements may have been lost instead. To prevent them from being transmitted twice, the system can use a second procedure known as *PDCP status reporting*. Figure 10.10 shows the combined effect of the two procedures. Note that the messages in this figure apply only to bearers that are using RLC acknowledged mode.

As part of the handover procedure described in Chapter 14, the old base station sends the new base station an X2-AP message known as *SN Status Transfer*, in which it lists the PDCP sequence numbers that it has received on the uplink. It also forwards any downlink PDCP service data units that the mobile has not yet acknowledged, as well as any uplink SDUs that it has received out of sequence.

The new base station can now send the mobile a PDCP control PDU known as a *PDCP Status Report* (step 1), in which it lists the sequence numbers that it has just received from the old base station. The mobile can delete these from its re-transmission buffer and only has to re-transmit the remainder (step 3). At the same time, the mobile can send a PDCP Status Report to the new base station (step 2), in which it lists the PDCP sequence numbers that it has received on the downlink. The new base station can delete these in the same way, before beginning its own re-transmission (step 4).

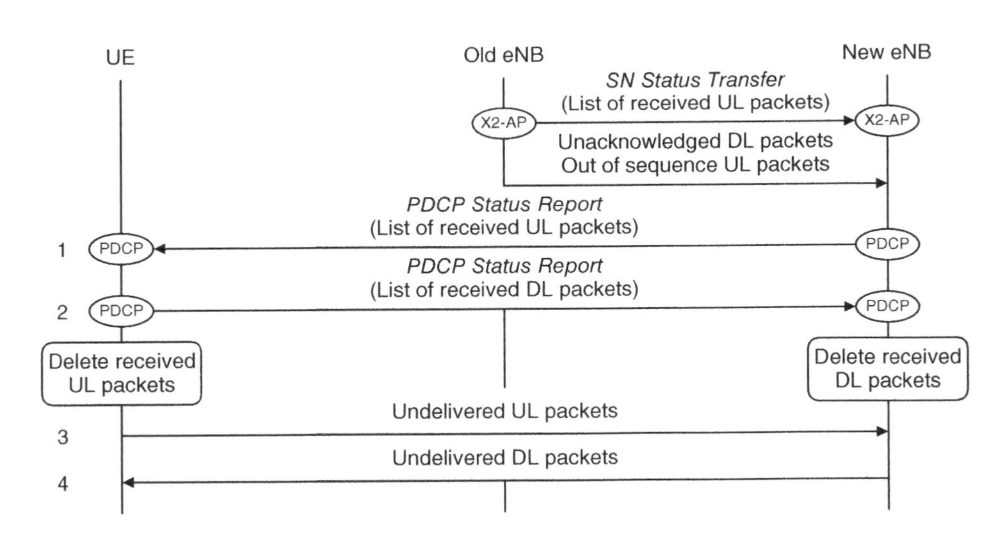

Figure 10.10 PDCP status reporting and prevention of packet loss after a handover

References

1. 3GPP TS 36.321 (2013) Medium Access Control (MAC) Protocol Specification, Release 11, July 2013.
2. 3GPP TS 36.331 (2013) Radio Resource Control (RRC) Protocol Specification, Release 11, Section 6.3.2 (Logical Channel Config, MAC-Main Config), September 2013.
3. 3GPP TS 36.213 (2013) Physical Layer Procedures, Release 11, Section 4.2.3, September 2013.
4. 3GPP TS 36.331 (2013) Radio Resource Control (RRC) Protocol Specification, Release 11, Section 6.3.2 (Time Alignment Timer), September 2013.
5. 3GPP TS 36.133 (2013) Evolved Universal Terrestrial Radio Access (E-UTRA); Requirements for Support of Radio Resource Management, Release 11, Section 9.1.8, September 2013.
6. 3GPP TS 36.322 (2012) Radio Link Control (RLC) Protocol Specification, Release 11, September 2012.
7. 3GPP TS 36.331 (2013) Radio Resource Control (RRC) Protocol Specification, Release 11, Section 6.3.2 (RLC-Config), September 2013.
8. 3GPP TS 36.323 (2013) Packet Data Convergence Protocol (PDCP) Specification, Release 11, March 2013.
9. 3GPP TS 36.331 (2013) Radio Resource Control (RRC) Protocol Specification, Release 11, Section 6.3.2 (PDCP-Config), September 2013.
10. 3GPP TS 26.101 (2012) Mandatory Speech Codec Speech Processing Functions; Adaptive Multi-Rate (AMR) Speech Codec Frame Structure, Release 11, Annex A, September 2012.
11. IETF RFC 4995 (2007) The RObust Header Compression (ROHC) Framework, July 2007.

11

Power-On and Power-Off Procedures

We have now completed our discussion of the LTE air interface. In the course of the next seven chapters, we will cover the signalling procedures that govern the high-level operation of LTE.

In this chapter, we describe the procedures that a mobile follows after it switches on, to select a cell and register its location with the network. We begin by reviewing the procedure at a high level and then continue with the three main steps of network and cell selection, RRC connection establishment and registration with the evolved packet core. A final section describes the detach procedure, through which the mobile switches off. As part of the chapter, we will refer to several of the low-level procedures that we have already encountered, notably the ones for cell acquisition and random access.

Network and cell selection are covered by two specifications, which describe the idle mode procedures for the non-access stratum [1] and the access stratum [2]. The signalling procedures in the rest of the chapter are summarized by the usual stage 2 specifications for LTE [3, 4]. Readers who require more detail about these procedures can find it by digging down into the relevant stage 3 signalling specifications [5–9], which define both the procedures on each individual interface and the contents of every signalling message.

11.1 Power-On Sequence

Figure 11.1 summarizes the procedures that the mobile follows after it switches on. The mobile begins by running the procedure for network and cell selection, which has three steps. In the first step, the mobile selects a public land mobile network (PLMN) that it will register with. In the second step, the mobile can optionally ask the user to select a closed subscriber group (CSG) for registration. In the third, the mobile selects a cell that belongs to the selected network and if necessary to the selected CSG. In doing so, it is said to *camp* on the cell.

The mobile then contacts the corresponding base station using the contention-based random access procedure from Chapter 9 and initiates the procedure for RRC connection

An Introduction to LTE: LTE, LTE-Advanced, SAE, VoLTE and 4G Mobile Communications, Second Edition.
Christopher Cox.

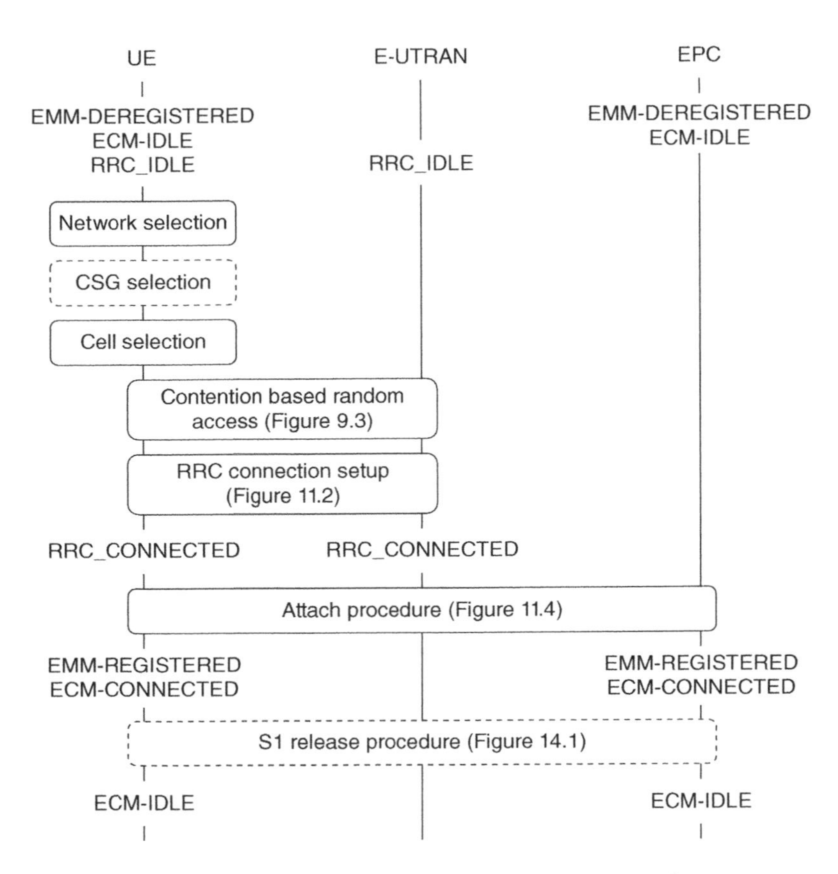

Figure 11.1 Overview of the mobile's power-on procedures

establishment. During the RRC procedure, the mobile establishes a signalling connection with the selected base station, configures signalling radio bearer 1 and moves from RRC_IDLE into RRC_CONNECTED. It also acquires a set of parameters through which it can communicate with the base station, such as a set of resources for the transmission of uplink control information on the physical uplink control channel (PUCCH).

In the final step, the mobile uses the attach procedure to contact the evolved packet core. As a result of that procedure, the mobile registers its location with a mobility management entity (MME) and moves to the states EMM-REGISTERED and ECM-CONNECTED. It also configures signalling radio bearer 2, acquires an IP address and establishes a default bearer through which it can communicate with the outside world.

The mobile is now in the states EMM-REGISTERED, ECM-CONNECTED and RRC_CONNECTED, and will stay in those states for as long as it is exchanging data with the network. If the user does nothing, then the network can transfer the mobile into ECM-IDLE and RRC_IDLE, using a procedure known as S1 release that is covered in Chapter 14.

11.2 Network and Cell Selection

11.2.1 Network Selection

In the network selection procedure [10–12], the mobile selects a public land mobile network (PLMN) that it will register with. To start the procedure, the mobile equipment interrogates the USIM and retrieves the globally unique temporary identity (GUTI) that it was using when last switched on, as well as the tracking area identity in which it was registered. From these quantities, it can identify the corresponding network, which is known as the *registered PLMN*. The mobile runs the CSG and cell selection procedures described below, in the hope of finding a suitable cell that belongs to the registered PLMN.

If the mobile cannot find the registered PLMN, then it scans all the LTE carrier frequencies that it supports and identifies the networks that it can actually find. To do this, the mobile uses the acquisition procedure from Chapter 7 to find the strongest LTE cell on each frequency, reads SIB 1 from its system information and identifies the network or networks that the cell belongs to. If the mobile also supports UMTS, GSM or cdma2000, then it runs a similar procedure to find networks that are using those radio access technologies.

There are then two network selection modes, automatic and manual. In *automatic mode*, the mobile runs in priority order through a list of networks that it should treat as home PLMNs, together with an associated list of radio access technologies. (These lists are both stored on the USIM.) When it encounters a network that it has previously found, the mobile runs the CSG and cell selection procedures in the manner described below.

If the mobile cannot find a home PLMN, then it repeats the procedure using first any user-defined list of networks and radio access technologies, and then any operator-defined list. If it cannot find any of those networks, then the mobile tries to select a cell from any network that is available. In this last case, it enters a limited service state, in which it can only make emergency calls and receive warnings from the earthquake and tsunami warning system.

In *manual mode*, the mobile presents the user with the list of networks that it has found, using the same priority order as in automatic mode. The user selects a preferred network and the mobile proceeds to the CSG and cell selection procedures as before.

11.2.2 Closed Subscriber Group Selection

A home base station is a base station that is controlling a femtocell, which may be accessible only by registered subscribers. To support this restriction, the base station can be associated with a closed subscriber group and a home eNB name, which it advertises in SIB 1 and SIB 9 respectively. Each USIM lists any closed subscriber groups that the subscriber is allowed to use [13], together with the identities of the corresponding networks.

If the USIM contains any closed subscriber groups, then the mobile has to run an additional procedure, known as *CSG selection* [14, 15]. The procedure has two modes of operation, automatic and manual, which are distinct from the network selection modes described above. In *automatic mode*, the mobile sends the list of allowed closed subscriber groups to the cell selection procedure, which selects either a non-CSG cell or a cell whose CSG is in the list. *Manual mode* is more restrictive. Here, the mobile identifies the CSG cells that it can find in the selected network. It presents this list to the user, indicates the corresponding home eNB names and indicates whether each CSG is in the list of allowed CSGs. The user selects a preferred closed subscriber group and the mobile selects a cell belonging to that CSG.

11.2.3 Cell Selection

During the *cell selection* procedure [16], the mobile selects a *suitable cell* that belongs to the selected network and, if necessary, to the selected closed subscriber group. It can do this in two ways. Usually, it has access to stored information about potential LTE carrier frequencies and cells, either from when it was last switched on or from the network selection procedure described above. If this information is unavailable, then the mobile scans all the LTE carrier frequencies that it supports and identifies the strongest cell on each carrier that belongs to the selected network.

A suitable cell is a cell that satisfies several criteria. The most important is the cell selection criterion:

$$S_{\text{rxlev}} > 0 \tag{11.1}$$

During initial network selection, the mobile calculates S_{rxlev} as follows:

$$S_{\text{rxlev}} = Q_{\text{rxlevmeas}} - Q_{\text{rxlevmin}} - P_{\text{compensation}} \tag{11.2}$$

In this equation, $Q_{\text{rxlevmeas}}$ is the cell's *reference signal received power* (RSRP), which is the average power per resource element that the mobile is receiving on the cell-specific reference signals [17]. Q_{rxlevmin} is a minimum value for the RSRP, which the base station advertises in SIB 1. These quantities ensure that a mobile will only select the cell if it can hear the base station's transmissions on the downlink. The final parameter, $P_{\text{compensation}}$, is calculated as follows:

$$P_{\text{compensation}} = \max(P_{\text{EMAX}} - P_{\text{PowerClass}}, 0) \tag{11.3}$$

Here, P_{EMAX} is an upper limit on the transmit power that a mobile is allowed to use, which the base station advertises as part of SIB 1. $P_{\text{PowerClass}}$ is the mobile's intrinsic maximum power, which the specifications limit to 23 dB relative to 1 mW (dBm), in other words about 200 mW. By combining these quantities, $P_{\text{compensation}}$ reduces the value of S_{rxlev} if the mobile cannot reach the power limit that the base station is assuming. It therefore ensures that a mobile will only select the cell if the base station can hear it on the uplink.

The cell selection procedure is enhanced in release 9 of the 3GPP specifications, so that a suitable cell also has to satisfy the following criterion:

$$S_{\text{qual}} > 0 \tag{11.4}$$

where

$$S_{\text{qual}} = Q_{\text{qualmeas}} - Q_{\text{qualmin}} \tag{11.5}$$

In this equation, Q_{qualmeas} is the *reference signal received quality* (RSRQ), which expresses the signal-to-interference plus noise ratio of the cell-specific reference signals. Q_{qualmin} is a minimum value for the RSRQ, which the base station advertises in SIB 1 as before. This condition prevents a mobile from selecting a cell on a carrier frequency that is subject to high levels of interference.

A suitable cell must also satisfy a number of other criteria. If the USIM contains a list of closed subscriber groups, then the cell has to meet the criteria for automatic or manual CSG selection that we defined above. If the USIM does not, then the cell must lie outside any closed subscriber groups. In addition, the network operator can bar a cell to all users or reserve it for operator use, by means of flags in SIB 1.

11.3 RRC Connection Establishment

11.3.1 Basic Procedure

Once the mobile has selected a network and a cell to camp on, it runs the contention-based random access procedure from Chapter 9. In doing so, it obtains a C-RNTI, an initial value for the timing advance and resources on the physical uplink shared channel (PUSCH) through which it can send a message to the network.

The mobile can then begin a procedure known as *RRC connection establishment* [18]. Figure 11.2 shows the message sequence. In step 1, the mobile's RRC protocol composes a message known as an *RRC Connection Request*. In this message, it specifies two parameters. The first is a unique non-access stratum (NAS) identity, either the S-TMSI (if the mobile was registered in the cell's tracking area when last switched on) or a randomly chosen value (otherwise). The second is the establishment cause, which can be mobile originated signalling (as in this example), mobile originated data, mobile terminated access (a response to paging), high priority access, or an emergency call.

The mobile transmits the message using signalling radio bearer 0, which has a simple configuration that the base station advertises in SIB 2. The message is sent on the common control channel, the uplink shared channel and the physical uplink shared channel.

The base station reads the message, takes on the role of serving eNB and composes a reply known as an *RRC Connection Setup* (step 2). In this message, it configures the mobile's physical layer and MAC protocols, as well as SRB 1. These configurations include several parameters that we have already seen. For example, the physical layer parameters include the PUCCH resources $n^{(1)}_{\text{PUCCH, SRI}}$ and $n^{(2)}_{\text{PUCCH}}$, the CQI reporting mode and the radio network temporary identities TPC-PUCCH-RNTI and TPC-PUSCH-RNTI. Similarly, the MAC parameters include the time alignment timer, the timer for periodic buffer status reports and the maximum number of hybrid ARQ transmissions on the uplink. Finally, the parameters for SRB 1 include the priorities and prioritized bit rates of the corresponding logical channels and

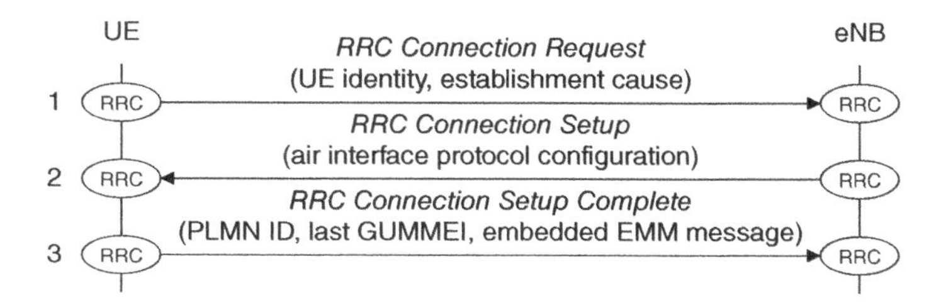

Figure 11.2 RRC connection establishment procedure. Source: TS 36.331. Reproduced by permission of ETSI

the parameters that govern polling and status reporting in the RLC. To reduce the size of the message, the base station can set several parameters to default values that are defined in the specifications. The base station transmits its message on SRB 0, as before, because the mobile does not yet understand the configuration of SRB 1.

The mobile reads the message, configures its protocols in the manner required and moves into RRC_CONNECTED. It then writes a confirmation message known as *RRC Connection Setup Complete* (step 3) and transmits it on SRB 1. In the message, the mobile includes three information elements. The first identifies the PLMN that it would like to register with. The second is the globally unique identity of the MME that was previously serving the mobile, which the mobile has extracted from its GUTI. The third is an embedded EPS mobility management message, which the base station will eventually forward to the MME. In this example, the embedded message is an attach request, but it can also be a detach request, a service request or a tracking area update request.

The RRC connection establishment procedure is also used later on, whenever a mobile in RRC_IDLE wishes to communicate with the network. We will see several examples in the chapters that follow.

11.3.2 Relationship with Other Procedures

As shown in Figure 11.3, the RRC connection establishment procedure overlaps with two other procedures, namely, the random access procedure that precedes it and the EPS mobility management (EMM) procedure that follows.

The mobile sends its RRC Connection Request in the third step of the contention-based random access procedure. The base station therefore uses the message in two ways: it echoes back the message during contention resolution and it replies to the message with its RRC Connection Setup. Similarly, the message RRC Connection Setup Complete is also the first step of the EMM procedure that follows. The base station accepts the RRC message as an acknowledgement of its RRC Connection Setup and forwards the embedded EMM message

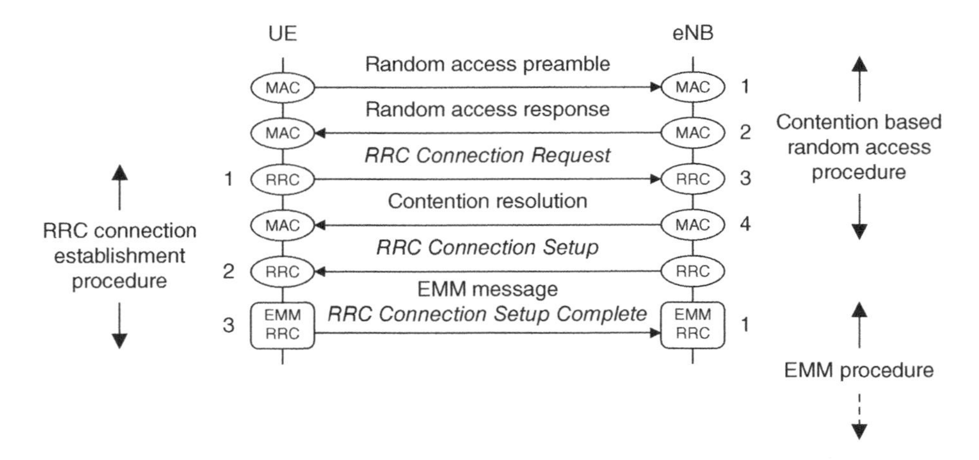

Figure 11.3 Relationships between RRC connection establishment and other procedures

to a suitable MME. Overlapping the procedures makes it harder to follow what is going on, but it brings a big advantage: it makes the signalling delays lower than in earlier systems, which helps the system to meet the latency requirements that were laid out in Chapter 1.

There is one final point to make. In Figures 11.2 and 11.3, we have only shown the high-level signalling messages that are transmitted on the PUSCH and PDSCH. We have omitted the lower-level control information on the PUCCH, PDCCH and PHICH, as well as the possibility of re-transmissions. We will follow this convention throughout the rest of the book, but it is worth remembering that the full sequence of air interface messages may be considerably longer than the resulting figures imply.

11.4 Attach Procedure

11.4.1 IP Address Allocation

During the attach procedure, the mobile acquires an IP version 4 address and/or an IP version 6 address, which it will subsequently use to communicate with the outside world. Before looking at the attach procedure itself, it is useful to discuss the methods that the network can use for IP address allocation [19].

IPv4 addresses are 32 bits long. In the usual technique, the PDN gateway allocates a dynamic IPv4 address to the mobile as part of the attach procedure. It can either allocate the IP address by itself or acquire a suitable IP address from a *dynamic host configuration protocol version 4* (DHCPv4) server. As an alternative, the mobile can itself use DHCPv4 to acquire a dynamic IP address after the attach procedure has completed. To do this, it contacts the PDN gateway over the user plane, with the PDN gateway acting as a DHCPv4 server towards the mobile.

Because of the shortage of IPv4 addresses, the allocated address is usually a private IP address that is invisible to the outside world. Using *network address translation* (NAT) [20], the system can map this address to a public IP address that is shared amongst several mobiles plus a mobile-specific TCP or UDP port number. The PDN gateway can either do the task of network address translation itself or delegate it to a separate physical device.

IPv6 addresses are 128 bits long and have two parts, namely a 64 bit network prefix and a 64 bit interface identifier. They are allocated using a procedure known as *IPv6 stateless address auto-configuration* [21]. In LTE's implementation of the procedure, the PDN gateway assigns the mobile a globally unique IPv6 prefix during the attach procedure, as well as a temporary interface identifier. It passes the interface identifier back to the mobile, which uses it to construct a temporary link-local IPv6 address. After the attach procedure has completed, the mobile uses the temporary address to contact the PDN gateway over the user plane and retrieve the IPv6 prefix, in a process known as router solicitation. It then uses the prefix to construct a full IPv6 address. Because the prefix is globally unique, the mobile can actually do this using any interface ID that it likes.

A mobile can also use a static IPv4 address or IPv6 prefix. The mobile does not store these permanently, however; instead, the network stores them in the home subscriber server or a DHCP server and sends them to the mobile during the attach procedure. Static IP addresses are unusual in the case of IPv4, due to the shortage of IPv4 addresses.

Subsequently, the mobile will use the same IP address for any dedicated bearers that it sets up with the same packet data network. If it establishes communications with another packet data network, then it will acquire another IP address using the same technique.

11.4.2 Overview of the Attach Procedure

The *attach procedure* has four main objectives. The mobile uses the procedure to register its location with a serving MME. The network configures signalling radio bearer 2, which carries subsequent non-access stratum signalling messages across the air interface. The network also gives the mobile an IP version 4 address and/or an IP version 6 address, using either or both of the techniques described above, and sets up a default EPS bearer, which provides the mobile with always-on connectivity to a default PDN.

Figure 11.4 summarizes the attach procedure. We will run through the individual steps of the procedure in the following sections, for the case where the S5/S8 interface is using the GPRS tunnelling protocol (GTP). In this figure and the ones that follow, solid lines show mandatory messages, while dashed lines indicate messages that are optional or conditional. The message numbers are the same as in TS 23.401 [22], a convention that we will follow for several of the other procedures in the book.

11.4.3 Attach Request

Figure 11.5 shows the first two steps of the procedure, which cover the mobile's attach request. The mobile starts by running the contention-based random access procedure and the first two steps of RRC connection establishment, in the manner described earlier.

The mobile then composes an EPS session management (ESM) message, *PDN Connectivity Request*, which asks the network to establish a default EPS bearer. The message includes a PDN type, which indicates whether the mobile supports IPv4, IPv6 or both. It can also include a set of protocol configuration options, which list any parameters that relate to the external network, such as a preferred access point name or a request to receive an IPv4 address over the

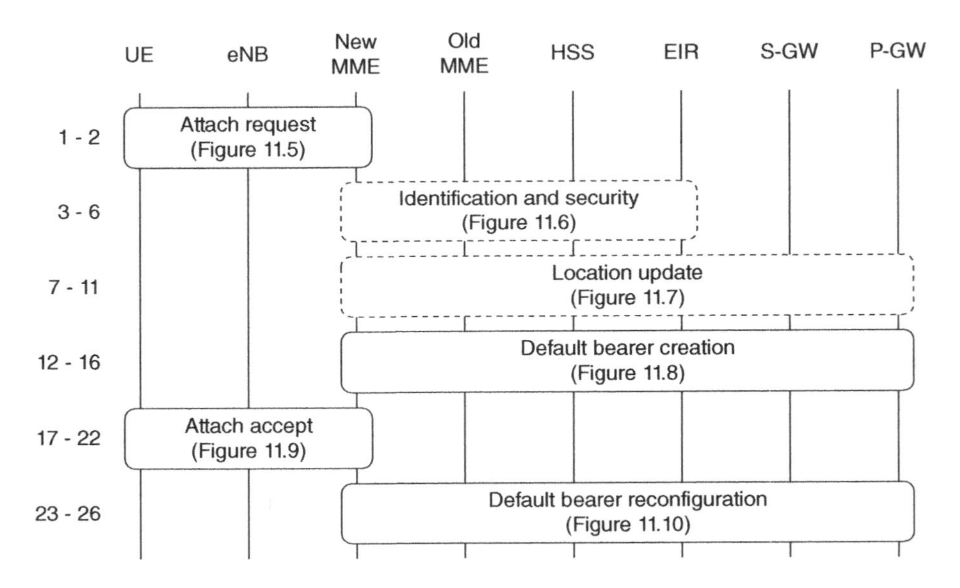

Figure 11.4 Overview of the attach procedure

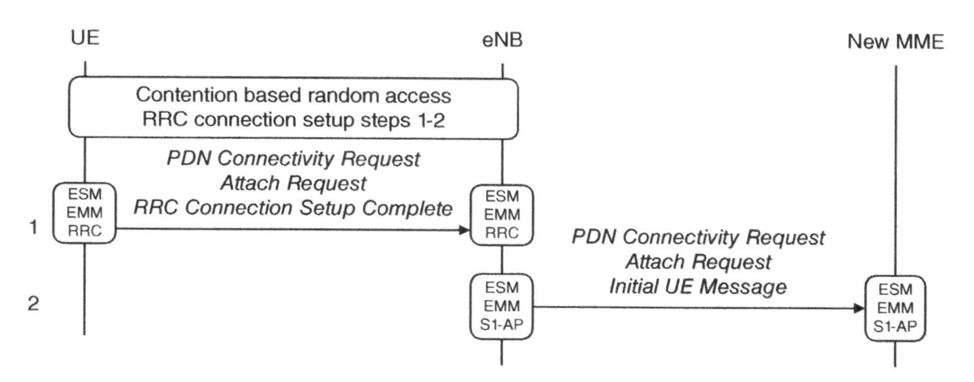

Figure 11.5 Attach procedure. (1) Attach request. Source: TS 23.401. Reproduced by permission of ETSI

user plane by means of DHCPv4. The mobile can either list its configuration options here or can set an ESM information transfer flag, which indicates a wish to send the options securely later on, after security activation. The mobile always uses the latter option if it wishes to indicate a preferred APN.

The mobile embeds the PDN connectivity request into an EMM *Attach Request*, in which it asks for registration with a serving MME. The message includes the globally unique temporary identity that the mobile was using when last switched on and the identity of the tracking area in which the mobile was last located. It also includes the mobile's non-access stratum capabilities, primarily the security algorithms that it supports.

In turn, the mobile embeds the Attach Request into the last message from the RRC connection establishment procedure, RRC Connection Setup Complete. As noted earlier, the RRC message also identifies the PLMN that the mobile would like to register with and the identity of its last serving MME. In step 1 of the attach procedure, the mobile sends this message to the serving eNB.

As described in Chapter 12, the mobile and MME can store their LTE security keys after the mobile switches off. If the mobile has a valid set of security keys, then it uses these to secure the attach request using a process known as integrity protection. This assures the MME that the request is coming from a genuine mobile and not from an intruder.

The base station extracts the EMM and ESM messages and embeds them into an S1-AP *Initial UE Message*, which requests the establishment of an S1 signalling connection for the mobile. As part of this message, the base station specifies the RRC establishment cause and the requested PLMN, which it received from the mobile during the RRC procedure, as well as the tracking area in which it lies.

The base station can now forward the message to a suitable MME (step 2). Usually, the chosen MME is the same one that the mobile was previously registered with. This can be done if two conditions are met: the base station has to lie in one of the old MME's pool areas and the old MME has to lie in the requested PLMN. If the mobile has changed the pool area since it was last switched on or if it is asking to register with a different network, then the base station selects another MME. It does so by choosing at random from the ones in its pool area, according to a load balancing algorithm [23].

11.4.4 Identification and Security Procedures

The MME receives the messages from the base station and can now run some procedures that relate to identification and security (Figure 11.6).

If the mobile has moved to a new MME since it was last switched on, then the MME has to find out the mobile's identity. To do this, it extracts the identity of the old MME from the mobile's GUTI and sends the GUTI to the old MME in a GTP-C *Identification Request* (3). The old MME's response includes the international mobile subscriber identity (IMSI) and the mobile's security keys. In exceptional cases, for example if the MME has purged its internal database, the mobile may be unknown to the old MME. If this happens, then the new MME asks the mobile for its IMSI using an EMM *Identity Request* (4), a message that is transported using the NAS information transfer procedure from Chapter 2.

The network can now run two security procedures (5a). In *authentication and key agreement*, the mobile and network confirm each other's identities and set up a new set of security keys. In *NAS security activation*, the MME activates those keys and initiates the secure protection of all subsequent EMM and ESM messages. These steps are mandatory if there was any problem with the integrity protection of the attach request and are optional otherwise. If the integrity

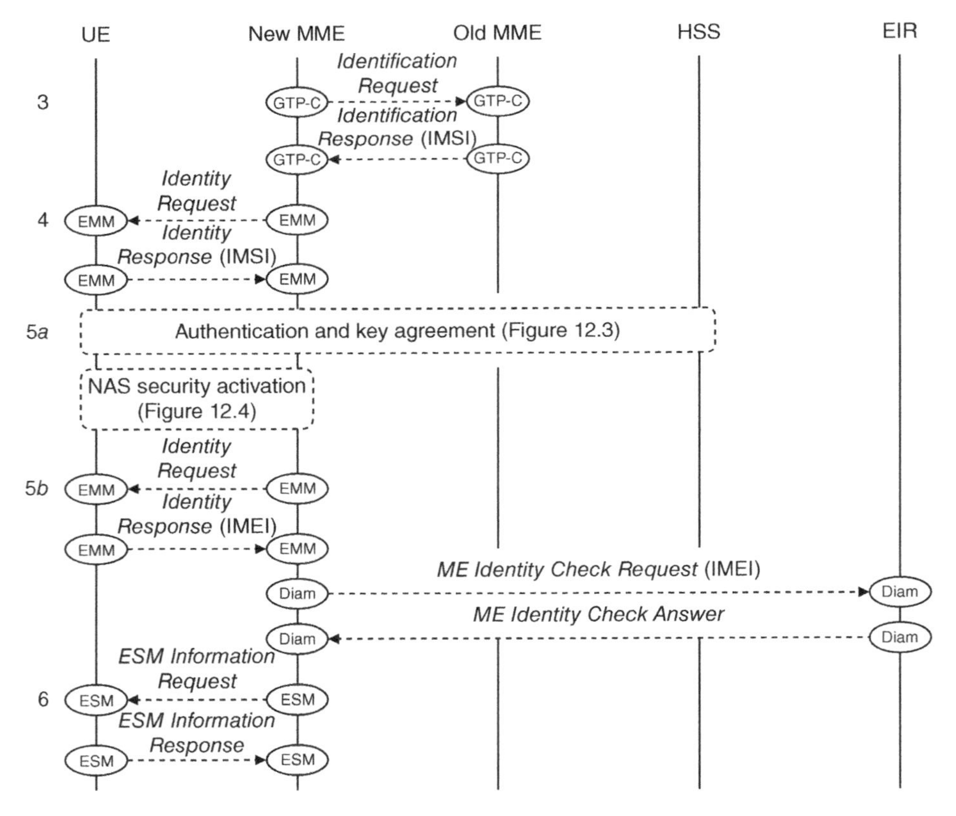

Figure 11.6 Attach procedure. (2) Identification and security procedures. Source: TS 23.401. Reproduced by permission of ETSI

check succeeded, then the MME can implicitly re-activate the mobile's old keys by sending it a signalling message that it has secured using those keys, thus skipping both of these procedures.

The MME then retrieves the international mobile equipment identity (IMEI) (5b). It can combine this message with NAS security activation to reduce the amount of signalling, but it is mandatory for the MME to retrieve the IMEI somehow. As a protection against stolen mobiles, the MME can optionally send the IMEI to the equipment identity register, which responds by either accepting or rejecting the device.

If the mobile sets the ESM information transfer flag in its PDN Connectivity Request, then the MME can now send it an *ESM Information Request* (6). The mobile sends its protocol configuration options in response; for example, any access point name that the mobile would like to request. Now that the network has activated NAS security, the mobile can send the message securely.

11.4.5 Location Update

The MME can now update the network's record of the mobile's location (Figure 11.7). If the mobile is re-attaching to its previous MME without having properly detached (for example, if its battery ran out), then the MME may still have some EPS bearers that are associated with the mobile. If this is the case then the MME deletes them (7), by following steps from the detach procedure that we will see later on.

If the MME has changed, then the new MME sends the mobile's IMSI to the home subscriber server (HSS), in a Diameter *Update Location Request* (8). The HSS updates its record of the mobile's location and tells the old MME to forget about the mobile (9). If the old MME has any EPS bearers that are associated with the mobile, then it deletes these as before (10).

In step 11, the HSS sends an *Update Location Answer* to the new MME, which includes the user's *subscription data* [24]. The subscription data list all the access point names (APNs) that the user has subscribed to, define each one using an *APN configuration* and identify one of the

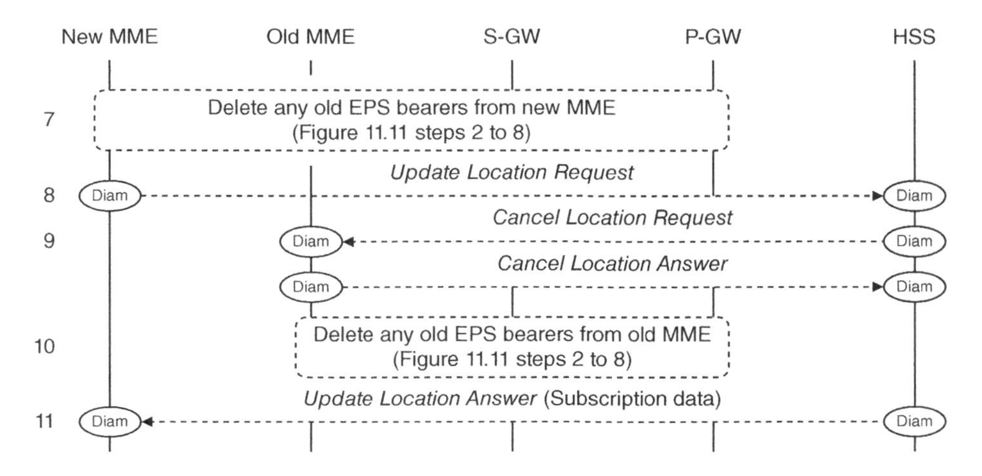

Figure 11.7 Attach procedure. (3) Location update. Source: TS 23.401. Reproduced by permission of ETSI

APN configurations as the default. In turn, each APN configuration identifies the access point name, states whether the corresponding packet data network supports IPv4, IPv6 or both, and defines the default EPS bearer's quality of service using parameters that we will see in Chapter 13. Optionally, it can also indicate a static IPv4 address or IPv6 prefix for the mobile to use when connecting to that APN.

11.4.6 Default Bearer Creation

The MME now has all the information that it needs to set up the default EPS bearer (Figure 11.8). It begins by selecting a suitable PDN gateway, using the mobile's preferred APN if it supplied one and the subscription data support it, or the default APN otherwise. It then selects a serving gateway and sends it a GTP-C *Create Session Request* (12). In this message, the MME includes the relevant subscription data and identifies the mobile's IMSI and the destination PDN gateway.

The serving gateway receives the message and forwards it to the PDN gateway (13). In the message, the serving gateway includes a GTP-U tunnel endpoint identifier (TEID), which the PDN gateway will eventually use to label the downlink packets that it sends across the S5/S8 interface.

If the message does not contain a static IP address, then the PDN gateway can allocate a dynamic IPv4 and/or IPv6 address for the mobile, using the methods we covered earlier. Alternatively, it can defer the allocation of an IPv4 address until later, if the mobile requested that in its protocol configuration options. The PDN gateway can also run a procedure known as *IP connectivity access network* (IP-CAN) *session establishment* (14), during which it receives authorization for the default bearer's quality of service. We will describe it along with the other procedures for managing quality of service, as part of Chapter 13.

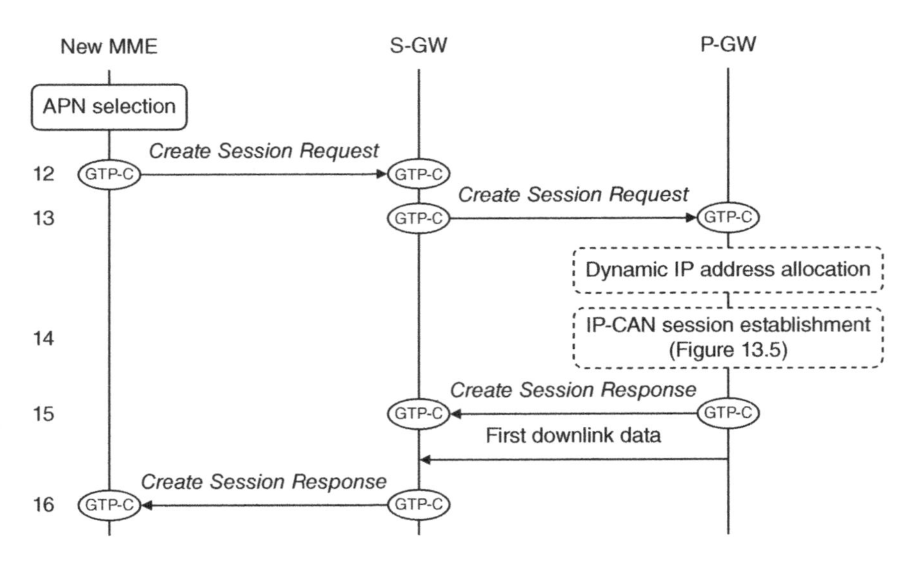

Figure 11.8 Attach procedure. (4) Default bearer creation. Source: TS 23.401. Reproduced by permission of ETSI

The PDN gateway now acknowledges the serving gateway's request by means of a GTP-C *Create Session Response* (15). In the message, it includes any IP address that the mobile has been allocated, as well as the quality of service of the default EPS bearer. The PDN gateway also includes a TEID of its own, which the serving gateway will eventually use to route uplink packets across S5/S8. The serving gateway forwards the message to the MME (16), except that it replaces the PDN gateway's tunnel endpoint identifier with an uplink TEID for the base station to use across S1-U.

11.4.7 Attach Accept

The MME can now reply to the mobile's attach request, as shown in Figure 11.9. It first initiates an ESM procedure known as *Default EPS bearer context activation*, which is a response to the mobile's PDN Connectivity Request and which starts with a message known as *Activate Default EPS Bearer Context Request*. The message includes the EPS bearer identity, the access point name, the quality of service and any IP address that the network has allocated to the mobile.

The MME embeds the ESM message into an EMM *Attach Accept*, which is a response to the mobile's original attach request. The message includes a list of tracking areas in which the MME has registered the mobile and a new globally unique temporary identity.

In turn, the MME embeds both messages into an S1-AP *Initial Context Setup Request*. This is the start of a procedure known as *Initial context setup*, which was triggered by the base

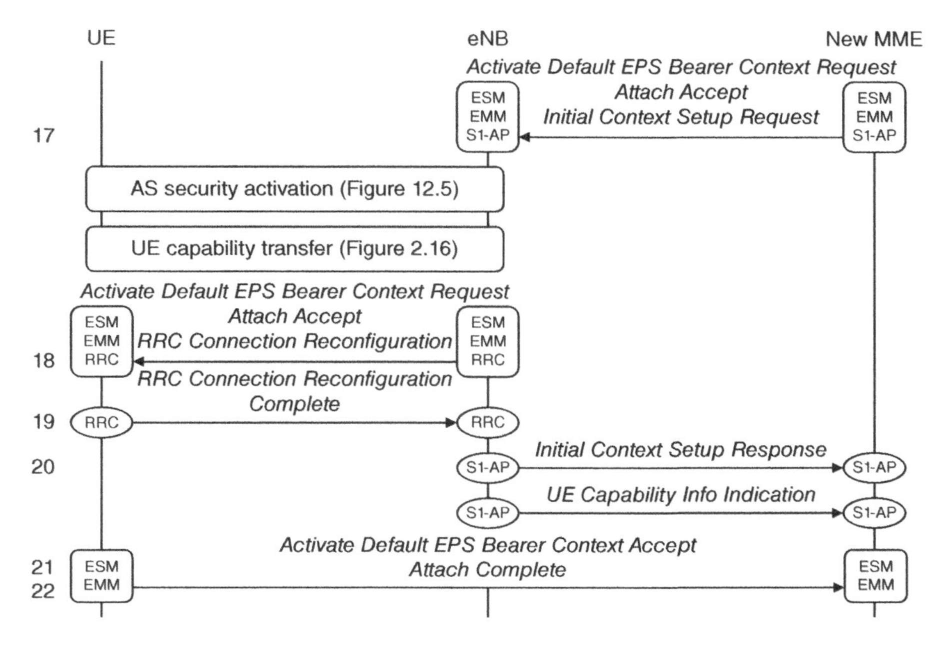

Figure 11.9 Attach procedure. (5) Attach accept. Source: TS 23.401. Reproduced by permission of ETSI

station's Initial UE Message. The procedure tells the base station to set up an S1 signalling connection for the mobile, and S1 and radio bearers that correspond to the default EPS bearer. The message includes the bearers' quality of service, the uplink TEID that the MME received from the serving gateway and a key for the activation of access stratum security. The MME sends all three messages to the base station, in step 17.

On receiving the message, the base station activates access stratum security using the secure key that the MME supplied. From this point, all the data and RRC signalling messages on the air interface are secured. It also retrieves the mobile's radio access capabilities using the procedure that we covered in Chapter 2 [25], so that it knows how to configure the mobile. The base station can then compose an *RRC Connection Reconfiguration* message, in which it modifies the mobile's RRC connection so as to set up two new radio bearers: a radio bearer that will carry the default EPS bearer and SRB 2. It sends this message to the mobile, along with the EMM and ESM messages that it has just received from the MME (18).

The mobile reconfigures its RRC connection as instructed and sets up the default EPS bearer. It then sends its acknowledgements to the network in two stages. Using SRB 1, the mobile first sends the base station an acknowledgement known as *RRC Connection Reconfiguration Complete*. This triggers an S1-AP *Initial Context Setup Response* to the MME (20), which includes a downlink TEID for the serving gateway to use across S1-U. It also triggers an S1-AP *UE Capability Info Indication*, in which the base station sends the mobile's capabilities back to the MME, where they are stored until the mobile detaches from the network.

The mobile then composes an ESM *Activate Default EPS Bearer Context Accept* and embeds it into an EMM *Attach Complete*, to acknowledge the ESM and EMM parts of message 18. It sends these messages to the base station on SRB 2 (21), using the NAS information transfer procedure, and the base station forwards the messages to the MME (22).

11.4.8 Default Bearer Update

The mobile can now send uplink data as far as the PDN gateway. However, we still need to tell the serving gateway about the identity of the selected base station and send it the tunnel endpoint identifier that the base station has just provided. To do this (Figure 11.10), the MME sends a GTP-C *Modify Bearer Request* to the serving gateway (23) and the serving gateway responds (24). From this point, downlink data packets can flow to the mobile.

The MME can also notify the HSS about the chosen PDN gateway and APN (25). It does this if the chosen PDN gateway is different from the one in the default APN configuration; for example, if the mobile requested an access point name of its own to connect to. The HSS stores the chosen PDN gateway, for use in any future handovers to non-3GPP systems, and responds (26).

Finally, the mobile may have to contact the PDN gateway across the user plane, to complete the allocation of its IP addresses. It does this when obtaining an IPv6 prefix using stateless auto-configuration and also when obtaining an IPv4 address using DHCPv4.

The mobile is now in the states EMM-REGISTERED, ECM-CONNECTED and RRC_CONNECTED and will stay in these states for as long as the user is actively communicating with the outside world. If the user does nothing, the network can transfer the mobile into ECM-IDLE and RRC_IDLE using a procedure known as S1 release. We will cover this procedure later, as part of Chapter 14.

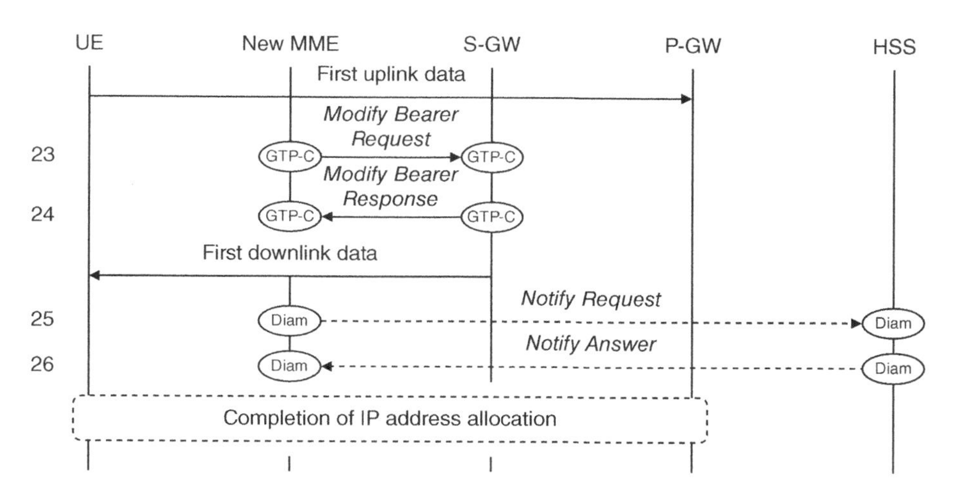

Figure 11.10 Attach procedure. (6) Default bearer reconfiguration. Source: TS 23.401. Reproduced by permission of ETSI

11.5 Detach Procedure

The last process to consider in this chapter is the *Detach procedure* [26]. This cancels the mobile's registration with the evolved packet core and is normally used when the mobile switches off, as shown in Figure 11.11.

We will assume that the mobile starts in ECM-CONNECTED and RRC_CONNECTED, consistent with its state at the end of the previous section. The user triggers the procedure by telling the mobile to shut down. In response, the mobile composes an EMM *Detach Request*, in which it specifies its GUTI, and sends the message to the MME (1). After sending the message, the mobile can switch off without waiting for a reply.

The MME now has to tear down the mobile's EPS bearers. To do this, it looks up the mobile's serving gateway and sends it a GTP-C *Delete Session Request* (2). The serving gateway forwards the message to the PDN gateway (3), which can run a procedure known as *IP-CAN session termination* (4) that undoes the earlier effect of IP-CAN session establishment. The PDN gateway then tears down all the mobile's bearers and replies to the serving gateway (5), which tears down its bearers in the same way and replies to the MME (6). If necessary, these steps are repeated for any other network that the mobile is connected to.

To finish the procedure, the MME tells the base station to tear down all the resources that are related to the mobile and indicates that the cause is a detach request (7). The base station does so and responds (8). The MME can now delete most of the information that it associated with the mobile. However, it keeps a record of the mobile's IMSI, GUTI and security keys, as it will need these next time the mobile switches on.

If the mobile starts in ECM-IDLE and RRC_IDLE, then it cannot send the detach request right away. Instead, it starts by running the contention-based random access procedure, followed by steps 1 and 2 of RRC connection establishment. It then embeds the detach request into the message RRC Connection Setup Complete, and the detach procedure continues as before.

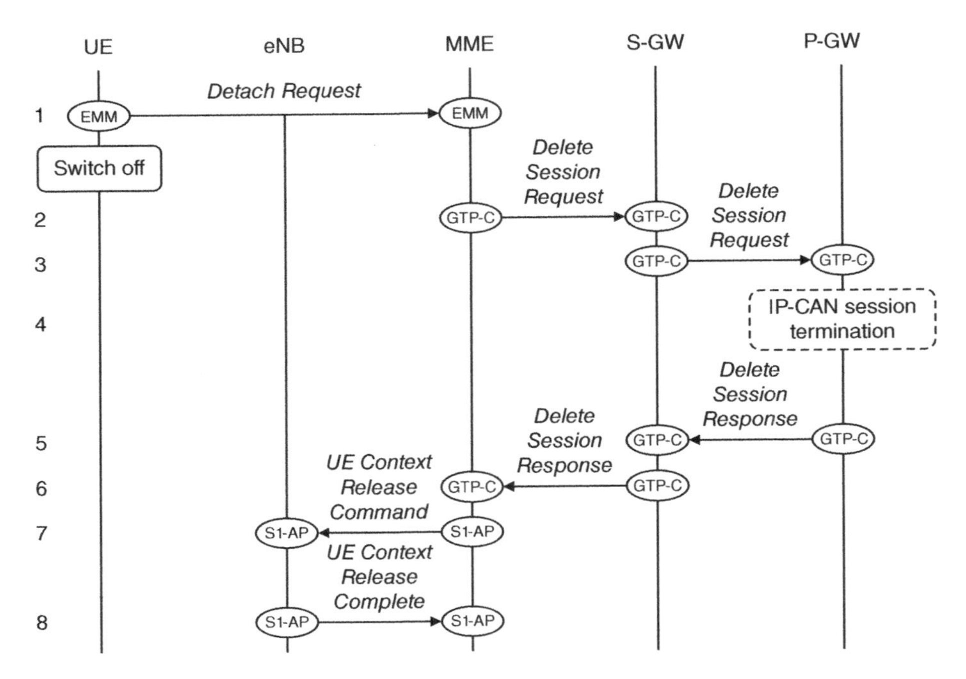

Figure 11.11 Detach procedure, triggered by the mobile switching off. Source: TS 23.401. Reproduced by permission of ETSI

References

1. 3GPP TS 23.122 (2012) Non-Access-Stratum (NAS) Functions Related to Mobile Station (MS) in Idle Mode, Release 11, December 2012.
2. 3GPP TS 36.304 (2013) User Equipment (UE) Procedures in Idle Mode, Release 11, September 2013.
3. 3GPP TS 23.401 (2013) General Packet Radio Service (GPRS) Enhancements for Evolved Universal Terrestrial Radio Access Network (E-UTRAN) Access, Release 11, September 2013.
4. 3GPP TS 36.300 (2013) Evolved Universal Terrestrial Radio Access (E-UTRA) and Evolved Universal Terrestrial Radio Access Network (E-UTRAN); Overall Description; Stage 2, Release 11, September 2013.
5. 3GPP TS 24.301 (2013) Non-Access-Stratum (NAS) Protocol for Evolved Packet System (EPS); Stage 3, Release 11, September 2013.
6. 3GPP TS 29.272 (2013) Evolved Packet System (EPS); Mobility Management Entity (MME) and Serving GPRS Support Node (SGSN) Related Interfaces Based on Diameter Protocol, Release 11, September 2013.
7. 3GPP TS 29.274 (2013) 3GPP Evolved Packet System (EPS); Evolved General Packet Radio Service (GPRS) Tunnelling Protocol for Control Plane (GTPv2-C); Stage 3, Release 11, September 2013.
8. 3GPP TS 36.331 (2013) Radio Resource Control (RRC); Protocol Specification, Release 11, September 2013.
9. 3GPP TS 36.413 (2013) Evolved Universal Terrestrial Radio Access Network (E-UTRAN); S1 Application Protocol (S1AP), Release 11, September 2013.
10. 3GPP TS 23.122 (2012) Non-Access-Stratum (NAS) Functions Related to Mobile Station (MS) in Idle Mode, Release 11, Sections 3.1, 4.3.1, 4.4, December 2012.
11. 3GPP TS 36.304 (2013) User Equipment (UE) Procedures in Idle Mode, Release 11, Sections 4, 5.1, September 2013.
12. 3GPP TS 31.102 (2013) Characteristics of the Universal Subscriber Identity Module (USIM) Application, Release 11, Sections 4.2.2, 4.2.5, 4.2.53, 4.2.54, 4.2.84, 4.2.91, September 2013.
13. 3GPP TS 31.102 (2013) Characteristics of the Universal Subscriber Identity Module (USIM) Application, Release 11, Section 4.4.6, September 2013.

14. 3GPP TS 23.122 (2012) Non-Access-Stratum (NAS) Functions Related to Mobile Station (MS) in Idle Mode, Release 11, Sections 3.1A, 4.4.3.1.3, December 2012.
15. 3GPP TS 36.304 (2013) User Equipment (UE) Procedures in Idle Mode, Release 11, Section 5.5, September 2013.
16. 3GPP TS 36.304 (2013) User Equipment (UE) Procedures in Idle Mode, Release 11, Sections 5.2.1, 5.2.2, 5.2.3, 5.3, September 2013.
17. 3GPP TS 36.214 (2012) Evolved Universal Terrestrial Radio Access (E-UTRA); Physical Layer; Measurements, Release 11, Section 5.1.1, December 2012.
18. 3GPP TS 36.331 (2013) Radio Resource Control (RRC); Protocol Specification, Release 11, Sections 5.3.3, 6.2.2 (RRCConnectionRequest, RRCConnectionSetup, RRCConnectionSetupComplete), September 2013.
19. 3GPP TS 23.401 (2013) General Packet Radio Service (GPRS) Enhancements for Evolved Universal Terrestrial Radio Access Network (E-UTRAN) Access, Release 11, Section 5.3.1, September 2013.
20. IETF RFC 2663 (1999) IP Network Address Translator (NAT) Terminology and Considerations, August 1999.
21. IETF RFC 4862 (2007) IPv6 Stateless Address, September 2007.
22. 3GPP TS 23.401 (2013) General Packet Radio Service (GPRS) Enhancements for Evolved Universal Terrestrial Radio Access Network (E-UTRAN) Access, Release 11, Section 5.3.2.1, September 2013.
23. 3GPP TS 23.401 (2013) General Packet Radio Service (GPRS) Enhancements for Evolved Universal Terrestrial Radio Access Network (E-UTRAN) Access, Release 11, Sections 4.3.7, 4.3.8, September 2013.
24. 3GPP TS 29.272 (2013) Evolved Packet System (EPS); Mobility Management Entity (MME) and Serving GPRS Support Node (SGSN) Related Interfaces Based on Diameter Protocol, Release 11, Sections 7.3.2, 7.3.34, 7.3.35, September 2013.
25. 3GPP TS 36.300 (2013) Evolved Universal Terrestrial Radio Access (E-UTRA) and Evolved Universal Terrestrial Radio Access Network (E-UTRAN); Overall Description; Stage 2, Release 11, Section 18, September 2013.
26. 3GPP TS 23.401 (2013) General Packet Radio Service (GPRS) Enhancements for Evolved Universal Terrestrial Radio Access Network (E-UTRAN) Access, Release 11, Section 5.3.8.2, September 2013.

12

Security Procedures

In this chapter, we review the security techniques that protect LTE against attacks from intruders. The most important issue is network access security, which protects the mobile's communications with the network across the air interface. In the first part of this chapter, we cover the architecture of network access security, the procedures that establish secure communications between the network and mobile, and the security techniques that are subsequently used. The system must also secure certain types of communication within the radio access network and the evolved packet core. This issue is known as network domain security and is the subject of the second part.

The 3GPP security procedures are covered by the 33 series specifications: those for LTE are summarized in TS 33.401 [1]. As in the last chapter, the details of the individual messages are in the specifications for the relevant signalling protocols [2–5]. For a detailed account of security in LTE, see Reference [6].

12.1 Network Access Security

12.1.1 Security Architecture

Network access security (Figure 12.1) protects the mobile's communications with the network across the air interface, which is the most vulnerable part of the system. It does this using four main techniques.

During *authentication*, the network and mobile confirm each other's identities. The evolved packet core (EPC) confirms that the user is authorized to use the network's services and is not using a cloned device. Similarly, the mobile confirms that the network is genuine and is not a spoof network set up to steal the user's personal data.

Confidentiality protects the user's identity. The international mobile subscriber identity (IMSI) is one of the quantities that an intruder needs to clone a mobile, so LTE avoids broadcasting it across the air interface wherever possible. Instead, the network identifies the user by means of temporary identities. If the EPC knows the MME pool area that the mobile is in (for example, during paging), then it uses the 40 bit S-TMSI. Otherwise (for example,

An Introduction to LTE: LTE, LTE-Advanced, SAE, VoLTE and 4G Mobile Communications, Second Edition.
Christopher Cox.

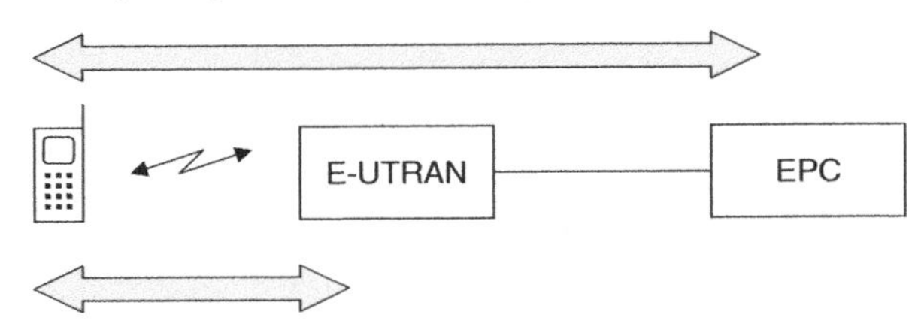

Figure 12.1 Network access security architecture

during the attach procedure), it uses the longer GUTI. Similarly, the radio access network uses the radio network temporary identifiers (RNTIs) that we introduced in Chapter 8.

Ciphering, also known as *encryption*, ensures that intruders cannot read the data and signalling messages that the mobile and network exchange. *Integrity protection* detects any attempt by an intruder to replay or modify signalling messages. It protects the system against problems such as man-in-the-middle attacks, in which an intruder intercepts a sequence of signalling messages and modifies and re-transmits them, in an attempt to take control of the mobile.

GSM and UMTS only implemented ciphering and integrity protection in the air interface's access stratum, to protect user plane data and RRC signalling messages between the mobile and the radio access network. As shown in Figure 12.1, LTE implements them in the non-access stratum as well, to protect EPS mobility and session management messages between the mobile and the MME. This brings two main advantages. In a wide-area network, it provides two cryptographically separate levels of encryption, so that even if an intruder breaks one level of security, the information is still secured on the other. It also eases the deployment of home base stations, whose access stratum security can be more easily compromised.

12.1.2 Key Hierarchy

Network access security is based on a hierarchy of keys [7] that is illustrated in Figure 12.2. Ultimately, it relies on the shared knowledge of a user-specific key, K, which is securely stored in the home subscriber server (HSS) and securely distributed within the universal integrated circuit card (UICC). There is a one-to-one mapping between a user's IMSI and the corresponding value of K, and the authentication process relies on the fact that cloned mobiles and spoof networks will not know the correct value of K.

From K, the HSS and UICC derive two further keys, denoted CK and IK. UMTS used those keys directly for ciphering and integrity protection, but LTE uses them differently, to derive an *access security management entity* (ASME) key, denoted K_{ASME}.

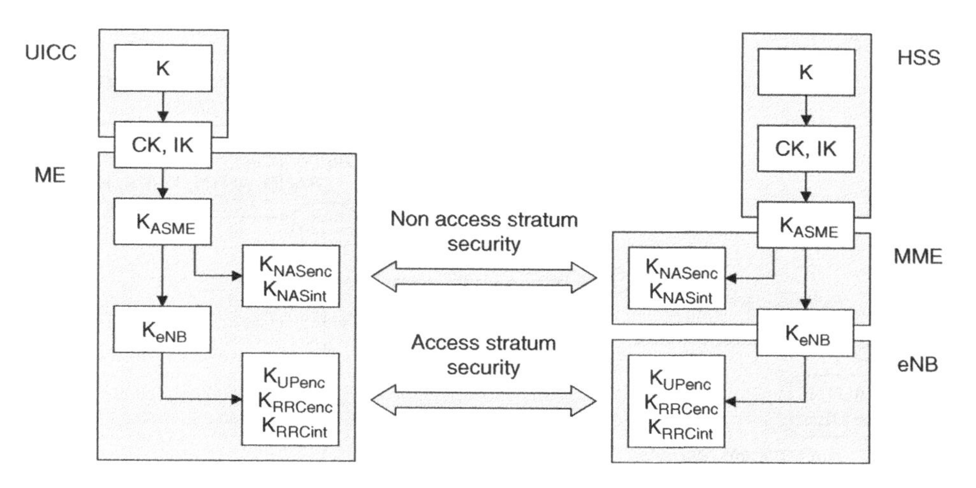

Figure 12.2 Hierarchy of the security keys used by LTE

From K_{ASME}, the MME and the mobile equipment derive three further keys, denoted K_{NASenc}, K_{NASint} and K_{eNB}. The first two are used for ciphering and integrity protection of non-access stratum (NAS) signalling messages between the mobile and the MME, while the last is passed to the base station. From K_{eNB}, the base station and the mobile equipment derive three access stratum keys, denoted K_{UPenc}, K_{RRCenc} and K_{RRCint}. These are respectively used for ciphering of data, ciphering of RRC signalling messages and integrity protection of RRC signalling messages in the access stratum (AS).

This set of keys is larger than the set used by GSM or UMTS, but it brings several benefits. Firstly, the mobile stores the values of CK and IK in its UICC after it detaches from the network, while the MME stores the value of K_{ASME}. This allows the system to apply integrity protection to the mobile's attach request when it next switches on, in the manner described in Chapter 11. The hierarchy also ensures that the AS and NAS keys are cryptographically separate, so that knowledge of one set of keys does not help an intruder to derive the other. At the same time, the hierarchy is backwards compatible with USIMs from 3GPP Release 99.

K, CK and IK contain 128 bits each, while the other keys all contain 256 bits. The current ciphering and integrity protection algorithms use 128 bit keys, which are derived from the least significant bits of the original 256 bit keys. If LTE eventually has to upgrade its algorithms to use 256 bit keys, then it will be able to do so with ease.

12.1.3 Authentication and Key Agreement

During *authentication and key agreement* (AKA) [8], the mobile and network confirm each other's identities and agree on a value of K_{ASME}. We have already seen this procedure used as part of the larger attach procedure; Figure 12.3 shows the full message sequence.

Before the procedure begins, the MME has retrieved the mobile's IMSI from its own records or from the mobile's previous MME, or exceptionally by sending an EMM Identity Request to

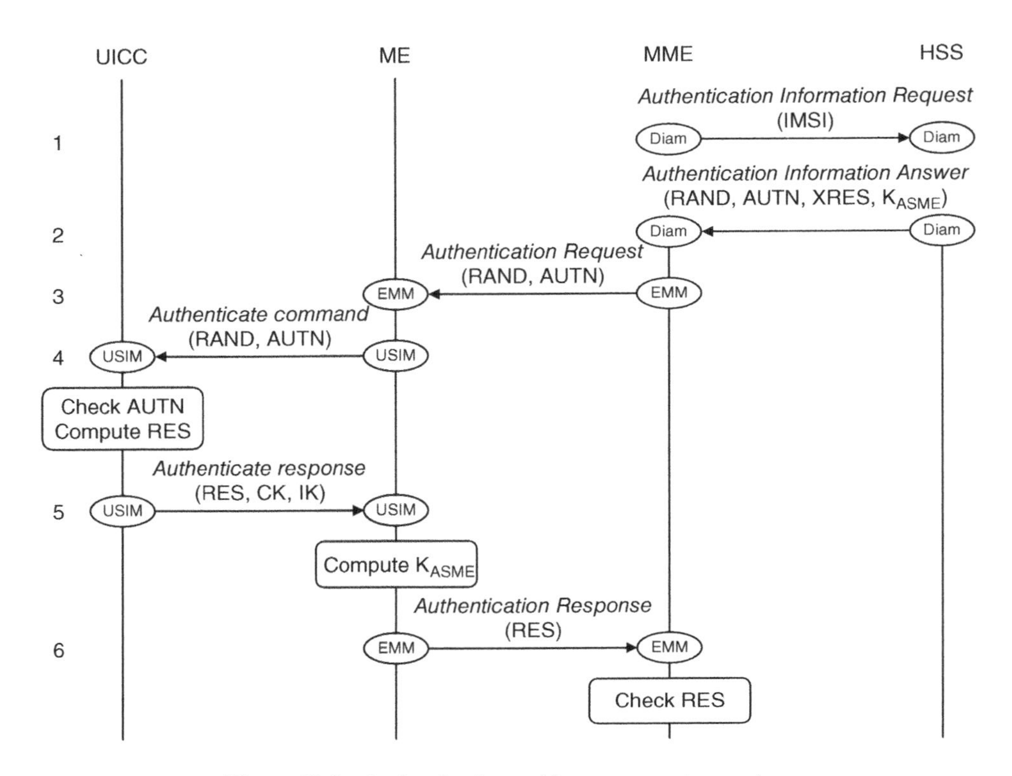

Figure 12.3 Authentication and key agreement procedure

the mobile itself. It now wishes to confirm the mobile's identity. To start the procedure, it sends a Diameter *Authentication Information Request* to the HSS (1), in which it includes the IMSI.

The HSS looks up the corresponding secure key K and calculates an *authentication vector* that contains four elements. RAND is a random number that the MME will use as an authentication challenge to the mobile. XRES is the expected response to that challenge, which can only be calculated by a mobile that knows the value of K. AUTN is an *authentication token*, which can only be calculated by a network that knows the value of K, and which includes a sequence number to prevent an intruder from recording an authentication request and replaying it. Finally, K_{ASME} is the access security management entity key, which is derived from CK and IK and ultimately from the values of K and RAND. In step 2, the HSS returns the authentication vector to the MME.

In GSM and UMTS, the HSS usually returns several authentication vectors at once, to minimize the number of separate messages that it has to handle. LTE actually discourages this technique, on the grounds that the storage of K_{ASME} has greatly reduced the number of messages that the HSS has to exchange.

The MME sends RAND and AUTN to the mobile equipment as part of an EMM *Authentication Request* (3) and the mobile equipment forwards them to the UICC (4). Inside the UICC, the USIM application examines the authentication token, to check that the network knows the value of K and that the enclosed sequence number has not been used before. If it is happy, then it calculates its response to the network's challenge, denoted RES, by combining RAND with its own copy of K. It also computes the values of CK and IK, and passes all three parameters back to the mobile equipment (5).

Using CK and IK, the mobile equipment computes the access security management entity key K_{ASME}. It then returns its response to the MME, as part of an EMM *Authentication Response* (6). In turn, the MME compares the mobile's response with the expected response that it received from the home subscriber server. If it is the same, then the MME concludes that the mobile is genuine. The system can then use the two copies of K_{ASME} to activate the subsequent security procedures, as described in the next section.

12.1.4 Security Activation

During security activation [9], the mobile and network calculate separate copies of the ciphering and integrity protection keys and start running the corresponding procedures. Security activation is carried out separately for the access and non-access strata.

The MME activates non-access stratum security immediately after authentication and key agreement, as shown in Figure 12.4. From K_{ASME}, the MME computes the ciphering and integrity protection keys K_{NASenc} and K_{NASint}. It then sends the mobile an EMM *Security Mode Command* (step 1), which tells the mobile to activate NAS security. The message is secured by integrity protection, but is not ciphered.

The mobile checks the integrity of the message in the manner described below. If the message passes the integrity check, then the mobile computes its own copies of K_{NASenc} and K_{NASint} from its stored copy of K_{ASME}, and starts ciphering and integrity protection. It then acknowledges the MME's command using an EMM *Security Mode Complete* (2). On receiving the message, the MME starts downlink ciphering.

If the mobile detaches from the MME, then the two devices delete their copies of K_{NASenc} and K_{NASint}. However, the MME retains its copy of K_{ASME}, while the mobile retains its copies of CK and IK. When the mobile switches on again, it re-computes its copies of K_{NASenc} and K_{NASint} and uses the latter to apply integrity protection to the subsequent attach request. The request is not ciphered, however, because of the risk that the network will not understand it.

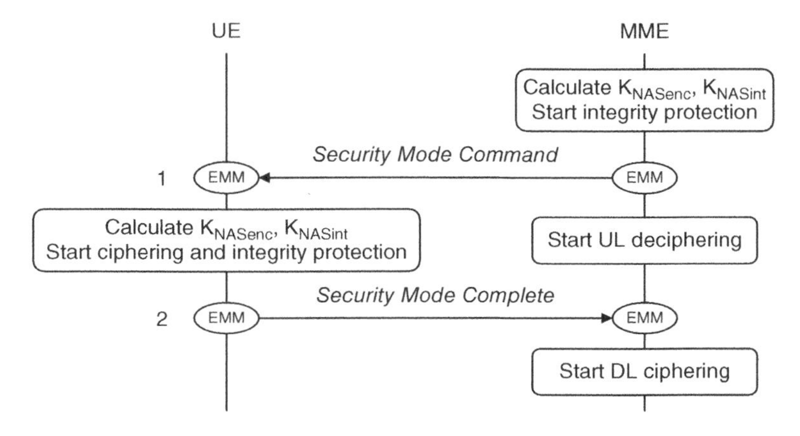

Figure 12.4 Activation of non-access stratum security. Source: TS 33.401. Reproduced by permission of ETSI

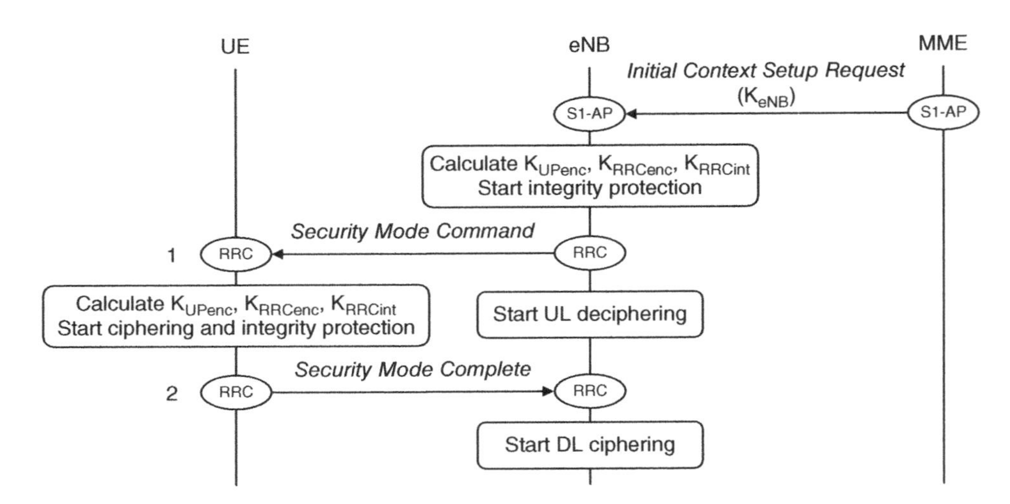

Figure 12.5 Activation of access stratum security. Source: TS 33.401. Reproduced by permission of ETSI

Access stratum security is activated later on, just before the network sets up the default radio bearer and signalling radio bearer 2. Figure 12.5 shows the message sequence. To trigger the process, the MME computes the value of K_{eNB}. It then passes K_{eNB} to the base station within the S1-AP Initial Context Setup Request, which we have already seen as step 17 of the attach procedure.

From here, the procedure is very like the activation of NAS security. Using K_{eNB}, the base station computes the ciphering and integrity protection keys K_{UPenc}, K_{RRCenc} and K_{RRCint}. It then sends the mobile an RRC *Security Mode Command* (step 1), whose integrity is protected using K_{RRCint}. The mobile checks the integrity of the message, computes its own keys and starts ciphering and integrity protection. It then acknowledges the base station's message using an RRC *Security Mode Complete* (step 2), at which point the base station can start downlink ciphering.

During the X2-based handover procedure that we cover in Chapter 14, the old base station derives a new secure key, denoted K_{eNB}^*. It can do this in two ways, either from another parameter known as *next hop* (NH) that it received from the MME at the end of the previous X2-based handover, or directly from K_{eNB}. The old base station then passes K_{eNB}^* to the new base station, which uses it as the new value of K_{eNB}. If the mobile moves to RRC_IDLE, then K_{eNB}, K_{UPenc}, K_{RRCenc} and K_{RRCint} are all deleted. However, the mobile and MME both retain K_{ASME} and use it to derive a new set of access stratum keys when the mobile returns to RRC_CONNECTED.

12.1.5 Ciphering

Ciphering ensures that intruders cannot read the information that is exchanged between the mobile and the network [10–12]. The packet data convergence protocol ciphers data and signalling messages in the air interface access stratum, while the EMM protocol ciphers signalling messages in the non-access stratum.

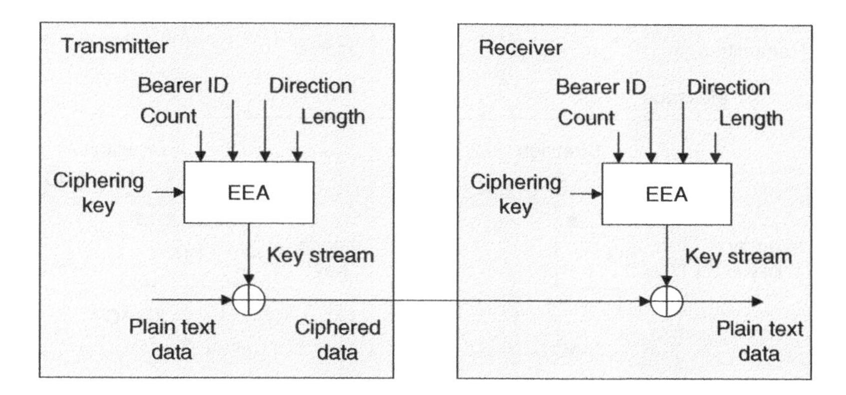

Figure 12.6 Operation of the ciphering algorithm. Source: TS 33.401. Reproduced by permission of ETSI

Figure 12.6 shows the ciphering process. The transmitter uses its ciphering key and other information fields to generate a pseudo-random key stream and mixes this with the outgoing data using an exclusive-OR operation. The receiver generates its own copy of the key stream and repeats the mixing process, so as to recover the original data. The algorithm is designed to be one-way, so that an intruder cannot recover the secure key from the transmitted message in a reasonable amount of computing time.

LTE currently supports four *EPS encryption algorithms* (EEAs). Two of them are SNOW 3G, which was originally used in the Release 7 standards for UMTS, and the *Advanced Encryption Standard* (AES). Release 11 adds a further algorithm, which is known as ZUC after the 5th century Chinese scientist Zu Chongzhi and is primarily intended for use in China. The final algorithm is a null ciphering algorithm, which means that, as in previous mobile communication systems, the air interface does not actually have to implement any ciphering at all. It is, however, mandatory for an LTE device to let the user know whether the air interface is using ciphering or not.

12.1.6 Integrity Protection

Integrity protection [13–15] allows a device to detect modifications to the signalling messages that it receives, as a protection against problems such as man-in-the-middle attacks. The packet data convergence protocol applies integrity protection to RRC signalling messages in the air interface's access stratum, while the EMM protocol applies integrity protection to its own messages in the non-access stratum.

Figure 12.7 shows the process. The transmitter passes each signalling message through an *EPS integrity algorithm* (EIA). Using the appropriate integrity protection key, the algorithm computes a 32-bit integrity field, denoted MAC-I, and appends it to the message. The receiver separates the integrity field from the signalling message and computes the expected integrity field XMAC-I. If the observed and expected integrity fields are the same, then it is happy. Otherwise, the receiver concludes that the message has been modified and discards it.

Integrity protection is mandatory for almost all of the signalling messages that the mobile and network exchange after security activation, and is based on SNOW 3G, the Advanced

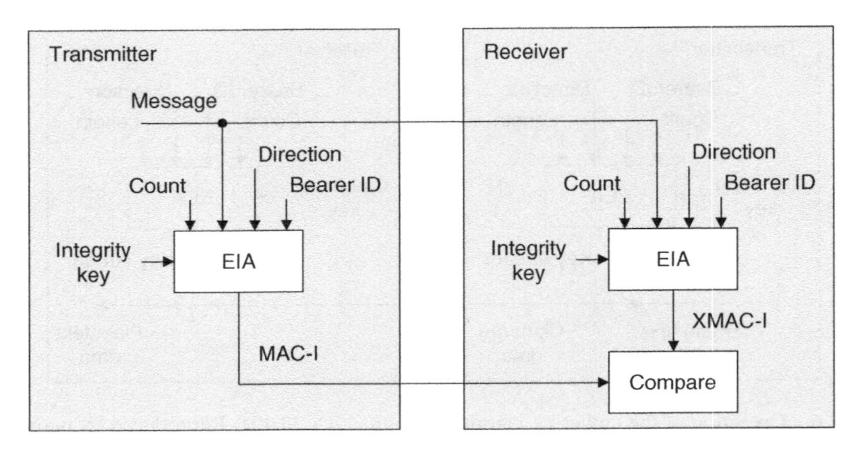

Figure 12.7 Operation of the integrity protection algorithm. Source: TS 33.401. Reproduced by permission of ETSI

Encryption Standard or ZUC as before. There is one exception, however: from Release 9, mobiles can use a null integrity protection algorithm for the sole purpose of making emergency voice calls without a UICC.

12.2 Network Domain Security

12.2.1 Security Protocols

Inside the fixed network, two devices often have to exchange information securely. Because the fixed network is based on IP, this can be done using standard IETF security protocols [16]. Devices first authenticate each other and establish a security association using a protocol known as *Internet Key Exchange version 2* (IKEv2) [17, 18]. This relies either on the use of a pre-shared secret, as in the air interface's use of the secure key K, or on public key cryptography.

Encryption and integrity protection are then implemented using the *Internet Protocol Security* (IPSec) *Encapsulating Security Payload* (ESP) [19, 20]. Depending on the circumstances, the network can use ESP *transport mode*, which just protects the payload of an IP packet, or *tunnel mode*, which protects the IP header as well. These techniques are used in two parts of the LTE network, in the manner described next.

12.2.2 Security in the Evolved Packet Core

In the evolved packet core, secure communications are required between networks that are run by different network operators, so as to handle roaming mobiles. To support this, the evolved packet core is modelled using *security domains*. A security domain usually corresponds to a network operator's EPC (Figure 12.8), but the operator can divide the EPC into more than one security domain if required.

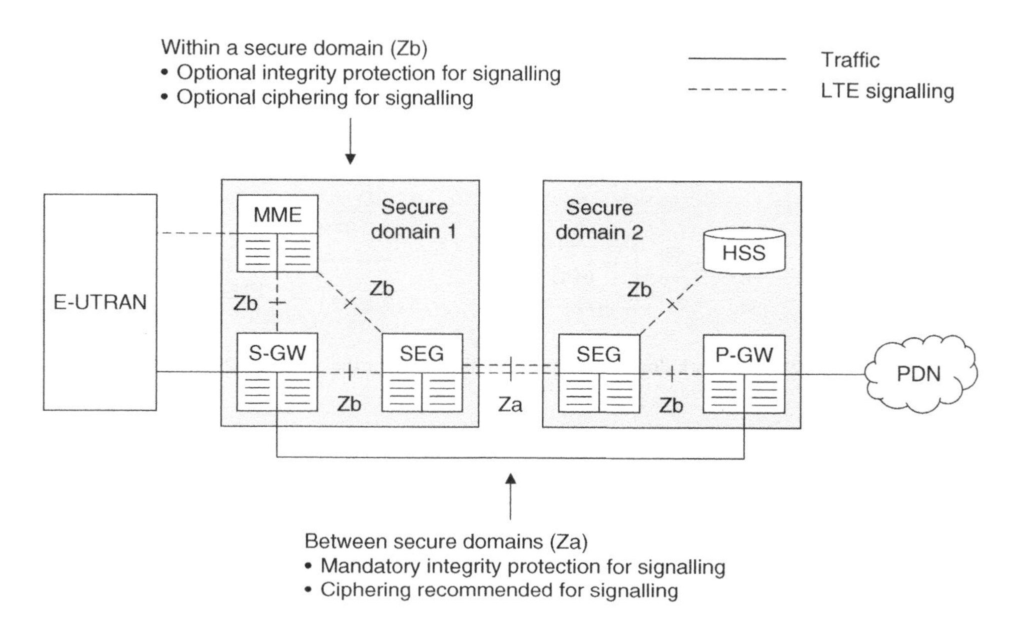

Figure 12.8 Example implementation of the network domain security architecture in the evolved packet core

From the viewpoint of the network domain security functions, different security domains are separated by the Za interface. On this interface, it is mandatory for LTE signalling messages to be protected using ESP tunnel mode. The security functions are implemented using *secure gateways* (SEGs), although operators can include the secure gateways' functions in the network elements themselves if they wish to. There is no protection for data, which will usually end up in an insecure public network anyway. If required, the data can be protected at the application layer.

Within a security domain, the network elements are separated by the Zb interface. This interface is usually under the control of a single network operator, so protection of LTE signalling messages across this interface is optional. If the interface is secured, then support of ESP tunnel mode is mandatory, while support of ESP transport mode is optional.

Security is the only difference between the S5 and S8 interfaces that were introduced in Chapter 2. Across the S5 interface, the serving and PDN gateways lie in the same security domain, so security functions are optional and are implemented using Zb. Across the S8 interface, the gateways lie in different security domains, so security functions are mandatory and are implemented using Za.

12.2.3 Security in the Radio Access Network

In the radio access network, it is usual for network operators to secure the X2 and S1 interfaces, which connect the base stations to each other and to the evolved packet core. This is especially important in a femtocell network, in which a home base station communicates with the EPC

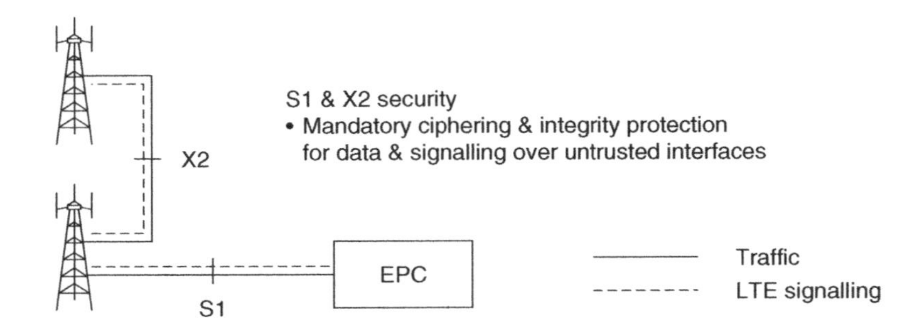

Figure 12.9 Network domain security architecture for the S1 and X2 interfaces

across a public IP backhaul, or in a network where the S1 and X2 interfaces are implemented using microwave links.

To handle these issues, it is mandatory for network operators to secure the S1 and X2 interfaces in the manner shown in Figure 12.9, unless the interfaces are already trusted by virtue of some other mechanism such as physical protection. If the interfaces are secured, then support of ESP tunnel mode is mandatory, while support of ESP transport mode is optional. The security functions are applied to the LTE signalling messages and to the user's data, which is a similar situation to the air interface's access stratum, but is different from the EPC.

References

1. 3GPP TS 33.401 (2013) 3GPP System Architecture Evolution (SAE): Security Architecture, Release 11, June 2013.
2. 3GPP TS 24.301 (2013) Non-Access-Stratum (NAS) Protocol for Evolved Packet System (EPS); Stage 3, Release 11, September 2013.
3. 3GPP TS 29.272 (2013) Evolved Packet System (EPS); Mobility Management Entity (MME) and Serving GPRS Support Node (SGSN) Related Interfaces Based on Diameter Protocol, Release 11, September 2013.
4. 3GPP TS 31.102 (2013) Characteristics of the Universal Subscriber Identity Module (USIM) Application, Release 11, September 2013.
5. 3GPP TS 36.331 (2013) Radio Resource Control (RRC) Protocol Specification, Release 11, September 2013.
6. Forsberg, D., Horn, G., Moeller, W.-D. and Niemi, V. (2012) *LTE Security*, 2nd edn, John Wiley & Sons, Ltd, Chichester.
7. 3GPP TS 33.401 (2013) 3GPP System Architecture Evolution (SAE): Security Architecture, Release 11, Section 6.2, June 2013.
8. 3GPP TS 33.401 (2013) 3GPP System Architecture Evolution (SAE): Security Architecture, Release 11, Section 6.1, June 2013.
9. 3GPP TS 33.401 (2013) 3GPP System Architecture Evolution (SAE): Security Architecture, Release 11, Section 7.2.4, June 2013.
10. 3GPP TS 33.401 (2013) 3GPP System Architecture Evolution (SAE): Security Architecture, Release 11, Annex B.1, June 2013.
11. 3GPP TS 36.323 (2013) Packet Data Convergence Protocol (PDCP) Specification, Release 11, Section 5.6, June 2013.
12. 3GPP TS 24.301 (2013) Non-Access-Stratum (NAS) Protocol for Evolved Packet System (EPS); Stage 3, Release 11, Section 4.4.5, September 2013.
13. 3GPP TS 33.401 (2013) 3GPP System Architecture Evolution (SAE): Security Architecture, Release 11, Annex B.2, June 2013.
14. 3GPP TS 36.323 (2013) Packet Data Convergence Protocol (PDCP) Specification, Release 11, Section 5.7, March 2013.

15. 3GPP TS 24.301 (2013) Non-Access-Stratum (NAS) Protocol for Evolved Packet System (EPS); Stage 3, Release 11, Section 4.4.4, September 2013.
16. 3GPP TS 33.401 (2013) 3GPP System Architecture Evolution (SAE): Security Architecture, Release 11, Sections 11, 12, June 2013.
17. 3GPP TS 33.310 (2012) Network Domain Security (NDS); Authentication Framework (AF); Network Domain Security (NDS); Authentication Framework (AF), Release 11, December 2012.
18. IETF RFC 4306 (2005) Internet Key Exchange (IKEv2) Protocol, December 2005.
19. 3GPP TS 33.210 (2012) 3G Security; Network Domain Security; IP Network Layer Security, Release 11, September 2012.
20. IETF RFC 4303 (2005) IP Encapsulating Security Payload (ESP), December 2005.

13

Quality of Service, Policy and Charging

In Chapters 2 and 11, we described how the evolved packet core transports data packets by the use of bearers and tunnels, and how it sets up a default EPS bearer for the mobile during the attach procedure. We also noted that each bearer was associated with a quality of service, which describes information such as the bearer's data rate, error rate and delay. However, we have not yet covered some important and related issues, namely how the network specifies and manages quality of service, and how it ultimately charges the user. These issues are the subject of this chapter.

We begin by defining the concept of policy and charging control, and by describing the architecture that is used for policy and charging in LTE. We continue by discussing the procedures for policy and charging control and for session management, through which an application can request a specific quality of service from the network. We conclude by discussing data transport in the evolved packet core, and offline and online charging.

Policy and charging control is described by two 3GPP specifications, namely TS 23.203 [1] and TS 29.213 [2]. The charging system is summarized in TS 32.240 [3] and TS 32.251 [4]. As usual, the details of the individual procedures are in the specifications for the relevant signalling protocols. Those for the evolved packet core are in References [5–8], but we will introduce some more protocols that are specifically related to policy and charging in the course of the chapter.

13.1 Policy and Charging Control

13.1.1 Quality of Service Parameters

In Chapter 2, we explained that LTE uses bidirectional data pipes known as EPS bearers, which transfer data between the mobile and the PDN gateway along the correct route and with the correct quality of service. To add some detail, let us look at how LTE specifies the quality of service of an EPS bearer [9, 10].

An Introduction to LTE: LTE, LTE-Advanced, SAE, VoLTE and 4G Mobile Communications, Second Edition.
Christopher Cox.

The most important parameter is the *QoS class identifier* (QCI). This is an 8-bit number that acts as a pointer into a look-up table and defines four other quantities. The first of these is the *resource type*, which divides bearers into two classes. A *GBR bearer* is associated with a *guaranteed bit rate*, which is the minimum long term average data rate that the mobile can expect to receive. GBR bearers are suitable for real-time services such as voice, in which the guaranteed bit rate might correspond to the lowest bit rate of the user's voice codec. A *non-GBR bearer* receives no such guarantees, so is suitable for non real-time services such as web browsing in which the data rate can fall to zero. A default bearer is always a non-GBR bearer, while a dedicated bearer can be either a GBR or a non-GBR bearer.

The *packet error/loss rate* is an upper bound for the proportion of packets that are lost because of errors in transmission and reception. The network should apply it reliably to GBR bearers, but non-GBR bearers can expect additional packet losses if the network becomes congested. The *packet delay budget* is an upper bound, with 98% confidence, for the delay that a packet receives between the mobile and the PDN gateway. Finally, the *QCI priority level* determines the order in which data packets are handled. Low numbers receive a high priority, and a congested network meets the packet delay budget of bearers with priority N, before moving on to bearers with priority $N+1$.

Some of the QoS class identifiers have been standardized and are associated with the parameters listed in Table 13.1. Bearers in these classes can expect to receive a consistent quality of service even if the mobile is roaming. Network operators can define other QoS classes for themselves, but these are only likely to work for non roaming mobiles.

The other QoS parameters are listed in Table 13.2. Each GBR bearer is associated with the guaranteed bit rate that we noted earlier and also with a *maximum bit rate* (MBR). The maximum bit rate is the highest data rate that the bearer will ever receive, so might correspond to the highest bit rate of a user's voice codec. Despite this distinction, the specifications only allow the maximum bit rate to be greater than the guaranteed bit rate from Release 10 onwards.

The data rates of non-GBR bearers are collectively limited by two other parameters. These are the *per APN aggregate maximum bit rate* (APN-AMBR), which limits the mobile's total data rate on the non-GBR bearers that are using a particular access point name, and the *per*

Table 13.1 Standardized QCI characteristics

QCI	Resource type	Packet error/ loss rate	Packet delay budget (ms)	QCI priority	Example services
1	GBR	10^{-2}	100	2	Conversational voice
2		10^{-3}	150	4	Real time video
3		10^{-3}	50	3	Real time games
4		10^{-6}	300	5	Buffered video
5	Non-GBR	10^{-6}	100	1	IMS signalling
6		10^{-6}	300	6	Web, email, FTP (high priority users)
7		10^{-3}	100	7	Voice, real time video, real time games
8		10^{-6}	300	8	Web, email, FTP (mid priority users)
9		10^{-6}	300	9	Web, email, FTP (low priority users)

Source: TS 23.303. Reproduced by permission of ETSI.

Table 13.2 Quality of service parameters

Parameter	Description	Use by GBR bearers	Use by non-GBR bearers
QCI	QoS class identifier	✓	✓
ARP	Allocation & retention priority	✓	✓
GBR	Guaranteed bit rate	✓	×
MBR	Maximum bit rate	✓	×
APN-AMBR	Per APN aggregate maximum bit rate	×	One field per APN
UE-AMBR	Per UE aggregate maximum bit rate	×	One field per UE

UE aggregate maximum bit rate (UE-AMBR), which limits a mobile's total data rate on all its non-GBR bearers.

Finally, the *allocation and retention priority* (ARP) contains three fields. The *ARP priority level* determines the order in which a congested network should satisfy requests to establish or modify a bearer, with level 1 receiving the highest priority. (Note that this parameter is different from the QCI priority level defined above.) The *pre-emption capability* field determines whether a bearer can grab resources from another bearer with a lower priority and might typically be set for emergency services. Similarly, the *pre-emption vulnerability* field determines whether a bearer can lose resources to a bearer with a higher priority.

We saw some of these parameters while discussing the medium access control protocol in Chapter 10. In the mobile's logical channel prioritization algorithm, the logical channel priority is derived from the QCI priority level in Table 13.1, while the prioritized bit rate is derived from the guaranteed bit rate in Table 13.2.

13.1.2 Service Data Flows

An EPS bearer is not restricted to carrying a single data stream. Instead, it can carry multiple data streams in the manner shown in Figure 13.1.

Each EPS bearer comprises one or more bidirectional *service data flows* (SDFs), each of which carries packets for a particular service such as a streaming video application. The service data flows in an EPS bearer have to share the same quality of service, specifically the same QCI and ARP, to ensure that they can be transported in the same way. For example, a user might be downloading two separate video streams, with each one implemented as a service data flow. The network can transport the streams using one EPS bearer if they share the same QCI and ARP, but has to use two EPS bearers otherwise. The mapping between service data flows and EPS bearers is governed by the PDN gateway.

In turn, each service data flow comprises one or more unidirectional *packet flows*, such as the audio and video streams that make up the service. As before, the packet flows in each service data flow have to share the same QCI and ARP. In the case of a video telephony service, for example, the network might wish to assign a lower allocation and retention priority to the video stream, so that it could drop the video stream in a congested cell but retain the audio. To do that, it would have to implement the packet flows using two service data flows with different

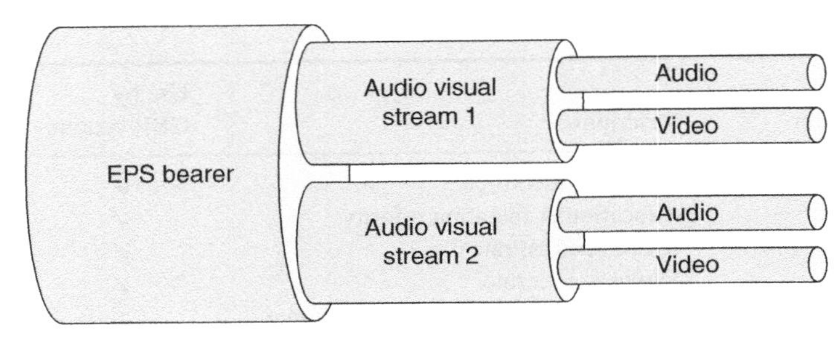

	EPS bearer	Service data flow	Packet flow
Role	Transports traffic over EPS	Basis for policy and charging control	Underlying data flow
Composition	Service data flows with same QCI and ARP	Packet flows from one service with same QCI and ARP	Packets with same IP addresses and port numbers
Scope	UE ⇔ PDN gateway	UE ⇔ PCRF	UE ⇔ AF
Identification	Traffic flow template	SDF template	Packet filter

Figure 13.1 Relationship between EPS bearers, service data flows and packet flows

allocation and retention priorities, and would ultimately have to use two EPS bearers. Packet flows are known to the application, while the mapping between packet flows and service data flows is governed by a network element, the policy and charging rules function (PCRF), which we will discuss shortly.

Adding some more detail, each packet flow is identified using a *packet filter*, which contains information such as the IP addresses of the source and destination devices, and the UDP or TCP port numbers of the source and destination applications. Each service data flow is identified using an *SDF template*, which is the set of packet filters that make it up. Similarly, each EPS bearer is identified using a *traffic flow template* (TFT), which is the set of SDF templates that make it up.

13.1.3 Charging Parameters

Each service data flow is associated with parameters that describe how the user will be charged. The *charging method* can be *offline charging*, which is suitable for simple monthly billing, or *online charging*, which supports more complex scenarios such as pre-paid services or monthly data limits. The *measurement method* determines whether the network should monitor the volume or duration of the data flow or both. Finally the *charging key*, also known as the *rating group*, indicates the tariff that the charging system will eventually use.

Once a service data flow has been configured, the serving and PDN gateways monitor the flow of traffic in accordance with these parameters and send information about it to an online or offline charging system. We will discuss the techniques that they use at the end of the chapter.

13.1.4 Policy and Charging Control Rules

Quality of service is associated with another concept known as *gating*, which determines whether or not packets are allowed to travel through a service data flow. Gating is important because the network often configures a service data flow in three stages, by authorizing its quality of service, configuring an EPS bearer and finally allowing packets to flow. To implement it, each service data flow is associated with a *gate status*, which can be either open or closed. Together, gating and QoS make up a concept known as *policy*.

In turn, the policy and charging parameters and the SDF template make up a dataset known as a *policy and charging control* (PCC) *rule* [11, 12]. Each service data flow is associated with a PCC rule, which describes how the network recognizes and implements the SDF. The network can also associate an SDF with multiple PCC rules, each of which includes a precedence level, an activation time and a deactivation time. This allows the network to apply different PCC rules at different times of day to support techniques such as time of day based charging. The network manages its PCC rules using a set of network elements which we have not previously seen, and which we will discuss in the next section.

13.2 Policy and Charging Control Architecture

13.2.1 Basic PCC Architecture

Figure 13.2 shows the basic architecture for policy and charging control in a network that is using a GTP based S5/S8 interface [13]. The architecture is applicable both to a non roaming mobile and to a mobile that is roaming using home routed traffic.

The most important device is the *policy and charging rules function* (PCRF), which authorizes the treatment that a service data flow will receive by specifying a suitable PCC rule. LTE uses two types of PCC rule, namely *pre-defined PCC rules*, which are permanently stored by the network and are only referenced by the PCRF, and *dynamic PCC rules*, which the PCRF composes on-the-fly. The network might use the first of these for the standardized QoS

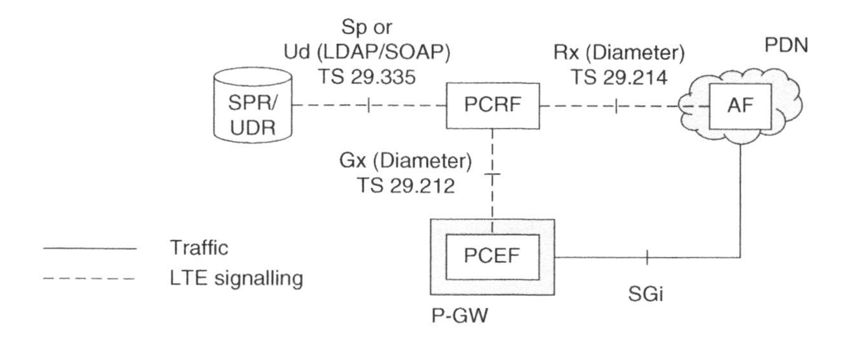

Figure 13.2 Policy control architecture, for a GTP based S5/S8 interface. Source: TS 23.203. Reproduced by permission of ETSI

characteristics that we introduced earlier, while reserving the second for more specialized applications.

The PDN gateway contains a *policy and charging enforcement function* (PCEF), which asks the PCRF to supply a PCC rule for a service data flow and which implements its decisions. An important part of the PCEF is a *bearer binding function* (BBF). This determines how the network will implement a service data flow using EPS bearers, for example by setting up a new EPS bearer or by folding the SDF into an existing bearer. Another part of the PCEF is an *event reporting function* (ERF). This can be configured to inform the PCRF when certain events occur, for example if the mobile's radio access technology changes, if radio communications with the mobile are lost or if the user runs out of credit.

To help it specify the policy and charging rules for a service data flow, the PCRF can communicate with various other devices. In Releases 8 and 9, details of the users' subscriptions are stored by a database known as the *subscription profile repository* (SPR), which contains similar information to the HSS but is a separate logical device. From Release 10, the users' subscription data can all be stored in a single centralized *user data repository* (UDR), which the HSS and PCRF both access by taking the role of application specific front ends. Finally, the *application function* (AF) runs an external application such as a voice over IP (VoIP) server and may be controlled by another party.

Using this architecture, an application can ask for a specific quality of service in two ways. Firstly, the application function can make a QoS request by sending a signalling message directly to the PCRF. Secondly, the mobile can make a QoS request by sending a signalling message to the MME, which forwards the request to the PDN gateway and hence to the PCRF. We will see how the network implements these techniques shortly.

To close this section, it is worth noting that LTE can operate without the use of a PCRF. Such a network uses a single pre-defined PCC rule for each access point name, which is the same for all users and for all packet flows. The network does not support dedicated bearers and omits most of the session management procedures that we cover later in the chapter. Despite its limited policy and charging functionality, such an architecture may still be useful for a network that only offers basic access to the internet, in which the user's quality of service is ultimately limited by what the internet can provide.

13.2.2 Local Breakout Architecture

If the mobile is roaming using local breakout then the PCRF's functions are split as shown in Figure 13.3, between a *home PCRF* (H-PCRF) and a *visited PCRF* (V-PCRF). When using this architecture, a QoS request from the mobile or application function is directed to the visited PCRF, which relays the request to the home PCRF. The home PCRF returns the PCC rule, which the visited PCRF can either accept or reject. In this way, the home and visited networks both have some control over the quality of service that the user will receive.

13.2.3 Architecture Using a PMIP Based S5/S8

If the S5/S8 interface is based on proxy mobile IP, then the EPS bearer only extends as far as the serving gateway. To support this architecture, the bearer binding process is delegated to a *bearer binding and event reporting function* (BBERF) that lies inside the serving gateway.

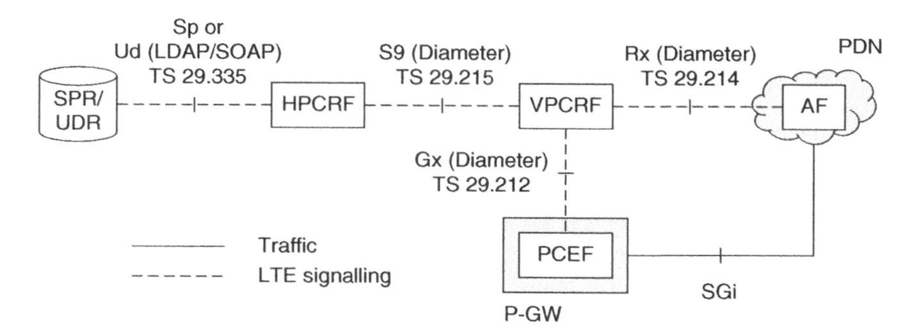

Figure 13.3 Policy and charging control architecture, for a GTP based S5/S8 interface and roaming using local breakout. Source: TS 23.203. Reproduced by permission of ETSI

Figure 13.4 shows the resulting architecture for a mobile in its home network. If the mobile is roaming, then the BBERF is controlled by the visited PCRF, while the PCEF is controlled by either the home or the visited PCRF as before.

13.2.4 Software Protocols

Most of the PCRF's signalling interfaces use Diameter applications, in a similar way to the interface between the MME and the home subscriber server. These include the Gx and Gxx interfaces to the PCEF and BBERF [14], the Rx interface to the application function [15] and the S9 interface between the home and visited PCRFs [16]. The Ud interface between the PCRF and the user data repository uses two non 3GPP protocols, namely the *lightweight access directory protocol* (LDAP) and the *simple object access protocol* (SOAP) [17]. The Sp interface to the stand-alone subscription profile repository has not yet been standardized.

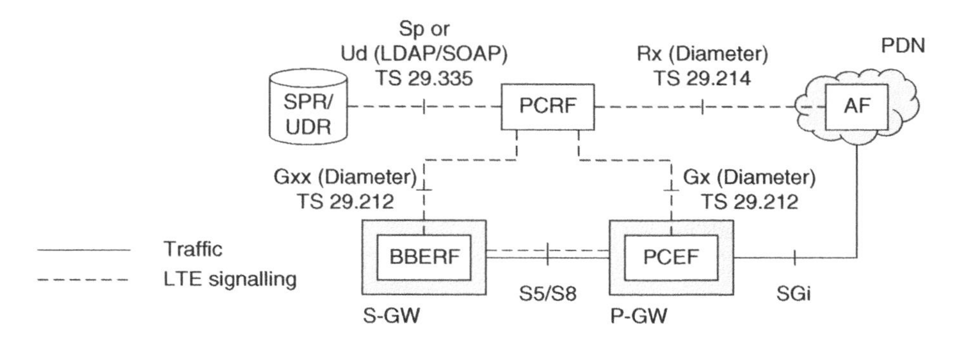

Figure 13.4 Policy and charging control architecture, for a PMIP based S5/S8 interface. Source: TS 23.203. Reproduced by permission of ETSI

13.3 Session Management Procedures

13.3.1 IP-CAN Session Establishment

During the attach procedure from Chapter 11, we skipped over a step known as IP connectivity access network (IP-CAN) session establishment, in which the PDN gateway configured the quality of service of the default EPS bearer. Figure 13.5 shows the details [18]. In this example, we have assumed that the mobile is not roaming and that the S5/S8 interface is based on GTP.

The procedure is triggered when the PDN gateway receives a request to establish a default EPS bearer. In response, the PCEF sends a Diameter *CC-Request* to the PCRF (1). The message includes the mobile's IMSI, the access point name that the mobile is connecting to, the corresponding quality of service from the user's subscription data and any IP address that the network has allocated for the mobile. The PCRF stores the information it receives, for use in later procedures.

If the PCRF requires the user's subscription details to define the default bearer's PCC rule, but does not yet have them, then it retrieves those details from the SPR or UDR (2).

The PCRF can now define the default bearer's quality of service and charging parameters, either by selecting a pre-defined PCC rule that the PCEF is already aware of, or by generating a dynamic rule (3). In doing so, it can either use the default quality of service that originated in the user's subscription data, or specify a new quality of service of its own. It then sends the information to the PCEF using a Diameter *CC-Answer* (4). If the PCC rule specifies the use of online charging, then the PCEF sends a credit request to the online charging system (5), in the manner described below. The PCEF can then proceed with the establishment of the default bearer.

If the user has suitable subscription details, then the PCRF can define multiple PCC rules with different qualities of service in step 3, and can send all those PCC rules to the PCEF in step 4. This triggers the establishment of one or more dedicated bearers to the same APN, as well as the usual default bearer.

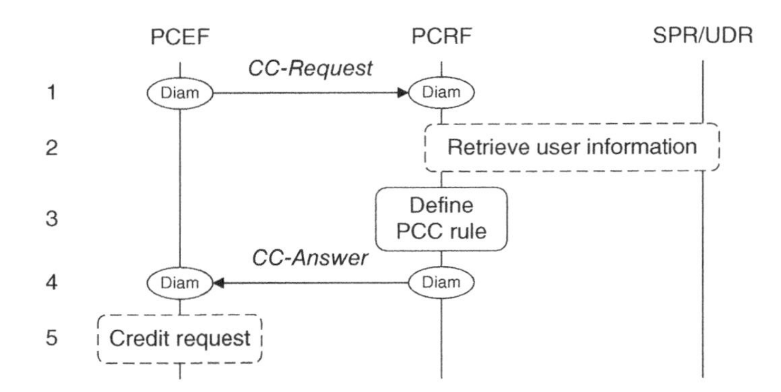

Figure 13.5 IP-CAN session establishment procedure. Source: TS 29.213. Reproduced by permission of ETSI

13.3.2 Mobile Originated SDF Establishment

After the attach procedure has completed, the PCRF can receive requests for a better quality of service from two sources, either the mobile itself or an external application function. The PDN gateway can satisfy those requests in two ways, either by triggering the establishment of a dedicated bearer that has a better quality of service or by modifying the quality of service of an existing bearer.

To illustrate this, Figure 13.6 shows how the mobile can ask for a new service data flow with an improved quality of service [19]. In this scenario, we assume that the mobile has previously made contact with an external VoIP server, using application-layer signalling messages that it has exchanged across a default EPS bearer. We also assume that the mobile's VoIP application would now like to set up a call with a better quality of service than the default EPS bearer can provide, typically with a guaranteed bit rate. Finally, we assume that the mobile is starting in ECM-CONNECTED state. When starting in ECM-IDLE, it first has to complete the service request procedure that we will describe in Chapter 14.

Inside the mobile, the VoIP application asks the LTE signalling protocols to set up a new service data flow with an improved quality of service. The protocols react by composing an ESM *Bearer Resource Allocation Request* and sending it to the MME (1). In the message, the mobile requests parameters such as the QoS class indicator, and the guaranteed and maximum bit rates for the uplink and downlink. It also specifies a traffic flow template that describes the service data flow, using parameters such as the source and destination IP addresses and the UDP port number of the VoIP application. The MME receives the mobile's request and forwards it to the serving gateway as a GTP-C *Bearer Resource Command* (2). In turn, the serving gateway forwards the message to the appropriate PDN gateway (3).

The PDN gateway reacts by triggering the procedure for *PCEF initiated IP-CAN session modification* [20] (Figure 13.7). As before, the PCEF sends the PCRF a Diameter CC-Request (4), in which it identifies the originating mobile and states the requested quality of service. The PCRF looks up the subscription details that it retrieved during IP-CAN session establishment, or contacts the SPR or UDR if required (5).

If the mobile has previously contacted the application function, then the application may have asked the PCRF to notify it about future events involving that mobile. If so, then the PCRF does so by using a Diameter *Re-Auth-Request* (6). To meet the mobile's request, the PCRF

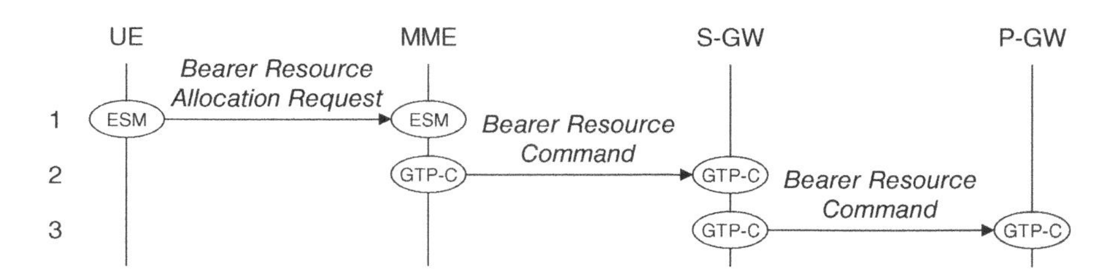

Figure 13.6 Mobile originated SDF establishment procedure. (1) Bearer resource allocation request. Source: TS 23.401. Reproduced by permission of ETSI

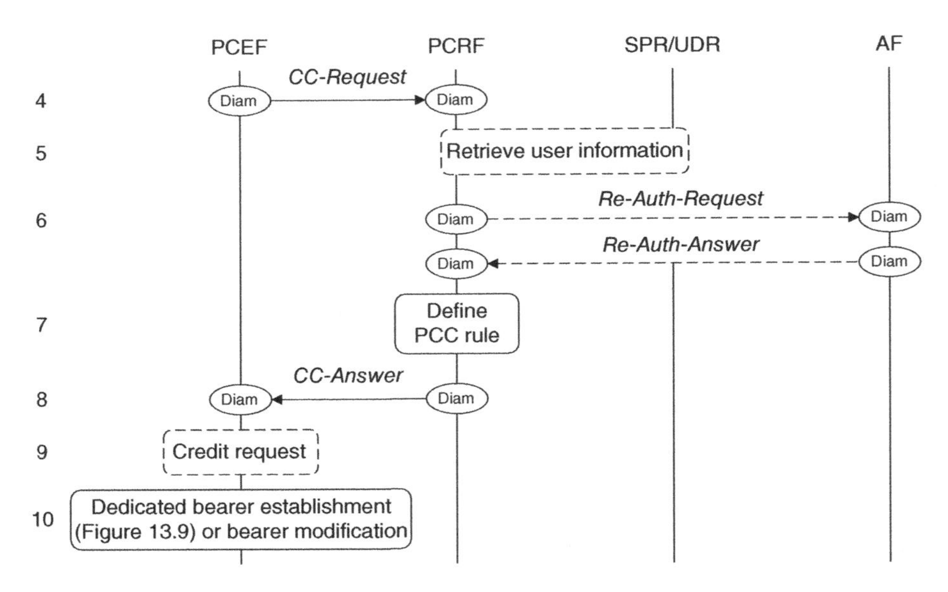

Figure 13.7 Mobile originated SDF establishment procedure. (2) PCEF initiated IP-CAN session modification. Source: TS 29.213. Reproduced by permission of ETSI

defines a new PCC rule (7). It then returns the corresponding QoS and charging parameters to the PCEF (8), which sends a credit request to the online charging system if required (9).

Usually, the PDN gateway reacts to the new PCC rule by establishing a new dedicated bearer (10). We will show how it does this below, after discussing how the application function can make a QoS request of its own. If, however, the mobile already has an EPS bearer with the same QCI and ARP, then the PDN gateway can modify that bearer to include the new service data flow, by increasing its data rate and adding the new packet filters.

13.3.3 Server Originated SDF Establishment

As an alternative to the mobile originated procedure described above, an application function can ask the PCRF for a new service data flow with an improved quality of service. The message sequence is shown in Figure 13.8 [21]. In the figure, we assume as before that the mobile has made contact with an application function such as a VoIP server, using application layer signalling communications over the default bearer. The server would now like to set up a call, using a better quality of service than the default bearer can provide.

In step 1, the application function sends a Diameter *AA-Request* to the PCRF. In the message, it identifies the mobile using its IP address, and describes the requested media using parameters such as the media type, codec and port number, and the maximum uplink and downlink data rates. The PCRF looks up the mobile's subscription details or retrieves them from the SPR or UDR (2), decides that it is willing to authorize the service data flow and returns an acknowledgement (3).

To meet the application function's request, the PCRF defines a new PCC rule (4). It then sends the information to the PCEF in a Diameter *Re-Auth-Request* (5, 6) and the PCEF sends a credit request to the online charging system if required (7). As before, the PDN gateway reacts

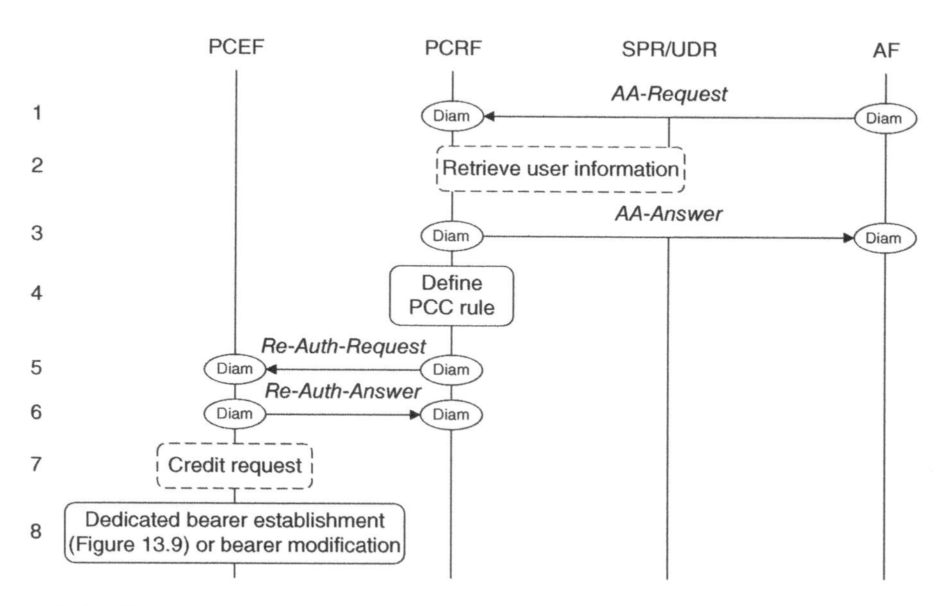

Figure 13.8 Server originated SDF establishment procedure. Source: TS 29.213. Reproduced by permission of ETSI

either by establishing a dedicated bearer in the manner described below or by modifying the quality of service of an existing bearer (8).

13.3.4 Dedicated Bearer Establishment

In the discussions above, we showed how the mobile and the application function could both ask the PCRF for a new service data flow with an improved quality of service. The PCRF responds by defining a new PCC rule, which usually leads to the establishment of a dedicated EPS bearer in the manner shown in Figure 13.9 [22].

The procedure is triggered by one of the quality of service requests from Figures 13.6 to 13.8 (1). In response, the PDN gateway tells the serving gateway to create a new EPS bearer for the mobile, defines its quality of service, and includes an uplink tunnel endpoint identifier for use over S5/S8 and an uplink traffic flow template for the mobile (2). The serving gateway receives the message and forwards it to the MME (3).

If the application function triggered the procedure, then the mobile may still be in ECM-IDLE. If it is, then the MME contacts the mobile using the paging procedure and the mobile responds with an EMM service request. We will discuss these procedures in Chapter 14.

The MME then composes an ESM *Activate Dedicated EPS Bearer Context Request*. This tells the mobile to set up a dedicated EPS bearer and includes the parameters that the MME received from the serving gateway. The MME embeds the message into an S1-AP *E-RAB Setup Request*, which tells the base station to set up the corresponding S1 and radio bearers and defines their qualities of service. In step 4, it sends both messages to the base station.

In response, the base station sends the mobile an RRC Connection Reconfiguration message (5), which tells the mobile how to configure the new radio bearer and which includes the ESM message that it received from the MME. The mobile configures the bearer as instructed and

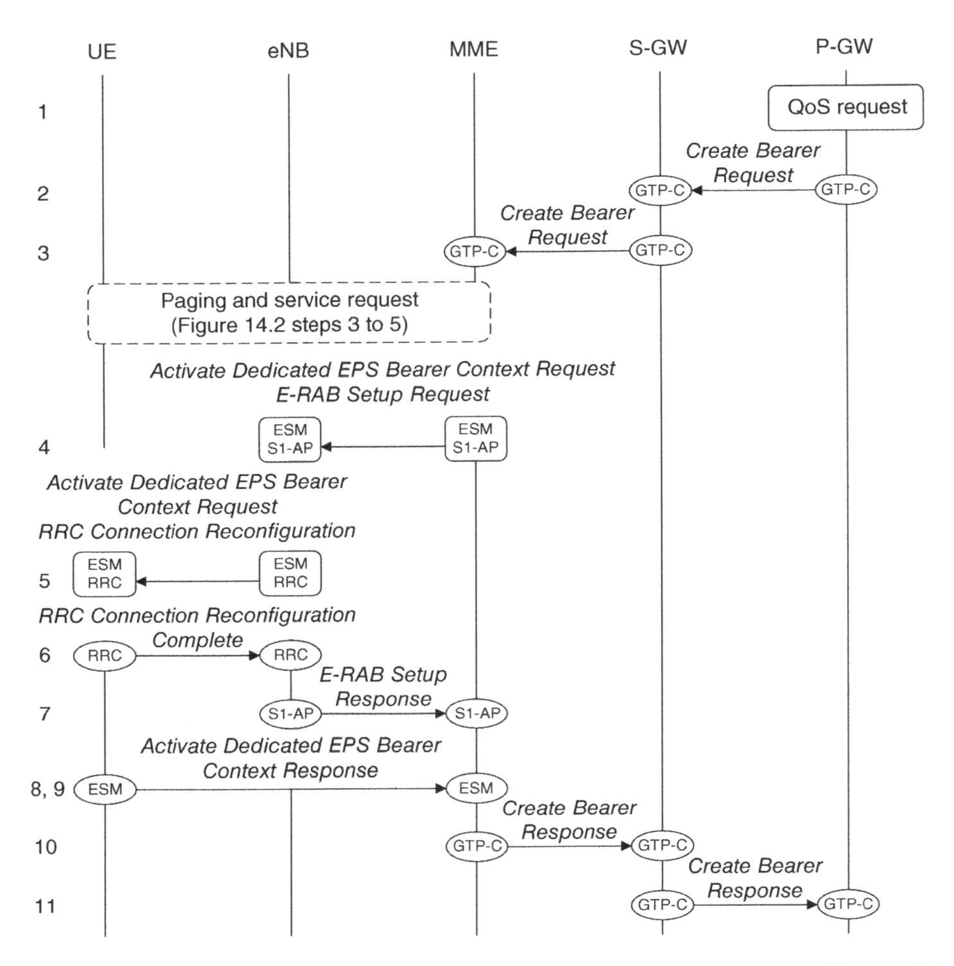

Figure 13.9 Dedicated bearer establishment procedure. Source: TS 23.401. Reproduced by permission of ETSI

acknowledges its RRC message (6), which triggers a further acknowledgement from the base station back to the MME (7). The mobile then acknowledges the ESM message (8, 9), in a reply that is transported using an uplink information transfer. The two final acknowledgements (10, 11) complete the configuration of the GTP tunnels and allow data to flow.

13.3.5 PDN Connectivity Establishment

By using a separate procedure known as *PDN connectivity establishment* [23], the mobile can ask for a connection to a second packet data network with a different access point name. The procedure gives the mobile a second default EPS bearer and a second IP address, and is illustrated in Figure 13.10. It is mandatory for the network to support connectivity to multiple access point names in this way, but optional for the mobile.

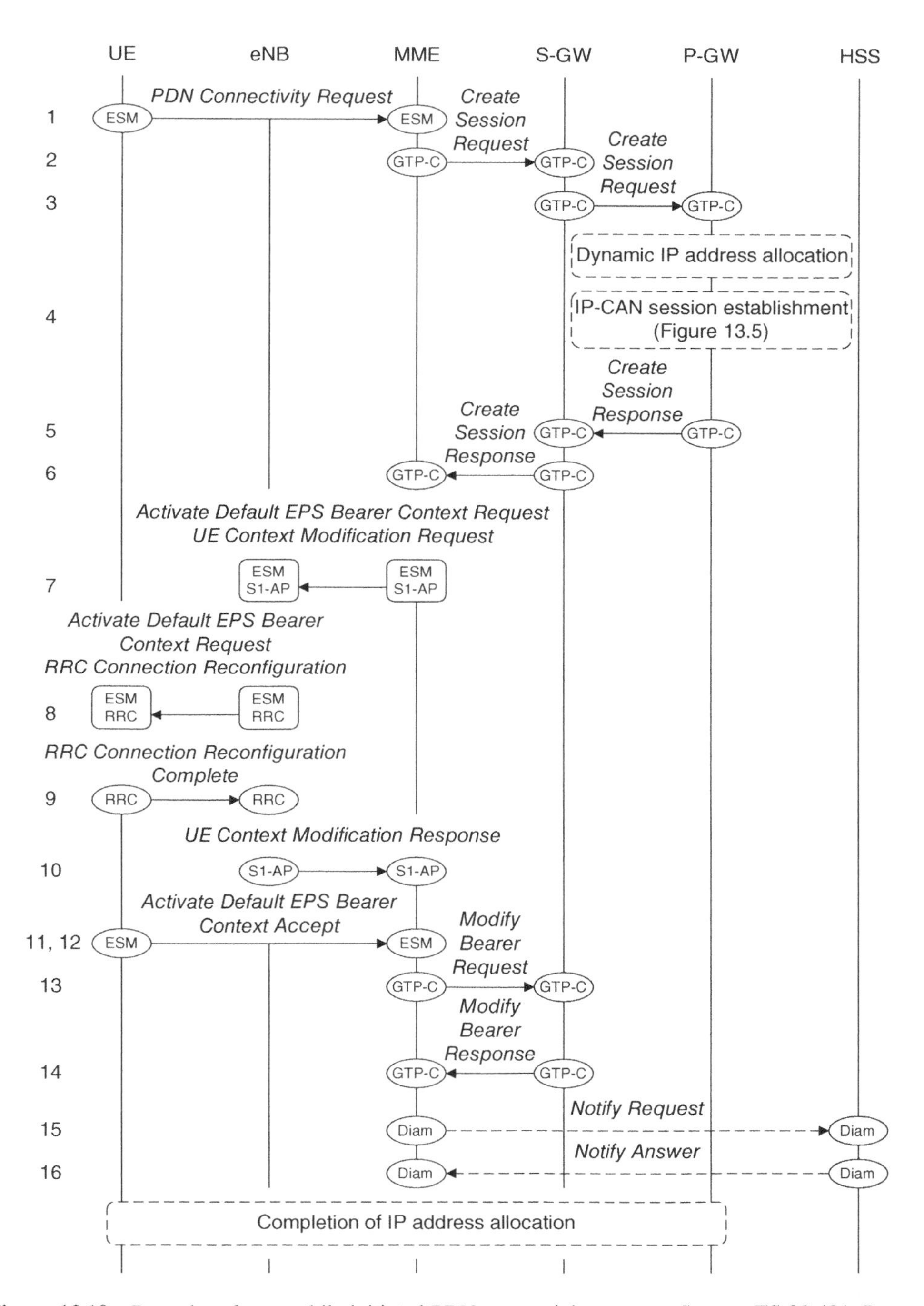

Figure 13.10 Procedure for a mobile initiated PDN connectivity request. Source: TS 23.401. Reproduced by permission of ETSI

The messages are very like the corresponding ones from the attach procedure, so we will only discuss them briefly. The mobile sends the MME a PDN Connectivity Request (1), in which the protocol configuration options include the requested access point name. In response (2–6), the evolved packet core gives the mobile a second IP address, gets approval for the new bearer's quality of service and configures the corresponding GTP-U tunnels. The MME passes the IP address and QoS parameters to the mobile (7–12) and instructs the base station to reconfigure the mobile's radio communications so as to carry the new data stream. The MME then informs the serving gateway about the base station's downlink tunnel endpoint identifier (13, 14), and informs the home subscriber server about the chosen PDN gateway if that gateway was different from the one in mobile's subscription data (15, 16). After the completion of IP address allocation, the mobile can communicate through the new access point name.

13.3.6 Other Session Management Procedures

After the establishment of a service data flow, the mobile and application function can both ask the PCRF to modify its quality of service, for example by increasing its data rate. In response, the PCRF modifies the corresponding PCC rule and sends the resulting parameters to the PDN gateway. In turn, the PDN gateway reacts by modifying the QoS of the corresponding bearer [24] or, if necessary, by extracting the modified service data flow into a new dedicated bearer. The procedures are very like the ones in Figures 13.6 to 13.9.

If the S5/S8 interface is based on PMIP, then it only uses one GRE tunnel per mobile and does not distinguish the qualities of service of the corresponding data streams. Instead, the serving gateway includes a bearer binding and event reporting function, which maps incoming downlink packets onto the correct EPS bearers. During the attach procedure, the BBERF runs a process known as *Gateway control session establishment* [25], in which it establishes communications with the PCRF, forwards the subscription data that it received from the home subscriber server and receives a PCC rule for the default EPS bearer in return. If the PCRF changes the PCC rule during IP-CAN session establishment, then it updates the BBERF using a further procedure known as *Gateway control and QoS rules provision* [26]. There are similar steps during dedicated bearer activation, bearer modification and UE requested PDN connectivity.

Finally, the detach procedure from Chapter 11 triggers a procedure known as *IP-CAN session termination* [27], in which the PCEF informs the PCRF that the mobile is detaching. As part of the procedure, the PCRF can notify the application function, while the PCEF can send a final credit report to the online charging system and return any remaining credit.

13.4 Data Transport in the Evolved Packet Core

13.4.1 Packet Handling at the PDN Gateway

Let us now consider how the evolved packet core delivers an incoming packet to a mobile and how it implements the QoS mechanisms that we have described. When a server sends a packet to a mobile, it addresses the packet to the mobile's IP address. That address lies in the address space of the PDN gateway that allocated it, so the network's incoming packets all arrive there.

Using a technique known as *deep packet inspection*, the PDN gateway examines the incoming packet's headers and compares them with the packet filters that it stored during bearer configuration. By doing so, the PDN gateway identifies the destination mobile, the service data flow and the EPS bearer, and looks up the corresponding S5/S8 bearer and GTP-U tunnel.

The PDN gateway then checks the data rate of the service data flow to ensure that the maximum bit rate is not being exceeded and updates its record of the SDF's traffic for charging purposes. If the evolved packet core is using network address translation, then the PDN gateway also maps the destination IP address and port number of the incoming packet to the private IP address that is uniquely assigned to the mobile.

13.4.2 Data Transport Using GTP

The PDN gateway can now send the packet to the serving gateway in the manner shown in Figure 13.11. The PDN gateway first adds a GTP-U header, which contains a 32 bit tunnel endpoint identifier for the outgoing GTP-U tunnel [28]. It then encapsulates the packet inside another IP packet by adding two further headers. The first is a UDP header, with UDP chosen rather than TCP because the incoming packets may be from a real time service such as voice over IP. The second is an IP header that includes the IP address of the mobile's serving gateway. Finally, the PDN gateway sends the packet into the underlying transport network, which delivers it to the serving gateway.

On arrival, the serving gateway strips off the extra headers, examines the incoming tunnel endpoint identifier and looks up the corresponding EPS bearer. It then identifies the destination mobile and the mobile's base station, adds GTP-U, UDP and IP headers of its own and sends the packet to the base station as before. (The tunnel endpoint identifiers are local to the S5/S8 and S1 interfaces, so the new TEID can be different from the old one.) In turn, the base station uses the transport mechanisms of the air interface to deliver the packet to the mobile.

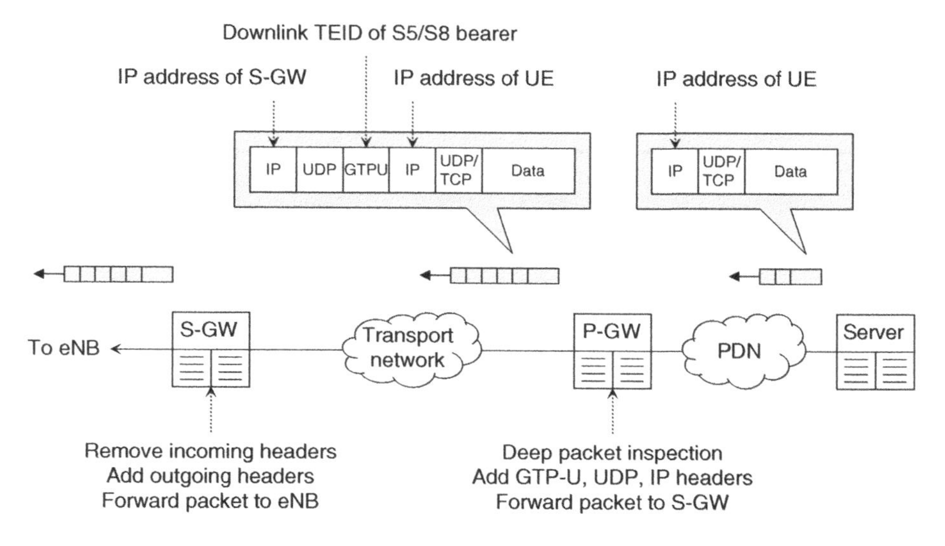

Figure 13.11 Packet forwarding on the S5/S8 interface using GTP

13.4.3 Differentiated Services

By itself, IP does not distinguish the qualities of service of different data streams. However the transport network can offer QoS guarantees using an enhancement to IP known as *differentiated services* (DiffServ) [29–32]. The 3GPP specifications mandate DiffServ on the S1-U and X2 interfaces [33, 34], and the protocol is also commonly used within the evolved packet core.

At the entry point to a DiffServ network, an ingress router examines the incoming packets, groups them into classes that are known as *per hop behaviours* (PHBs) and labels them using a six bit *differentiated services code point* (DSCP) field in the IP header. Inside the network, internal routers use the DSCP field to support their algorithms for queuing, packet dropping and packet forwarding.

There are three types of per hop behaviour. The default behaviour uses *best effort* (BE) forwarding in the same way as a normal IP router and offers no quality of service guarantees. At the other extreme is *expedited forwarding* (EF), an emulation of circuit switching that guarantees a minimum aggregate data rate for that traffic class.

Assured forwarding (AF) offers softer guarantees about prioritization and packet loss. It encompasses four priority levels, each of which is assigned a certain maximum aggregate data rate, together with three levels of drop precedence. A router handles packets from the assured forwarding class in priority order. If the number of buffered packets within a priority level is sufficiently low then they can be forwarded unconditionally; if not, then some of the packets are discarded in accordance with the drop precedence.

When the PDN gateway adds the IP header that we noted earlier, it computes the DSCP field by inspecting the QCI of the overlying bearer. The 3GPP specifications do not mandate any particular mapping from QCI to DSCP, but the GSM Association's specifications for the IP packet exchange do [35]. Table 13.3 shows this mapping, in order to illustrate how the PDN gateway might typically configure the DSCP field. The highest assured forwarding priority is number 4, so packets from QCI 4 are delayed less than those from QCI 5. Within each priority level, packets with drop precedence 1 are the least likely to be discarded, so packets from QCI 5 have a lower packet loss rate than those from QCI 6.

Table 13.3 Mapping from QoS class indicator to DSCP field in the IP exchange

QCI	Per hop behaviour	AF priority	AF drop precedence	DSCP	Example services
1	EF			101 110	Conversational voice
2	EF			101 110	Real time video
3	EF			101 110	Real time games
4	AF	4	1	100 010	Buffered video
5	AF	3	1	011 010	IMS signalling
6	AF	3	2	011 100	Web, email, FTP (high priority users)
7	AF	2	1	010 010	Voice, real time video, real time games
8	AF	1	1	001 010	Web, email, FTP (mid priority users)
9	BE			000 000	Web, email, FTP (low priority users)

13.4.4 Multiprotocol Label Switching

The use of IP for packet routing has some disadvantages. For example, the routers have to make routing decisions using IP addresses that contain 32 or 128 bits, which can be slow. Furthermore, different traffic classes reach a particular destination device along the same traffic path and cannot be segregated from each other.

To overcome these issues, network operators often use an additional protocol known as *multi-protocol label switching* (MPLS) [36, 37]. At the entry point to an MPLS network, an ingress router examines the destination IP address and DSCP field of the incoming packet and adds an MPLS header that includes a 20 bit label. Inside the network, internal routers forward the packet by inspecting the MPLS label rather than the destination IP address. The MPLS label can change from one router to the next in a similar way to the GTP-U tunnel endpoint identifier, so does not have to be globally unique.

The technique brings various advantages. Firstly, the routers can make their routing decisions faster than before, using the simpler MPLS header and the shorter MPLS label. Secondly, the transport network can assign different labels to different DiffServ traffic classes that share the same destination IP address, so can segregate the traffic classes and can improve the network's response to congestion. There is no single mechanism for mapping from the DSCP field to the MPLS header, but various options are possible and can be configured by the network operator [38].

13.4.5 Data Transport Using GRE and PMIP

If the S5/S8 interface is implemented using GRE and PMIP, then that interface only uses one GRE tunnel per mobile. The tunnel handles all the data packets that the mobile is transmitting or receiving, without any quality of service guarantees.

When a downlink packet arrives at the PDN gateway, that device still carries out deep packet inspection so as to direct incoming packets to the correct GRE tunnel and hence to the correct serving gateway. When the packet arrives at the serving gateway, the bearer binding and event reporting function carries out a second process of deep packet inspection to handle the one-to-many mapping from GRE tunnels to EPS bearers. From that point on, the packet is transported in the same way as before.

13.5 Charging and Billing

13.5.1 High Level Architecture

LTE supports a flexible charging model, in which the cost of a session can be calculated from information such as the data volume or duration, and can depend on issues such as the user's quality of service and radio access technology. It uses the same charging architecture as the packet switched domains of UMTS and GSM [39], as shown in Figure 13.12.

There are two different charging systems: the *offline charging system* (OFCS) is suitable for basic post-paid services for which the subscriber receives a regular bill, while the *online charging system* (confusingly abbreviated to OCS) is suitable for more complex charging models

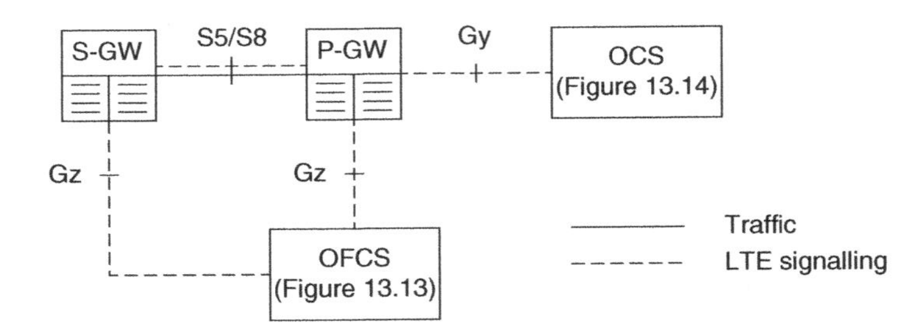

Figure 13.12 High level charging architecture

such as monthly data limits and pre-pay. From a technical point of view, the online charging system can affect a session in real time, for example by terminating the session if the subscriber runs out of credit, while the offline charging system cannot. The charging systems use much the same protocols, which are defined in References [40–44]. We will discuss the two systems in turn below.

13.5.2 Offline Charging

Figure 13.13 shows the internal architecture of the offline charging system. There are four components. *Charging trigger functions* (CTFs) monitor the subscriber's use of resources and generate *charging events*, which describe activities that the system will charge for. *Charging data functions* (CDFs) receive charging events from one or more charging trigger functions and collect them into *charging data records* (CDRs). The *charging gateway function* (CGF) post-processes the charging data records and collects them into *CDR files*, while the *billing domain* (BD) determines how much the resources have cost and sends an invoice to the user.

In LTE, the serving and PDN gateways both contain charging trigger functions. The charging data functions can be separate devices, as shown in Figure 13.13, or they can be integrated into the serving and PDN gateways. Depending on which choice is made, the Gz interface from Figure 13.12 corresponds to either the Rf or the Ga interface from Figure 13.13.

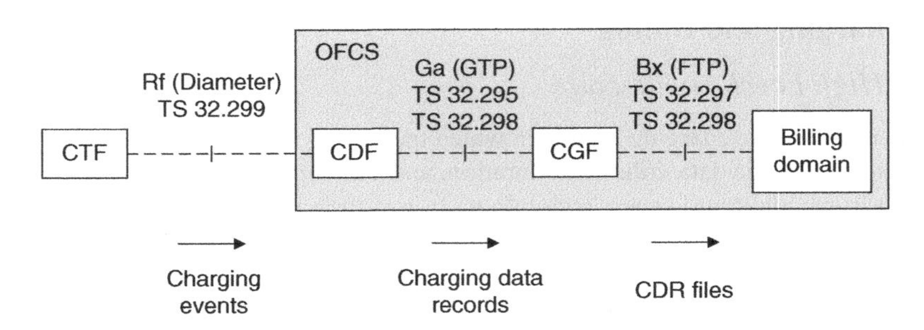

Figure 13.13 Architecture of the offline charging system. Source: TS 32.240. Reproduced by permission of ETSI

If the CDF is a separate device, then the CTF periodically sends it information about a subscriber's use of resources by means of a Diameter *Accounting Request* over Rf. The message contains information such as the rating group, the uplink and downlink traffic volumes, the QoS parameters, the current radio access technology and the user's IMSI. The information makes its way to the CGF and billing domain by way of the Ga and Bx interfaces, and the billing domain uses the information to bill the subscriber.

If the user is roaming, then the billing domain and charging gateway function are in the home network, while the charging data function is in the same network as the PDN gateway or serving gateway. In the case of home routed traffic, for example, the PDN gateway sends charging events to the home network's CDF, which processes them in the usual way. Meanwhile, the serving gateway sends charging events to the visited network's CDF, which uses them in two ways. Firstly, the visited CDF sends its charging data records to the home CGF, which uses them to invoice the subscriber. Secondly, the visited network uses its charging data records to invoice the home network for the subscriber's use of resources.

13.5.3 Online Charging

Figure 13.14 shows the internal architecture of the online charging system. In this architecture, the charging trigger function seeks permission to begin a session by sending a Diameter *Credit Control Request* to the *online charging function* (OCF). This retrieves the cost of the requested resources from the *rating function* (RF) and retrieves the balance of the subscriber's account from the *account balance management function* (ABMF). It then replies to the CTF with a Diameter *Credit Control Answer*, which typically specifies how long the session can last or how much data the user can transfer. The CTF can then allow the session to proceed.

As the session continues, the charging trigger function monitors the subscriber's use of resources. If the subscriber approaches the end of the original allocation, then the CTF sends a new Credit Control Request to the online charging function, to ask for additional resources. At the end of the session, the CTF notifies the online charging function about any remaining credit, and the online charging function returns the credit to the account balance management function.

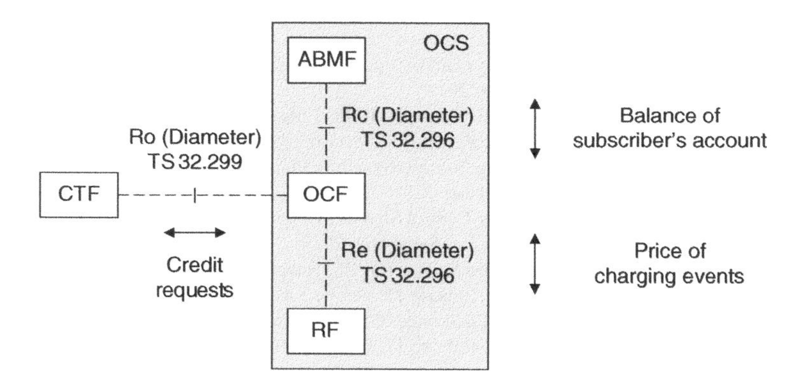

Figure 13.14 Architecture of the online charging system. Source: TS 32.240. Reproduced by permission of ETSI

In LTE, the charging trigger function lies inside the PDN gateway and is actually handled by the PCEF. The Gy interface from Figure 13.12 is identical to the Ro interface from Figure 13.14.

The online charging system is always in the subscriber's home network. If the user is roaming, then the visited network simultaneously creates charging data records using its offline charging system and uses these to invoice the home network for the subscriber's use of resources as before.

References

1. 3GPP TS 23.203 (2013) Policy and Charging Control Architecture, Release 11, September 2013.
2. 3GPP TS 29.213 (2013) Policy and Charging Control Signalling Flows and Quality of Service (QoS) Parameter Mapping, Release 11, September 2013.
3. 3GPP TS 32.240 (2013) Telecommunication Management; Charging Management; Charging Architecture and Principles, Release 11, March 2013.
4. 3GPP TS 32.251 (2013) Telecommunication Management; Charging Management; Packet Switched (PS) Domain Charging, Release 11, June 2013.
5. 3GPP TS 24.301 (2013) Non-Access-Stratum (NAS) Protocol for Evolved Packet System (EPS); Stage 3, Release 11, September 2013.
6. 3GPP TS 29.274 (2013) 3GPP Evolved Packet System (EPS); Evolved General Packet Radio Service (GPRS) Tunnelling Protocol for Control Plane (GTPv2-C); Stage 3, Release 11, September 2013.
7. 3GPP TS 36.331 (2013) Radio Resource Control (RRC); Protocol Specification, Release 11, September 2013.
8. 3GPP TS 36.413 (2013) Evolved Universal Terrestrial Radio Access Network (E-UTRAN); S1 Application Protocol (S1AP), Release 11, September 2013.
9. 3GPP TS 23.203 (2013) Policy and Charging Control Architecture, Release 11, Section 6.1.7, September 2013.
10. 3GPP TS 23.401 (2013) General Packet Radio Service (GPRS) Enhancements for Evolved Universal Terrestrial Radio Access Network (E-UTRAN) Access, Release 11, Section 4.7.3, September 2013.
11. 3GPP TS 23.203 (2013) Policy and Charging Control Architecture, Release 11, Sections 4, 6.3, September 2013.
12. 3GPP TS 29.212 (2013) Policy and Charging Control over Gx/Sd Reference Point, Release 11, Sections 4.3, 5.3.4, September 2013.
13. 3GPP TS 23.203 (2013) Policy and Charging Control Architecture, Release 11, Section 5, September 2013.
14. 3GPP TS 29.212 (2013) Policy and Charging Control over Gx/Sd Reference Point, 3rd Generation Partnership Project, Release 11, September 2013.
15. 3GPP TS 29.214 (2013) Policy and Charging Control over Rx Reference Point, Release 11, September 2013.
16. 3GPP TS 29.215 (2013) Policy and Charging Control (PCC) Over S9 Reference Point; Stage 3, Release 11, September 2013.
17. 3GPP TS 29.335 (2012) User Data Repository Access Protocol Over the Ud interface; Stage 3, Release 11, December 2012.
18. 3GPP TS 29.213 (2013) Policy and Charging Control Signalling Flows and Quality of Service (QoS) Parameter Mapping, Release 11, Section 4.1, September 2013.
19. 3GPP TS 23.401 (2013) General Packet Radio Service (GPRS) Enhancements for Evolved Universal Terrestrial Radio Access Network (E-UTRAN) Access, Release 11, Section 5.4.5, September 2013.
20. 3GPP TS 29.213 (2013) Policy and Charging Control Signalling Flows and Quality of Service (QoS) Parameter Mapping, Release 11, Section 4.3.2.1, September 2013.
21. 3GPP TS 29.213 (2013) Policy and Charging Control Signalling Flows and Quality of Service (QoS) Parameter Mapping, Release 11, Sections 4.3.1.1, 4.3.1.2.1, September 2013.
22. 3GPP TS 23.401 (2013) General Packet Radio Service (GPRS) Enhancements for Evolved Universal Terrestrial Radio Access Network (E-UTRAN) Access, Release 11, section 5.4.1, September, 2013.
23. 3GPP TS 23.401 (2013) General Packet Radio Service (GPRS) Enhancements for Evolved Universal Terrestrial Radio Access Network (E-UTRAN) Access, Release 11, Section 5.10.2, September 2013.
24. 3GPP TS 23.401 (2013) General Packet Radio Service (GPRS) Enhancements for Evolved Universal Terrestrial Radio Access Network (E-UTRAN) Access, Release 11, Section 5.4.2.1, September 2013.
25. 3GPP TS 29.213 (2013) Policy and Charging Control Signalling Flows and Quality of Service (QoS) Parameter Mapping, Release 11, Section 4.4.1, September 2013.

26. 3GPP TS 29.213 (2013) Policy and Charging Control Signalling Flows and Quality of Service (QoS) Parameter Mapping, Release 11, Section 4.4.3, September 2013.
27. 3GPP TS 29.213 (2013) Policy and Charging Control Signalling Flows and Quality of Service (QoS) Parameter Mapping, Release 11, section 4.2.1, September 2013.
28. 3GPP TS 29.281 (2013) General Packet Radio System (GPRS) Tunnelling Protocol User Plane (GTPv1-U), Release 11, Section 5, March 2013.
29. IETF RFC 2474 (1998) Definition of the Differentiated Services Field (DS Field in the IPv4 and IPv6 Headers, December 1998.
30. IETF RFC 2475 (1998) An Architecture for Differentiated Services, December 1998.
31. IETF RFC 2597 (1999) Assured Forwarding PHB Group, June 1999.
32. IETF RFC 3246 (2002) An Expedited Forwarding PHB (Per-Hop Behavior), March 2002.
33. 3GPP TS 36.414 (2012) Evolved Universal Terrestrial Radio Access Network (E-UTRAN); S1 Data Transport, Release 11, Section 5.4, September 2012.
34. 3GPP TS 36.424 (2012) Evolved Universal Terrestrial Radio Access Network (E-UTRAN); X2 Data Transport, Release 11, Section 5.4, September 2012.
35. GSM Association IR.34 (2013) Guidelines for IPX Provider networks, Version 9.1, Section 6.2, May 2013.
36. IETF RFC 3031 (2001) Multiprotocol Label Switching Architecture, January 2001.
37. IETF RFC 3032 (2001) MPLS Label Stack Encoding, January 2001.
38. IETF RFC 3270 (2002) Multi-Protocol Label Switching (MPLS) Support of Differentiated Services, May 2002.
39. 3GPP TS 32.240 (2013) Telecommunication Management; Charging Management; Charging Architecture and Principles, Release 11, Section 4, March 2013.
40. 3GPP TS 32.295 (2012) Charging Data Record (CDR) Transfer, Release 11, September 2012.
41. 3GPP TS 32.296 (2013) Online Charging System (OCS): Applications and Interfaces, Release 11, March 2013.
42. 3GPP TS 32.297 (2012) Charging Data Record (CDR) File Format and Transfer, Release 11, September 2012.
43. 3GPP TS 32.298 (2013) Charging Data Record (CDR) Parameter Description, Release 11, March 2013.
44. 3GPP TS 32.299 (2013) Diameter Charging Applications, Release 11, March 2013.

14

Mobility Management

In this chapter, we discuss the mobility management procedures that the network uses to keep track of the mobile's location.

The choice of mobility management procedures depends on the state that the mobile is in. Mobiles in RRC_IDLE use a mobile-triggered procedure known as cell reselection, whose objective is to maximize the mobile's battery life and minimize the load on the network. In contrast, mobiles in RRC_CONNECTED use the network-triggered procedures of measurements and handover, to give the base station the control it requires over mobiles that are actively transmitting and receiving. We begin the chapter by covering the procedures that switch a mobile between these states in response to changes in the user's activity, namely S1 release, paging and service requests, and continue by describing the mobility management procedures themselves.

Several specifications are relevant for this chapter. TS 36.304 [1] defines the mobility management procedures that the mobile should follow in RRC_IDLE, while TS 23.401 [2] and TS 36.300 [3] describe the signalling procedures in RRC_CONNECTED and the procedures that switch the mobile between states. As usual, the relevant stage 3 specifications [4–8] define the details of the individual signalling messages. Another specification [9] defines the measurements that the mobile has to make in both RRC states and the corresponding performance requirements.

14.1 Transitions between Mobility Management States

14.1.1 S1 Release Procedure

After the attach procedure from Chapter 11, the mobile had completed its power-on procedures and was in the states EMM-REGISTERED, ECM-CONNECTED and RRC_CONNECTED. The user was able to communicate with the outside world, using the default EPS bearer. If the user does nothing, then the network can use a procedure known as *S1 release* [10] to take the mobile into ECM-IDLE and RRC_IDLE. As part of this procedure, the network tears down signalling radio bearers 1 and 2 and deletes all the user's data radio bearers and S1 bearers.

An Introduction to LTE: LTE, LTE-Advanced, SAE, VoLTE and 4G Mobile Communications, Second Edition.
Christopher Cox.

Using the cell reselection procedure that we cover later, the mobile can then move from one cell to another without the need to re-route the bearers, so only has a minimal impact on the network.

Figure 14.1 shows the message sequence. After a period of user inactivity, a timer expires in the base station and triggers the procedure. In response, the base station asks the MME to move the mobile into ECM-IDLE (1).

On receiving the request, the MME tells the serving gateway to tear down the S1 bearers that it was using for the mobile (2). The serving gateway does so and responds (3), but the PDN gateway is not involved, so the S5/S8 bearers remain intact. From this point, any downlink data can only travel as far as the serving gateway, where they trigger the paging procedure described below.

Using messages that we have already seen in the detach procedure, the MME tells the base station to tear down its signalling communications with the mobile (4), and the base station sends a similar message across the air interface (5). In response, the mobile tears down SRB 1, SRB 2 and all its data radio bearers, and moves into ECM-IDLE and RRC_IDLE. There is no need for it to reply. At the same time, the base station tears down its own records of the mobile's S1 and radio bearers, and sends an acknowledgement to the MME (6).

The S1 release procedure has several variations. The base station can trigger the procedure for other reasons, such as a repeated failure of the integrity check or a loss of radio communications with the mobile. The MME can trigger the procedure starting from step 2; for example, after a failure of the authentication procedure. The MME can also trigger the procedure by sending an S1-AP message to the mobile before step 1; for example, during circuit switched fallback. In this last variation, the base station knows in advance that the MME will accept its UE Context Release Request, so it can release the mobile's RRC connection right away instead of delaying that message until step 5.

On completion of the S1 release procedure, the MME will sometimes tear down the mobile's GBR bearers on the grounds that it can no longer support them. This typically happens if the S1 connection is being released because of a problem such as a loss of radio communications, but not if it is being released for a benign reason such as user inactivity or a redirection to another radio access technology.

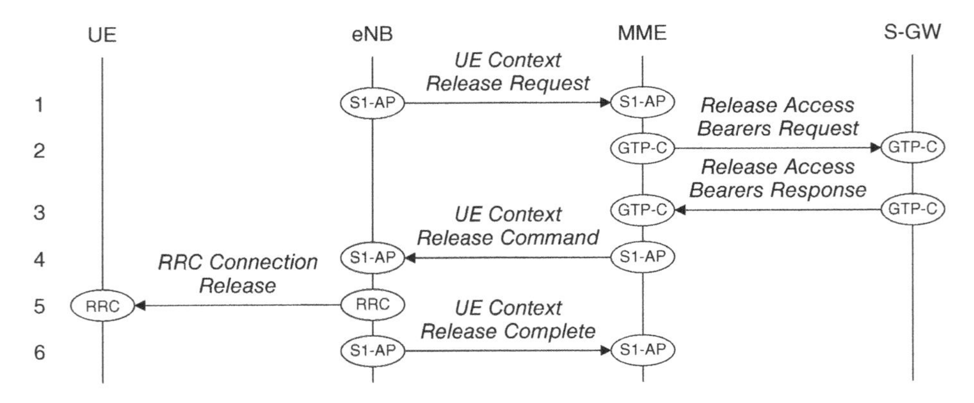

Figure 14.1 S1 release procedure, triggered by user inactivity in the base station. Source: TS 23.401. Reproduced by permission of ETSI

14.1.2 Paging Procedure

At the end of the S1 release procedure, the mobile is in ECM-IDLE and RRC_IDLE, and its S1 and radio bearers have all been torn down. What happens if downlink data arrive at the serving gateway?

Figure 14.2 shows the answer [11]. The PDN gateway forwards an incoming data packet to the serving gateway (1), but the serving gateway is unable to send the data any further. Instead, it sends a GTP-C *Downlink Data Notification* message to the MME (2), to tell the MME what has happened. The MME acknowledges, which stops the serving gateway from sending further notifications if more data packets arrive.

The MME now composes an S1-AP *Paging* message and sends it to the base stations in all of the mobile's tracking areas (3). In the message, it specifies the mobile's S-TMSI and all the information that the base stations will need to calculate the paging frame and the paging occasion. In response, each base station sends a single RRC *Paging* message to the mobile (4), in accordance with the discontinuous reception procedure from Chapter 8. The mobile receives the message from one of the base stations and responds using the service request procedure described below (5).

14.1.3 Service Request Procedure

The mobile runs the *service request* procedure [12] if it is in ECM-IDLE state but wishes to communicate with the network. The procedure can be triggered in two ways, either by the paging procedure described above or internally; for example, if the user tries to contact a server while the mobile is idle. During the procedure, the network moves the mobile into ECM-CONNECTED, re-establishes SRB 1 and SRB 2, and re-establishes the mobile's data radio bearers and S1 bearers. The mobile can then exchange data with the outside world.

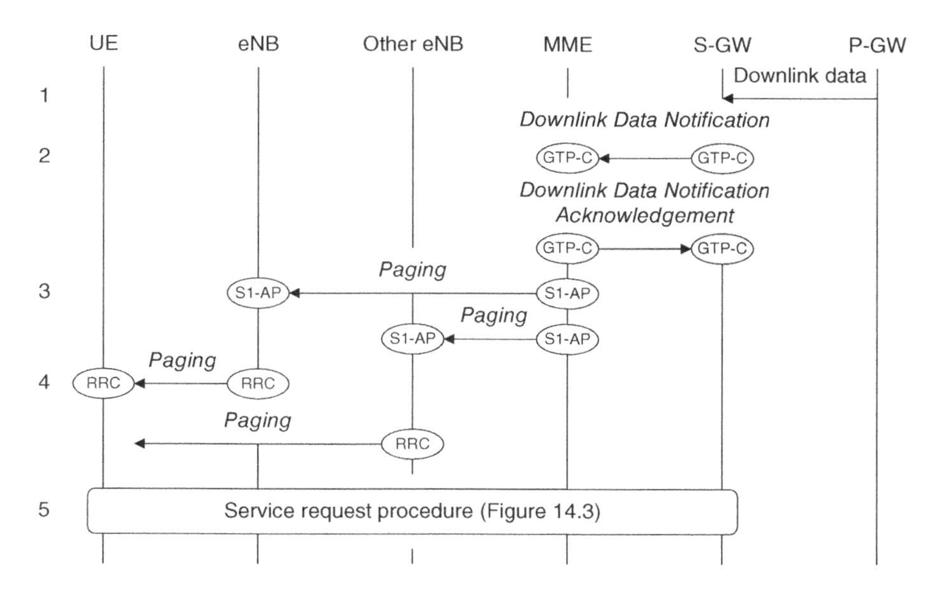

Figure 14.2 Paging procedure. Source: TS 23.401. Reproduced by permission of ETSI

The message sequence is shown in Figure 14.3. There is actually little new material here, as most of the messages have already appeared within the procedures for registration and session management.

The mobile starts by establishing a signalling connection with a serving eNB, using the procedures for contention based random access and RRC connection establishment. It then composes an EMM *Service Request*, which asks the serving MME to move it into ECM-CONNECTED. It embeds the request into its RRC Connection Setup Complete and sends both messages to the base station (1).

The base station extracts the mobile's request and forwards it to the MME (2). Optionally, the MME can now authenticate the mobile and can update non access stratum security using the resulting keys (3). These steps are mandatory if the EMM message failed the integrity check and are optional otherwise.

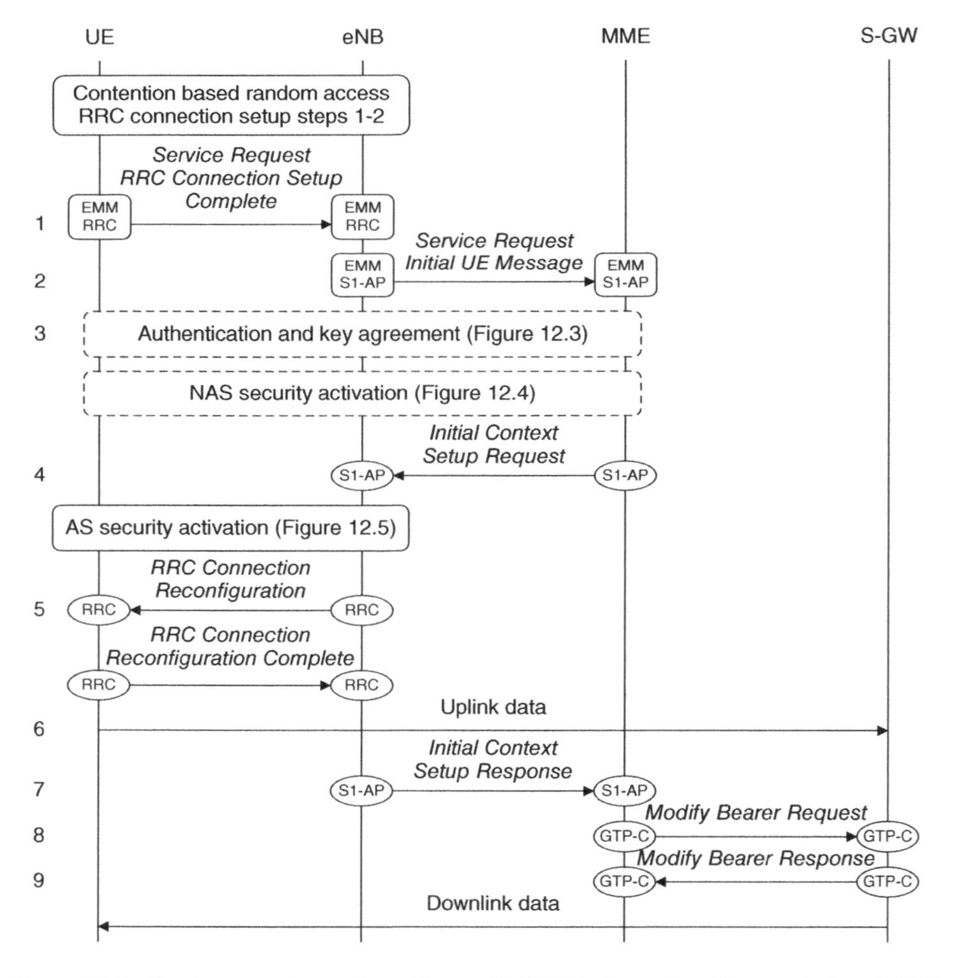

Figure 14.3 Service request procedure. Source: TS 23.401. Reproduced by permission of ETSI

The MME now tells the base station to set up the mobile's S1 and radio bearers (4). The message includes several parameters that the MME stored after the attach procedure, such as the identity of the serving gateway, the tunnel endpoint identifiers for the uplink S1 bearers, the mobile's radio access capabilities and the security key K_{eNB}. The base station reacts to the last of these by activating access stratum security.

The base station can now send an RRC message to the mobile (5), which configures SRB 2 and the mobile's data radio bearers. The mobile treats the message as an implicit acceptance of its EMM service request, returns an acknowledgement, and can now send any uplink data to the outside world (6). In turn, the base station sends an acknowledgement to the MME (7) and includes tunnel endpoint identifiers (TEIDs) for the serving gateway to use on the downlink.

The MME now forwards the downlink TEIDs to the serving gateway and identifies the target base station (8). The serving gateway responds (9) and can now send downlink data to the mobile, notably any data that triggered the paging procedure in Figure 14.2.

14.2 Cell Reselection in RRC_IDLE

14.2.1 Objectives

For mobiles in RRC_IDLE and ECM-IDLE, the mobility management procedures have two main objectives. The first is to maximize the mobile's battery life and the second is to minimize the signalling load on the network.

LTE achieves these objectives using four techniques. Firstly, the mobile usually wakes up only once in every discontinuous reception cycle to monitor the network for paging messages and to make the measurements described below. Using this technique, the mobile can spend most of its time in a low power state. Secondly, the mobile decides by itself whether to stay with the previous cell or move to a new one, by following a procedure known as *cell reselection* that does not require any explicit RRC signalling. Thirdly, the mobile does not inform the network every time it changes cell; instead, it only does so if it moves into a tracking area in which it was not previously registered. Finally, the mobile does not have to camp on the cell with the strongest signal; instead, it only has to camp on a cell whose signal lies above a predefined threshold.

In this section, we discuss the mobility management procedures that these mobiles use. We will mainly follow the steps used in Release 8, but will also note some additional features that were introduced in Release 9.

14.2.2 Measurement Triggering on the Same LTE Frequency

We start the cell reselection procedure [13, 14] by assuming that the network is only using a single LTE carrier frequency. If this is the case, then the mobile wakes up once every discontinuous reception cycle, in the same subframes that it is already monitoring for paging messages. In Release 8, the mobile uses those subframes to measure the reference signal received power (RSRP) from the serving cell. If the RSRP is high enough, then the mobile can continue camping on that cell and does not have to measure any neighbouring cells at all. This technique minimizes the number of measurements that the mobile performs and the time for which it is awake, so maximizes its battery life.

This situation continues until the RSRP falls below the following threshold:

$$S_{\text{rxlev}} \leq S_{\text{IntraSearchP}} \tag{14.1}$$

In this equation, $S_{\text{IntraSearchP}}$ is a threshold that the serving cell advertises as part of SIB 3. S_{rxlev} depends on the RSRP of the serving cell and is calculated using Equation 11.2. If the above condition is met, then the mobile starts to measure neighbouring cells that are on the same LTE carrier frequency as the serving cell. To do this, it runs the first three steps of the acquisition procedure from Chapter 7, so as to receive the primary and secondary synchronization signals, discover the location and content of the downlink reference signals and measure their RSRP.

From Release 9, a mobile can also start measurements of neighbouring cells if the reference signal received quality (RSRQ) falls below the following threshold:

$$S_{\text{qual}} \leq S_{\text{IntraSearchQ}} \tag{14.2}$$

Here, $S_{\text{IntraSearchQ}}$ is another threshold that the base station advertises as part of SIB 3. S_{qual} depends on the serving cell's RSRQ and is calculated using Equation 11.5.

Unlike in earlier systems, the mobile can find neighbouring LTE cells by itself: the base station does not have to advertise an LTE neighbour list as part of its system information. This brings three benefits. Firstly, the network operator can configure the radio access network more easily in LTE than before. Secondly, there is no risk of a mobile missing nearby cells due to errors in the neighbour list. Thirdly, it is easier for an operator to introduce home base stations, which the user can install in locations that are unknown to the surrounding macrocell network. However the base station can still identify individual neighbouring cells in SIB 4 using their physical cell identities, and can describe them using optional cell-specific parameters that we will see next.

14.2.3 Cell Reselection to the Same LTE Frequency

After finding and measuring the neighbouring cells, the mobile computes the following ranking scores:

$$\begin{aligned} R_{\text{s}} &= Q_{\text{meas, s}} + Q_{\text{hyst}} \\ R_{\text{n}} &= Q_{\text{meas, n}} - Q_{\text{offset, s, n}} \end{aligned} \tag{14.3}$$

Here, R_{s} and R_{n} are the ranking scores of the serving cell and one of its neighbours, while $Q_{\text{meas, s}}$ and $Q_{\text{meas, n}}$ are the corresponding reference signal received powers. Q_{hyst} is a hysteresis parameter that the base station advertises in SIB 3, which discourages the mobile from bouncing back and forth between cells as the signal levels fluctuate. $Q_{\text{offset, s, n}}$ is an optional cell-specific offset, which the serving cell can advertise in SIB 4 to encourage or discourage the mobile to or from individual neighbours.

The mobile then switches to the best ranked cell, provided that three conditions are met. Firstly, the mobile must have been camped on the serving cell for at least one second. Secondly, the new cell must be suitable, according to the criteria laid out in Chapter 11. Finally, the new cell must be better ranked than the serving cell for a time of at least $T_{\text{reselection, EUTRA}}$, which is advertised in SIB 3 and has a value of 0 to 7 s. The mobile uses the same procedure if any of

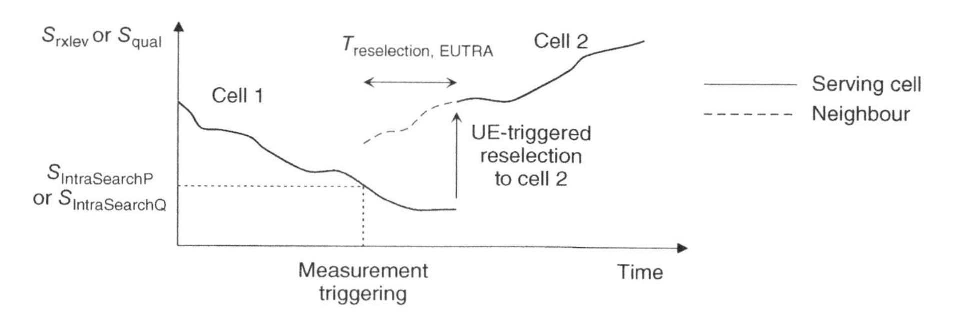

Figure 14.4 Intra frequency cell reselection

the neighbouring cells belongs to a closed subscriber group, except that the mobile must also belong to the group in order to camp on a CSG cell.

Figure 14.4 shows the end result. In this diagram, the mobile is initially camping on cell 1, but is moving away from that cell and towards cell 2. When S_{rxlev} falls below $S_{\mathrm{IntraSearchP}}$ or S_{qual} falls below $S_{\mathrm{IntraSearchQ}}$, the mobile starts to measure neighbouring cells and discovers cell 2. Subject to the conditions noted above, it can then carry out a cell reselection.

14.2.4 Measurement Triggering on a Different LTE Frequency

If the network is using more than one LTE carrier frequency, then the serving cell advertises the other carriers as part of SIB 5. As before, it may include offsets for the individual neighbouring cells, but is not obliged to do so. However, the serving cell does associate each carrier frequency with a priority from 0 to 7, where 7 is the highest priority. The network can use these priorities to encourage or discourage the mobile to or from individual carriers, a feature that is particularly useful in layered networks, as microcells are usually on a different carrier frequency from macrocells and usually have a higher priority.

The measurement triggering procedure depends on the relative priorities of the two carrier frequencies. The mobile always measures cells on higher priority carriers, no matter how strong the signal from the serving cell. It makes the measurements separately from the discontinuous reception cycle, as the mobile cannot look for paging messages on one carrier and measure cells on another at the same time. However the mobile only has to measure one carrier frequency every minute, so the load on the mobile is small.

In Release 8, the mobile starts to measure cells on equal or lower priority carriers if the following condition is satisfied:

$$S_{\mathrm{rxlev}} \leq S_{\mathrm{NonIntraSearchP}} \tag{14.4}$$

Here, $S_{\mathrm{NonIntraSearchP}}$ is a threshold that the base station advertises in SIB 3, while S_{rxlev} depends on the serving cell's RSRP as before. From Release 9, the mobile can also start to measure these carriers if the RSRQ falls below the following threshold:

$$S_{\mathrm{qual}} \leq S_{\mathrm{NonIntraSearchQ}} \tag{14.5}$$

14.2.5 Cell Reselection to a Different LTE Frequency

The cell reselection procedure is also affected by the relative priorities of the two carrier frequencies. The mobile moves to a new cell on a higher priority carrier if three conditions are met. We have already seen the first two: the mobile must have been camped on the serving cell for at least 1 s, and the new cell must be suitable according to the criteria from Chapter 11. In Release 8, the new cell's RSRP must also meet the following condition, for a time of at least $T_{\text{reselection, EUTRA}}$:

$$S_{\text{rxlev}, x, n} > \text{Thresh}_{x, \text{HighP}} \tag{14.6}$$

Here, $\text{Thresh}_{x, \text{HighP}}$ is a threshold for frequency x that the serving cell advertises in SIB 5. $S_{\text{rxlev}, x, n}$ depends on the new cell's RSRP and is calculated using Equation 11.2. The mobile does not measure the RSRP from the serving cell in making this decision, so it moves to a higher priority frequency whenever it finds a cell that is good enough. From Release 9, the base station can optionally replace this last condition with a similar one based on the RSRQ.

The mobile moves to a new cell on a lower priority carrier if five conditions are met. The first two are the same as before: the mobile must have been camped on the serving cell for at least 1 s and the new cell must be suitable according to the criteria from Chapter 11. In addition, the mobile must be unable to find a satisfactory cell on the original frequency or on a frequency with an equal or higher priority. In Release 8, the RSRPs of the serving and neighbouring cells must also satisfy the following conditions, for a time of at least $T_{\text{reselection, EUTRA}}$:

$$S_{\text{rxlev}} < \text{Thresh}_{\text{Serving, LowP}}$$

$$S_{\text{rxlev}, x, n} > \text{Thresh}_{x, \text{LowP}} \tag{14.7}$$

As before, $\text{Thresh}_{\text{Serving, LowP}}$ and $\text{Thresh}_{x, \text{LowP}}$ are thresholds that the base station advertises in SIB 5, while S_{rxlev} and $S_{\text{rxlev}, x, n}$ depend on the RSRP of the serving and neighbouring cells. From Release 9, the base station can optionally replace these last conditions with similar ones based on the RSRQ.

The mobile moves to a new cell on an equal priority carrier using nearly the same criteria that it did for the same carrier frequency. The only difference is in Equation 14.3, where the serving cell can optionally add a frequency-specific offset $Q_{\text{offset, frequency}}$ to the cell-specific offset $Q_{\text{offset, s}, n}$.

Finally, there is one adjustment for mobiles that belong to a closed subscriber group. If a mobile is camped on a non-CSG cell and detects a suitable CSG cell that is the highest ranked on another carrier, then it moves to that CSG cell, irrespective of the new carrier's priority.

14.2.6 Fast Moving Mobiles

In the above algorithms, the mobile can only move to a neighbouring cell whose received signal power has been above a suitable threshold for a time of at least $T_{\text{reselection, EUTRA}}$. Usually, the value of $T_{\text{reselection, EUTRA}}$ is fixed. For fast moving mobiles, however, the use of a fixed value can introduce unwanted delays into the procedure, and can even prevent a mobile from moving to a new cell altogether.

To deal with this problem, the mobile measures the rate at which it is making cell reselections, ignoring any reselections that cause it to bounce back and forth between neighbouring cells.

Depending on the result, it places itself either in a normal mobility state or in a state of medium or high mobility.

In the medium and high mobility states, the mobile makes two adjustments. Firstly, it reduces the reselection time, $T_{\text{reselection, EUTRA}}$, using a state-dependent scaling factor that lies between 0.25 and 1. This reduces the delays in the cell reselection procedure and allows the mobile to move to a neighbouring cell more quickly. Secondly, the mobile reduces the hysteresis parameter Q_{hyst} from Equation 14.3. This makes the mobile less likely to stick in the current cell, so also eases the process of cell reselection. The network specifies all the necessary thresholds and adjustments as part of SIB 3.

14.2.7 Tracking Area Update Procedure

After the mobile reselects to a new cell, it reads the cell's system information and examines the tracking area code. If the mobile has moved into a tracking area in which it was not previously registered, then it tells the evolved packet core using a procedure known as a *tracking area update* [15].

Figure 14.5 shows a basic version of the procedure. The diagram assumes that the mobile is starting in ECM-IDLE and RRC_IDLE and also that the mobile has stayed in the same MME pool area and S-GW service area, so that the MME and serving gateway can remain unchanged.

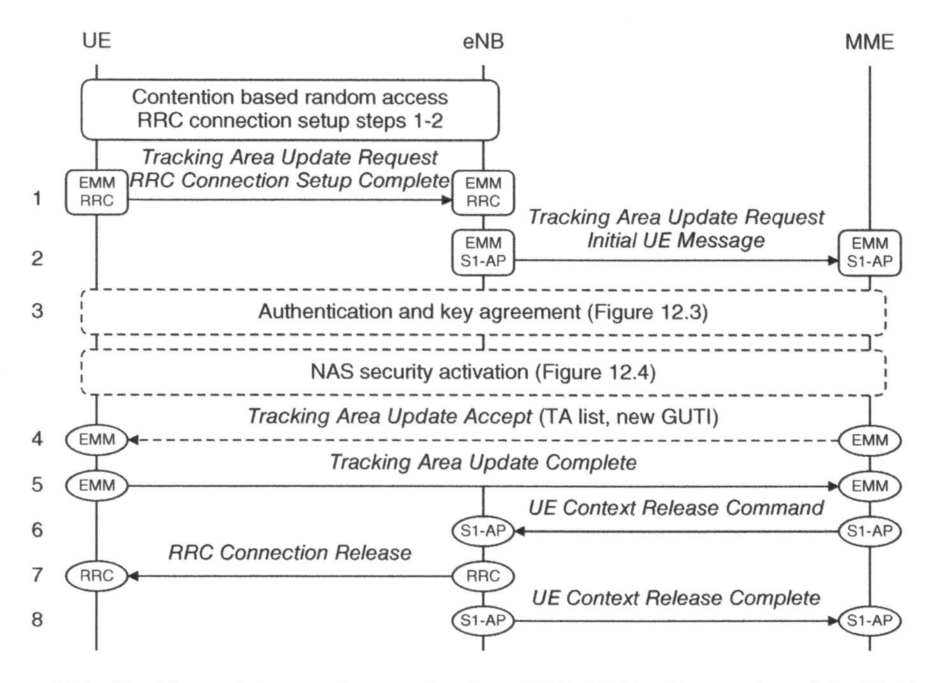

Figure 14.5 Tracking update procedure, starting from RRC_IDLE, with retention of the MME and serving gateway. Source: TS 23.401. Reproduced by permission of ETSI

The procedure begins in a similar way to the service request that we saw earlier. The mobile runs the random access procedure and steps 1 and 2 of RRC connection establishment, so as to move temporarily into RRC_CONNECTED. It then composes an EMM *Tracking Area Update Request* and sends it to the base station by embedding it into its RRC Connection Setup Complete (1). In turn, the base station forwards the message to the serving MME (2). As before, the MME can authenticate the mobile and update non access stratum security (3).

The MME examines the new tracking area and decides in this example that the serving gateway can remain unchanged. In response, it sends the mobile an EMM *Tracking Area Update Accept* (4), in which it lists the tracking areas in which the mobile is now registered and optionally gives the mobile a new globally unique temporary identity. If the GUTI does change, then the mobile sends an acknowledgement (5). The MME can now tell the base station to move the mobile back into RRC_IDLE, in messages that we saw as part of S1 release (steps 6–8).

The mobile also runs the tracking area update procedure periodically, even if it stays in the same tracking area, to tell the MME that it is still switched on and in an area of LTE coverage. The timer has a default value of 54 minutes, but the MME can choose a different value in its Attach Accept or Tracking Area Update Accept messages.

There are two complications. If the mobile has moved into a new MME pool area, then the new base station will not be connected to the old MME. Instead, the base station chooses a new MME after step 1, using the same technique that we saw during the attach procedure. The new MME retrieves the mobile's details from the old MME and contacts the home subscriber server to update its record of the mobile's location. It also tells the serving gateway that it is now looking after the mobile.

Independently, the mobile may have moved into a new S-GW service area. In place of the last interaction from the paragraph above, the MME tells the new serving gateway to set up a new set of EPS bearers for the mobile and to redirect the S5/S8 bearers by contacting the PDN gateway. The MME also tells the old serving gateway to tear down its bearers. If the MME has changed as well, then these steps are performed by the new and old MME respectively.

14.2.8 Network Reselection

If a mobile is configured for automatic network selection and is roaming in a visited network, then it periodically runs a procedure for *network reselection* whenever it is in RRC_IDLE, to search for networks that have the same country code but a higher priority [16, 17]. The search period is stored in the USIM and has a default value of 60 minutes [18].

The procedure is the same as the earlier one for network and cell selection, except that the conditions for a suitable cell (Equations 11.2 and 11.5) are modified as follows:

$$S_{\text{rxlev}} = Q_{\text{rxlevmeas}} - Q_{\text{rxlevmin}} - Q_{\text{rxlevminoffset}} - P_{\text{compensation}} \tag{14.8}$$

$$S_{\text{qual}} = Q_{\text{qualmeas}} - Q_{\text{qualmin}} - Q_{\text{qualminoffset}} \tag{14.9}$$

In the first equation, the original serving cell specifies the parameter $Q_{\text{rxlevminoffset}}$ as part of SIB 1. This parameter increases the minimum RSRP that is required in the destination cell and prevents a mobile from selecting a high priority network that only contains poor cells. The same applies to the parameter $Q_{\text{qualminoffset}}$ from Release 9.

14.3 Measurements in RRC_CONNECTED

14.3.1 Objectives

If a mobile is in the states of ECM-CONNECTED and RRC_CONNECTED, then the mobility management procedures are completely different. In these states, the mobile is already transmitting and receiving with a high data rate, so the procedures do not significantly increase its power consumption or signalling load. Instead, the objective is to maximize the mobile's data rate and the overall capacity of the system.

LTE achieves this objective using three techniques, which are very different from the ones used in ECM-IDLE and RRC_IDLE. Firstly, the base station always knows which cell is serving the mobile. Secondly, the base station chooses the serving cell by means of mobile-specific RRC signalling messages. Thirdly, the base station can ensure that the serving cell is the one with the strongest signal of all, not just a cell whose signal lies above a threshold.

The result is a two-step procedure. In the first step, the mobile measures the signal levels from the serving cell and its nearest neighbours and sends measurement reports to the serving eNB. In the second step, the serving eNB can use those measurements to request a handover to a neighbouring cell.

14.3.2 Measurement Procedure

The measurement procedure [19, 20] is shown in Figure 14.6. To start the procedure, the serving eNB sends the mobile an RRC Connection Reconfiguration message (step 1). We have already seen this message being used to reconfigure a mobile's radio bearers in places such as the attach procedure, but it can also be used to specify the measurements that a mobile should be making.

In the RRC message, the measurements are specified using an information element known as a *measurement configuration*. This contains a list of *measurement objects*, each of which describes an LTE carrier frequency to measure and the corresponding downlink bandwidth. It also contains a list of *reporting configurations*, which tell the mobile when to report the results. Finally, it defines each individual measurement using a *measurement identity*, which simply pairs up a measurement object with a reporting configuration.

As before, the mobile can identify the neighbouring cells by itself, so the base station does not have to list them. The measurement object can, however, contain several optional fields, such

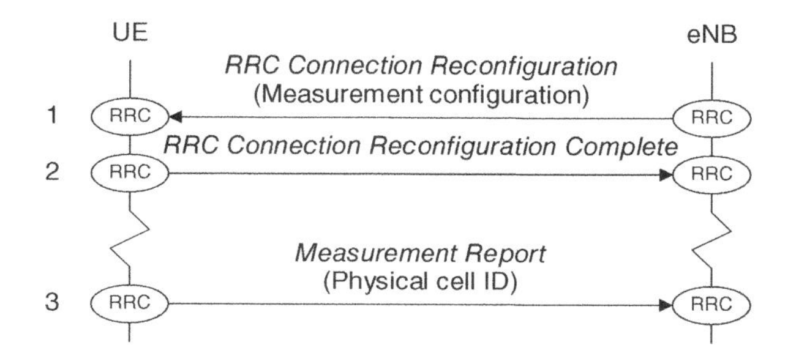

Figure 14.6 Measurement reporting procedure

as a list of cells that the mobile should ignore when making measurements, and frequency- and cell-specific offsets for use in measurement reporting. In these fields, the base station identifies each cell using its physical cell identity.

The mobile acknowledges (step 2) and makes the measurements. Eventually, the mobile sends an RRC *Measurement Report* to the serving eNB (step 3), and identifies the cell that triggered the report using its physical cell identity. In response, the serving eNB can run the handover procedure that we will cover shortly.

14.3.3 Measurement Reporting

The base station tells the mobile when to return a measurement report using the reporting configuration that we introduced above. The most general reporting mechanism in LTE is event-triggered periodic reporting, in which a mobile starts to send periodic measurement reports if a signal level crosses over a threshold, and stops if the signal crosses back. By simplifying this mechanism, the specifications also support one-off event-triggered reports, as well as unconditional periodic and one-off reports.

Table 14.1 lists the events that are used for measurements of other LTE cells, and notes some possible applications. We will see some other measurement events when we discuss inter-system operation in Chapter 15.

As an example, let us consider measurement event A3, which triggers the majority of LTE handovers. The mobile enters the event A3 reporting state when the following condition is met for a time of at least *TimeToTrigger* (0 to 5120 milliseconds):

$$M_n + \mathrm{Of}_n + \mathrm{Oc}_n > M_s + \mathrm{Of}_s + \mathrm{Oc}_s + \mathrm{Off} + \mathrm{Hys} \tag{14.10}$$

Whilst in this state, the mobile sends measurement reports to the base station with a period of *ReportInterval* (20 milliseconds to 60 minutes), up to a maximum of *ReportAmount* reports (1 to 64 or unlimited). The mobile leaves the event A3 reporting state when the following condition is met for a time of at least TimeToTrigger:

$$M_n + \mathrm{Of}_n + \mathrm{Oc}_n < M_s + \mathrm{Of}_s + \mathrm{Oc}_s + \mathrm{Off} - \mathrm{Hys} \tag{14.11}$$

Table 14.1 Measurement events used to report neighbouring LTE cells

Event	Description	Possible application
A1	Serving cell > Threshold	Remove measurement gaps and stop measuring other carriers Add secondary cell (Release 10)
A2	Serving cell < Threshold	Insert measurement gaps and start measuring other carriers RRC connection release with redirection to 2G/3G carrier Remove secondary cell (Release 10)
A3	Neighbour > Serving cell + Offset	Handover to LTE cell on same carrier Handover to LTE cell on equal priority carrier
A4	Neighbour > Threshold	Handover to LTE cell on higher priority carrier
A5	Serving cell < Threshold 1 & Neighbour > Threshold 2	Handover to LTE cell on lower priority carrier
A6	Neighbour > SCell + Offset	Change of secondary cell (Release 10)

In these equations, M_s and M_n are the mobile's measurements of the serving and neighbouring cells respectively. As part of the reporting configuration, the base station can tell the mobile to measure either the cells' reference signal received power or their reference signal received quality [21]. In the case of measurement event A3, the neighbour is typically on the same carrier frequency as the serving cell or on another carrier that has the same priority.

Hys is a hysteresis parameter for measurement reporting. If the mobile enters the event A3 reporting state, then Hys prevents it from leaving that state until the signal levels have changed by 2 × Hys. Similarly, Off is a hysteresis parameter for handovers. If the measurement report triggers a handover, then Off prevents the mobile from moving back to the original cell until the signal levels have changed by 2 × Off. Of_s and Of_n are the optional frequency-specific offsets noted earlier, while Oc_s and Oc_n are the cell-specific offsets.

Figure 14.7 shows the end result. As in the previous diagram, the mobile is initially being served by cell 1, but is moving away from that cell and towards cell 2. When the RSRP or RSRQ satisfies the terms of Equation 14.10, the mobile starts sending periodic measurement reports for event A3. The base station can then use these reports to trigger a handover, in the manner described below.

14.3.4 Measurement Gaps

If a neighbouring cell is on the same carrier frequency as the serving eNB, then the mobile can measure it at any time. If it is on a different frequency, then things are more difficult. Unless the mobile has an expensive dual frequency receiver, it cannot transmit and receive on one frequency and make measurements on another at the same time.

To deal with this problem, the base station can define *measurement gaps* (Figure 14.8) as part of the measurement configuration. Measurements gaps are subframes in which the base station promises not to schedule any transmissions to or from the mobile. During these subframes, the mobile can move to another carrier frequency and make a measurement, confident that it will not miss any downlink data or uplink transmission opportunities. Each gap has a *measurement gap length* (MGL) of six subframes, enough for a single measurement of the primary and

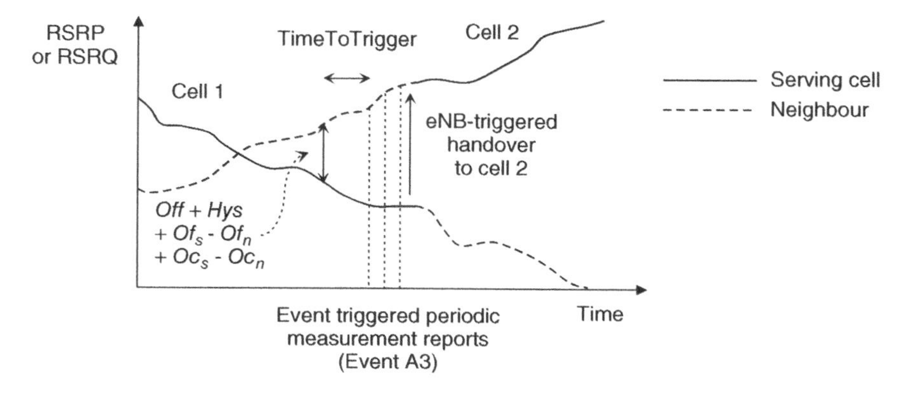

Figure 14.7 Measurement reporting and handover using measurement event A3

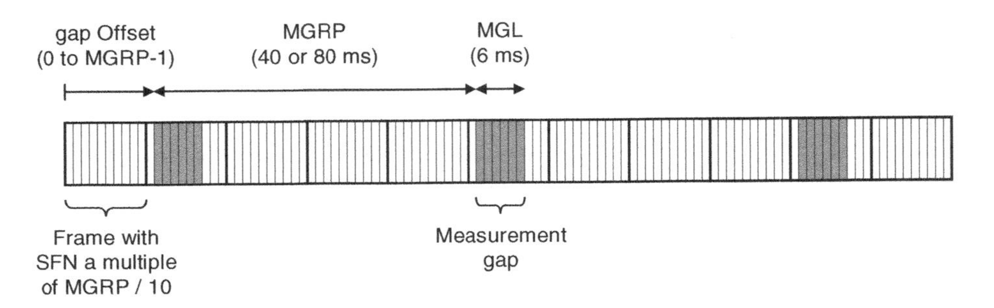

Figure 14.8 Measurement gaps

secondary synchronization signals, and a *measurement gap repetition period* (MGRP) of either 40 or 80 subframes. A mobile receives an individual offset within this period as part of its measurement configuration.

14.4 Handover in RRC_CONNECTED

14.4.1 X2 Based Handover Procedure

After receiving a measurement report, the serving eNB may decide to hand the mobile over to another cell. There are a few scenarios, but the most common is the basic *X2-based handover procedure*. This involves a change of base station using signalling messages over the X2 interface, but no change of serving gateway or MME.

The procedure begins in Figure 14.9, which follows the numbering scheme from TS 36.300 [22]. The mobile identifies a neighbouring cell in a measurement report (1, 2) and the old base station decides to hand the mobile over (3). Using an X2-AP *Handover Request* (4), it asks the new base station to take control of the mobile and includes the new cell's global ID, the identity of the mobile's serving MME, the new security key $K_{eNB}{}^{*}$ and the mobile's radio access capabilities. It also identifies the bearers that it would like to transfer and describes their qualities of service.

The new base station examines the list of bearers and identifies the bearers that it is willing to accept (5). It might reject some if the new cell is overloaded, for example, or if the new cell has a smaller bandwidth than the old one. It then composes an RRC Connection Reconfiguration message, which tells the mobile how to communicate with the new cell. In the message, it gives the mobile a new C-RNTI and includes the configurations of SRB 1, SRB 2 and the data radio bearers that it is willing to accept. As described in Chapter 9, it can optionally include a preamble index for the non contention based random access procedure. The new base station embeds its RRC message into an X2-AP *Handover Request Acknowledge*, which acknowledges the old base station's request and lists the bearers that it will accept. It then sends both messages to the old base station (6).

The old base station extracts the RRC message and sends it to the mobile (7). At the same time, it sends an X2-AP *SN Status Transfer* to the new base station (8), which identifies the PDCP service data units that it has successfully received on the uplink, on bearers that are using RLC acknowledged mode. It also forwards any uplink packets that it has received out

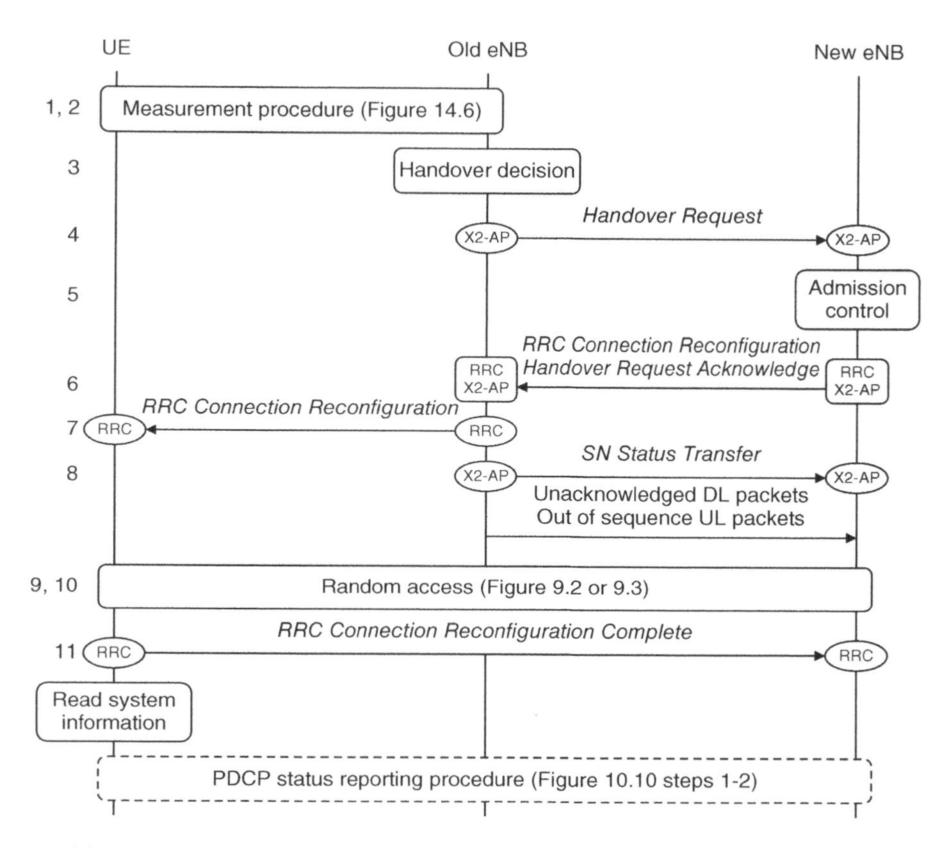

Figure 14.9 X2 based handover procedure. (1) Handover decision, preparation and execution. Source: TS 23.401 and TS 36.300. Reproduced by permission of ETSI

of sequence, any downlink packets that the mobile has not yet acknowledged and any more downlink packets that arrive from the serving gateway.

On receiving the RRC message, the mobile reconfigures itself for the new cell and runs either the non contention based or the contention based random access procedure (9, 10), depending on whether or not it received a preamble index. It then acknowledges the RRC message (11) and reads the new cell's system information. Optionally, the base station and mobile can also run the PDCP status reporting procedure from Chapter 10, to minimize the amount of duplicate packet re-transmission.

There is one task remaining, shown in Figure 14.10. The serving gateway is still sending downlink packets to the old base station, so we need to contact it and change the downlink path. To do this, the new base station sends an S1-AP *Path Switch Request* to the MME (12), in which it lists the bearers that it has accepted and includes tunnel endpoint identifiers for the serving gateway to use on the downlink. The MME forwards the TEIDs to the serving gateway (13), along with the IP address of the new base station.

On receiving the message, the serving gateway redirects the GTP-U tunnels for the bearers that the base station accepted (14) and deletes the ones that were rejected. To indicate the end

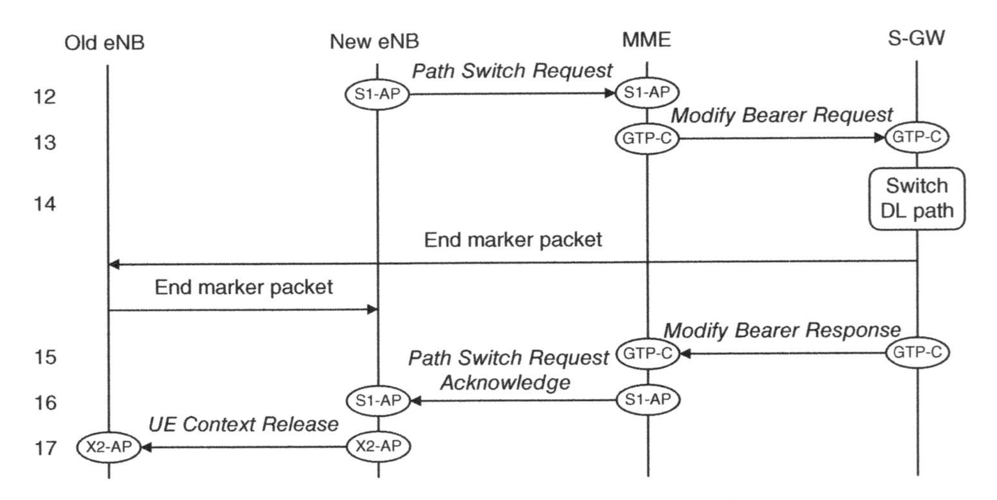

Figure 14.10 X2 based handover procedure. (2) Handover completion. Source: TS 23.401 and TS 36.300. Reproduced by permission of ETSI

of the data stream, it also sends a GTP-U *End marker* packet to the old base station, which forwards it to the new one. Once the new base station has sent all the forwarded packets to the mobile and has reached the end marker, it can be confident that no more will arrive and can start transmitting any new packets that arrive direct from the serving gateway.

To conclude the procedure, the serving gateway sends an acknowledgement to the MME (15), in which it includes TEIDs for the base station to use on the uplink. The MME sends an acknowledgement to the base station (16), which includes the serving gateway's TEIDs and the security parameter known as next hop (NH). In turn, the new base station tells the old one that the handover has completed successfully (17). Once the old base station has received the message and forwarded the end marker packet, it can delete all the resources that were associated with the mobile.

If it wishes to do so, the new base station can use NH to calculate a new set of security keys right away and can bring them into use by means of an RRC connection reconfiguration procedure. This has the benefit that the old base station will not know the access stratum security keys that the new one is using. Otherwise, the new base station can keep NH in reserve until the next X2 based handover, in which case key separation is completed after two handovers rather than one.

As a result of the handover procedure, the mobile can move into a tracking area in which it was not previously registered. If so, then it runs the tracking area update procedure after the handover has completed. There are a few simplifications, as the mobile is starting and finishing in RRC_CONNECTED state and there is no need to change the MME or the serving gateway.

14.4.2 Handover Variations

There are several variations to the basic procedure described above [23]. The first leads to a simplification. If the same base station controls both cells, then we can leave out all the X2 messages from the sequence above and can keep the downlink path unchanged.

If the mobile changes base station and moves into a new S-GW serving area, then the MME has to change the serving gateway in place of steps 13 to 15. To do this, the MME tells the old serving gateway to tear down the bearers that it was using for the mobile and tells the new serving gateway to set up a new set of bearers. In turn, the new serving gateway contacts the PDN gateway, to change the traffic path on S5/S8.

As we noted in Chapter 2, the X2 interface is optional. If there is no X2 interface between the two base stations, then the procedure described above is unsuitable, so the handover is instead carried out using the *S1-based handover procedure*. In that procedure, the base stations communicate by way of the MME, using messages that are exchanged on the S1 interface instead of on X2.

If the mobile moves into a new MME pool area, then the S1-based handover procedure is mandatory. As before, the old base station requests a handover by contacting the old MME. The old MME hands control of the mobile over to a new MME, and the new MME forwards the handover request to the new base station.

References

1. 3GPP TS 36.304 (2013) User Equipment (UE) Procedures in Idle Mode, Release 11, September 2013.
2. 3GPP TS 23.401 (2013) General Packet Radio Service (GPRS) Enhancements for Evolved Universal Terrestrial Radio Access Network (E-UTRAN) Access, Release 11, September 2013.
3. 3GPP TS 36.300 (2013) Evolved Universal Terrestrial Radio Access (E-UTRA) and Evolved Universal Terrestrial Radio Access Network (E-UTRAN); Overall Description; Stage 2, Release 11, September 2013.
4. 3GPP TS 24.301 (2013) Non-Access-Stratum (NAS) Protocol for Evolved Packet System (EPS); Stage 3, Release 11, September 2013.
5. 3GPP TS 29.274 (2013) 3GPP Evolved Packet System (EPS); Evolved General Packet Radio Service (GPRS) Tunnelling Protocol for Control Plane (GTPv2-C); Stage 3, Release 11, September 2013.
6. 3GPP TS 36.331 (2013) Radio Resource Control (RRC); Protocol Specification, Release 11, September 2013.
7. 3GPP TS 36.413 (2013) Evolved Universal Terrestrial Radio Access Network (E-UTRAN); S1 Application Protocol (S1AP), Release 11, September 2013.
8. 3GPP TS 36.423 (2013) Evolved Universal Terrestrial Radio Access Network (E-UTRAN); X2 Application Protocol (X2AP), Release 11, September 2013.
9. 3GPP TS 36.133 (2013) Requirements for Support of Radio Resource Management, Release 11, September 2013.
10. 3GPP TS 23.401 (2013) General Packet Radio Service (GPRS) Enhancements for Evolved Universal Terrestrial Radio Access Network (E-UTRAN) Access, Release 11, Section 5.3.5, September 2013.
11. 3GPP TS 23.401 (2013) General Packet Radio Service (GPRS) Enhancements for Evolved Universal Terrestrial Radio Access Network (E-UTRAN) Access, Release 11, Section 5.3.4.3, September 2013.
12. 3GPP TS 23.401 (2013) General Packet Radio Service (GPRS) Enhancements for Evolved Universal Terrestrial Radio Access Network (E-UTRAN) Access, Release 11, Section 5.3.4.1, September 2013.
13. 3GPP TS 36.304 (2013) User Equipment (UE) Procedures in Idle Mode, Release 11, Section 5.2.4, September 2013.
14. 3GPP TS 36.133 (2013) Requirements for Support of Radio Resource Management, Release 11, Section 4.2, September 2013.
15. 3GPP TS 23.401 (2013) General Packet Radio Service (GPRS) Enhancements for Evolved Universal Terrestrial Radio Access Network (E-UTRAN) Access, Release 11, Section 5.3.3, September 2013.
16. 3GPP TS 23.122 (2012) Non-Access-Stratum (NAS) Functions Related to Mobile Station (MS) in Idle Mode, Release 11, Section 4.4.3.3, December 2012.
17. 3GPP TS 36.304 (2013) User Equipment (UE) Procedures in Idle Mode, Release 11, Section 5.2.3, September 2013.
18. 3GPP TS 31.102 (2013) Characteristics of the Universal Subscriber Identity Module (USIM) Application, Release 11, Section 4.2.6, September 2013.

19. 3GPP TS 36.331 (2013) Radio Resource Control (RRC); Protocol Specification, Release 11, Sections 5.5, 6.3.5, September 2013.
20. 3GPP TS 36.133 (2013) Requirements for Support of Radio Resource Management, Release 11, Sections 8.1.2.1, 8.1.2.2, 8.1.2.3, September 2013.
21. 3GPP TS 36.214 (2012) Physical Layer; Measurements, Release 11, Sections 5.1.1, 5.1.3, December 2012.
22. 3GPP TS 36.300 (2013) Evolved Universal Terrestrial Radio Access (E-UTRA) and Evolved Universal Terrestrial Radio Access Network (E-UTRAN); Overall Description; Stage 2, Release 11, Section 10.1.2, September 2013.
23. 3GPP TS 23.401 (2013) General Packet Radio Service (GPRS) Enhancements for Evolved Universal Terrestrial Radio Access Network (E-UTRAN) Access, Release 11, Section 5.5.1, September 2013.

15

Inter-operation with UMTS and GSM

In the early stages of rolling out the technology, LTE has been available only in large cities and isolated hotspots. In other areas, network operators have continued to use older technologies such as GSM, UMTS and cdma2000. Similarly, most LTE mobiles are actually multiple mode devices that also support some or all of those other technologies. To handle this situation, LTE has been designed so that it can inter-operate with other mobile communication systems, particularly by handing mobiles over if they move outside the coverage area of LTE.

In this chapter, we discuss the most important issue, namely inter-operation with the earlier 3GPP technologies of UMTS and GSM. There are two possible inter-operation architectures: one requires enhancements to the 2G/3G packet switched domain to make it compatible with LTE, while the other requires extra functions in the evolved packet core that make it backwards compatible with the older systems. The specifications support mobility between LTE and UMTS or GSM in both RRC_IDLE and RRC_CONNECTED, and include the option for optimized handovers that transfer mobiles with no packet loss and with a minimal break in communications. We delay any discussion of non-3GPP technologies until the next chapter and any discussion of the 2G/3G circuit switched domain until Chapter 21.

This chapter uses similar specifications to the previous one. The most important ones cover the procedures that a mobile should follow in RRC_IDLE [1] and inter-operation with UMTS and GSM in RRC_CONNECTED [2]. Other specifications define the measurements that the mobile makes in both RRC states [3] and the individual signalling procedures [4–7]. There are several detailed accounts of UMTS and GSM for readers who would like further information about those technologies [8–11].

15.1 System Architecture

15.1.1 Architecture of the 2G/3G Packet Switched Domain

In Chapter 1, we briefly introduced the architecture of the 2G/3G packet switched domain. Figure 15.1 reviews that architecture and adds some further detail. The diagram omits the 2G/3G circuit switched domain, which we will cover in Chapter 21.

An Introduction to LTE: LTE, LTE-Advanced, SAE, VoLTE and 4G Mobile Communications, Second Edition.
Christopher Cox.

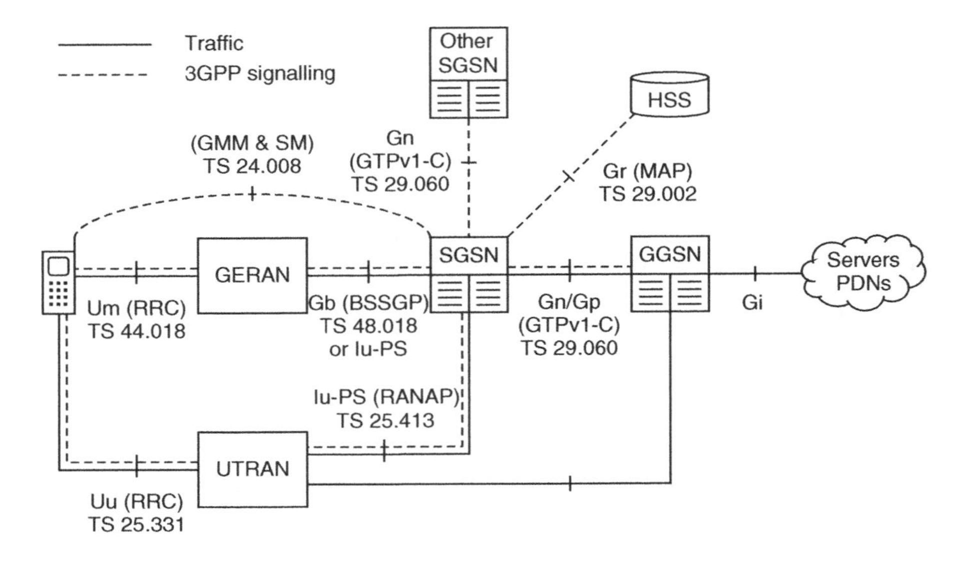

Figure 15.1 Architecture of the packet switched domain of UMTS and GSM

The core network's packet switched domain has two main components. The gateway GPRS support node (GGSN) is the point of contact with external packet data networks and behaves very like the PDN gateway. The serving GPRS support node (SGSN) communicates with the UMTS terrestrial radio access network (UTRAN) and the GSM EDGE radio access network (GERAN), and combines the functions of the MME and the serving gateway. Using an optional technique known as direct tunnelling, data packets can travel directly between the UTRAN and the GGSN, by-passing the SGSN. Although not shown in the diagram, the core network can also interact with the network elements for policy and charging control that we introduced in Chapter 13.

There are several signalling protocols. The SGSN communicates with the home subscriber server (HSS) using the *mobile application part* (MAP) [12], while the core network's other signalling interfaces use the *GPRS tunnelling protocol control part version 1* (GTPv1-C) [13]. The SGSN controls the UMTS radio access network using the *radio access network application part* (RANAP) [14] and controls the GSM radio access network using either RANAP or the *base station system GPRS protocol* (BSSGP) [15]. The SGSN controls the mobile's non access stratum using two signalling protocols [16], namely the *GPRS mobility management* (GMM) protocol, which looks after the network's internal bookkeeping, and the *session management* (SM) protocol, which manages the mobile's data streams. Finally, the radio access network controls the mobile's access stratum using the radio resource control (RRC) protocol for UMTS or GSM [17, 18], as appropriate.

The core network is organized into *routing areas* (RAs), which are similar to tracking areas but do not have any equivalent to mobile-specific tracking area lists. The core network identifies each user using a *packet temporary mobile subscriber identity* (P-TMSI), while the core and radio access networks keep their own records of the mobile's state in a similar way to the state diagrams for LTE.

The core network transports data packets using data pipes known as *UMTS bearers*, which are described using data structures known as *packet data protocol* (PDP) *contexts*. UMTS bearers transport data packets in the same way as EPS bearers and can provide them with quality of service guarantees. However, they are described using different QoS parameters and are configured using different signalling messages.

A UMTS bearer is set up by a process known as *PDP context activation*. This differs from EPS bearer establishment in two ways. Firstly, PDP context activation is initiated by the mobile, whereas EPS bearer establishment is initiated by the PDN gateway. Secondly, a mobile does not activate a PDP context during the 2G/3G attach procedure; instead, it only does so later on, when it wishes to communicate with the outside world.

15.1.2 S3/S4-Based Inter-operation Architecture

LTE can inter-operate with 2G/3G core networks using two possible architectures. The long term solution [19] is shown in Figure 15.2. The diagram only shows the components and interfaces that are actually relevant to inter-operation: it omits, for example, the S11 signalling interface between the MME and the serving gateway.

When a mobile hands over from LTE to UMTS or GSM, the MME transfers control of the mobile to an SGSN. In contrast, the PDN gateway and serving gateway both stay in the data path. The use of the PDN gateway allows the mobile to retain its IP address and EPS bearers after the handover and to maintain its communications with the outside world. The use of the serving gateway allows that device to be a common point of contact for roaming mobiles, whichever technology they are using, and eases the implementation of certain procedures such as paging. The architecture also retains the home subscriber server (HSS), which acts as a common database for the 2G, 3G and 4G networks.

Data packets usually flow through the SGSN as well. However, the optional S12 interface also allows packets to flow directly between the serving gateway and the radio access network

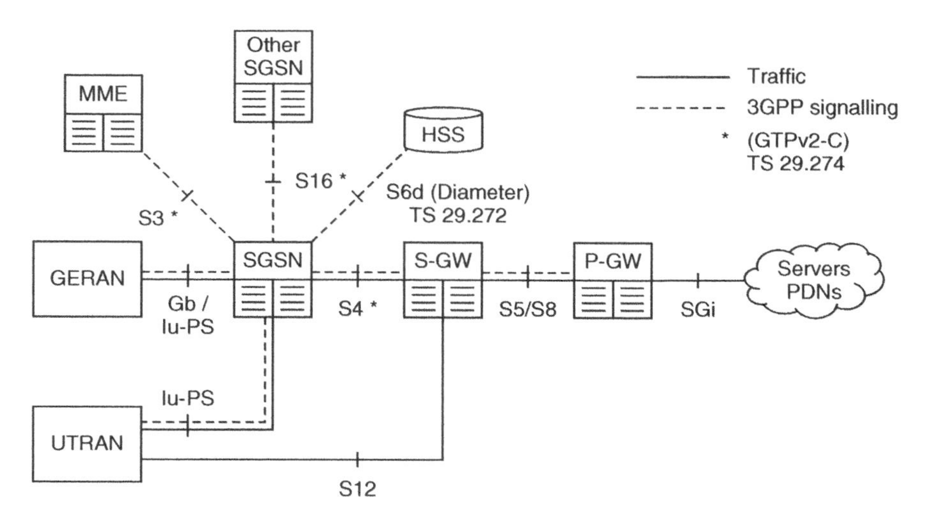

Figure 15.2 Architecture for inter-operation with UMTS and GSM, using an enhanced S3/S4-based SGSN

of UMTS, so as to replicate the UMTS direct tunnelling technique. If the network is using direct tunnelling, the SGSN only handles the mobile's signalling messages, so it bears a close resemblance to the MME.

To use the architecture in Figure 15.2, the SGSN must be enhanced so that it also supports GTPv2-C signalling messages on the S3, S4 and S16 interfaces, and Diameter communications on S6d. If the system is to support cell reselection and handover back to LTE, then the UMTS and GSM radio access networks have to be enhanced as well, so that they can tell their mobiles about the neighbouring LTE carrier frequencies and cells.

When using the S3/S4-based architecture, the evolved packet core refers to EPS bearers using signalling messages across the S4 and S5/S8 interfaces, while the mobile and radio access networks refer to PDP contexts. The SGSN handles the conversion between the two using QoS mapping rules, which are defined in the specifications [20]. Despite this issue, the data packets are transported in exactly the same way, so there is no need for any conversion in the user plane.

In UMTS and GSM, a mobile can explicitly request the creation of an additional PDP context, using a procedure known as secondary PDP context activation. This is different from the situation in LTE, in which the mobile requests a new service data flow, which the PDN gateway implements using either a new EPS bearer or an existing one. To handle this distinction, an S3/S4-based network will always create a dedicated EPS bearer in response to a successful secondary PDP context activation request by the mobile [21] and will not attempt to fold the new SDF into an existing bearer.

15.1.3 Gn/Gp-Based Inter-operation Architecture

Figure 15.3 shows an alternative interim architecture [22]. Here, the SGSN does not have to be enhanced at all: it has to only handle GTPv1-C signalling messages on the Gn and Gp interfaces and MAP signalling messages on Gr. Instead, the MME and PDN gateway are made backwards

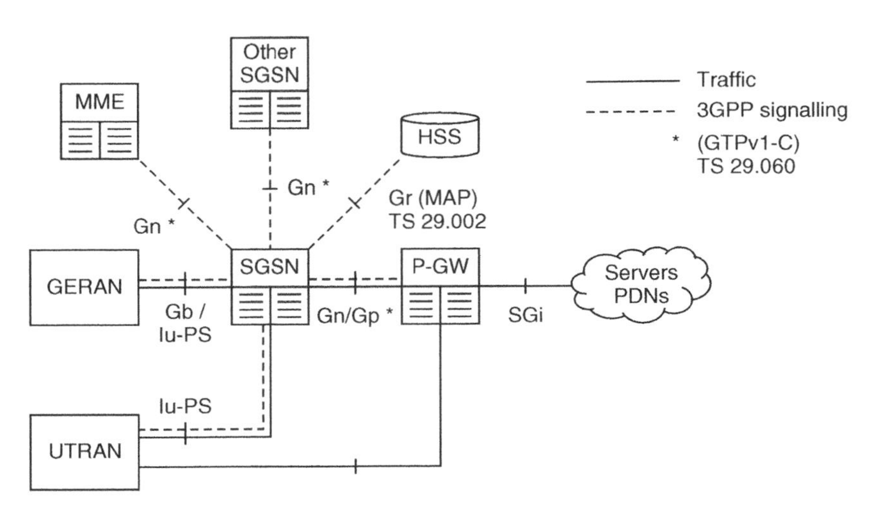

Figure 15.3 Architecture for inter-operation with UMTS and GSM, using a legacy Gn/Gp-based SGSN

compatible with the 2G/3G packet switched domain so that they can handle messages written using GTPv1-C.

The Gn/Gp architecture uses PDP contexts alone, so the MME converts a mobile's EPS bearers to PDP contexts when it moves the mobile to 2G or 3G. If the SGSN dates from Release 7 or before, then it can handle a PDP context that is associated with an IPv4 address or with an IPv6 address, but not one that is associated with both. To handle this difficulty, a PDN gateway will never establish an EPS bearer that uses both IPv4 and IPv6 if there is a risk that the mobile will move to a Release 7 SGSN later on. Instead, it will set up one EPS bearer that uses IPv4 and one that uses IPv6.

The Gn/Gp architecture has a couple of advantages: it allows a network operator to introduce LTE without upgrading its legacy SGSNs, and it allows a mobile to roam in a visited network that has not yet been upgraded for LTE. It does, however, have a few limitations. Firstly, it only supports handovers to and from LTE networks whose S5/S8 interfaces are based on GTP, not PMIP. Secondly, it does not support a technique known as idle mode signalling reduction, which we will cover below.

In discussing the signalling procedures later in this chapter, we will mainly focus on the S3/S4-based inter-operation architecture from Figure 15.2. The procedures for the Gn/Gp-based architecture are, however, very similar.

15.2 Power-On Procedures

As we noted in Chapter 11, the USIM contains prioritized lists of home networks and of any networks that the user or network operator have specified. Each network can be associated with a prioritized list of radio access technologies, including not just LTE, but also UMTS, GSM and cdma2000. During the procedures for network and cell selection, the mobile can use these lists to look for cells that belong to those other radio access technologies [23]. It does this using the procedures that are appropriate for each technology: those for UMTS and GSM are in [24] and [25] respectively.

If the mobile selects a UMTS or GSM cell, then it runs the appropriate attach procedure and registers with an SGSN [26]. It does not activate a PDP context right away; instead, it only does so later on, when it needs to communicate with the outside world. If the network is using the S3/S4-based architecture from Figure 15.2, then the mobile always reaches the outside world through a PDN gateway, even if it has no LTE capability. In the case of the Gn/Gp-based architecture from Figure 15.3, an LTE capable mobile uses a PDN gateway, while other mobiles can continue using a gateway GPRS support node (GGSN) as before.

15.3 Mobility Management in RRC_IDLE

15.3.1 Cell Reselection

In RRC_IDLE state, a mobile can switch to a 2G or 3G cell using the cell reselection procedure from Chapter 14 [27, 28]. To support this procedure, the base station lists the frequencies of neighbouring UMTS and GSM carriers using information in SIB 6 and SIB 7 respectively. Each carrier frequency is associated with a priority level, which has to be different from the priority of the serving LTE cell. The neighbour lists do not include any information about individual UMTS or GSM cells, so the mobile identifies the neighbouring cells by itself.

The actual algorithm is the same one that we saw in Chapter 14 for reselection to another LTE frequency with a higher or lower priority. The only difference is that some of the quantities are re-interpreted to ones that are suitable for the target radio access technology.

15.3.2 Routing Area Update Procedure

After a cell reselection, the mobile reads the new cell's system information, which includes the 2G/3G routing area identity. This triggers a procedure known as a *routing area update* [29, 30], which is similar to the tracking area update procedure from LTE. Figure 15.4 shows the start of the procedure, for the case where the mobile is moving from an MME to an S3/S4-based SGSN and there is no change of serving gateway.

The mobile first runs the 3G procedure for RRC connection establishment (1) so as to establish signalling communications with a radio network controller and move temporarily into RRC_CONNECTED. The mobile then sends a GMM *Routing Area Update Request* to the RNC (2), which forwards the message to a suitable SGSN. In the message, the mobile

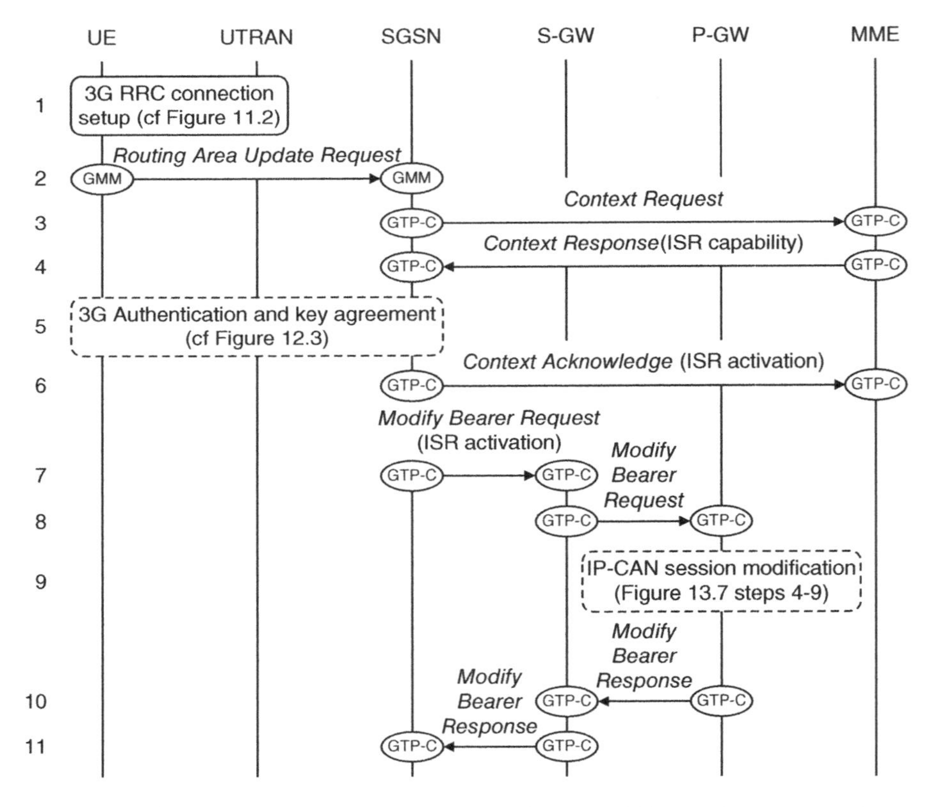

Figure 15.4 Routing area update from an MME to an S3/S4 based SGSN, with retention of the serving gateway. (1) Routing area update request. Source: TS 23.401. Reproduced by permission of ETSI

identifies itself by mapping its globally unique temporary identity (GUTI) to a 3G packet TMSI and a 3G routing area identity, and tells the SGSN that it derived those identities from a GUTI.

The SGSN reconstructs the mobile's GUTI, extracts the identity of the serving MME and requests the mobile's IMSI, the identity of its serving gateway and the descriptions of its EPS bearers using a GTPv2-C *Context Request* (3). The MME returns the requested information using a *Context Response* (4). After an optional authentication procedure (5), the SGSN maps the EPS bearers onto PDP contexts and confirms its acceptance of the mobile using a *Context Acknowledge* (6). As part of messages 4 and 6, the SGSN and MME also decide whether to activate a technique known as idle mode signalling reduction (ISR), which we discuss below.

The SGSN now informs the serving gateway that it is in control of the mobile (7). As part of this message, the SGSN tells the serving gateway about the change of radio access technology: the serving gateway relays this information to the PDN gateway and the PCRF for possible use by the charging system (8–10), and sends an acknowledgement to the SGSN (11).

To complete the procedure (Figure 15.5), the SGSN tells the home subscriber server that it is in control of the mobile and asks for the mobile's subscription data (12). The HSS cancels the mobile's registration with any previous SGSN (13, 14) and returns the subscription data as requested (15). The HSS does not cancel the mobile's registration with the MME; instead, it can simultaneously maintain details of one MME and one S3/S4-based SGSN for each mobile, to help support idle mode signalling reduction.

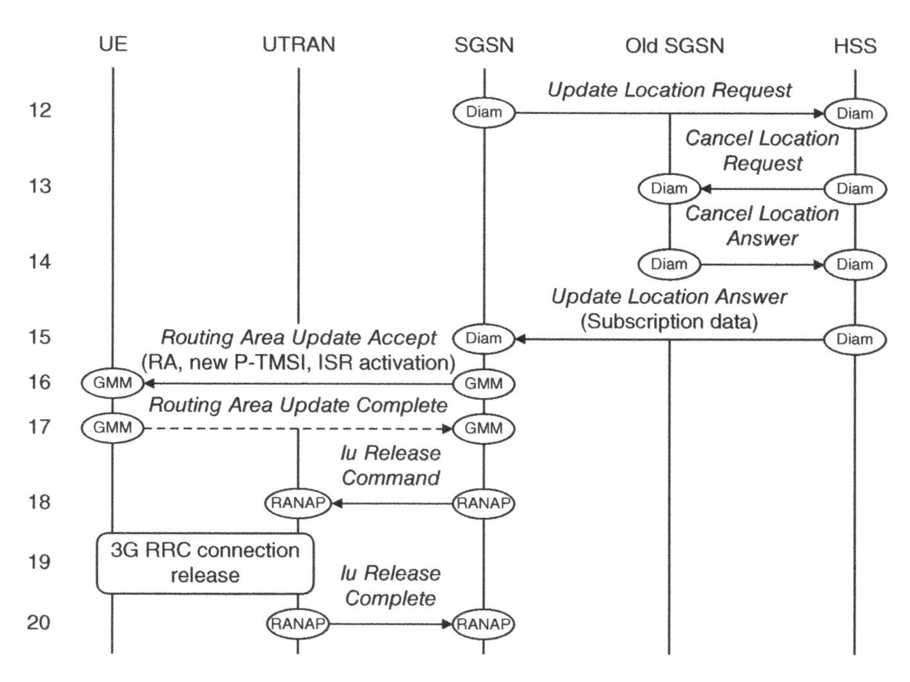

Figure 15.5 Routing area update from an MME to an S3/S4 based SGSN, with retention of the serving gateway. (2) Routing area update accept. Source: TS 23.401. Reproduced by permission of ETSI

The SGSN accepts the mobile's routing area update request and can optionally send the mobile a new P-TMSI (16). If it does so, then the mobile acknowledges receipt (17). Finally, the SGSN moves the mobile back into RRC_IDLE (18–20), using messages that are similar to steps 4, 5 and 6 of the S1 release procedure from Chapter 14.

Later on, the mobile can move back into a region of LTE coverage. If the 2G/3G network has been enhanced to support LTE and is broadcasting information about neighbouring LTE frequencies in its system information, then the mobile can carry out a cell reselection back to LTE. After the reselection, the mobile requests a tracking area update, and the MME takes control back from the SGSN.

15.3.3 Idle Mode Signalling Reduction

If the mobile is near the edge of the LTE coverage area, then it can easily bounce back and forth between cells that are using LTE and cells that are using UMTS or GSM. There is then a risk that the mobile will execute a large number of routing and tracking area updates, leading to excessive signalling.

To avoid this problem, S3/S4-based networks can optionally implement a technique known as *idle mode signalling reduction* (ISR) [31]. This technique behaves very like the registration of a mobile in multiple tracking areas. When using idle mode signalling reduction, the network can simultaneously register the mobile in a routing area that is served by an S3/S4-based SGSN and in one or more tracking areas that are served by an MME. The mobile can then freely reselect between cells that are using the three radio access technologies and only has to inform the network if it moves into a routing or tracking area in which it is not currently registered.

The network activates ISR either during the routing area update from Figures 15.4 and 15.5 or during a tracking area update back to LTE. In the case of the routing area update, the MME tells the SGSN whether it supports ISR as part of its Context Response (step 4). The SGSN decides whether or not to activate ISR, informs the MME as part of its Context Acknowledge (6) and informs the serving gateway as part of its Modify Bearer Request (7). Finally, the SGSN informs the mobile about the activation of ISR as part of its Routing Area Update Accept (16). The mobile is now registered with both the SGSN and the MME.

If downlink data arrive from the PDN gateway, then the paging procedure is modified so that the serving gateway contacts both the MME and the SGSN. The MME pages the mobile in all the tracking areas in which it is registered, as before, while the SGSN pages the mobile throughout its routing area. After the mobile's response, the network moves it back into connected mode using the appropriate radio access technology.

15.4 Mobility Management in RRC_CONNECTED

15.4.1 RRC Connection Release with Redirection

Early LTE networks have generally transferred active mobiles to UMTS or GSM using a procedure known as *RRC connection release with redirection*. In this procedure, the network moves the mobile into RRC_IDLE but gives it details of a 2G/3G carrier frequency on which it should look for a suitable cell. The mobile can then select a 2G/3G cell and establish an RRC

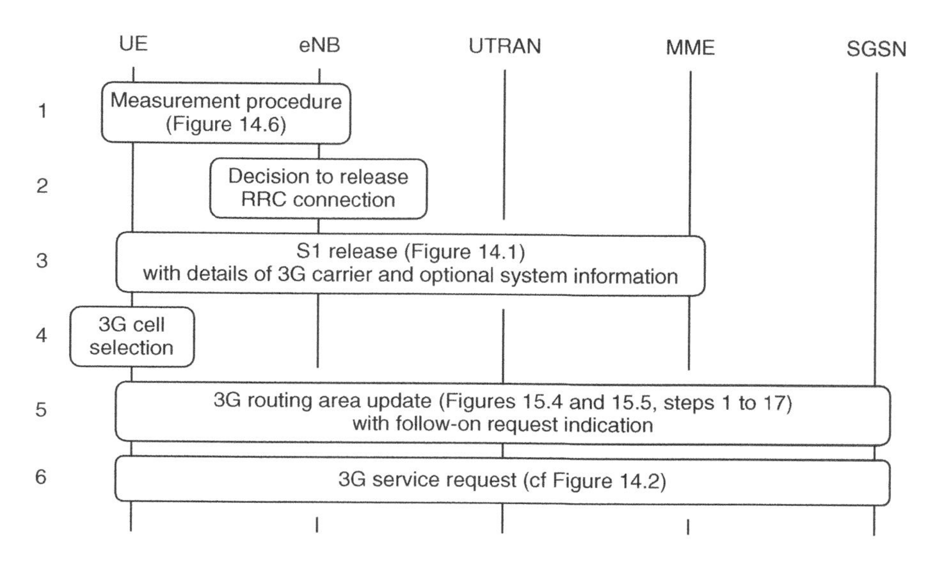

Figure 15.6 RRC connection release with redirection to a UMTS carrier frequency

connection in the usual way. Figure 15.6 shows the procedure for the case where the target radio access technology is 3G.

The message sequence begins when the mobile sends a measurement report to the base station (1). In this procedure, the mobile does not have to report any information about neighbouring 3G cells. Instead, it typically sends a measurement report when the signal from the serving LTE cell becomes too weak and there are no other satisfactory LTE cells nearby, using measurement event A2.

In response, the base station decides to release the mobile's RRC signalling connection (2) and initiates the S1 release procedure from Chapter 14 (3). In the message RRC Connection Release, the base station sends the mobile details of a 3G carrier frequency on which to look for a suitable cell. The mobile moves to the requested carrier frequency, selects a 3G cell (4) and runs the 3G routing area update procedure (5). In its Routing Area Update Request, the mobile can indicate that it has another message to send, and the SGSN reacts by maintaining the mobile's RRC signalling connection. The mobile can then initiate the procedure for a 3G service request so as to re-activate its bearers (6).

By the end of this procedure, the mobile is communicating successfully through a 3G cell. There has been a break in data communications as the mobile travelled through RRC_IDLE, typically a few seconds, so some data packets are lost or are retransmitted from end to end using TCP. However, the procedure does have several advantages. Firstly, the mobile does not have to make any measurements of nearby 3G cells. Secondly, the signalling messages are more straightforward than they are in an optimized handover. Thirdly, for reasons we discuss at the end of the chapter, the procedure can be more reliable than an optimized handover and can lead to fewer dropped connections.

There is one last point to make. In the basic procedure described here, the mobile has to read the new cell's system information in step (4), as part of cell selection. This can take a few seconds. To reduce those delays, the Release 9 specifications allow the base station to pass the system information of nearby 3G cells to the mobile within the RRC Connection Release message in step (3). After finding the nearby cells, the mobile can look up their system information from the details that the base station supplied. This speeds up the reselection process, but begs the question of how the base station discovers the system information in the first place. We will address that question when discussing self optimizing networks in Chapter 17.

15.4.2 Measurement Procedures

To support optimized handovers, the base station can tell the mobile to make measurements of neighbouring 2G/3G cells using the measurement procedure from Chapter 14 [32]. This time, the base station has to tell the mobile about the individual neighbours as part of the measurement procedure, because it is considered too hard for the mobile to find UMTS or GSM cells by itself during measurement gaps. As before, we will see how the base station might discover the neighbouring cells when discussing self optimizing networks in Chapter 17.

The mobile's measurement reports can be triggered by two measurement events, B1 and B2, which are listed in Table 15.1. To illustrate their behaviour, the mobile enters the event B2 reporting state when both of the following conditions are met for a time of at least TimeToTrigger. Firstly, the signal from the LTE serving cell must lie below one threshold:

$$M_s < \text{Thresh}_1 - \text{Hys} \tag{15.1}$$

Secondly, the signal from the non-LTE neighbour must lie above another threshold:

$$M_n + \text{Of}_n > \text{Thresh}_2 + \text{Hys} \tag{15.2}$$

Whilst in this state, the mobile sends periodic measurement reports to the base station in the manner described in Chapter 14. The mobile leaves the event B2 reporting state when either of the following conditions is met for a time of at least TimeToTrigger:

$$M_s > \text{Thresh}_1 + \text{Hys} \tag{15.3}$$

$$M_n + \text{Of}_n < \text{Thresh}_2 - \text{Hys} \tag{15.4}$$

In these equations, M_s is the reference signal received power or reference signal received quality from the serving LTE cell. M_n is the measurement of the neighbouring cell, namely the *received signal strength indicator* (RSSI) in the case of GSM and either the *received signal code power* (RSCP) or SINR per chip (Ec/No) in the case of UMTS [33]. Thresh_1 and Thresh_2

Table 15.1 Measurement events used to report neighbouring UMTS, GSM and cdma2000 cells

Event	Description	Possible application
B1	Neighbour > Threshold	Handover to 2G/3G cell on higher priority carrier
B2	Serving cell < Threshold 1 & Neighbour > Threshold 2	Handover to 2G/3G cell on lower priority carrier

are thresholds, Hys is a hysteresis parameter for measurement reporting and Of_n is an optional frequency-specific offset.

Using this measurement event, the mobile can inform the serving eNB if the signal from the serving LTE cell is sufficiently weak, while the signal from a neighbouring UMTS or GSM cell is sufficiently strong. The neighbouring cell is typically on a lower priority carrier frequency in the case of measurement event B2, or on a higher priority carrier in the case of the other measurement event, B1. On the basis of this measurement report, the base station can hand the mobile over to UMTS or GSM.

15.4.3 Optimized Handover

Optimized handovers between LTE and UMTS or GSM are similar to handovers within LTE. They have two main features. Firstly, the MME and SGSN exchange signalling messages before the actual handover takes place, to prepare the 3G network, minimize the gap in communications and minimize the risk of packet loss. Secondly, the network hands the mobile over to a specific 2G/3G cell.

To illustrate the procedure, Figures 15.7 and 15.8 show the procedure that is typically used for a handover from LTE to UMTS [34]. Although it looks complex at first glance, the procedure is actually very like the one for a normal S1-based handover. In these figures, we have assumed that the network is using a GTP-based S5/S8 interface and an S3/S4-based SGSN, and is not using the S12 direct tunnelling interface. We also assume that the radio access networks are not in direct communication, so that the handover uses a technique known as indirect data forwarding in which the eNB forwards unacknowledged downlink packets to the target RNC by way of the serving gateway. Finally, we assume that the serving gateway remains unchanged.

The procedure begins in Figure 15.7. On the basis of a measurement report, usually for measurement event B1 or B2, the eNB decides to hand the mobile over to a specific UMTS cell. It tells the MME that a handover is required (1) and identifies the target Node B, together with the corresponding radio network controller and routing area. The MME identifies a suitable SGSN, asks it to relocate the mobile and describes the EPS bearers that the mobile is currently using (2).

The SGSN reacts by mapping the EPS bearers' QoS parameters onto the parameters for the corresponding PDP contexts. It then asks the RNC to establish resources for the mobile and describes the bearers that it would like to set up (3), using a *Relocation Request* that is written using the UMTS Radio access network application part (RANAP). The RNC configures the Node B (4) and returns an acknowledgement to the SGSN (5). In its acknowledgement, the RNC lists the bearers that it is willing to accept and includes an embedded *Handover to UTRAN Command*, which tells the mobile how to communicate with the target cell. This last message is written using the UMTS RRC protocol, which the diagram denotes as 3GRRC.

The SGSN acknowledges the MME's request, tells it which bearers have been accepted and attaches the UMTS RRC message that it received from the RNC (6). It also includes a tunnel endpoint identifier (TEID), which the serving gateway can use to forward any downlink packets that the mobile has not yet acknowledged. The MME sends the TEID to the serving gateway (7), which creates the tunnel as requested. The MME can then tell the eNB to hand the mobile over to UMTS (8). In the message, the MME states which bearers should be retained and which should be released, and includes the UMTS RRC message that it received from the SGSN.

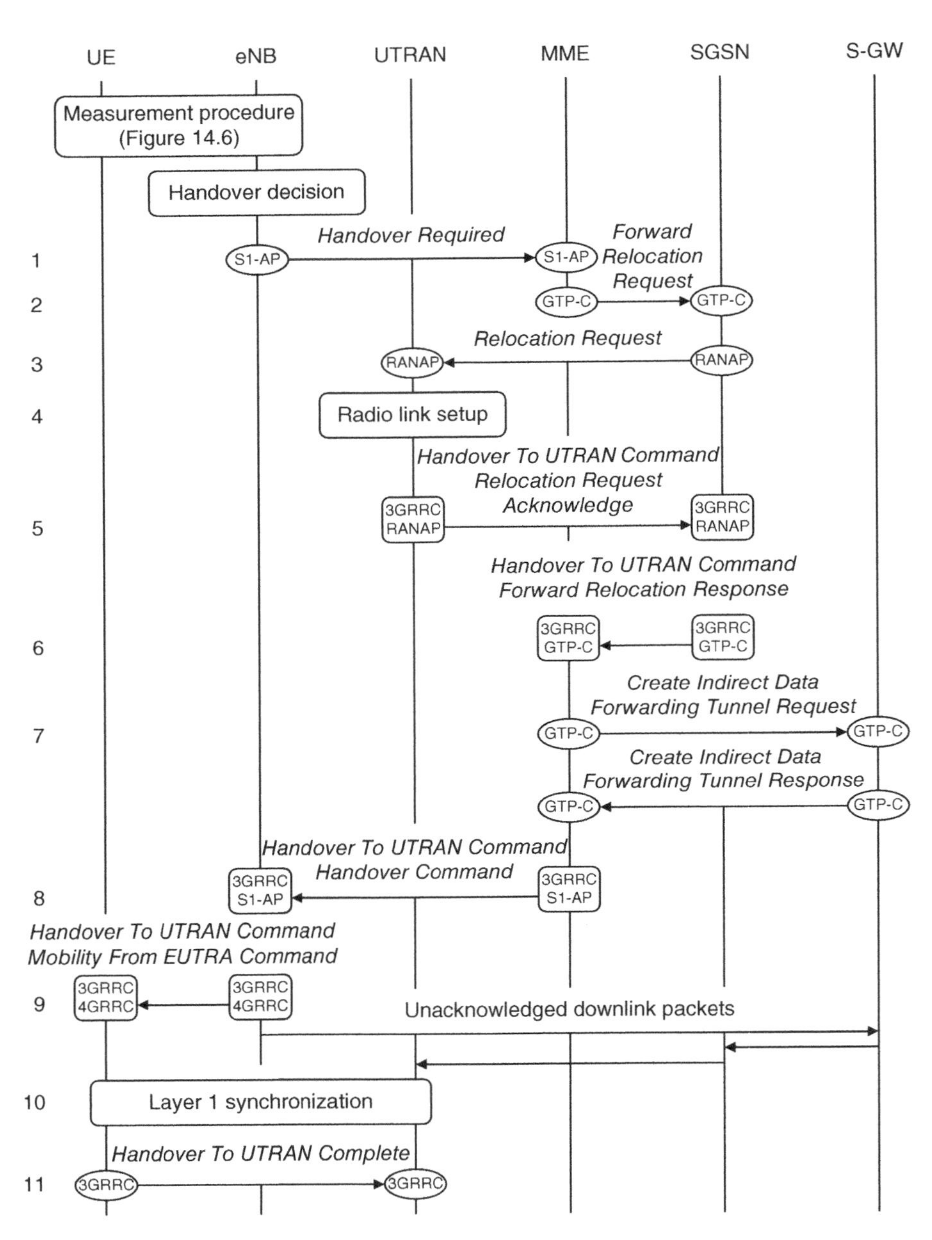

Figure 15.7 Handover from LTE to UMTS. (1) Handover preparation and execution. Source: TS 23.401. Reproduced by permission of ETSI

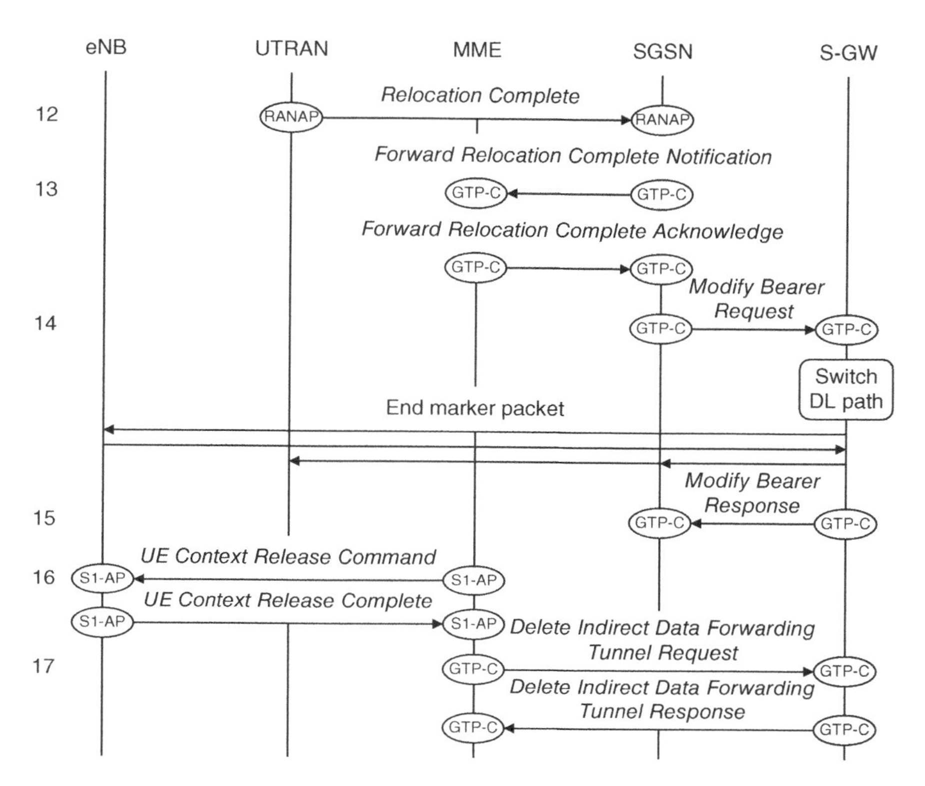

Figure 15.8 Handover from LTE to UMTS. (2) Handover completion. Source: TS 23.401. Reproduced by permission of ETSI

To trigger the handover, the eNB sends the mobile a *Mobility from EUTRA Command* (9). This message is written using the LTE RRC protocol, which the diagram denotes as 4GRRC. Embedded in the message is the UMTS handover command that was originally written by the target RNC. The mobile reads both RRC messages, switches to UMTS, synchronizes with the new cell (10) and sends an acknowledgement to the RNC (11). At the same time, the eNB starts to return any unacknowledged downlink packets to the serving gateway, as well as any new packets that continue to arrive. The serving gateway forwards the packets to the SGSN using the tunnel that it has just created, and the SGSN forwards the packets to the target RNC.

The network still has to release the old resources and redirect the data path from the serving gateway (Figure 15.8). To achieve this, the RNC tells the SGSN that the relocation procedure is complete (12) and the SGSN forwards this information to the MME (13). The SGSN also tells the serving gateway to redirect its downlink path (14) so as to send future downlink packets to the SGSN. The serving gateway does so, indicates the end of the data stream by sending an end marker packet on the old downlink path to the eNB and returns an acknowledgement to the SGSN (15).

On the expiry of a timer, the MME tells the eNB to release the mobile's resources (16), and tells the serving gateway to tear down the indirect forwarding tunnel that it created earlier (17). At the same time, the mobile notices that it has moved into a new 2G/3G routing area.

It responds by running a routing area update, unless it is using idle mode signalling reduction and is already registered there.

The handover procedure from LTE to GSM is very similar. The specifications also support handovers from UMTS and GSM back to LTE, but these are less important, as the network will have few regions where there is coverage from LTE but not from other technologies.

This procedure does not cause any significant gap in the data communications, but does have a few disadvantages. Firstly, the signalling messages are rather complex, so the network operator may prefer not to implement it. Secondly, the network has to supply the mobile with a list of neighbouring 2G/3G cells. Thirdly, there may be a significant delay between the mobile's measurement and the handover command, during which the signal received from the target cell can fall so far that the handover fails. For these reasons, RRC connection release with redirection is often preferred.

References

1. 3GPP TS 36.304 (2013) User Equipment (UE) Procedures in Idle Mode, Release 11, September 2013.
2. 3GPP TS 23.401 (2013) General Packet Radio Service (GPRS) Enhancements for Evolved Universal Terrestrial Radio Access Network (E-UTRAN) Access, Release 11, September 2013.
3. 3GPP TS 36.133 (2013) Requirements for Support of Radio Resource Management, Release 11, September 2013.
4. 3GPP TS 24.301 (2013) Non-Access-Stratum (NAS) Protocol for Evolved Packet System (EPS); Stage 3, Release 11, September 2013.
5. 3GPP TS 29.274 (2013) 3GPP Evolved Packet System (EPS); Evolved General Packet Radio Service (GPRS) Tunnelling Protocol for Control Plane (GTPv2-C); Stage 3, Release 11, September 2013.
6. 3GPP TS 36.331 (2013) Radio Resource Control (RRC); Protocol Specification, Release 11, September 2013.
7. 3GPP TS 36.413 (2013) Evolved Universal Terrestrial Radio Access Network (E-UTRAN); S1 Application Protocol (S1AP), Release 11, September 2013.
8. Sauter, M. (2010) *From GSM to LTE: An Introduction to Mobile Networks and Mobile Broadband*, John Wiley & Sons, Ltd, Chichester.
9. Eberspächer, J., Vögel, H.-J., Bettstetter, C. and Hartmann, C. (2008) *GSM: Architecture, Protocols and Services*, 3rd edn, John Wiley & Sons, Ltd, Chichester.
10. Johnson, C. (2008) *Radio Access Networks for UMTS: Principles and Practice*, John Wiley & Sons, Ltd, Chichester.
11. Kreher, R. and Ruedebusch, T. (2007) *UMTS Signaling: UMTS Interfaces, Protocols, Message Flows and Procedures Analyzed and Explained*, 2nd edn, John Wiley & Sons, Ltd, Chichester.
12. 3GPP TS 29.002 (2013) Mobile Application Part (MAP) Specification, Release 11, September 2013.
13. 3GPP TS 29.060 (2013) General Packet Radio Service (GPRS); GPRS Tunnelling Protocol (GTP) Across the Gn and Gp Interface, Release 11, September 2013.
14. 3GPP TS 25.413 (2013) Radio Access Network Application Part (RANAP) Signalling, Release 11, June 2013.
15. 3GPP TS 48.018 (2013) BSS GPRS Protocol (BSSGP), Release 11, September 2013.
16. 3GPP TS 24.008 (2013) Mobile Radio Interface Layer 3 Specification; Core Network Protocols; Stage 3, Release 11, September 2013.
17. 3GPP TS 25.331 (2013) Radio Resource Control (RRC), Release 11, September 2013.
18. 3GPP TS 44.018 (2013) Radio Resource Control (RRC) Protocol, Release 11, September 2013.
19. 3GPP TS 23.401 (2013) General Packet Radio Service (GPRS) Enhancements for Evolved Universal Terrestrial Radio Access Network (E-UTRAN) Access, Release 11, Sections 4.2, 4.4, September 2013.
20. 3GPP TS 23.401 (2013) General Packet Radio Service (GPRS) Enhancements for Evolved Universal Terrestrial Radio Access Network (E-UTRAN) Access, Release 11, Annex E, September 2013.
21. 3GPP TS 23.060 (2013) General Packet Radio Service (GPRS); Service Description; Stage 2, Release 11, Section 9.2.2.1.1A, September 2013.
22. 3GPP TS 23.401 (2013) General Packet Radio Service (GPRS) Enhancements for Evolved Universal Terrestrial Radio Access Network (E-UTRAN) Access, Release 11, Annex D, September 2013.

23. 3GPP TS 23.122 (2012) Non-Access-Stratum (NAS) Functions Related to Mobile Station (MS) in Idle Mode, Release 11, Section 4.4, December 2012.
24. 3GPP TS 25.304 (2013) User Equipment (UE) Procedures in Idle Mode and Procedures for Cell Reselection in Connected Mode, Release 11, Section 5.2.6, September 2013.
25. 3GPP TS 43.022 (2012) Functions Related to Mobile Station (MS) in Idle Mode and Group Receive Mode, Release 11, Section 4.5, September 2012.
26. 3GPP TS 23.401 (2013) General Packet Radio Service (GPRS) Enhancements for Evolved Universal Terrestrial Radio Access Network (E-UTRAN) Access, Release 11, Section 5.3.2.2, September 2013.
27. 3GPP TS 36.304 (2013) User Equipment (UE) Procedures in Idle Mode, Release 11, Section 5.2.4.5, September 2013.
28. 3GPP TS 36.133 (2013) Requirements for Support of Radio Resource Management, Release 11, Section 4.2.2.5, September 2013.
29. 3GPP TS 23.401 (2013) General Packet Radio Service (GPRS) Enhancements for Evolved Universal Terrestrial Radio Access Network (E-UTRAN) Access, Release 11, Sections 5.3.3.3, 5.3.3.6, September 2013.
30. 3GPP TR 25.931 (2012) UTRAN Functions, Examples on Signalling Procedures, 3rd Generation Partnership Project, Release 11, Sections 7.3, 7.4, September 2012.
31. 3GPP TS 23.401 (2013) General Packet Radio Service (GPRS) Enhancements for Evolved Universal Terrestrial Radio Access Network (E-UTRAN) Access, Release 11, Annex J, September 2013.
32. 3GPP TS 36.331 (2013) Radio Resource Control (RRC); Protocol Specification, Release 11, Sections 5.5, 6.3.5, September 2013.
33. 3GPP TS 36.214 (2012) Physical Layer; Measurements, Release 11, Section 5.1, December 2012.
34. 3GPP TS 23.401 (2013) General Packet Radio Service (GPRS) Enhancements for Evolved Universal Terrestrial Radio Access Network (E-UTRAN) Access, Release 11, Section 5.5.2.1, September 2013.

16

Inter-operation with Non-3GPP Technologies

In the last chapter, we saw how LTE can inter-operate with the older 3GPP technologies of UMTS and GSM. This is not the limit of LTE's capabilities, as it can also inter-operate with communication networks that are defined by other standards bodies. These capabilities help operators offload traffic onto other communication technologies such as wireless local area networks and help operators of other technologies such as cdma2000 upgrade to LTE.

We begin by discussing the architectural options and signalling procedures through which LTE can inter-operate with a generic non-3GPP communication network. The specifications allow a mobile to transfer between LTE and another technology while maintaining its IP addresses and its connections with any external servers, but do not include any support for optimized handovers. We then move to the special case of inter-operation between LTE and cdma2000. Here, the specifications do support optimized handovers, so allow a mobile to transfer between the two technologies with no packet loss and with a minimal break in communications.

The most important specification for this chapter is TS 23.402 [1], which defines the inter-operation architecture and presents a high-level view of the signalling procedures. Other 3GPP specifications define the layer 3 signalling messages on the individual interfaces, while two 3GPP2 specifications [2, 3] define inter-operation between LTE and cdma2000 from the viewpoint of the latter system. There are several detailed accounts of the other technologies that we cover in this chapter, notably cdma2000 [4–6], WiMAX [7, 8] and wireless LAN [9].

16.1 Generic System Architecture

16.1.1 Network-Based Mobility Architecture

LTE can inter-operate with non-3GPP technologies using two main architectures [10], which route data packets and handle mobility and roaming using functions in the network and the mobile respectively.

An Introduction to LTE: LTE, LTE-Advanced, SAE, VoLTE and 4G Mobile Communications, Second Edition.
Christopher Cox.

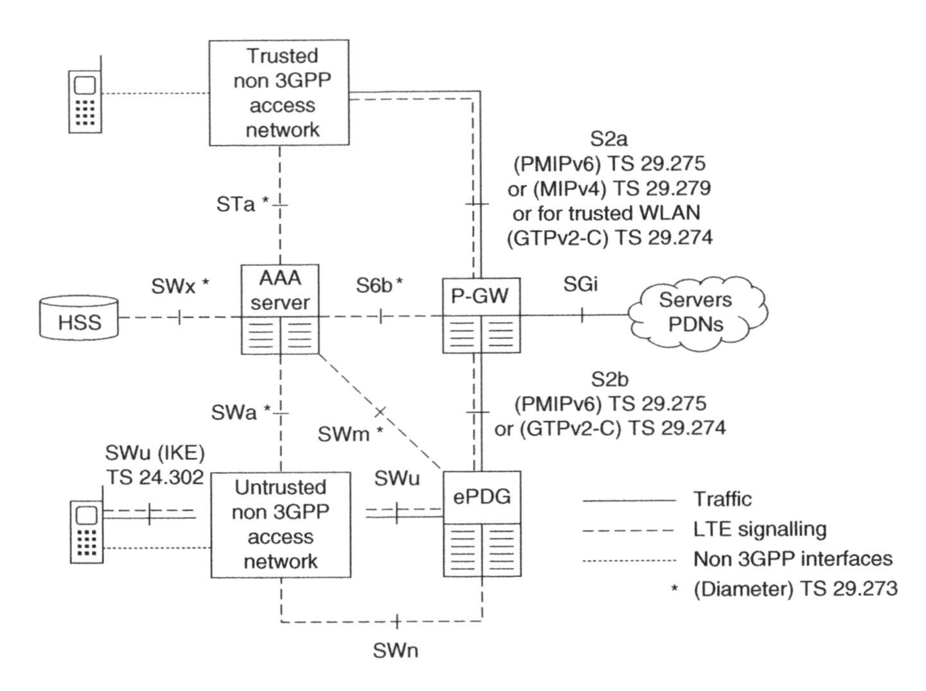

Figure 16.1 Architecture for inter-operation with non-3GPP access technologies, using network-based mobility

Figure 16.1 shows the architecture when using network-based mobility. The architecture retains the home subscriber server and the PDN gateway, which respectively act as a common subscriber database and a common point of contact with the outside world. The mobile is controlled using devices inside the non-3GPP access network, while the *3GPP authentication, authorization and accounting* (AAA) *server* authenticates the mobile and grants it access to services through the PDN gateway. The signalling interfaces to the AAA server use protocols that are based on Diameter [11].

When using the network-based mobility architecture, the mobile receives an IP address from the PDN gateway. The PDN gateway delivers a mobile's downlink packets to the access network using similar tunnelling procedures to those in the evolved packet core, and the access network extracts the packets and transmits them to the mobile. The process requires no extra functionality in the mobile.

The tunnels are usually implemented using generic routing encapsulation (GRE), which we have already seen as one of the options for the S5/S8 interface of the evolved packet core. The GRE tunnels are managed using signalling messages between the access network and the PDN gateway, which are written using either proxy mobile IP version 6 (PMIPv6) [12] or an older protocol known as *mobile IP version 4* (MIPv4) [13] in the case of S2a. Release 10 introduced support for GTP over the S2b interface, a feature that is useful if the network operator is using PDN gateways that do not support PMIP. Release 11 introduced limited support for *S2a mobility based on GTP* (SaMOG) for the sole case of trusted wireless local area networks, with more extensive support introduced in Release 12 [14].

The architecture also makes a distinction between *trusted* and *untrusted* access networks. A trusted network provides adequate security for the mobile's radio communications using techniques such as ciphering and integrity protection, while an untrusted network does not. It is the responsibility of the AAA server to decide whether to treat a particular network as trusted or untrusted.

The untrusted network architecture is suitable for scenarios such as insecure wireless local area networks and includes an extra component known as the *evolved packet data gateway* (ePDG). Together, the mobile and gateway secure the data that they exchange by means of ciphering and integrity protection, using IPSec in tunnel mode [15]. The access network is forced to send uplink traffic through the gateway by means of signalling messages over the SWn interface, for which the signalling protocol lies outside the 3GPP specifications.

The architecture also retains the policy and charging rules function (PCRF), which acts as a common source of policy and charging control with the evolved packet core. The PDN gateway contains a policy and charging enforcement function (PCEF), as before, while the trusted access network and the evolved packet data gateway both contain a bearer binding and event reporting function (BBERF).

If the mobile is roaming, then the architecture supports both home routed traffic and local breakout. The trusted access network and ePDG both communicate with the visited PCRF, while the PDN gateway communicates with either the home or the visited PCRF depending upon its location. The AAA server is also split between an AAA server in the home network and a visited AAA proxy.

The mobile identifies itself to the AAA server using a *network access identifier* (NAI) with the format `username@realm` [16]. The realm part is constructed from the mobile network code and mobile country code, so it identifies the home network operator. The username part is constructed either from the international mobile subscriber identity (IMSI) or from a temporary pseudonym supplied by the AAA server. In the latter case, the network access identifier has a similar role to a 3GPP temporary mobile subscriber identity.

16.1.2 Host-Based Mobility Architecture

Figure 16.2 shows an alternative architecture, in which the mobility management functions lie in the mobile. When using this architecture, the mobile receives an IP address from the PDN gateway in the same way as before. This time, however, the mobile also receives a local IP address from the access network, and informs the PDN gateway using signalling messages over the S2c interface. The PDN gateway can then deliver downlink packets using a tunnel that runs all the way to the mobile. The tunnel is managed using a signalling protocol known as *dual stack mobile IP version 6* (DSMIPv6) [17], which is implemented in the mobile and the PDN gateway. There is no mobility management functionality in the access network, which simply acts as a router.

The choice of mobility architecture and protocol can be configured either statically within the mobile and the network or dynamically during the non-3GPP attach procedure. We will see an example of dynamic configuration shortly.

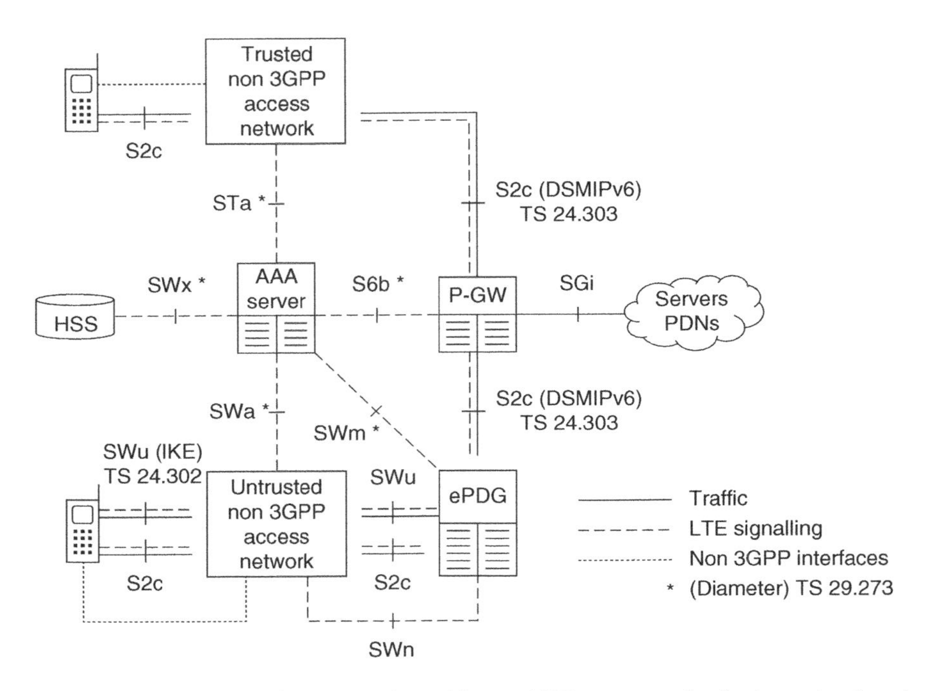

Figure 16.2 Architecture for inter-operation with non-3GPP access technologies, using host-based mobility

16.1.3 Access Network Discovery and Selection Function

There is one more network element that is important for non-3GPP radio access, the *access network discovery and selection function* (ANDSF) [18, 19]. The ANDSF provides the mobile with two main sets of information, namely details of nearby radio access networks and operator policies that influence which access network the mobile should choose.

Figure 16.3 shows the architecture. The mobile communicates with the ANDSF across the S14 interface, which delivers information using *management objects* (MOs) that comply with the *Open Mobile Alliance* (OMA) specifications for device management [20, 21]. The communications are transparent to the evolved packet system and can use any type of radio access technology. If the mobile is roaming, then it can contact an ANDSF in the home network, the visited network or both, with precedence given to any information it receives from the visited ANDSF.

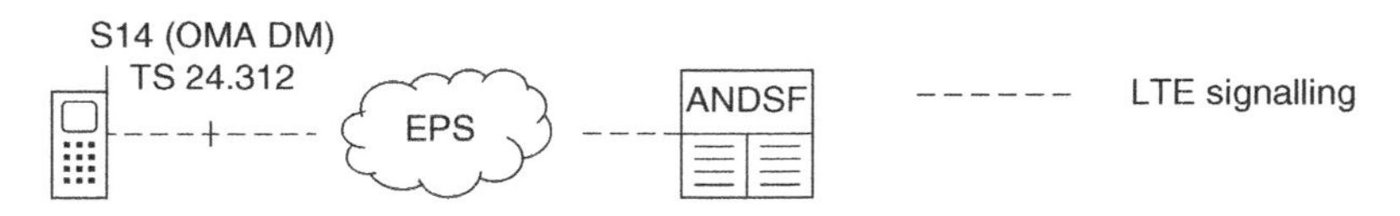

Figure 16.3 Architecture for the access network discovery and selection function. Source: TS 23.402. Reproduced by permission of ETSI

After attaching to LTE in the usual way, the mobile uses the mobile network code and mobile country code to construct the ANDSF's domain name [22] and looks up the corresponding IP address in a domain name server. It then contacts the ANDSF and sends details of its location such as the E-UTRAN cell global identifier (ECGI).

In response, the ANDSF sends two sets of information to the mobile. Firstly, the *access network discovery information* contains details of nearby access networks such as WiFi *service set identifiers* (SSIDs) and WiMAX *network access provider identities* (NAP-IDs). Secondly, the *inter-system mobility policy* contains a prioritized list of policy rules, only one of which is active at any one time. Each rule contains a prioritized list of radio access technologies such as 3GPP, WiFi and WiMAX, together with optional information describing the rule's validity in terms of the mobile's location, the date and the time of day.

On receiving this information, the mobile inspects the validity and priority of each policy rule and establishes which rule it should activate. With the assistance of the discovery information, the mobile can establish which radio access networks are nearby. The mobile then implements the active policy rule and, if necessary, moves to another radio access technology using the reselection procedure that we describe below. The mobile can also store the policy rules and discovery information for use the next time it switches on.

16.2 Generic Signalling Procedures

16.2.1 Overview of the Attach Procedure

To illustrate the procedures for inter-operation with non-3GPP networks, let us consider the case of network-based mobility across a trusted access network, with the signalling messages written using PMIPv6. This option might be suitable for inter-operation with mobile WiMAX or with a trusted wireless local area network and is also a useful prerequisite for the discussion of cdma2000 below. The message sequence is summarized in Figure 16.4.

The procedure begins when the mobile switches on in a region of non-3GPP coverage. Guided by any inter-system mobility policy that it has previously downloaded from the ANDSF, the mobile establishes radio communications with a non-3GPP access network using

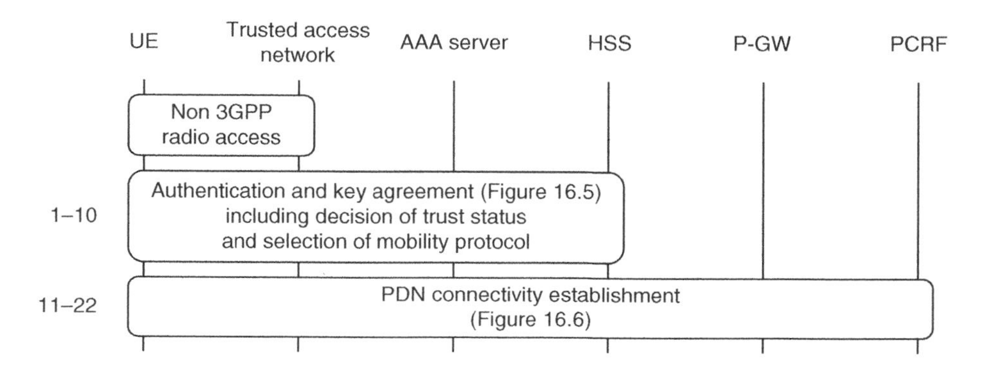

Figure 16.4 Overview of the attach procedure across a non-3GPP access network

techniques that correspond to the LTE procedures of network selection, cell selection and RRC connection establishment.

The access network then retrieves the mobile's identity and contacts the AAA server, which authenticates the mobile. During this second stage, the AAA server decides whether to treat the access network as trusted or untrusted, transfers the subscription data to the access network and decides the mobility protocol that it will use. In the third stage, the mobile attaches to the access network, acquires an IP address and connects to an external packet data network. We will look at the second and third stages in more detail below.

16.2.2 Authentication and Key Agreement

The security procedures for non-3GPP radio access are very like those for LTE [23]. The mobile and network authenticate each other using an IETF protocol known as the *Improved Extensible Authentication Protocol Method for 3G Authentication and Key Agreement* (EAP-AKA′) [24]. This improves upon an earlier protocol known as EAP-AKA [25], by computing secure keys that include the access network's identity and are unusable elsewhere. In turn, EAP-AKA is an implementation of the IETF extensible authentication protocol (EAP) [26].

Figure 16.5 shows the procedure for authentication and key agreement across a trusted access network [27]. The diagram assumes that the network is using dynamic configuration of the IP mobility protocol and that the mobile is not roaming.

At the start of the procedure, the mobile has established radio communications with a non-3GPP access network. Using an *EAP Request/Identity* message, the access network requests the mobile's identity (2), and the mobile returns its network access identifier using a previous pseudonym if one is available or its IMSI otherwise. The access network extracts the 3GPP network operator's domain name from the NAI and looks up the AAA server's IP address. It then forwards the mobile's response to the AAA server by embedding it in a *Diameter EAP Request* (2), which includes the access network's identity and the type of radio access being used.

On receiving the message, the AAA server looks up the mobile's IMSI and asks the home subscriber server for a set of authentication data (3). The HSS replies with the authentication parameters RAND, AUTN, XRES, CK′ and IK′. We saw the first three of these when looking at the LTE security procedures, while CK′ and IK′ are respectively derived from CK and IK using the identity of the access network. If the AAA server does not have a copy of the user's subscription data then it can also retrieve them here (4).

Using CK′ and IK′, the AAA server computes a *master session key* (MSK), which has a role similar to the LTE key K_{ASME} and is used to derive the actual keys used for ciphering and integrity protection. It also uses the information that the access network supplied in step (2) to decide whether to treat the access network as trusted, as in this example, or untrusted. The AAA server then sends an authentication challenge to the mobile in which it includes the parameters RAND and AUTN, indicates the network's trust status and asks the mobile for the mobility management protocols that it supports (5). The message is forwarded to the access network by embedding it in a *Diameter EAP Response*, in a similar way to step 2. The access network then extracts the EAP message and forwards it to the mobile.

On receiving the message, the mobile checks the authentication token, computes its own copy of the master session key and computes its response to the network's challenge. The mobile then returns its response to the AAA server along with a list of its supported

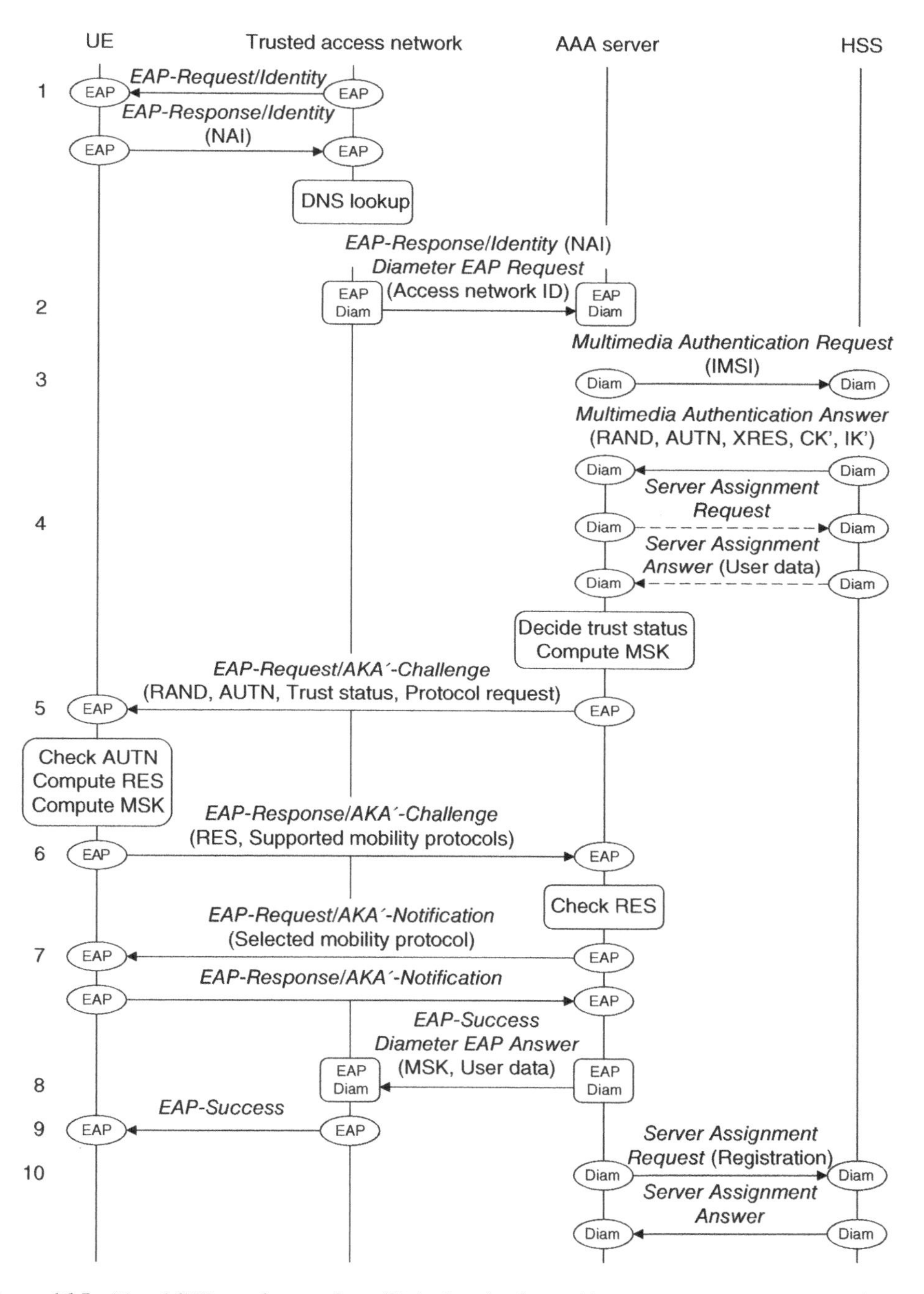

Figure 16.5 Non-3GPP attach procedure. (1) Authentication and key agreement across a trusted access network. Source: TS 33.402. Reproduced by permission of ETSI

mobility management protocols (6). The AAA server checks the mobile's response, selects a mobility management protocol and indicates its selection (7). In this example, we assume that the AAA server selects network-based mobility using PMIPv6, which requires no additional features within the mobile.

To complete the procedure, the AAA server passes the subscription data and master session key to the access network and informs the mobile that the authentication procedure has been completed successfully (8, 9). The access network can use the master session key for its own air interface security procedures. The AAA server also informs the home subscriber server that it has successfully registered the mobile (10).

There is a similar procedure for the case of an untrusted access network. Instead of relying on the security features of the access network itself, the procedure sets up a tunnel between the mobile and the evolved packet data gateway, across which they can exchange information securely using IPSec.

16.2.3 PDN Connectivity Establishment

In the last stage of the attach procedure, the mobile acquires an IP address and establishes a connection to an external packet data network. Figure 16.6 shows the steps that are involved, assuming that the mobile is not roaming and that we are using network-based mobility across a trusted access network with the signalling messages written using PMIPv6 [28].

At the start of the diagram, the network has just informed the mobile that the authentication and key agreement procedure has completed successfully. In response, the mobile sends a layer 3 attach request to the access network (11), using messages that correspond to steps 1 and 2 from the LTE attach procedure. The mobile may be able to request a preferred access point name as part of these messages, depending on the type of access network that it is using.

The access network selects a suitable access point name and a corresponding PDN gateway, using either any APN that the mobile requested or the default APN from the subscription data. In step 12, the access network's bearer binding and event reporting function runs the procedure of gateway control session establishment that we noted in Chapter 13. In this procedure, it sets up signalling communications with the PCRF, forwards the default quality of service that it received in the subscription data and receives a provisional policy and charging control rule in return.

The access network then sends a PMIPv6 *Proxy Binding Update* to the chosen PDN gateway (13), which identifies the mobile, requests the establishment of a GRE tunnel across the S2a interface and includes a downlink GRE key to identify the tunnel. The PDN gateway can then allocate a dynamic IP address for the mobile and can run the usual procedure of IP-CAN session establishment (14). The PDN gateway also tells the AAA server about its identity, so that the AAA server can store this information in the HSS for use in future reselections of the radio access technology (15–18). It then acknowledges the access network's request with a PMIPv6 *Proxy Binding Acknowledgement* (19), which includes the mobile's IP address and an uplink GRE key.

If the PCRF changed the provisional policy control and charging rule during IP-CAN session establishment, then it informs the access network using the procedure of gateway control and QoS rules provision (20). The access network then accepts the mobile's attach request, and includes the selected access point name and any IP address that the mobile has been allocated (21). The mobile carries out any steps that it needs for the completion of IP address

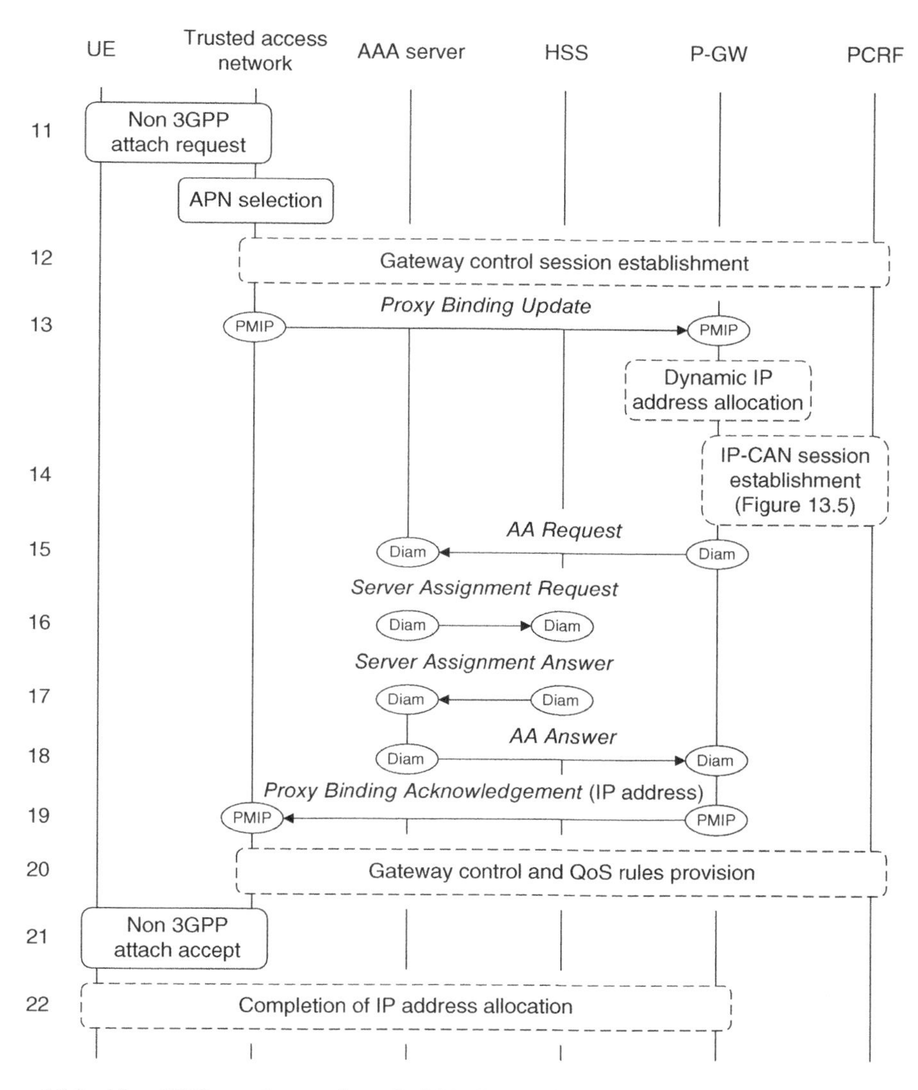

Figure 16.6 Non-3GPP attach procedure. (2) PDN Connectivity Establishment across a trusted access network, with network-based mobility using PMIPv6. Source: TS 23.402. Reproduced by permission of ETSI

allocation (22) and can then communicate through the PDN gateway using the functions of the non-3GPP access network.

There are two differences for the case of a trusted wireless local area network. Firstly, the PMIP messages can be replaced by a GTPv2-C Create Session Request and Response, in which case the gateway control procedures are not required. Secondly, the Release 11 solution has no impact on the mobile, which cannot specify an access point name of its own in step 11 and cannot use the handovers with IP address preservation that we cover below. These limitations are removed as part of Release 12, by means of a new signalling protocol between the mobile and the access network [29].

There is also a similar procedure for the case of host-based mobility over S2c [30]. In this procedure, the access network skips steps 13 to 19, and instead accepts the mobile's attach request, identifies the PDN gateway and gives the mobile an IP address that is local to the access network (21). In response, the mobile establishes a security association with the PDN gateway and sends it a DSMIPv6 *Binding Update* which includes the IP address that the access network assigned. The PDN gateway allocates the mobile an IP address from its own address space, which it will use to identify the mobile to the outside world. It then runs the usual procedure for IP-CAN session establishment and replies with a DSMIPv6 *Binding Acknowledgement*. The PDN gateway can now tunnel any incoming data packets to the mobile's local IP address, across the access network.

16.2.4 Radio Access Network Reselection

A mobile can switch from LTE to a non-3GPP access network and back again, possibly guided by any information that it has downloaded from the ANDSF. However, the generic non-3GPP architecture does not include any signalling interfaces to the MME, or any interfaces that would allow it to exchange data packets with the serving gateway. As a result, it does not support the optimized handovers that we have previously seen.

This has two implications. Firstly, the MME cannot send the mobile any advance information about the non-3GPP radio access network, such as the parameters that the mobile requires to access the target cell or even the carrier frequency. Instead, the mobile has to reselect the access network by itself, by detaching from LTE and attaching to the non-3GPP technology. The resulting procedure [31] is similar to the one from Figures 16.4 to 16.6 and can lead to a significant gap in communications. Secondly, there is no mechanism for forwarding unacknowledged downlink packets from LTE to the non-3GPP network. This implies that packets are likely to be lost during the communication gap.

The reselection procedure does, however, offer more functions than the basic attach procedure. During the authentication process (Figure 16.5), the AAA server retrieves the mobile's current access point name and PDN gateway from the home subscriber server, which stored them at the end of the LTE attach procedure in Chapter 11. It then forwards these to the access network. As part of the non-3GPP attach request (step 11 in Figure 16.6), the mobile may be able to ask the network to preserve its original session and IP address, depending on the type of access network that it is using. The access network responds by contacting the original PDN gateway in step 13, and the PDN gateway can continue using the mobile's old IP address instead of allocating a new one. As a result, the mobile retains its session and IP address during the handover, so it can maintain its communications with any external servers.

16.3 Inter-Operation with cdma2000 HRPD

16.3.1 System Architecture

As we saw in Chapter 1, most cdma2000 operators are planning to migrate their systems to LTE. In the early stages of this process, network operators will only have LTE cells in hotspots and large cities, leading to frequent handovers between the two systems. It is desirable that

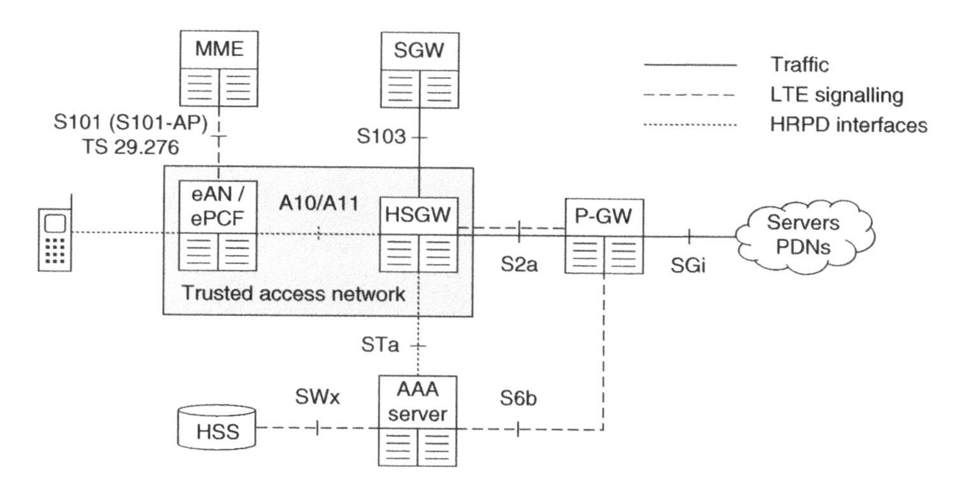

Figure 16.7 Architecture for inter-operation with cdma2000 HRPD

these handovers should be as efficient as possible. To support this, the LTE specifications define an optimized variant of the non-3GPP architecture for use with cdma2000 high-rate packet data (HRPD), also known as cdma2000 evolution data optimized (EV-DO). An HRPD network that supports this architecture is known as *evolved HRPD* (eHRPD).

Figure 16.7 shows the architecture that is used [32–34]. The architecture is based on the one for network-based mobility across a trusted non-3GPP access network, but with two new interfaces. Firstly, the MME can exchange signalling messages with an HRPD *evolved access network/evolved packet control function* (eAN/ePCF) across the S101 interface. The interface uses the *S101 application protocol* (S101-AP) [35], which simply transports HRPD signalling messages between the two devices. Secondly, the serving gateway can exchange data packets with an *HRPD serving gateway* (HSGW) across the S103 interface, so as to minimize the risk of packet loss during a handover.

16.3.2 Preregistration with cdma2000

If a mobile switches on in a region of LTE coverage, then it attaches to LTE in the usual way. However, the network can then tell it to run a procedure known as *HRPD preregistration* [36–38]. Using this procedure, the mobile establishes a dormant session in the HRPD network, which can be used later on to speed up any subsequent reselection or handover. The procedure is summarized in Figure 16.8.

After running the usual LTE attach procedure (1), the mobile reads SIB 8, which contains the parameters it will need for reselection to cdma2000 (2). One of these parameters is a preregistration trigger, which tells the mobile whether it should preregister with an HRPD network. By setting this trigger on a cell-by-cell basis, the network operator can tell a mobile to preregister with HRPD if it is near the edge of the LTE coverage area.

If the trigger is set, then the mobile registers with the HRPD access network using HRPD messages that are tunnelled over the air interface and over S1 and S101 (3). The network

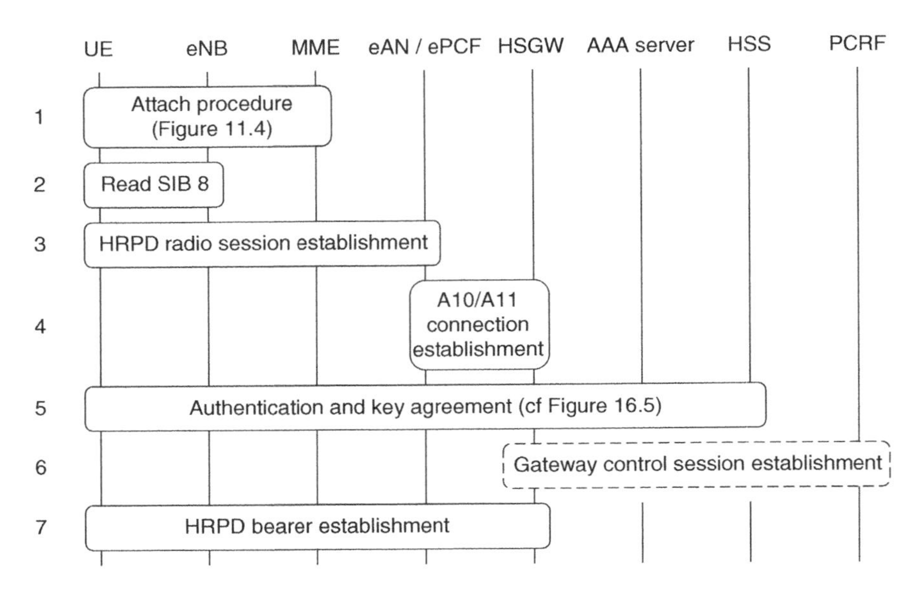

Figure 16.8 Procedure for preregistration with cdma2000 HRPD. Source: TS 23.402. Reproduced by permission of ETSI

establishes a signalling connection for the mobile with the HSGW (4) and the mobile authenticates itself to the 3GPP AAA server (5). During the authentication process, the AAA server retrieves the mobile's identity from the home subscriber server, along with information such as the mobile's access point name and PDN gateway, and sends these to the eAN/ePCF.

Inside the HSGW, the bearer binding and event reporting function runs the procedure of gateway control session establishment (6), to establish communications with the PCRF. Finally, the HSGW exchanges a set of HRPD messages with the mobile (7), to establish a dormant set of HRPD bearers that mirror the ones in the evolved packet core. The existence of these bearers will speed up any subsequent reselection or handover to HRPD.

Later on, the evolved packet core may set up new EPS bearers for the mobile, or may modify the quality of service of existing bearers. If this happens, then the mobile exchanges additional messages with the HSGW, so as to keep the HRPD bearers in step.

16.3.3 Cell Reselection in RRC_IDLE

In RRC_IDLE state, a mobile can reselect to a cdma2000 HRPD cell using the same procedure that it used for UMTS and GSM. The only difference is that SIB 8 contains a list of neighbouring cdma2000 HRPD cells, so the mobile is not expected to find these by itself.

Once reselection is complete, there are two possibilities. If the mobile has preregistered with HRPD, then it contacts the HRPD access network to indicate that it has arrived and the network sets up a new set of resources for the mobile. The effect is similar to the cdma2000 handover procedure described below, although with fewer steps. If the mobile has not preregistered, then it has to run the generic non-3GPP handover procedure that we covered earlier.

Note that there is no analogue to idle mode signalling reduction for cdma2000. Network operators may therefore wish to configure the cell reselection parameters so as to minimize the risk of mobiles bouncing back and forth between cdma2000 and LTE.

16.3.4 Measurements and Handover in RRC_CONNECTED

In RRC_CONNECTED state, the base station can release the mobile's RRC signalling connection and redirect it to a cdma2000 HRPD carrier, in the same way that it did for UMTS and GSM. Alternatively, it can tell the mobile to measure the signal-to-interference ratios of neighbouring cdma2000 cells and can use measurement events B1 and B2 to trigger a handover to a cdma2000 HRPD cell. To do this, it uses the procedure shown in the figures that follow [39–41]. In this procedure, we assume that the mobile has already preregistered with the HRPD network. If it has not done so, then it instead has to run the generic non-3GPP handover procedure that we covered earlier.

To begin the procedure (Figure 16.9), the mobile identifies a neighbouring HRPD cell in a measurement report (1) and the base station decides to hand the mobile over (2). In step 3, the base station tells the mobile to contact the HRPD network, using the signalling path that it established during preregistration. In response, the mobile composes an HRPD *Connection Request*, in which it requests the parameters of a physical *traffic channel* that it can use for communication across the HRPD air interface. In steps 4, 5 and 6, the message is forwarded across the Uu, S1 and S101 interfaces to the eAN/ePCF.

On receiving the message, the eAN/ePCF retrieves the mobile's access point name and PDN gateway identity, which it stored during the preregistration procedure, and sends these to the HSGW (7). In response, the HSGW returns an IP address to which the serving gateway can forward any unacknowledged downlink data packets. The eAN/ePCF can then allocate a set of radio resources for the mobile and can compose an HRPD *Traffic Channel Assignment*, which provides the mobile with the details that it requested in step 2. It sends the message to the MME by embedding it in an S101-AP direct transfer (8), which also contains the forwarding address that the HSGW sent earlier. The MME sends the forwarding address to the serving gateway (9), which responds by creating the requested tunnel.

The MME can now send the traffic channel assignment to the base station (10). In turn, the base station sends the message to the mobile (11) and also starts to return any unacknowledged downlink packets to the serving gateway. Using the forwarding address that it received earlier, the serving gateway can send these packets across the S103 interface to the HSGW, along with any other packets that arrive on the downlink later on. On receiving the base station's message, the mobile switches to cdma2000, acquires the specified traffic channel (12) and acknowledges the HRPD network's message using an HRPD *Traffic Channel Completion* (13).

In step 14 (Figure 16.10), the HRPD access network tells the HSGW to activate the bearers that it set up during preregistration. The HSGW asks the PDN gateway to establish a GRE tunnel to carry the data, which triggers the procedure of IP-CAN session modification in which the PDN gateway retrieves the corresponding policy and charging control rule. After the acknowledgements, traffic can flow on the uplink and downlink between the HRPD access network, the HSGW and the PDN gateway. If necessary, the network repeats these steps for every access point name that the mobile is using.

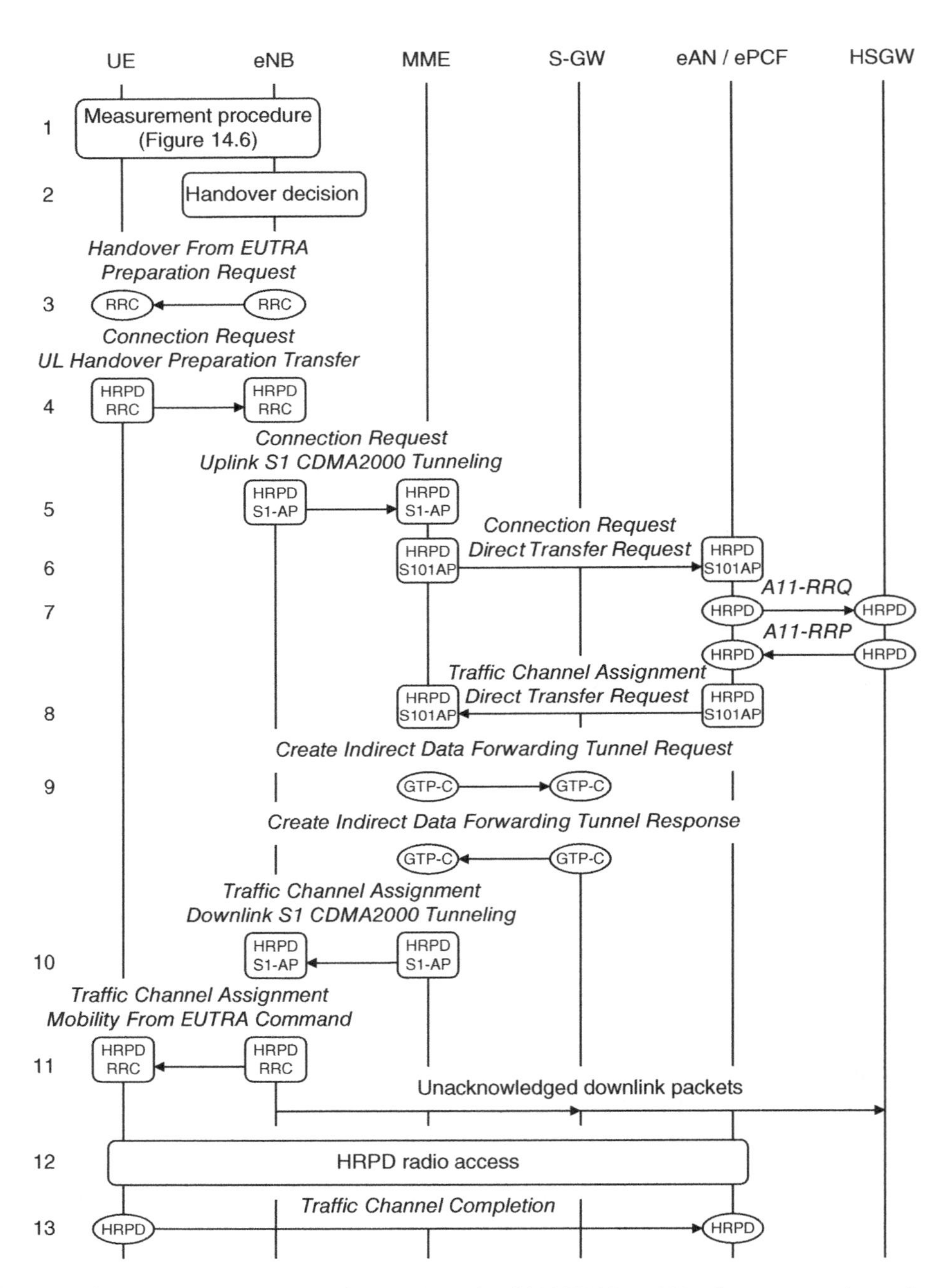

Figure 16.9 Inter-system handover from LTE to cdma2000 HRPD. (1) Handover preparation and execution. Source: TS 23.402. Reproduced by permission of ETSI

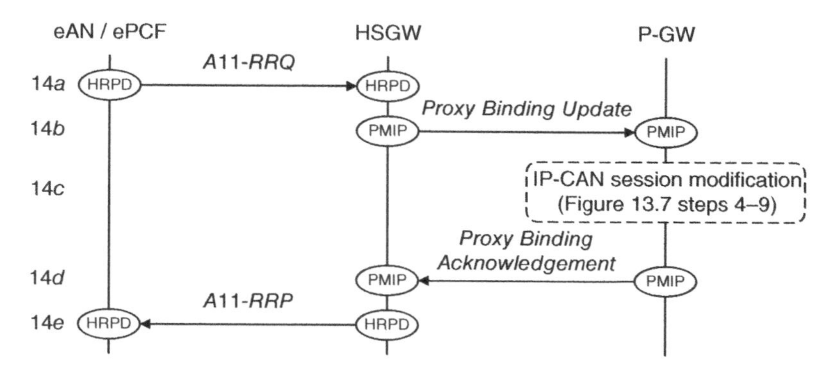

Figure 16.10 Inter-system handover from LTE to cdma2000 HRPD. (2) Bearer activation. Source: TS 23.402. Reproduced by permission of ETSI

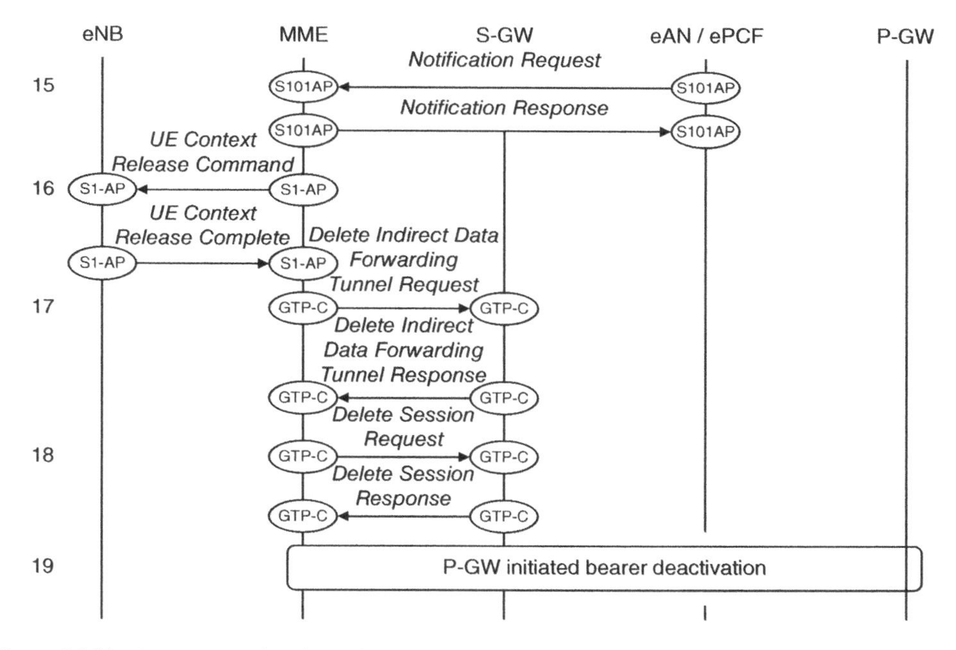

Figure 16.11 Inter-system handover from LTE to cdma2000 HRPD. (3) Handover completion. Source: TS 23.402. Reproduced by permission of ETSI

We still need to tear down the resources that the mobile was using in LTE (Figure 16.11). To achieve this, the HRPD access network tells the MME that the handover has completed (15). In response, the MME tells the base station to tear down the mobile's resources (16) and sends similar messages to the serving gateway (17, 18). At about the same time, the PDN gateway runs a procedure known as *PDN GW initiated bearer deactivation* (19) [42], which releases the remaining resources in the evolved packet core.

References

1. 3GPP TS 23.402 (2013) Architecture Enhancements for Non-3GPP Accesses, Release 11, June 2013.
2. 3GPP2 A.S0022-0 (2010) Interoperability Specification (IOS) for Evolved High Rate Packet Data (eHRPD) Radio Access Network Interfaces and Interworking with Enhanced Universal Terrestrial Radio Access Network (EUTRAN), Version 2.0, April 2010.
3. 3GPP2 X.S0057-0 (2010) E-UTRAN – eHRPD Connectivity and Interworking: Core Network Aspects, Version 3.0, September 2010.
4. Etemad, K. (2004) *CDMA2000 Evolution: System Concepts and Design Principles*, John Wiley & Sons, Ltd, Chichester.
5. Vanghi, V., Damnjanovic, A. and Vojcic, B. (2004) *The cdma2000 System for Mobile Communications: 3G Wireless Evolution*, Prentice Hall.
6. Yang, S. (2004) *3G CDMA2000 Wireless System Engineering*, Artech.
7. Ahmadi, S. (2010) *Mobile WiMAX: A Systems Approach to Understanding IEEE 802.16 m Radio Access Technology*, Academic Press.
8. Andrews, J.G., Ghosh, A. and Muhamed, R. (2007) *Fundamentals of WiMAX: Understanding Broadband Wireless Networking*, Prentice Hall.
9. Gast, M. (2011) *802.11 Wireless Networks: The Definitive Guide*, 2nd edn, O'Reilly.
10. 3GPP TS 23.402 (2013) Architecture Enhancements for Non-3GPP Accesses, Release 11, Sections 4.2.2, 4.2.3, 16.1.1, June 2013.
11. 3GPP TS 29.273 (2013) Evolved Packet System (EPS); 3GPP EPS AAA Interfaces, Release 11, September 2013.
12. 3GPP TS 29.275 (2013) Proxy Mobile IPv6 (PMIPv6) Based Mobility and Tunnelling Protocols; Stage 3, Release 11, June 2013.
13. 3GPP TS 29.279 (2012) Mobile IPv4 (MIPv4) Based Mobility Protocols; Stage 3, Release 11, September 2012.
14. 3GPP TR 23.852 (2013) Study on S2a Mobility based on GPRS Tunnelling Protocol (GTP) and Wireless Local Area Network (WLAN) Access to the Enhanced Packet Core (EPC) network (SaMOG); Stage 2, Release 12, September 2013.
15. 3GPP TS 24.302 (2013) Access to the 3GPP Evolved Packet Core (EPC) via Non-3GPP Access networks, Release 11, Section 7, June 2013.
16. 3GPP TS 23.003 (2013) Numbering, Addressing and Identification, Release 11, Section 19.3, September 2013.
17. 3GPP TS 24.303 (2013) Mobility Management Based on Dual-Stack Mobile IPv6, Release 11, June 2013.
18. 3GPP TS 23.402 (2013) Architecture Enhancements for Non-3GPP Accesses, Release 11, Section 4.8, June 2013.
19. 3GPP TS 24.302 (2013) Access to the 3GPP Evolved Packet Core (EPC) via Non-3GPP Access Networks, Release 11, Sections 5.1, 6.8, June 2013.
20. 3GPP TS 24.312 (2013) Access Network Discovery and Selection Function (ANDSF) Management Object (MO), Release 11, March 2013.
21. Open Mobile Alliance OMA-ERELD-DM-V1_2 (2008) Enabler Release Definition for OMA Device Management, Version 1.2.1, June 2008.
22. 3GPP TS 23.003 (2013) Numbering, Addressing and Identification, Release 11, Section 22, September 2013.
23. 3GPP TS 33.402 (2012) Security Aspects of Non-3GPP Accesses, Release 11, June 2012.
24. IETF RFC 5448 (2009) Improved Extensible Authentication Protocol Method for 3rd Generation Authentication and Key Agreement (EAP-AKA′), May 2009.
25. IETF RFC 4187 (2006) Extensible Authentication Protocol Method for 3rd Generation Authentication and Key Agreement (EAP-AKA), January 2006.
26. IETF RFC 3748 (2004) Extensible Authentication Protocol (EAP), June 2004.
27. 3GPP TS 33.402 (2012) Security Aspects of Non-3GPP Accesses, Release 11, Section 6.2, June 2012.
28. 3GPP TS 23.402 (2013) Architecture Enhancements for Non-3GPP Accesses, Release 11, Section 6.2.1, June 2013.
29. 3GPP TS 24.244 (2014) Wireless LAN control plane protocol for trusted WLAN access to EPC, Release 12, March 2014.
30. 3GPP TS 23.402 (2013) Architecture Enhancements for Non-3GPP Accesses, Release 11, Section 6.3, June 2013.
31. 3GPP TS 23.402 (2013) Architecture Enhancements for Non-3GPP Accesses, Release 11, Section 8.2.2, June 2013.
32. 3GPP TS 23.402 (2013) Architecture Enhancements for Non-3GPP Accesses, Release 11, Section 9.1, June 2013.

33. 3GPP2 A.S0022-0 (2010) Interoperability Specification (IOS) for Evolved High Rate Packet Data (eHRPD) Radio Access Network Interfaces and Interworking with Enhanced Universal Terrestrial Radio Access Network (EUTRAN), Version 2.0, Section 1.4, April 2010.
34. 3GPP2 X.S0057-0 (2010) E-UTRAN – eHRPD Connectivity and Interworking: Core Network Aspects, Version 3.0, Section 4, September 2010.
35. 3GPP TS 29.276 (2012) Optimized Handover Procedures and Protocols between E-UTRAN Access and cdma2000 HRPD Access, Release 11, September 2012.
36. 3GPP TS 23.402 (2013) Architecture Enhancements for Non-3GPP Accesses, Release 11, Section 9.3.1, June 2013.
37. 3GPP2 A.S0022-0 (2010) Interoperability Specification (IOS) for Evolved High Rate Packet Data (eHRPD) Radio Access Network Interfaces and Interworking with Enhanced Universal Terrestrial Radio Access Network (EUTRAN), Version 2.0, Section 3.2.1, April 2010.
38. 3GPP2 X.S0057-0 (2010) E-UTRAN - eHRPD Connectivity and Interworking: Core Network Aspects, Version 3.0, Section 13.1.1, September 2010.
39. 3GPP TS 23.402 (2013) Architecture Enhancements for non-3GPP Accesses, Release 11, Section 9.3.2, June 2013.
40. 3GPP2 A.S0022-0 (2010) Interoperability Specification (IOS) for Evolved High Rate Packet Data (eHRPD) Radio Access Network Interfaces and Interworking with Enhanced Universal Terrestrial Radio Access Network (EUTRAN), Version 2.0, Section 3.2.2.2, April 2010.
41. 3GPP2 X.S0057-0 (2010) E-UTRAN – eHRPD Connectivity and Interworking: Core Network Aspects, Version 3.0, Section 13.1.2, September 2010.
42. 3GPP TS 23.401 (2013) General Packet Radio Service (GPRS) Enhancements for Evolved Universal Terrestrial Radio Access Network (E-UTRAN) Access, Release 11, Section 5.4.4.1, September 2013.

17

Self-Optimizing Networks

In common with other mobile telecommunication technologies, an LTE network is controlled by a network management system. This has a wide range of functions; for example, it sets the parameters that the network elements are using, manages their software, and detects and corrects any faults in their operation. Using such a management system, an operator can remotely configure and optimize every base station in the radio access network and every component of the core network. However the process requires manual intervention, which can make it time-consuming, expensive and prone to error. To deal with this issue, 3GPP has gradually introduced a technique known as *self-optimizing* or *self-organizing networks* (SON) into LTE.

In this chapter, we cover the main self-optimization features that have been added to LTE in each of its releases. These fall into four broad categories, namely self-configuration of an LTE base station, interference coordination, mobility management and drive test minimization. We also discuss a technique known as radio access network information management (RIM), through which an LTE base station can exchange self-optimization data with the radio access networks of UMTS and GSM.

Self-optimizing networks are summarized in TR 36.902 [1] and TS 36.300 [2]. Their main impact is on the radio access network's signalling procedures, notably the ones on the X2 interface [3]. For some more detailed accounts of the use of self-optimizing networks in LTE, see References [4–6].

17.1 Self-Configuration of an eNB

17.1.1 Automatic Configuration of the Physical Cell Identity

LTE has been designed so that a network operator can set up a new base station with minimal knowledge of the outside world, which might include the domain name of the network management system, and the domain names of its MMEs and serving gateways. The base station can acquire the other information it needs by a process of self-configuration [7]. During this process, the base station contacts the management system and downloads the software it will require for its operation. It also downloads a set of configuration parameters [8], such as a tracking area code, a list of PLMN identities, and the global cell identity and maximum transmit power of each cell.

An Introduction to LTE: LTE, LTE-Advanced, SAE, VoLTE and 4G Mobile Communications, Second Edition. Christopher Cox.

In the configuration parameters, the management system can explicitly assign a physical cell identity to each of the base station's cells. However, this places an unnecessary burden on the network planner, as every cell must have a different identity from any other cells that are nearby. It also causes difficulties in networks that contain home base stations, which can be sited without any knowledge of their neighbours at all.

As an alternative, the management system can simply give the base station a short list of allowed physical cell identities. If the base station has a suitable downlink receiver, then it can listen for other LTE cells that are nearby, reject their physical cell identities and choose an identity at random from the ones that remain. For the configuration of subsequent cells, the base station can also reject any identities that it receives during measurement reports and any that nearby base stations list during the X2 setup procedure described below.

As part of the self-configuration process, the base station also runs a procedure known as *S1 setup* [9], to establish communications with each of the MMEs that it is connected to. In this procedure, the base station tells the MME about the tracking area code and PLMN identities of each of its cells, as well as any closed subscriber groups that it belongs to. The MME replies with a message that indicates its globally unique identity, and it can now communicate with the base station over the S1 interface.

17.1.2 Automatic Neighbour Relations

During the configuration process described above, there is no need for a base station to find out anything about its individual neighbouring cells and no need for it to set up a neighbour list. This removes a large burden from the network operator and a large potential source of error. Instead, a mobile can identify a neighbouring cell by itself and can tell the base station about it later on using the RRC measurement reports that we covered in Chapter 14. The base station can then establish communications with its neighbour using the *automatic neighbour relation* procedure shown in Figure 17.1 [10].

The procedure is triggered when the base station receives a measurement report containing a physical cell identity that it was not previously aware of (1). The base station cannot contact the new cell right away, so it sends the mobile a second measurement configuration to ask for more information about that cell (2). In response, the mobile completes the acquisition procedure for the neighbouring cell, reads system information block 1 and returns the cell's global identity, tracking area code and PLMN list in a second measurement report (3). The base station now has enough information to initiate an S1-based handover to the new cell.

To support X2-based handovers, the base station sends the global cell ID to the MME and asks it to return an IP address that the neighbouring base station is using for communications over X2 (4). The MME is already communicating with the neighbouring base station over S1, so it can send the request onwards (5) and can return the neighbour's reply (6). The two base stations can now establish communications across the X2 interface (7), using a procedure known as *X2 setup* [11]. During this procedure, the base stations exchange information about all the cells they are controlling, including their global cell identities, physical cell identities and carrier frequencies. This last field might include frequencies that the original base station was not previously aware of, which it can use to populate the list of neighbouring frequencies in SIB 5.

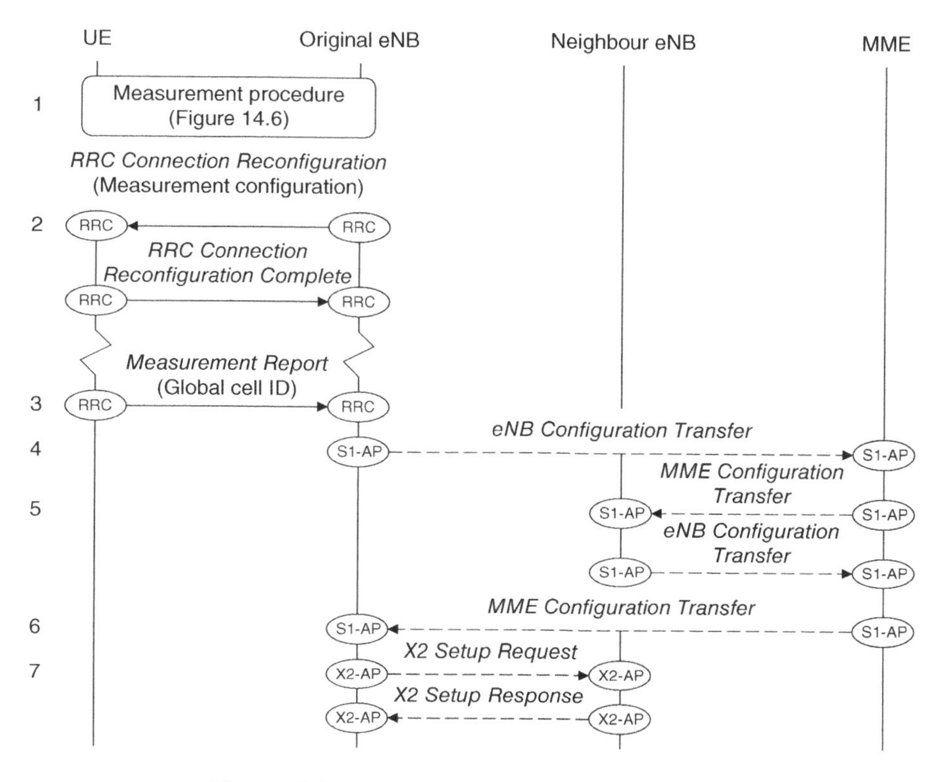

Figure 17.1 Automatic neighbour relation procedure

17.1.3 Random Access Channel Optimization

Manual configuration of the random access channel is another potential burden for the network operator. From Release 9, base stations can exchange information about the parameters they are using for the random access channel during the X2 setup procedure. The information includes the PRACH frequency offset and PRACH configuration index, which determine the resource blocks that the channel is using, and the zero correlation zone configuration, root sequence index and high-speed flag, which determine the cell's choice of random access preambles. Using this information, the base stations can minimize the interference between random access transmissions in nearby cells by allocating them different sets of resource blocks and different preambles.

Subsequently, the base station can use an RRC message known as a *UE Information Request* to retrieve information about a mobile's last successful random access attempt. The information includes the number of preambles that the mobile sent before receiving a reply and an indication of whether the contention resolution procedure failed at any stage. Using this information, the base station can adjust the random access channel's power settings and resource block allocations, so as to minimize the load that the channel makes on the air interface.

17.2 Inter-Cell Interference Coordination

The X2-AP *Load Indication* procedure [12] helps a network to minimize the interference between neighbouring base stations and to implement the fractional frequency re-use schemes that we introduced in Chapter 4. To use the procedure, a base station sends an X2-AP *Load Information* message to one of its neighbours. In the message, the base station can include three information elements for each cell that it is controlling. The first is the *relative narrowband Tx power*, which states whether the transmitted power in each downlink resource block lies above or below a threshold. The neighbour can use this information in its scheduling procedure, by avoiding downlink transmissions to distant mobiles in resource blocks that are subject to high levels of downlink interference.

The second information element is the *UL high interference indication*, which lists the uplink resource blocks in which the base station intends to schedule distant mobiles. The neighbour can use this in a similar way, so that it does not schedule uplink transmissions from distant mobiles in resource blocks that will be subject to high uplink interference. The third is the *UL interference overload indication*, which states whether the received interference in each uplink resource block is high, moderate or low. Here, the neighbour could again avoid scheduling uplink transmissions from distant mobiles in overloaded resource blocks, this time to ensure that it does not generate high levels of uplink interference.

From Release 10, the base station can include a fourth information element known as the *ABS* (almost blank subframe) *information*. This supports the enhanced inter-cell interference coordination techniques that we will cover in Chapter 19.

17.3 Mobility Management

17.3.1 Mobility Load Balancing

We will now discuss three self-optimization techniques that relate to mobility management within LTE. Figure 17.2 shows the first of these, which is known as *mobility load balancing* or *resource status reporting* [13]. Using this procedure, nearby base stations can cooperate to even out the load in the radio access network and to maximize the total capacity of the system.

Using an X2-AP *Resource Status Request* (1), a Release 8 base station can ask one of its neighbours to report three items of information. The first is the percentage of resource blocks that the neighbour is using in each of its cells, for both GBR and non-GBR traffic. The second is the load on the S1 interface, while the third is the hardware load. The neighbour returns an

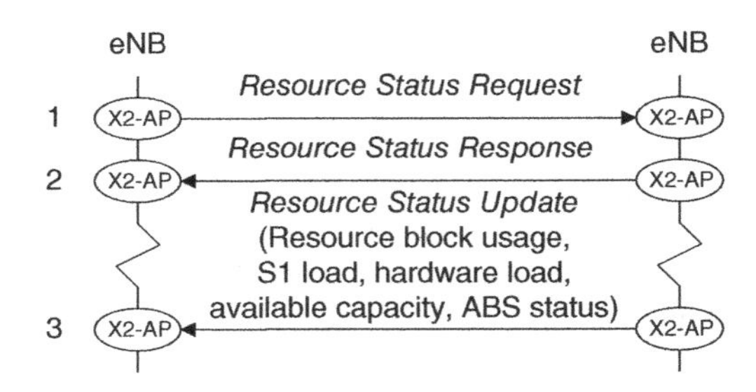

Figure 17.2 Resource status reporting procedure

acknowledgement (2) and then reports each item periodically for both the uplink and downlink, using an X2-AP *Resource Status Update* (3). As a result of this information, a congested base station can hand over a mobile to a neighbouring cell that has enough spare capacity and can even out the load in the radio access network.

From Release 9, the neighbour can report a fourth field in its Resource Status Update, the *composite available capacity group*, which indicates the capacity that it has available for load balancing purposes on the uplink and downlink. The original base station can use this information to assist its handover decision. From Release 10, the neighbour can report a fifth field, the *ABS status*, which we will cover in Chapter 19.

After such a handover, there is a risk that the new base station will hand the mobile straight back to the old one. To prevent this from happening, Release 9 introduces another X2 procedure, known as *mobility settings change* [14]. Using this procedure, a base station can ask a neighbour to adjust the thresholds that it is using for measurement reporting, by means of the cell-specific offsets that we introduced in Chapter 14. After the adjustment, the mobile should stay in the target cell, instead of being handed back.

17.3.2 Mobility Robustness Optimization

Mobility robustness optimization [15] is a self-optimization technique that first appears in Release 9. Using this technique, a base station can gather information about any problems that have arisen due to the use of unsuitable measurement reporting thresholds. It can then use the information to adjust the thresholds it is using and to correct the problem.

There are three main causes of trouble, the first of which is shown in Figure 17.3. Here, the base station has started a handover to a new cell (1) but it has done this too late, because its

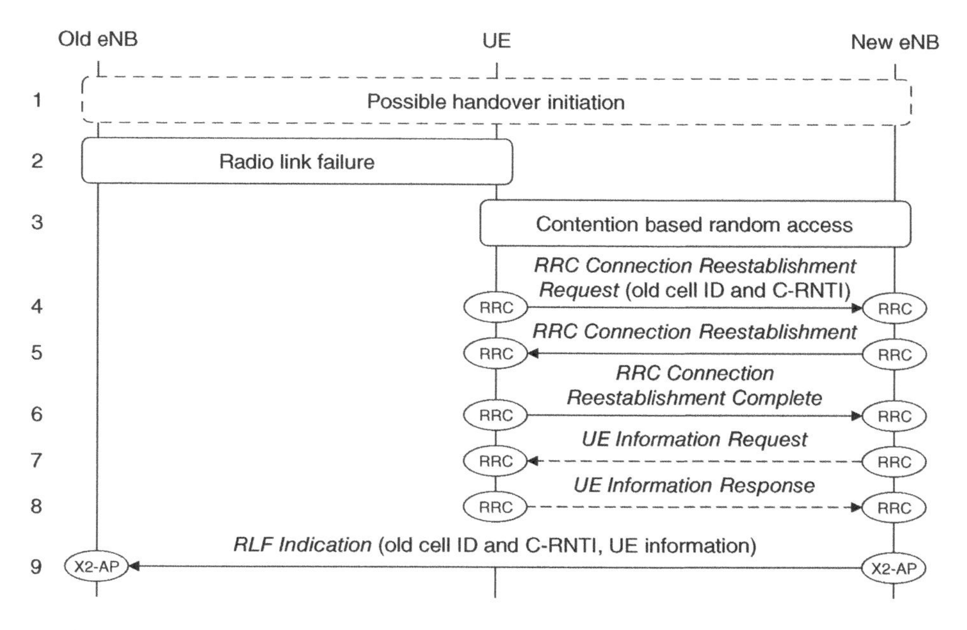

Figure 17.3 Mobility robustness optimization, triggered by a handover that was too late

measurement reporting thresholds have been poorly set. Alternatively, it may not have started the handover at all. Before any handover is executed, the mobile's received signal power falls below a threshold and its radio link fails (2). In response, the mobile runs the cell selection procedure and discovers the cell that it should have been handed to. It contacts the new cell using the random access procedure (3) and a procedure known as *RRC connection reestablishment* (4, 5, 6), in which it identifies itself using the old cell's physical cell ID and its old C-RNTI. In step 6, the mobile can also indicate that it has measurements from immediately before the radio link failure of the power received from the old cell and its neighbours. If it does, then the base station retrieves this information using an RRC *UE Information* procedure (7, 8).

The new base station can now tell the old base station about the problem, using a Release 9 message known as an X2-AP *Radio Link Failure* (RLF) *Indication* (9). After a series of such reports, the old base station can take action by adjusting its measurement reporting thresholds, using proprietary optimization software that is outside the scope of the 3GPP specifications.

The next problem is in Figure 17.4. Here, the base station has carried out a handover too early (1), perhaps in an isolated area where the mobile is briefly receiving line-of-sight coverage from the new cell. The mobile completes the handover, but its radio link soon fails (2). On running the cell selection procedure, the mobile rediscovers the old cell, re-establishes an RRC connection (3), and identifies itself using its new physical cell ID and C-RNTI. The old base station notifies the new one, as before (4). However, the new base station notices that it had just received the mobile in a handover from the old one, so it tells the old base station using another Release 9 message, an X2-AP *Handover Report* (5). Once again, the old base station can use the information to adjust its measurement thresholds.

The final problem is in Figure 17.5. Here, the base station has handed the mobile over to the wrong cell, perhaps due to an incorrect cell-specific measurement offset (1). The mobile's radio link fails as before (2) and it re-establishes an RRC connection with a third cell, the one it should have been handed to in the first place (3). In response, the third base station sends a radio link failure indication to the second (4), which notifies the original base station using a handover report as before (5).

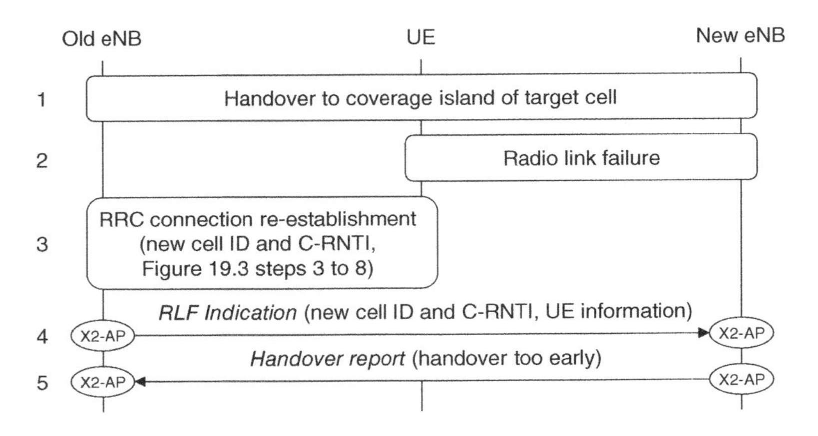

Figure 17.4 Mobility robustness optimization, triggered by a handover that was too early

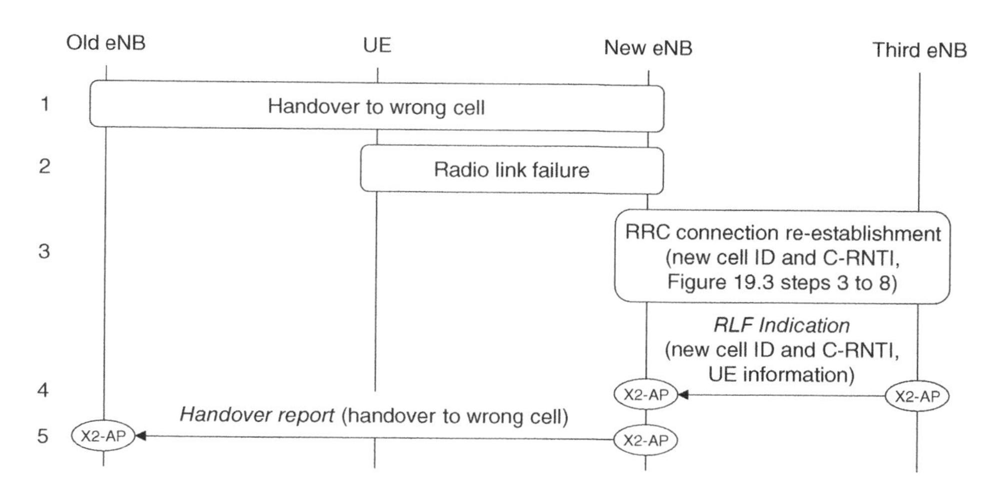

Figure 17.5 Mobility robustness optimization, triggered by a handover to the wrong cell

17.3.3 Energy Saving

The aim of the final mobility management procedure, first introduced in Release 9 [16], is to save energy by switching off cells that are not being used. A typical situation is the use of picocells in a shopping centre, in which a cell can be switched off outside shopping hours if it only contributes to the network's capacity, but not to its coverage.

If a base station supports this feature, then it can decide to switch the cell off after a long period of low load. To do this, it hands any remaining mobiles over to cells that have overlapping coverage, tells them about the change using an X2-AP *eNB Configuration Update* and switches the cell off. The base station itself remains switched on, so, at a later time, a neighbour can ask the base station to switch the cell on again using an X2-AP *Cell Activation Request*.

17.4 Radio Access Network Information Management

17.4.1 Introduction

So far we have only discussed the transfer of information within the radio access network of LTE. *Radio access network information management* (RIM) [17–19] was originally developed for information transfers within the radio access network of GSM. It was later extended to information transfers between the radio access networks of GSM, UMTS and LTE, specifically between the GSM base station controller, the UMTS radio network controller and the eNB.

At the lowest level, the core network transports information between the three radio access networks by embedding it into signalling messages on the intervening interfaces. Higher up, the end-to-end information transfer is managed by a separate RIM protocol, which supports single and event-driven reporting, and which guarantees reliable delivery by means of acknowledgements and retransmissions. At the highest level, the specifications define various RIM applications that handle different types of information. These applications allow an LTE base station to discover the system information of nearby UMTS and GSM cells, and extend the capabilities of self-optimizing networks to cover inter-system optimization.

17.4.2 Transfer of System Information

Figure 17.6 shows a signalling procedure that uses this technique. There are two stages. In the first stage, a new LTE base station discovers one of its UMTS neighbours and adds the neighbour to its list of 3G measurement objects. In the second stage, the base station uses an RIM procedure to request the neighbour's system information.

The procedure begins when the base station receives a measurement report about a UMTS neighbour that it was not previously aware of. The mobile originally found the cell by itself whilst in RRC_IDLE, but has subsequently moved to RRC_CONNECTED and has continued to measure the cell. The mobile describes the cell using its 3G physical cell identity, but the base station does not recognize it and asks for more information. In response, the mobile completes the 3G acquisition procedure, reads the cell's system information and returns its global cell identity, location area code, routing area code and PLMN identity (1). The base station can now hand the mobile over to the new cell. It can also add the cell to its list of 3G measurement objects, so that the cell can be measured by other mobiles that are already in RRC_CONNECTED.

The base station can now compose an RIM *RAN Information Request* in which it asks for multiple event-driven reporting of the new cell's system information (2). The network transports the message to the destination radio network controller by embedding it into an S1-AP *eNB Direct Information Transfer*, a GTPv1-C or GTPv2-C *RAN Information Relay* and an Iu-PS *Direct Information Transfer*. After the RNC's reply (3), the base station can send the system information to a mobile during the procedure of RRC connection release with

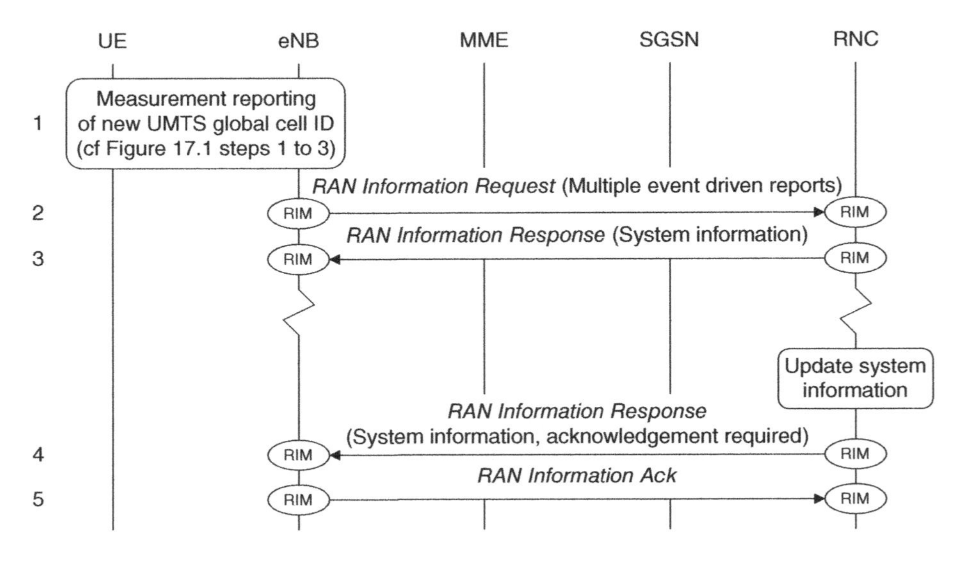

Figure 17.6 Transfer of system information from a new UMTS cell to LTE

redirection from Chapter 15, so as to speed up that procedure. Later on, the RNC may update the system information. Because of the previous request for multiple event-driven reporting, it alerts the LTE base station (4) and requests an acknowledgement to ensure reliable delivery (5).

17.4.3 Transfer of Self-Optimization Data

We can apply similar procedures to transfer three types of self-optimization data. Firstly, the radio access networks can exchange cell load information to support inter-system mobility load balancing. An LTE cell reports the composite available capacity group, while UMTS and GSM cells report similar information known as the *cell load information group*.

Secondly, the radio access networks can exchange information in support of mobility robustness optimization, to help them detect unnecessary inter-system handovers. After a handover from LTE to UMTS or GSM, the new radio access network can ask the mobile to continue measuring the signal power that it is receiving from nearby LTE cells. If the signal power is sufficiently high, then the network can tell the LTE base station that it triggered the handover unnecessarily. The base station can use a series of such reports to adjust its measurement reporting thresholds.

Thirdly, the radio access networks can exchange information to support inter-system energy saving. This works in the same way as the LTE energy-saving procedures from earlier, by allowing a cell to report when it has switched off or ask a neighbour to switch on.

17.5 Drive Test Minimization

Network operators have traditionally assessed the coverage of a radio access network by transporting measurement devices around its intended coverage area, in a technique known as drive testing. As well as being time-consuming and expensive, this technique provides coverage data that are limited to the route of the drive test and supplies little or no information about coverage indoors. Network operators do, however, have another ready supply of measurement devices in the form of the users' mobiles. In a Release 10 technique known as *minimization of drive tests* (MDT) [20–22], an operator can ask its mobiles to return measurements that supplement or even replace the ones obtained from traditional drive testing.

As part of the customer care process, the operator is obliged to obtain the users' consent for using their mobiles in drive test minimization. The network stores the relevant information in the home subscriber server and checks it before measurement activation.

If the user does consent, then two measurement modes are available: immediate measurements for mobiles in RRC_CONNECTED state and logged measurements for mobiles in RRC_IDLE. Immediate measurements follow the same reporting procedure that we saw in Chapter 14. The mobile measures the downlink RSRP or RSRQ and reports these quantities to the base station; it also reports any location data that it has available, but does not make any location measurements just for the purpose of MDT. The base station can then return the information to the management system, using the existing network management procedures for trace reporting.

A base station can also send an RRC message known as *Logged Measurement Configuration* to an active mobile, to configure it for logged measurements once it enters RRC_IDLE. In idle mode, the mobile makes its measurements with a period that is a multiple of the discontinuous reception cycle. It then stores the information in a log, along with time stamps and any location data that it has available. When the mobile next establishes an RRC connection, it can signal the availability of its measurement log using a field in the message RRC Connection Setup Complete. The base station can then retrieve the logged measurements from the mobile using the RRC UE Information procedure and can forward them to the management system as before.

The Release 11 specifications enhance drive test minimization in two ways. Firstly, the mobile returns additional measurements to the base station. The uplink and downlink data throughput indicate the user's quality of service, the uplink and downlink data volume indicate where the traffic is concentrated, the uplink power headroom indicates the uplink coverage, while the number of radio link and RRC connection establishment failures provide further information about any coverage holes. Secondly, the network can instruct the mobile to measure and return its location for the purpose of drive test minimization alone. This makes the information more valuable to the network operator but can increase the mobile's power consumption. It might typically be managed using an extra subscription option within the home subscriber server.

References

1. 3GPP TR 36.902 (2011) Evolved Universal Terrestrial Radio Access Network (E-UTRAN); Self-Configuring and Self-Optimizing Network (SON) Use Cases and Solutions, Release 9, April 2011.
2. 3GPP TS 36.300 (2013) Evolved Universal Terrestrial Radio Access (E-UTRA) and Evolved Universal Terrestrial Radio Access Network (E-UTRAN); Overall Description; Stage 2, Release 11, Section 22, September 2013.
3. 3GPP TS 36.423 (2013) Evolved Universal Terrestrial Radio Access Network (E-UTRAN); X2 Application Protocol (X2AP), Release 11, September 2013.
4. Hämäläinen, S., Sanneck, H. and Sartori, C. (2011) *LTE Self-Organizing Networks: Network Management Automation for Operational Efficiency*, John Wiley & Sons, Ltd, Chichester.
5. Ramiro, J. and Hamied, K. (2011) *Self-Organizing Networks: Self Planning, Self Optimization and Self Healing for GSM, UMTS and LTE*, John Wiley & Sons, Ltd, Chichester.
6. 4G Americas (2011) Self-Optimizing Networks: The Benefits of SON in LTE, July 2011.
7. 3GPP TS 32.501 (2012) Telecommunication Management; Self-Configuration of Network Elements; Concepts and Requirements, Release 11, Section 6.4.2, September 2012.
8. 3GPP TS 32.762 (2013) Telecommunication Management; Evolved Universal Terrestrial Radio Access Network (E-UTRAN) Network Resource Model (NRM) Integration Reference Point (IRP); Information Service (IS), Release 11, Section 6, March 2013.
9. 3GPP TS 36.413 (2013) Evolved Universal Terrestrial Radio Access Network (E-UTRAN); S1 Application Protocol (S1AP), Release 11, Section 8.7.3, September 2013.
10. 3GPP TS 36.300 (2013) Evolved Universal Terrestrial Radio Access (E-UTRA) and Evolved Universal Terrestrial Radio Access Network (E-UTRAN); Overall Description; Stage 2, Release 11, Section 22.3.3, September 2013.
11. 3GPP TS 36.423 (2013) Evolved Universal Terrestrial Radio Access Network (E-UTRAN); X2 Application Protocol (X2AP), Release 11, Section 8.3.3, September 2013.
12. 3GPP TS 36.423 (2013) Evolved Universal Terrestrial Radio Access Network (E-UTRAN); X2 Application Protocol (X2AP), Release 11, Section 8.3.1, September 2013.
13. 3GPP TS 36.423 (2013) Evolved Universal Terrestrial Radio Access Network (E-UTRAN); X2 Application Protocol (X2AP), Release 11, Section 8.3.6, 8.3.7, September 2013.
14. 3GPP TS 36.423 (2013) Evolved Universal Terrestrial Radio Access Network (E-UTRAN); X2 Application Protocol (X2AP), Release 11, Sections 8.3.8, September 2013.
15. 3GPP TS 36.423 (2013) Evolved Universal Terrestrial Radio Access Network (E-UTRAN); X2 Application Protocol (X2AP), Release 11, Sections 8.3.9, 8.3.10, September 2013.
16. 3GPP TS 36.423 (2013) Evolved Universal Terrestrial Radio Access Network (E-UTRAN); X2 Application Protocol (X2AP), Release 11, Sections 8.3.5, 8.3.11, September 2013.

17. 3GPP TS 23.401 (2013) General Packet Radio Service (GPRS) Enhancements for Evolved Universal Terrestrial Radio Access Network (E-UTRAN) Access, Release 11, Section 5.15, September 2013.
18. 3GPP TS 48.018 (2013) General Packet Radio Service (GPRS); Base Station System (BSS) – Serving GPRS Support Node (SGSN); BSS GPRS Protocol (BSSGP), Release 11, Section 8c, September 2013.
19. 3GPP TS 36.413 (2013) Evolved Universal Terrestrial Radio Access Network (E-UTRAN); S1 Application Protocol (S1AP), Sections 8.13, 8.14, Release 11, Annex B, September 2013.
20. 3GPP TS 37.320 (2013) Universal Terrestrial Radio Access (UTRA) and Evolved Universal Terrestrial Radio Access (E-UTRA); Radio Measurement Collection for Minimization of Drive Tests (MDT); Overall Description; Stage 2, Release 11, March 2013.
21. 3GPP TS 36.331 (2013) Radio Resource Control (RRC); Protocol Specification, Release 11, Sections 5.6.6, 5.6.7, 5.6.8, September 2013.
22. 3GPP TS 32.422 (2013) Telecommunication Management; Subscriber and Equipment Trace; Trace Control and Configuration Management, Release 11, Sections 4.2.8, 6, July 2013.

18

Enhancements in Release 9

At the time when the specifications for LTE Release 8 were being written, there was competition between LTE and WiMAX for support from network operators and equipment vendors. In view of this competition, 3GPP understandably wanted to finalize the specifications for Release 8 as soon as they could. To help achieve this, they delayed some of the peripheral features of the system until Release 9, which was frozen in December 2009.

This chapter covers the capabilities that were first introduced in Release 9, which include the multimedia broadcast/multicast service, location services and dual layer beamforming. The contents of Release 9 are described in the relevant 3GPP release summary [1] and there is a useful review in a regularly updated white paper by 4G Americas [2].

18.1 Multimedia Broadcast/Multicast Service

18.1.1 Introduction

Mobile cellular networks are normally used for one-to-one communication services such as phone calls and web browsing, but they can also be used for one-to-many services such as mobile television. There are two types of one-to-many service: *broadcast services* are available to anyone, while *multicast services* are only available to users who have subscribed to a multicast group. Broadcast and multicast services require several different techniques from traditional unicasting. For example, the network has to distribute the data using IP multicast, while the encryption techniques have to be modified to ensure that all subscribing users can receive the information stream.

UMTS implements these techniques using the *multimedia broadcast/multicast service* (MBMS) [3], which was introduced in 3GPP Release 6. Although it is not widely used, it has proved popular in a few markets such as Japan and South Korea. The LTE multimedia broadcast/multicast service was introduced in Release 9 and is often known as *evolved MBMS* (eMBMS). Despite the service's name, LTE currently only supports broadcast services, which do not require the user to subscribe to a multicast group.

To transmit MBMS data streams, LTE uses an air interface technique known as *multicast/broadcast over a single frequency network* (MBSFN). MBSFN was fully specified in

---

An Introduction to LTE: LTE, LTE-Advanced, SAE, VoLTE and 4G Mobile Communications, Second Edition.
Christopher Cox.

Release 8, but we will cover both MBMS and MBSFN in this chapter so as to keep the two related issues together.

18.1.2 Multicast/Broadcast over a Single Frequency Network

When delivering a broadcast or multicast service, the radio access network transmits the same information stream from several nearby cells. This is different from the usual situation in a mobile telecommunication system, in which nearby cells are transmitting completely different information. LTE exploits this feature to improve the transmission of broadcast and multicast services, using the technique of MBSFN (Figure 18.1).

When using MBSFN, nearby base stations are synchronized so that they broadcast the same content at the same time and on the same sub-carriers. The mobile receives multiple copies of the information, which are identical except for their different arrival times, amplitudes and phases. The mobile can then process the information streams using exactly the same techniques that we introduced in Chapter 4 for handling multipath. It is not even aware that the information is coming from multiple cells.

Because the extra cells are transmitting the same information stream, they do not cause any interference to the mobile: instead, they contribute to the received signal power. This increases the mobile's SINR and maximum data rate, particularly at the edge of the cell where interference is usually high. In turn, this helps LTE to reach a target spectral efficiency of 1 bit s^{-1} Hz^{-1} for the delivery of MBMS [4], equivalent to 16 mobile TV channels in a 5 MHz bandwidth at a rate of 300 kbps each.

18.1.3 Implementation of MBSFN in LTE

In LTE, the synchronized base stations that we introduced above lie in a geographical region known as an *MBSFN area*. MBSFN areas can overlap, so that one base station can transmit multiple sets of content from multiple MBSFN areas.

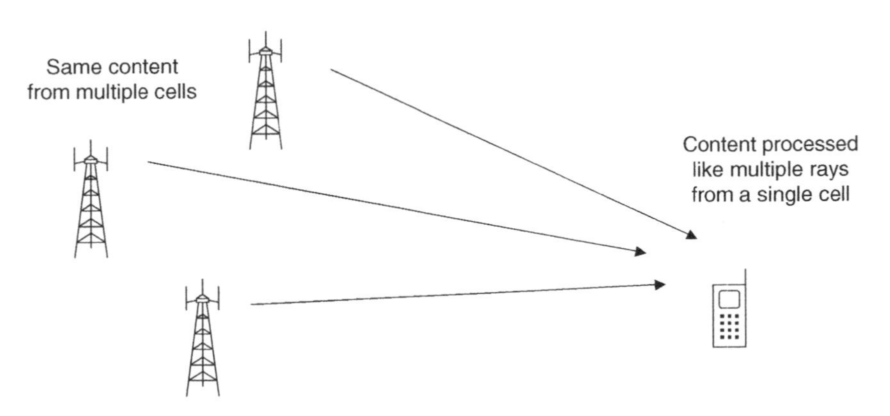

Figure 18.1 Multicast/broadcast over a single frequency network

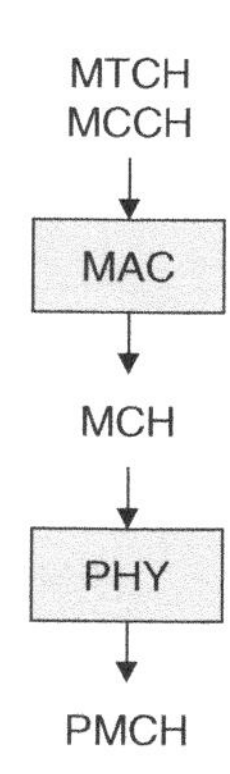

Figure 18.2 Channels used for MBMS and MBSFN

In each MBSFN area, the LTE air interface delivers MBMS using the channels shown in Figure 18.2. There are two logical channels. The multicast traffic channel (MTCH) carries broadcast traffic such as a television station, while the multicast control channel (MCCH) carries RRC signalling messages that describe how the traffic channels are being transmitted. Each MBSFN area contains one multicast control channel and multiple instances of the multicast traffic channel.

The multicast traffic and control channels are transported using the multicast channel (MCH) and the physical multicast channel (PMCH). Each MBSFN area contains multiple instances of the PMCH, each of which carries either the multicast control channel, or one or more multicast traffic channels.

There are a few differences between the transmission techniques used for the PMCH and the PDSCH [5]. The PMCH supports the MBSFN techniques described above, and always uses the extended cyclic prefix to handle the long delay spreads that result from the use of multiple base stations. It is transmitted on antenna port 4, to keep it separate from the base station's other transmissions, and does not use transmit diversity, spatial multiplexing or hybrid ARQ. Each instance of the channel uses a fixed modulation scheme and coding rate, which are configured by means of RRC signalling.

The PMCH uses a different set of reference signals from usual, known as *MBSFN reference signals*. These are tagged with the MBSFN area identity instead of the physical cell identity, to ensure that the mobile can successfully combine the reference signals that it receives from different cells.

To date, MBMS is only delivered on carriers that are shared with unicast traffic. This is achieved using time division multiplexing. In any one cell, each downlink subframe is allocated either to unicast traffic on the PDSCH, or to broadcast traffic on the PMCH, according to a mapping that is defined in SIB 2. Thus a particular subframe contains either the PDSCH, or the PMCH, but not both. Furthermore, each broadcast subframe is allocated to a single MBSFN area. Within an MBSFN area, each broadcast subframe carries either signalling messages on the multicast control channel, or broadcast traffic on a single instance of the multicast traffic channel.

A broadcast subframe still starts with a PDCCH control region, but this is only used for uplink scheduling and uplink power control commands, and is only one or two symbols long.

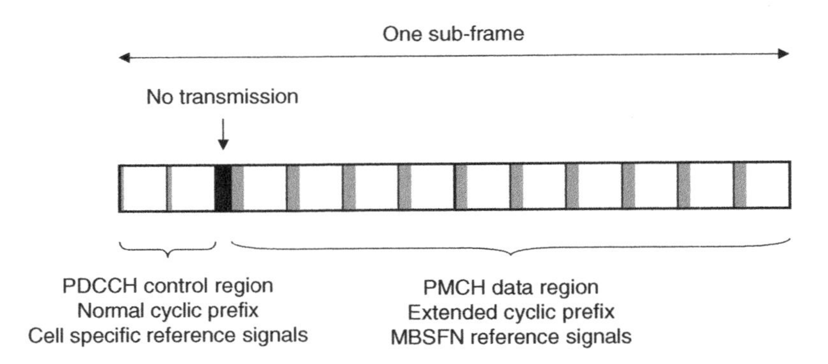

Figure 18.3 Example structure of an MBSFN subframe, for a 2-symbol control region using the normal cyclic prefix

The control region uses the cell's usual cyclic prefix duration (normal or extended) and the cell-specific reference signals. The rest of the subframe is occupied by the PMCH, which uses the extended cyclic prefix and the MBSFN reference signals. Figure 18.3 shows the resulting slot structure, for the case where the control region uses the normal cyclic prefix and contains two symbols. Note the gap in the downlink transmission, as the cell changes from one cyclic prefix duration to the other.

Eventually, it is envisaged that LTE will also support cells that are dedicated to MBMS. These cells will only use the downlink reference and synchronization signals, the physical broadcast channel and the physical multicast channel and will not support any uplink transmissions. They have not yet been fully specified, but are worth mentioning because they can optionally use a special slot structure in support of very large cells. In this slot structure, the sub-carrier spacing is reduced to 7.5 kHz, which increases the symbol duration to 133.3 μs and the cyclic prefix duration to 33.3 μs. This option appears in the air interface specifications from Release 8 [6], but is not yet usable and can generally be ignored.

18.1.4 Architecture of MBMS

Figure 18.4 shows the architecture that is used for the delivery of MBMS over LTE [7, 8]. The *broadcast/multicast service centre* (BM-SC) receives MBMS content from a content provider. The *MBMS gateway* (MBMS-GW) distributes the content to the appropriate base stations, while the *multicell/multicast coordination entity* (MCE) schedules the transmissions from all the base stations in a single MBSFN area.

The BM-SC indicates the start of each MBMS session by sending a signalling message across the SGmb interface. The interface is not defined by the 3GPP specifications, but the message describes the session's quality of service and tells the MBMS gateway to reserve resources for it. The message is propagated across the Sm [9], M3 [10] and M2 [11] interfaces, the last of which also defines the modulation scheme, coding rate and subframe allocation that the base stations in the MBSFN area should use.

The BM-SC then broadcasts the data across the SGi-mb interface using IP multicast. Each MBMS gateway forwards the data to the appropriate base stations across the M1 interface,

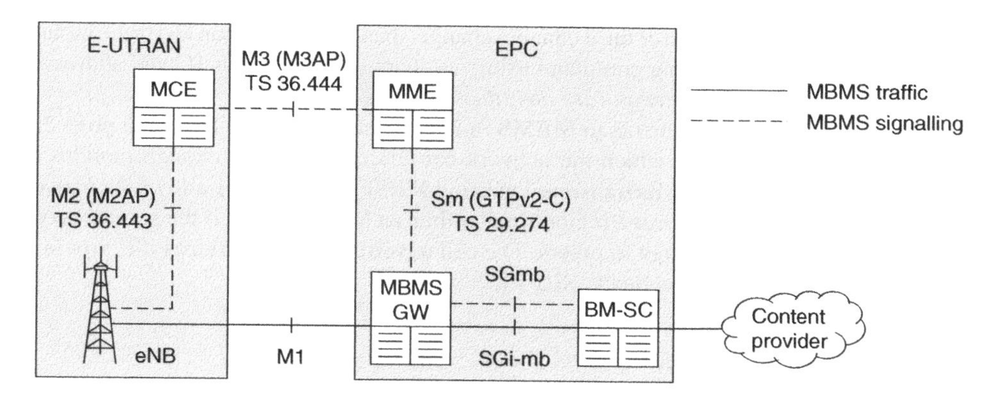

Figure 18.4 Architecture for MBMS in LTE

along with a header that indicates each packet's transmission time with an accuracy of 10 ms [12]. By combining this with the scheduling information that it receives from the multicell/multicast coordination entity, the base station can establish the exact transmission time for each packet.

18.1.5 Operation of MBMS

The best way to understand the operation of MBMS is to look at the procedures within the mobile [13–16]. After it switches on, the mobile reads SIB 2. By doing this, it discovers which subframes have been reserved for MBSFN transmissions on the PMCH and which have been reserved for unicast transmissions on the PDSCH. The MBSFN subframes repeat with a period of 1 to 32 frames. They do not clash with the subframes used by the synchronization signals or the physical broadcast channel, or with paging subframes.

If the user wishes to receive broadcast services, then the mobile continues by reading a new system information block, SIB 13. This lists the MBSFN areas that the cell belongs to. For each MBSFN area, it also defines the subframes that carry the multicast control channel, and the modulation scheme and coding rate that those transmissions will use.

The mobile can now receive the multicast control channel, which carries a single RRC signalling message, *MBSFN Area Configuration*. This message lists the physical multicast channels that the MBSFN area is using. For each PMCH, the message lists the corresponding multicast traffic channels and defines the subframes that will be used, the modulation scheme, the coding rate and a parameter known as the *MCH scheduling period* that lies between 8 and 1024 frames.

The mobile can now receive each instance of the PMCH. It still has to discover one more piece of information, namely the way in which the PMCH subframes are shared amongst the various multicast traffic channels that it carries. The base station signals this information in a new MAC control element known as *MCH scheduling information*, which it transmits at the beginning of each MCH scheduling period. Once it has read this control element, the mobile can receive each instance of the MTCH.

If the contents of the multicast control channel change, then the base station alerts the mobiles by writing a PDCCH scheduling command using a variant of DCI format 1C and addressing it to the *MBMS radio network temporary identifier* (M-RNTI).

There are two main enhancements to MBMS in later releases. Release 10 adds a procedure known as *MBMS counting*, in which the network can discover whether enough mobiles are interested in a service to justify its transmission using MBSFN. Release 11 allows the network to transmit an MBMS service over a region smaller than an MBSFN area, if the service is only relevant to some locations but not to others. The cell advertises which services it is providing using a new system information block, SIB 15.

18.2 Location Services

18.2.1 Introduction

Location services (LCS) [17, 18], also known as *location-based services* (LBS), allow an application to find out the geographical location of a mobile. UMTS supported location services from Release 99, but they were only introduced into LTE from Release 9.

The biggest single motivation is emergency calls. This issue is especially important in the USA, where the Federal Communications Commission requires network operators to localize an emergency call to an accuracy between 50 and 300 m, depending on the type of positioning technology used [19]. Location services are also of increasing importance to the user, for applications such as navigation and interactive games. Other applications include lawful interception by the police or security services and the use of a mobile's location to support network-based functions such as handover.

18.2.2 Positioning Techniques

In common with other mobile communication systems, LTE can calculate a mobile's position using three different techniques. The most accurate and increasingly common technique is the use of a *global navigation satellite system* (GNSS), a collective term for satellite navigation systems such as the *Global Positioning System* (GPS). There are two variants. With *UE-based positioning*, the mobile has a complete satellite receiver and calculates its own position. The network can send it information to assist this calculation, such as an initial position estimate and a list of visible satellites. With *UE assisted positioning*, the mobile has a more basic satellite receiver, so it sends a basic set of measurements to the network and the network calculates its position. Whichever method is adopted, the measurement accuracy is typically around 10 m.

The second technique is known as *downlink positioning* or *observed time difference of arrival* (OTDOA). Here, the mobile measures the times at which signals arrive from its serving cell and the nearest neighbours, and reports the time differences to the network. The network can then calculate the mobile's position by triangulation. The timing measurements are made on a new set of reference signals, known as *positioning reference signals* [20], which are transmitted on a new antenna port, number 6. The positioning accuracy is limited by multipath, typically to around 100 m [21], so the technique has difficulty in meeting the requirements of the US FCC. It is often used as a backup to satellite positioning, as a mobile may not be able to receive a satisfactory satellite signal if it is surrounded by tall buildings or is indoors. Release 11 introduces a variant known as *uplink time difference of arrival* (UTDOA), in which the network measures the mobile's position by timing the arrival of the sounding reference signal at different base stations.

The last technique is known as *enhanced cell ID positioning*. Here, the network estimates the mobile's position from its knowledge of the serving cell identity and additional information such as the mobile's timing advance. The positioning accuracy depends on the cell size, being excellent in femtocells (provided that the base station's position is actually known), but very poor in macrocells.

18.2.3 Location Service Architecture

Figure 18.5 shows the main hardware components that LTE uses for location services [22, 23]. The *gateway mobile location centre* (GMLC) receives location requests from external clients across the Le interface. It retrieves the identity of the mobile's serving MME from the home subscriber server and forwards the location request to the MME. In turn, the MME delegates responsibility for calculating a mobile's position to the *evolved serving mobile location centre* (E-SMLC).

Two other components can be separate devices or can be integrated into the GMLC. The *privacy profile register* (PPR) contains the users' privacy details, which determine whether a location request from an external client will actually be accepted. The *pseudonym mediation device* (PMD) retrieves a mobile's IMSI using the identity supplied by the external client.

The architecture uses several signalling protocols. The GMLC communicates with the home subscriber server and the MME using Diameter applications [24, 25], while the MME communicates with the E-SMLC using the *LCS application protocol* (LCS-AP) [26]. The MME can also send positioning-related information to the mobile using *supplementary service* (SS) messages that are embedded into EMM messages on the air interface [27]. The E-SMLC communicates with the mobile and the base station using the *LTE positioning protocol* (LPP) [28], the messages being transported by embedding them into lower-level LCS-AP and EMM messages. The GMLC can communicate with the external client using a few different techniques, such as the *open service architecture* (OSA).

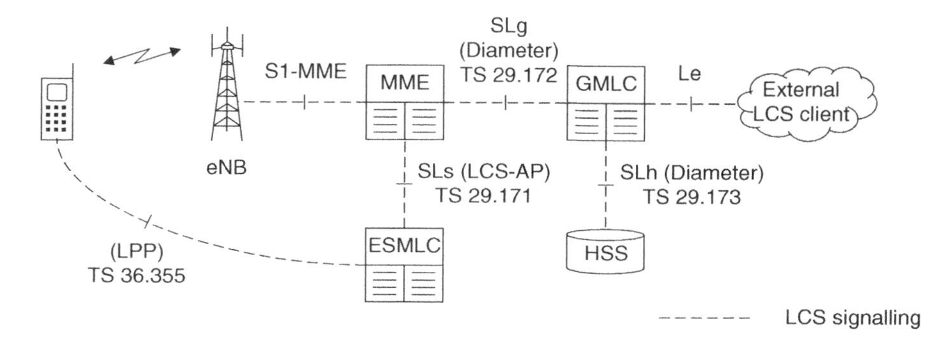

Figure 18.5 Architecture for location services in LTE

18.2.4 Location Service Procedures

To illustrate the operation of location services, Figure 18.6 shows how the network might respond to a location request from an external client [29]. The diagram assumes that the mobile is not roaming and that the GMLC can communicate directly with the mobile's serving MME. It also assumes the use of a mobile-assisted or mobile-based positioning technique such as GNSS or OTDOA.

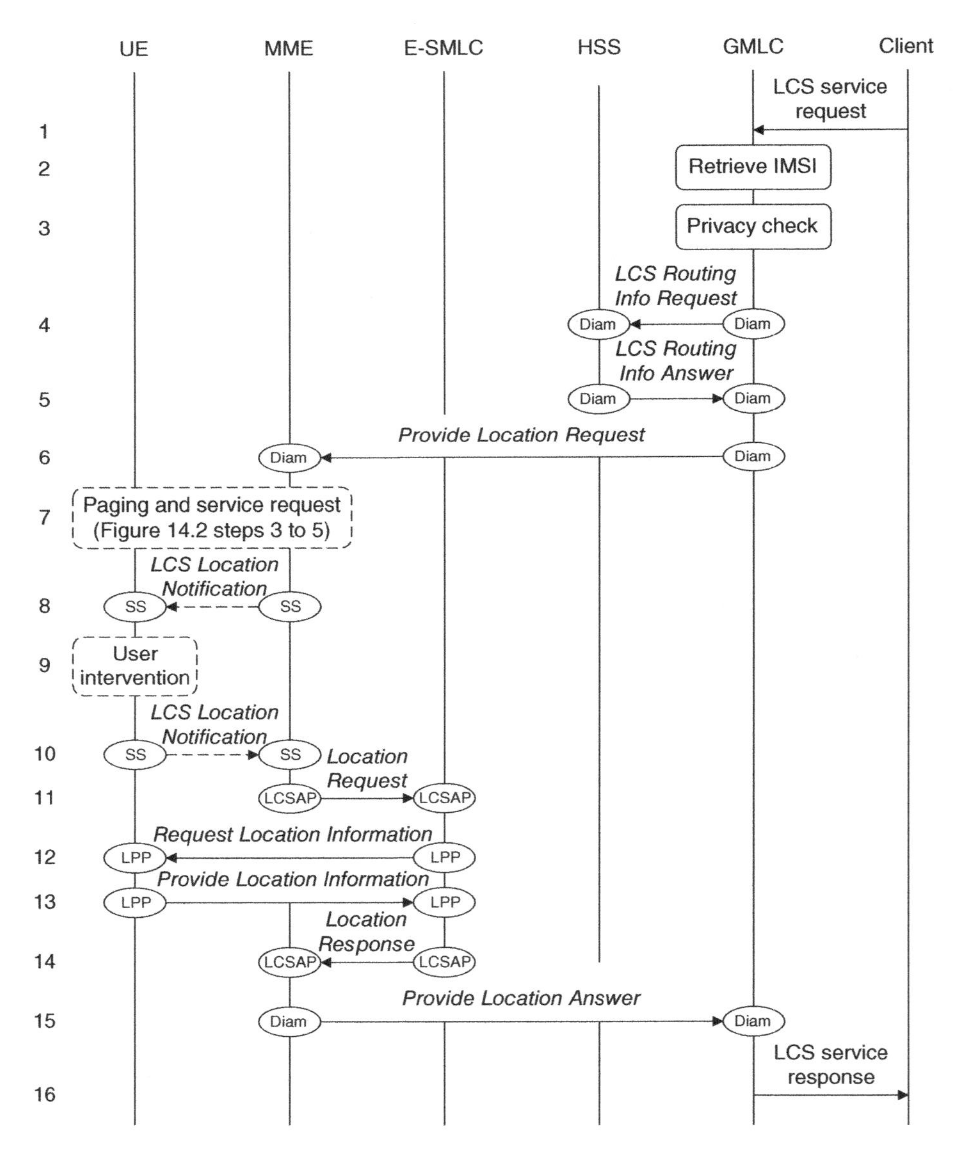

Figure 18.6 Mobile-terminated location request procedure. Source: TS 23.271. Reproduced by permission of ETSI

The procedure begins when the external client asks the GMLC for the mobile's position and optionally velocity (1). The client typically identifies the mobile using its IP address. The GMLC retrieves the mobile's IMSI from the pseudonym mediation device (2) and interrogates the privacy profile register to establish whether the location request can be accepted (3). It then retrieves the identity of the mobile's serving MME from the home subscriber server (4, 5) and forwards the location request there (6). As part of that message, it can specify the location estimate's quality of service in terms of the positional accuracy and response time, the information being obtained either from the client or from the mobile's subscription data.

If the mobile is in ECM-IDLE state, then the MME wakes it up using the paging procedure, to which the mobile responds by initiating a service request (7). If the privacy information indicates that the user should be notified of an incoming location request, then the MME sends a notification message to the mobile (8). The mobile asks the user whether the request can be accepted (9) and indicates the response to the MME (10). Unless the user withholds permission, the MME selects an E-SMLC and forwards the location request there (11). The message includes the mobile's location capabilities, which are supplied as part of its non-access stratum capabilities during the attach procedure.

Using the mobile's capabilities and the requested quality of service, the E-SMLC decides the positioning technique that it will use. Assuming the use of a mobile-assisted or mobile-based technique, it sends the mobile a location request (12). In its message, the E-SMLC specifies the selected positioning technique and supplies supporting information such as the satellites that should be visible or the positioning reference signals that nearby base stations are transmitting. The message is transported to the mobile by embedding it into lower-level LCS-AP and EMM messages.

The mobile makes the measurements that have been requested and sends a response to the E-SMLC (13). In turn, the E-SMLC returns its position estimate to the client (14, 15, 16).

18.3 Other Enhancements in Release 9

18.3.1 Dual Layer Beamforming

In LTE Release 8, the base station can use multiple antennas in two distinct configurations, namely spatial multiplexing, in which the antennas are widely spaced or have different polarizations, and beamforming, in which the antennas are close together. However, in practical deployments, the base station often mixes these configurations by using two arrays of closely spaced antennas with two different polarizations.

LTE exploits this arrangement in Release 9 with the introduction of *dual layer beamforming* [30–32]. In this technique, the base station constructs two independent beams using two antenna arrays that have different polarizations, and sends a different data stream on each one. The mobile receives the beams using two receive antennas and can reconstruct the transmitted data using the normal techniques of spatial multiplexing. Dual layer beamforming can therefore double the mobile's received data rate, provided that the signal-to-interference plus noise ratio is sufficiently high.

To use dual layer beamforming, the base station configures the mobile into a new transmission mode, mode 8, and schedules it using a new DCI format, 2B. It then transmits to the mobile using either or both of the two new antenna ports, numbers 7 and 8. The antenna ports use a new set of UE-specific reference signals, which are only transmitted in the physical resource blocks that the mobile is using and which behave in the same way as the reference signals that are used for single layer beamforming on port 5. As before, the base station processes the reference signals using the same antenna weights that it applies to the PDSCH so that the weights are transparent to the mobile and are removed as a side-effect of channel estimation.

18.3.2 Commercial Mobile Alert System

The US Federal Communications Commission established the *commercial mobile alert system* (CMAS) in response to the US Warning Alert and Response Network act of 2006. Using this system, participating network operators can transmit three types of emergency message: Presidential alerts about local, regional or national emergencies, imminent threat alerts about natural disasters such as hurricanes, and child abduction emergency alerts.

In Release 9, LTE supports CMAS by generalizing its earthquake and tsunami warning system to a *public warning system* (PWS) that covers both types of information [33]. The base station continues to send earthquake and tsunami warnings on SIBs 10 and 11, and transmits commercial mobile alerts on a new system information block, SIB 12.

References

1. 3rd Generation Partnership Project (2013) FTP Directory, ftp://ftp.3gpp.org/Information/WORK_PLAN/Description_Releases/ (accessed 15 October 2013).
2. 4G Americas (2012) 4G Mobile Broadband Evolution: Release 10, Release 11 and Beyond: HSPA+, SAE/LTE and LTE-Advanced, October 2012.
3. 3GPP TS 23.246 (2012) Multimedia Broadcast/Multicast Service (MBMS); Architecture and Functional Description, Release 11, March 2012.
4. 3GPP TR 25.913 (2009) Requirements for Evolved UTRA (E-UTRA) and Evolved UTRAN (E-UTRAN), Release 9, Section 7.5, January 2009.
5. 3GPP TS 36.211 (2013) Physical Channels and Modulation, Release 11, Sections 6.5, 6.10.2, September 2013.
6. 3GPP TS 36.211 (2013) Physical Channels and Modulation, Release 11, Sections 6.2.3, 6.10.2.2, 6.12, September 2013.
7. 3GPP TS 23.246 (2012) Multimedia Broadcast/Multicast Service (MBMS); Architecture and Functional Description, Release 11, Sections 4.2.2, 4.3.3, 5, March 2012.
8. 3GPP TS 36.440 (2013) Evolved Universal Terrestrial Radio Access Network (E-UTRAN); General Aspects and Principles for Interfaces Supporting Multimedia Broadcast Multicast Service (MBMS) within E-UTRAN, Release 11, Section 4, March 2013.
9. 3GPP TS 29.274 (2013) 3GPP Evolved Packet System (EPS); Evolved General Packet Radio Service (GPRS) Tunnelling Protocol for Control Plane (GTPv2-C); Stage 3, Release 11, Section 7.13, September 2013.
10. 3GPP TS 36.444 (2013) Evolved Universal Terrestrial Radio Access Network (E-UTRAN); M3 Application Protocol (M3AP), M3 Application Protocol (M3AP), Release 11, June 2013.
11. 3GPP TS 36.443 (2013) Evolved Universal Terrestrial Radio Access Network (E-UTRAN); M2 Application Protocol (M2AP), M2 Application Protocol (M2AP), Release 11, June 2013.
12. 3GPP TS 25.446 (2012) MBMS Synchronisation Protocol (SYNC), Release 11, September 2012.
13. 3GPP TS 36.331 (2013) Radio Resource Control (RRC); Protocol Specification, Release 11, Sections 5.8, 6.2.2 (MBSFNAreaConfiguration), 6.3.1, 6.3.7, September 2013.
14. 3GPP TS 36.321 (2013) Medium Access Control (MAC) Protocol Specification, Release 11, Sections 5.12, 6.1.3.7, July 2013.
15. 3GPP TS 36.213 (2013) Physical Layer Procedures, Release 11, Section 11, September 2013.
16. 3GPP TS 36.212 (2013) Multiplexing and Channel Coding, Release 11, Section 5.3.3.1.4, June 2013.
17. 3GPP TS 23.271 (2013) Functional Stage 2 Description of Location Services (LCS), Release 11, March 2013.
18. 3GPP TS 36.305 (2013) Evolved Universal Terrestrial Radio Access Network (E-UTRAN); Stage 2 Functional Specification of User Equipment (UE) Positioning in E-UTRAN, Release 11, March 2013.

19. Federal Communications Commission (2013) Wireless 911 Services, http://www.fcc.gov/guides/wireless-911-services (accessed 5 November 2013).
20. 3GPP TS 36.211 (2013) Physical Channels and Modulation, Release 11, Section 6.10.4, September 2013.
21. 3GPP R1-090768 (2009) Performance of DL OTDOA with Dedicated LCS-RS, Release 11, February 2009.
22. 3GPP TS 23.271 (2013) Functional Stage 2 Description of Location Services (LCS), Release 11, Section 5, March 2013.
23. 3GPP TS 36.305 (2013) Evolved Universal Terrestrial Radio Access Network (E-UTRAN); Stage 2 Functional Specification of User Equipment (UE) Positioning in E-UTRAN, Release 11, Section 6, March 2013.
24. 3GPP TS 29.173 (2012) Location Services (LCS); Diameter-Based SLh Interface for Control Plane LCS, Release 11, December 2012.
25. 3GPP TS 29.172 (2013) Location Services (LCS); Evolved Packet Core (EPC) LCS Protocol (ELP) between the Gateway Mobile Location Centre (GMLC) and the Mobile Management Entity (MME); SLg Interface, Release 11, September 2013.
26. 3GPP TS 29.171 (2013) Location Services (LCS); LCS Application Protocol (LCS-AP) between the Mobile Management Entity (MME) and Evolved Serving Mobile Location Centre (E-SMLC); SLs Interface, Release 11, June 2013.
27. 3GPP TS 24.171 (2012) Control Plane Location Services (LCS) Procedures in the Evolved Packet System (EPS), Release 11, September 2012.
28. 3GPP TS 36.355 (2013) Evolved Universal Terrestrial Radio Access (E-UTRA); LTE Positioning Protocol (LPP), Release 11, September 2013.
29. 3GPP TS 23.271 (2013) Functional Stage 2 Description of Location Services (LCS), Release 11, Sections 9.1.1, 9.1.15, 9.3a, March 2013.
30. 3GPP TS 36.211 (2013) Physical Channels and Modulation, Release 11, Section 6.10.3, September 2013.
31. 3GPP TS 36.212 (2013) Multiplexing and Channel Coding, Release 11, Section 5.3.3.1.5B, June 2013.
32. 3GPP TS 36.213 (2013) Physical Layer Procedures, Release 11, Section 7.1.5A, September 2013.
33. 3GPP TS 22.268 (2012) Public Warning System (PWS) Requirements, Release 11, December 2012.

19

LTE-Advanced and Release 10

Release 10 enhances the capabilities of LTE, to make the technology compliant with the International Telecommunication Union's requirements for IMT-Advanced. The resulting system is known as LTE-Advanced. This chapter covers the new features of the LTE-Advanced air interface, by focussing on carrier aggregation, relaying and enhancements to multiple antenna transmission on the uplink and downlink. We also discuss the techniques that are used to control interference in heterogeneous networks, in which the cells are on the same carrier frequency but have different sizes. Finally, we cover techniques for offloading traffic from the LTE network and for controlling the load from machine-type communications.

For the most part, the Release 10 enhancements are designed to be backwards compatible with Release 8. Thus a Release 10 base station can control a Release 8 mobile, normally with no loss of performance, while a Release 8 base station can control a Release 10 mobile. In the few cases where there is a loss of performance, the degradation has been kept to a minimum.

TR 36.912 [1] is 3GPP's submission to the ITU for LTE-Advanced and is a useful summary of the new features of the system. The contents of Release 10 are also described in the relevant 3GPP release summary [2] and are reviewed in a regularly updated white paper by 4G Americas [3].

19.1 Carrier Aggregation

19.1.1 Principles of Operation

The ultimate goal of LTE-Advanced is to support a maximum bandwidth of 100 MHz. This is an extremely large bandwidth, which is most unlikely to be available as a contiguous allocation in the foreseeable future. To deal with this problem, LTE-Advanced allows a mobile to transmit and receive on up to five *component carriers* (CCs), each of which has a maximum bandwidth of 20 MHz. This technique is known as *carrier aggregation* (CA) [4].

There are three scenarios, shown in Figure 19.1. In *inter-band aggregation*, the component carriers are located in different frequency bands and are separated by a multiple of 100 kHz, which is the usual LTE carrier spacing. In *non-contiguous intra-band aggregation*, the carriers are in the same band, while in *contiguous intra-band aggregation* the carriers are in the same

An Introduction to LTE: LTE, LTE-Advanced, SAE, VoLTE and 4G Mobile Communications, Second Edition.
Christopher Cox.

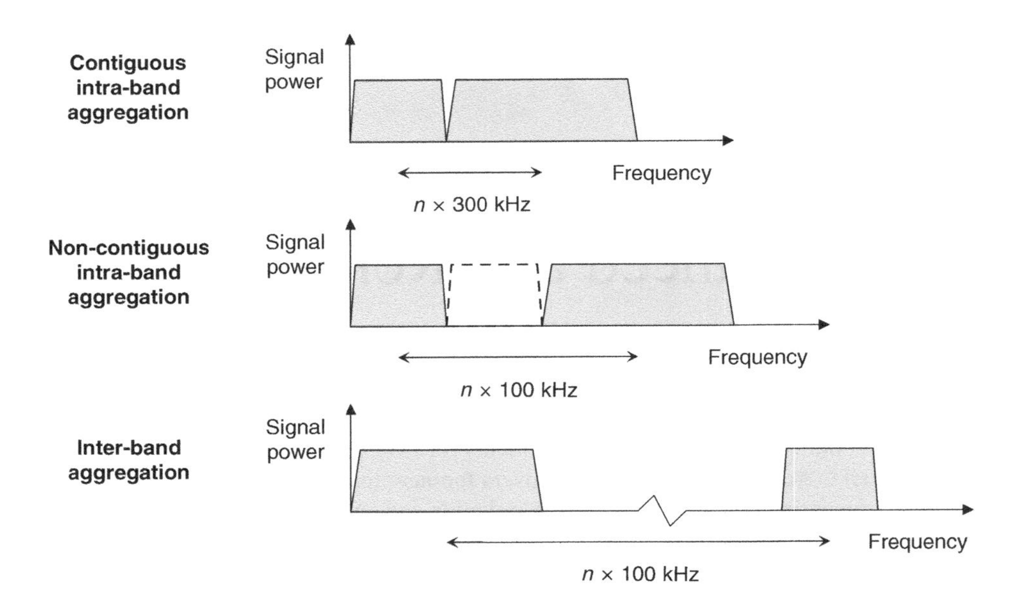

Figure 19.1 Carrier aggregation scenarios

band and are adjacent to each other. In this last scenario, the carriers are separated by a multiple of 300 kHz, which is consistent with the orthogonality requirement described in Chapter 4, so that the different sets of sub-carriers are orthogonal to each other and do not interfere.

There are a few more restrictions. In FDD mode, the allocations on the uplink and downlink can be different, but the number of downlink component carriers is always greater than or equal to the number used on the uplink. In TDD mode, each component carrier must have the same TDD configuration in Release 10, but that restriction was removed as part of Release 11. Finally, the component carriers must have the same mode of operation (FDD or TDD) up to Release 11, but that restriction is itself removed in Release 12.

The component carriers are organized into one *primary cell* (PCell) and up to four *secondary cells* (SCells). The primary cell contains one component carrier in TDD mode, or one downlink CC and one uplink CC in FDD mode. It is used in exactly the same way as a cell in Release 8. Secondary cells are only used by mobiles in RRC_CONNECTED and are added or removed by means of mobile-specific signalling messages. Each secondary cell contains one component carrier in TDD mode, or one downlink CC and optionally one uplink CC in FDD mode.

Carrier aggregation only affects the physical layer and the MAC protocol on the air interface, and the RRC, S1-AP and X2-AP signalling protocols. There is no impact on the RLC or PDCP and no impact on data transport in the fixed network.

19.1.2 UE Capabilities

Ultimately, carrier aggregation will allow a mobile to transmit and receive using five component carriers in a variety of frequency bands. It is handled by a single UE category, category 8, which supports a peak data rate of 3000 Mbps in the downlink and 1500 Mbps in the uplink.

Despite this, a category 8 mobile does not have to support every feature of category 8 and does not have to support such a high peak data rate; instead, the mobile declares its support for individual features as part of its UE capabilities. Furthermore, 3GPP has not introduced full support for category 8 right away.

There are two main aspects [5]. Firstly, the maximum number of downlink component carriers in Releases 10 and 11 is two rather than five, while the maximum number of uplink component carriers is two when using contiguous intra-band aggregation and one otherwise. Secondly, the specifications only support carrier aggregation in a limited number of frequency bands. This limits the complexity of the specifications, because some of the radio frequency requirements have to be individually defined for each band or band combination.

Figure 19.2 shows the band combinations that support carrier aggregation in Release 11 and indicates their approximate carrier frequencies and intended areas of use. Several new combinations are scheduled for Release 12, including inter-band aggregation using three component carriers on the downlink and one on the uplink, and inter-band aggregation using two component carriers on the downlink and two on the uplink.

A mobile declares which bands and band combinations it supports as part of its radio access capabilities [6, 7]. The mobile also declares a capability known as the *CA bandwidth class* for each individual band or combination, which states the number of component carriers that the mobile supports and the total number of resource blocks that it can handle. Table 19.1 lists the classes that are used by LTE-Advanced. Classes A, B and C are supported in Releases 10 and 11, while classes D, E and F are reserved for future releases.

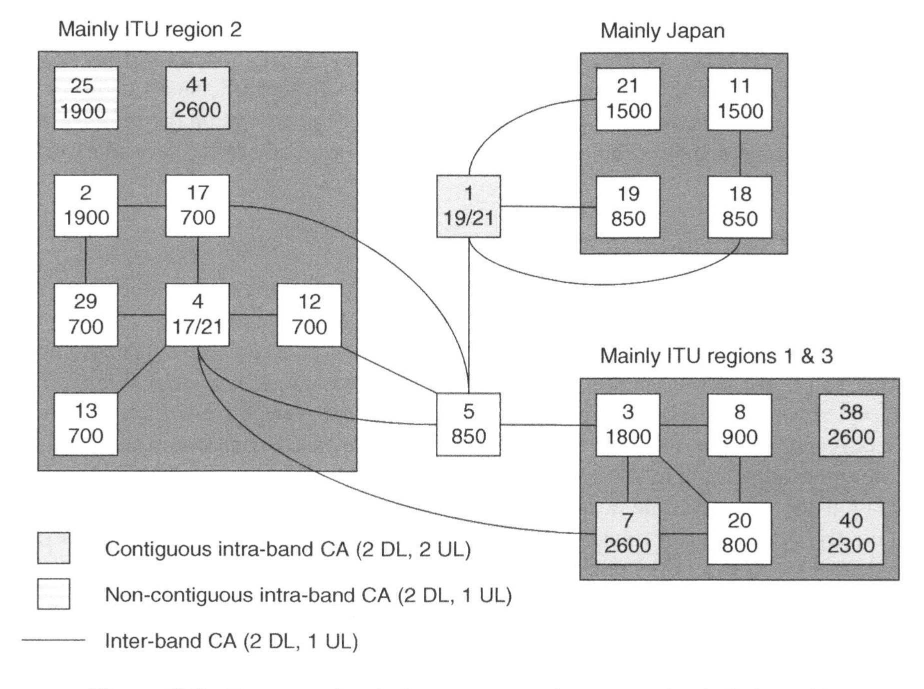

Figure 19.2 Frequency bands that support carrier aggregation in Release 11

Table 19.1 Carrier aggregation bandwidth classes

Carrier aggregation bandwidth class	Release	Maximum number of component carriers	Maximum number of resource blocks
A	R10	1	100
B	R10	2	100
C	R10	2	200
D	R12 or beyond	Not yet determined	Not yet determined
E	R12 or beyond	Not yet determined	Not yet determined
F	R12 or beyond	Not yet determined	Not yet determined

Source: TS 36.101. Reproduced by permission of ETSI.

19.1.3 Scheduling

In Chapter 8, we described how a mobile looks for PDCCH scheduling messages in common and UE-specific search spaces, which lie within the downlink control region at the start of every subframe. Release 10 continues to use this process, but with a few modifications [8, 9].

Each component carrier is independently scheduled and generates an independent set of hybrid ARQ feedback bits. The system does, however, support cross carrier scheduling: the base station can trigger an uplink or downlink transmission on one component carrier using a scheduling message on another. Release 10 implements cross carrier scheduling by adding a *carrier indicator field* (CIF) to each DCI format, which indicates the carrier to be used for the subsequent transmission.

Using cross carrier scheduling, a base station can transmit its scheduling messages on the component carrier that has the greatest coverage, so as to maximize the reliability of successful reception. It can also use the technique to balance the loads from traffic and scheduling across the different component carriers. The common search space is always on the primary cell, but the UE-specific search spaces can be on the primary cell or on any of the secondary cells.

19.1.4 Data Transmission and Reception

Carrier aggregation does not affect data transmission in the downlink, but it does lead to some changes in the uplink. In Release 8, a mobile uses SC-FDMA, which assumes that the mobile is transmitting on a single contiguous block of sub-carriers. In Release 10, this assumption is no longer valid: instead, the mobile uses a more general technique known as *discrete Fourier transform spread orthogonal frequency division multiple access* (DFT-S-OFDMA). This multiple access technique is the same as SC-FDMA, except that it supports transmission on a non-contiguous allocation of sub-carriers.

To exploit the new multiple access technique, the specifications are relaxed in two other ways. Firstly, a mobile can transmit on each component carrier using sub-carriers that are grouped into two blocks, rather than one. These transmissions are scheduled using a new uplink resource allocation scheme, known as type 1. Secondly, a mobile can transmit on the PUCCH and PUSCH at the same time. Both features are optional for the mobile, which declares support for them as part of its capabilities [10].

A mobile's peak output power is higher when using DFT-S-OFDMA than when it is using SC-FDMA. This puts greater demands on the mobile's power amplifier, which increases the cost of the amplifier and the uplink power consumption.

19.1.5 Uplink and Downlink Feedback

Carrier aggregation leads to a few changes in the transmission of uplink control information [11–13]. The most important is that the mobile only transmits the PUCCH on the primary cell. However, it can send uplink control information using the PUSCH on the primary cell or on any of the secondary cells.

If the mobile needs to send hybrid ARQ acknowledgements to the base station, then it groups them together onto a single component carrier. When using the PUCCH, it can send the acknowledgements in two ways. The first is to transmit on multiple PUCCH resources using PUCCH format 1b, in a similar way to the use of ACK/NACK multiplexing in TDD mode. The second way is to use a new PUCCH format, number 3. This format handles the simultaneous transmission of up to 10 hybrid ARQ bits in FDD mode and 20 in TDD mode, together with an optional scheduling bit, using resource block pairs that are shared amongst five mobiles.

There are no significant changes to the procedure for uplink transmission and reception. In particular, a base station sends its PHICH acknowledgements on the same cell (primary or secondary) that the mobile used for its uplink data transmission.

19.1.6 Other Physical Layer and MAC Procedures

Carrier aggregation introduces a few other changes to the physical layer and MAC procedures from Chapters 8 to 10. As noted earlier, the base station adds and removes secondary cells using mobile-specific RRC Connection Reconfiguration messages. In addition, it can quickly activate and deactivate a secondary cell by sending a MAC *Activation/Deactivation* control element to the target mobile [14].

As before, the base station can control the power of a mobile's PUSCH transmissions using DCI formats 0, 3 and 3A. In LTE-Advanced, each component carrier has a separate power control loop. When using DCI format 0, the base station identifies the component carrier using the carrier indicator field from earlier. In the case of formats 3 and 3A, it assigns a different value of TPC-PUSCH-RNTI to each component carrier and uses that value as the target of the power control command.

19.1.7 RRC Procedures

Carrier aggregation introduces a few changes to the RRC procedures from Chapters 11 to 15, but not many. In RRC_IDLE state, the mobile carries out cell selection and reselection using one cell at a time, as before. The RRC connection setup procedure is unchanged too: at the end of the procedure, the mobile is only communicating with a primary cell. Once the mobile is in RRC_CONNECTED state, the base station can add or remove secondary cells using mobile-specific RRC Connection Reconfiguration messages.

In RRC_CONNECTED state, the mobile measures individual neighbouring cells in much the same way as before. The serving cell corresponds to the primary cell in measurement events A3, A5 and B2, and to either the primary or a secondary cell in measurement events A1 and A2. There is also a new measurement event, A6, which the mobile reports if the power from a neighbour cell rises sufficiently far above the power from a secondary cell. The base station might use this measurement report to trigger a change of secondary cell.

During a handover, the new base station tells the mobile about the new secondary cells using its RRC Connection Reconfiguration command, in the same way that it conveyed the random access preamble index in Release 8. This allows the network to change all the secondary cells as part of the handover procedure and also to hand a mobile over between base stations with differing support for Releases 8 and 10.

19.2 Enhanced Downlink MIMO

19.2.1 Objectives

Release 10 extends LTE's support for downlink multiple antenna transmission using a new technique known as *eight layer spatial multiplexing*. The technique has three main objectives. Firstly, it supports single user MIMO transmissions with a maximum of eight layers. Secondly, it supports multiple user MIMO transmissions to a maximum of four mobiles and includes the accurate feedback that MU-MIMO requires. Thirdly, it allows the base station to switch a mobile between the two techniques every subframe without the need for additional RRC signalling.

We have already seen that the peak downlink data rate in Release 10 is 1200 Mbps. This is four times greater than in Release 8 and results from the use of two component carriers, each of which carries eight transmission layers rather than four. Eventually, LTE should support a peak downlink data rate of 3000 Mbps, through the use of five component carriers.

19.2.2 Downlink Reference Signals

Reference signals have two functions: they provide an amplitude and phase reference in support of channel estimation and demodulation, and they provide a power reference in support of channel quality measurements and frequency-dependent scheduling. In the Release 8 downlink, the cell-specific reference signals support both of these functions, at least in transmission modes 1 to 6.

In principle, the designers of LTE-Advanced could have supported eight antenna MIMO in the same way as four antenna MIMO, by adding four new antenna ports that each carried cell-specific reference signals. However, this approach would have led to a few difficulties. The reference signals would occupy more resource elements, which would increase the overhead for Release 10 mobiles that recognized them and increase the interference for Release 8 mobiles that did not. They would also do nothing to improve the performance of multiple user MIMO.

Instead, Release 10 introduces some new downlink reference signals [15], in which the two functions are split. The UE-specific reference signals (Figure 19.3) support channel estimation and demodulation, in a similar way to the demodulation reference signals on the uplink, and

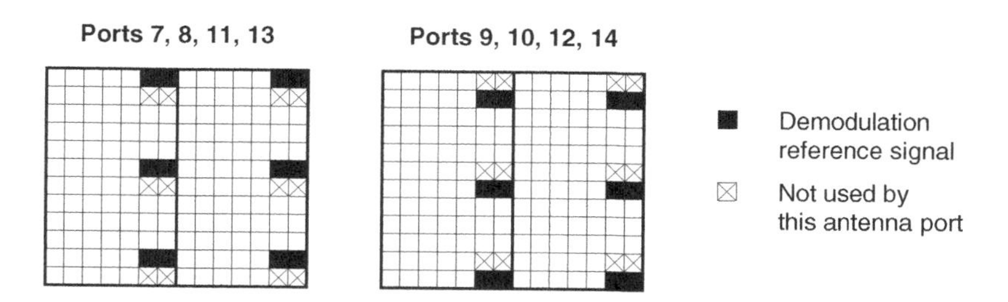

Figure 19.3 Resource element mapping for the UE-specific demodulation reference signals, using FDD mode and a normal cyclic prefix. Source: TS 36.211. Reproduced by permission of ETSI

are transmitted on antenna ports 7 to 14. The signals on ports 7 and 8 are the same ones used by dual-layer beamforming, while those on ports 9 to 14 support single user MIMO with a maximum of eight antenna ports. (A base station can still use the techniques described in this section even if it only has two or four antenna ports rather than eight.)

As shown in the figure, each individual reference symbol is actually shared amongst four antenna ports by means of orthogonal code division multiplexing. There is no cell-specific frequency offset; instead, the reference symbols are at fixed locations with respect to their surrounding resource blocks. This feature helps the introduction of coordinated multipoint transmission in Release 11.

The base station precodes the UE-specific reference signals using the same precoding matrix that it applies to the PDSCH. This makes the precoding operation transparent to the mobile, so that the base station can apply any precoding matrix it likes, and improves the performance of multiple user MIMO, which requires a free choice of precoding matrix to ensure that the signals reach the mobiles with the correct constructive or destructive interference. Furthermore, the base station only transmits the UE-specific reference signals in the physical resource blocks that the target mobile is actually using. As a result, the reference signals do not cause any overhead or interference for the other mobiles in the cell. However, it also means that the reference signals are unsuitable for channel quality measurements, which are across the whole of the downlink band.

To deal with this issue, the base station also transmits *CSI reference signals* on eight more antenna ports, numbered from 15 to 22. (The signals are not precoded, so the antenna ports are different from the ones used by the UE-specific reference signals mentioned earlier.) The CSI reference signals support channel quality measurements and frequency-dependent scheduling, in a similar way to the sounding reference signals on the uplink.

A cell can transmit the CSI reference signal using two, four, or eight resource elements per resource block pair, depending on the number of antenna ports that it has. The cell chooses these resource elements from a larger set of 40, with nearby cells choosing different resource elements so as to minimize the interference between them. Figure 19.4 shows an example transmission, for a cell with eight antenna ports.

The base station then supplies each mobile with a *CSI reference signal configuration*. This defines the subframes in which the mobile should measure the signal and the resource elements that it should inspect, with a measurement interval of 5 to 80 ms that depends on the mobile's speed. The base station can also supply the mobile with additional *zero power CSI reference signal configurations*. Despite their name, these simply define other resource elements that

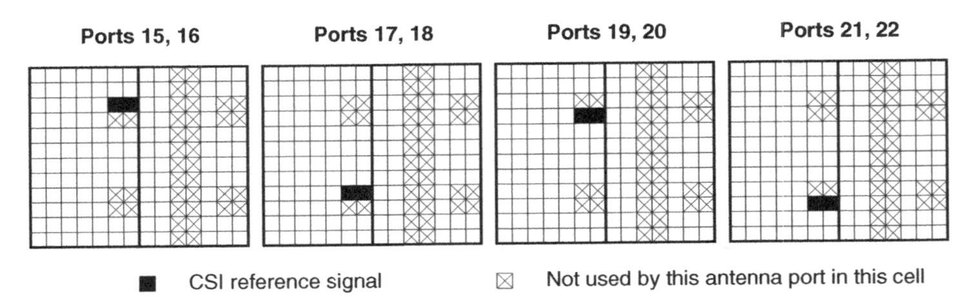

Figure 19.4 Example resource element mapping for the CSI reference signals, using FDD mode and a normal cyclic prefix. Source: TS 36.211. Reproduced by permission of ETSI

the base station has assigned to the CSI reference signal, typically for other mobiles that are measuring it more often, and which the mobile should skip when receiving the PDSCH.

The CSI reference signals cause some overheads for Release 10 mobiles, but the long transmission interval implies that the overheads are acceptably small. They can also cause CRC failures for Release 8 mobiles that do not recognize them, but the base station can avoid these by scheduling Release 8 mobiles in different resource blocks.

19.2.3 Downlink Transmission and Feedback

To use eight layer spatial multiplexing [16–18], the base station configures the mobile into a new transmission mode, mode 9 and schedules it using a new DCI format, 2C. In its scheduling command, the base station can allocate each resource block in two distinct ways. Using single user MIMO, the base station can transmit a maximum of eight layers on antenna ports 7 to 7 + n, where n is the number of layers that the mobile is receiving. Using multiple-user MIMO, the base station transmits a maximum of two layers on each of antenna ports 7 and 8, giving four layers altogether. The base station can then use these in various ways. In one scenario, four mobiles can each receive one of the transmitted layers, with the mobiles separated using beamforming. Alternatively, two mobiles can each receive one layer on port 7 and one layer on port 8 by means of the dual-layer beamforming techniques that we introduced in Chapter 18.

As the radio propagation changes, the base station can dynamically switch each mobile between the two techniques without the need to change transmission mode. It might, for example, schedule a mobile using SU-MIMO if it is in a rich multipath environment with uncorrelated channel conditions, but change the mobile to MU-MIMO if it moves into line-of-sight communications.

Optionally, the base station can tell the mobile to feed back a precoding matrix indicator, which signals the discrepancy between the precoding that the base station is transparently providing and the precoding that the mobile would ideally like to receive. To help achieve this, the specifications introduce an improved feedback technique, which is restricted to base stations with eight antenna ports in Release 10 but is extended to base stations with four antenna ports in Release 12.

When using this technique, the mobile feeds back two indices, denoted i_1 and i_2, from which the base station reconstructs the requested precoding matrix. Both indices can vary from 0 to

15, which provides more finely grained feedback than the PMI and improves the performance of multiple user MIMO. Their exact definition is targeted at a common deployment scenario in which the base station has two cross polarized arrays, each with four closely spaced antennas. The index i_1 captures the relationship between the signals received from the different antennas within each array, which varies slowly and is only reported occasionally. The index i_2 captures the relationship between the two polarizations, which varies more quickly and is reported more often.

19.3 Enhanced Uplink MIMO

19.3.1 Objectives

The only multiple antenna scheme supported by the Release 8 uplink was multiple user MIMO. This increased the cell capacity while only requiring the mobile to have a single transmit power amplifier and was far easier to implement than on the downlink. However, it did nothing for the peak data rate of a single mobile.

In LTE-Advanced, the uplink is enhanced to support single user MIMO, using up to four transmit antennas and four transmission layers. The mobile declares how many layers it supports as part of its uplink capabilities [19]. The peak uplink data rate in Release 10 is 600 Mbps. This is eight times greater than in Release 8, and results from the use of four transmission layers and two component carriers. Eventually, LTE should support a peak uplink data rate of 1500 Mbps, through the use of five component carriers.

19.3.2 Implementation

To support single user MIMO, the base station configures a Release 10 mobile into one of the transmission modes listed in Table 19.2 [20]. These are used in a similar way to the transmission modes on the downlink. Mode 1 corresponds to single antenna transmission, while mode 2 corresponds to single user MIMO, specifically the use of closed loop spatial multiplexing.

Once a mobile has been configured into mode 2, the base station sends it a scheduling grant for closed loop spatial multiplexing using a new DCI format, number 4 [21]. As part of the scheduling grant, the base station specifies the number of layers that the mobile should use for its transmission and the precoding matrix that it should apply. The maximum number of uplink codewords is increased to two, the same as on the downlink.

The PUSCH transmission process is then modified to include the additional steps of layer mapping and precoding [22], which work in the same way as the corresponding steps on the downlink. The antenna ports (Table 19.3) are numbered in an unexpected way. Port 10 is used

Table 19.2 Uplink transmission modes in 3GPP Release 10

Mode	Purpose
1	Single antenna transmission
2	Closed loop spatial multiplexing

Table 19.3 Uplink antenna ports in 3GPP Release 10

Antenna port	Channels	Application
10	PUSCH, SRS	Single antenna transmission
20–21		2 antenna closed loop spatial multiplexing
40–43		4 antenna closed loop spatial multiplexing
100	PUCCH	Single antenna transmission
200–201		2 antenna open loop transmit diversity

for single antenna transmission of the PUSCH, ports 20 and 21 for dual antenna transmission and ports 40 to 43 for transmission on four antennas, while the same antenna ports are also used by the sounding reference signal. The PUCCH can be transmitted from a single antenna on port 100, or from two antennas using open loop diversity on ports 200 and 201.

19.4 Relays

19.4.1 Principles of Operation

Repeaters and *relays* (Figure 19.5) are devices that extend the coverage area of a cell. They are useful in sparsely populated areas in which the performance of a network is limited by coverage rather than capacity. They can also increase the data rate at the edge of a cell, by improving the signal-to-interference plus noise ratio there.

A repeater receives a radio signal from the transmitter, and amplifies and rebroadcasts it, so appears to the receiver as an extra source of multipath. Unfortunately the repeater amplifies the incoming noise and interference as well as the received signal, which ultimately limits its performance. FDD repeaters were fully specified in Release 8, with the sole specification [23] referring to the radio performance requirements. TDD repeaters are harder to implement, because of the increased risk of interference between uplink and downlink, and have not yet been specified.

A relay takes things a step further, by decoding the received radio signal, before re-encoding and rebroadcasting it. By doing this, it removes the noise and interference from the retransmitted signal, so can achieve a higher performance than a repeater. Relays are first specified in Release 10, for both FDD and TDD modes.

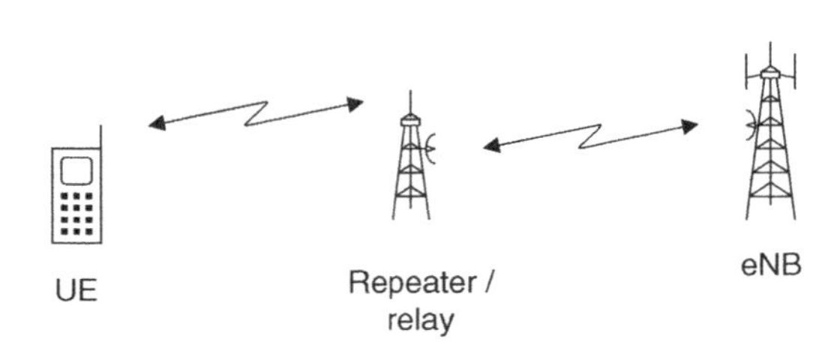

Figure 19.5 Operation of repeaters and relays

19.4.2 Relaying Architecture

Figure 19.6 shows the architecture that is used for relaying [24–26]. There are two parts. In the inner part, the *relay node* (RN) acts as a mobile towards the network. The relay node's non-access stratum is controlled by an MME, while its access stratum is controlled by a *donor eNB* (DeNB). The donor eNB also incorporates the functions of a PDN gateway and a serving gateway, which allocate an IP address for the relay node and handle its traffic.

In the outer part, the relay node acts as an eNB towards the mobile and controls the mobile's access stratum. To carry out this task, the relay node has one or more physical cell identities of its own, broadcasts its own synchronization signals and system information, and schedules all of its transmissions on the Uu interface. It can also communicate with other base stations across an X2 interface to support handovers. The mobile's non-access stratum is controlled by an MME, which can be different from the MME that is controlling the relay node.

The donor eNB includes one last function, a *relay gateway*. This is similar to a home eNB gateway, and shields the core network and the other base stations from the need to know anything about the relay nodes themselves.

The Un interface is the air interface between the relay node and the donor eNB. It might be implemented as a normal air interface or as a point-to-point microwave link, and in the latter case would have a far greater range than the air interface to a mobile.

The Un and Uu interfaces can use either the same carrier frequency, or different ones. If the carrier frequencies are different, then the Un interface can be implemented in exactly the same way as a normal air interface. For example, the relay node acts like a base station on the Uu interface towards the mobile and independently acts like a mobile on the Un interface towards the donor eNB. If the carrier frequencies are the same, then the Un interface requires some extra functions, covered in the next section, to share the resources of the air interface with Uu.

There are a couple of restrictions on the use of relaying in Release 10. Relay nodes are assumed to be stationary, so a relay node cannot be handed over from one donor eNB to another. In addition, multi-hop relaying is not supported, so that one relay node cannot control another

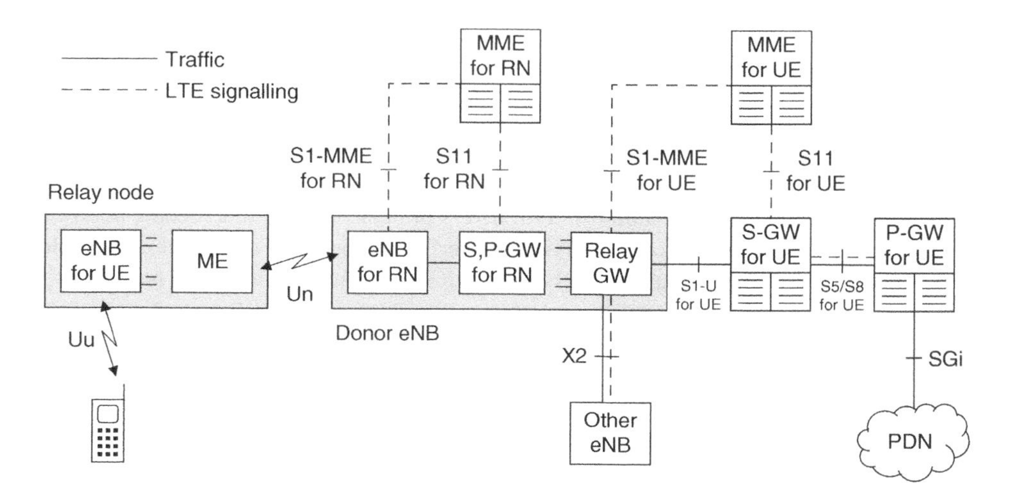

Figure 19.6 Relaying architecture in LTE

relay node. However, there is no impact on the mobile, which is completely unaware that it is being controlled by a relay. This implies that Release 8 mobiles support relaying in just the same way as Release 10 mobiles.

19.4.3 Enhancements to the Air Interface

If the Uu and Un interfaces use the same carrier frequency, then some enhancements are required to Un so that the resources of the air interface can be shared. The physical layer enhancements are ring-fenced in a single specification [27], while some extra RRC signalling messages are required as well [28]. There is no impact on the Uu interface, to ensure backwards compatibility with Release 8 mobiles, or on the layer 2 protocols.

The air interface resources are shared using time division multiplexing, with individual subframes allocated to either Un or Uu. This is implemented in two stages. Firstly, the donor eNB tells the relay node about the allocation using an RRC *RN Reconfiguration* message. Secondly, the relay node configures the Un subframes as MBSFN subframes on Uu, but does not transmit any downlink MBSFN data in them and does not schedule any data transmissions on the uplink. This is a little ugly, but is backwards compatible with mobiles that only support Release 8.

Unfortunately, the start of an MBSFN subframe is used by PDCCH transmissions on the Uu interface, typically scheduling grants for uplink transmissions that will occur a few subframes later. This prevents the use of the PDCCH on the Un interface. Instead, the specification introduces the *relay physical downlink control channel* (R-PDCCH), which takes the place of the PDCCH on Un. R-PDCCH transmissions look much the same as normal PDCCH transmissions, but occur in reserved resource element groups in the part of the subframe that is normally used by data. Transmissions in the first slot of a subframe are used for downlink scheduling commands, while transmissions in the second slot are used for uplink scheduling grants.

The Un interface does not use the physical hybrid ARQ indicator channel: instead, the donor eNB acknowledges the relay node's uplink transmissions implicitly, using scheduling grants on the R-PDCCH. In the absence of the PDCCH and PHICH, the physical control format indicator channel is not required either.

19.5 Heterogeneous Networks

19.5.1 Introduction

Radio access networks often contain different layers of cells, such as macrocells, microcells and picocells. A network operator can minimize interference between these layers by deploying them on different carrier frequencies. However this technique is not always feasible as the operator may not have enough carriers available, and it does not make the most efficient use of the operator's total bandwidth. As an alternative, a network operator can deploy different layers on the same carrier frequency, so that they occupy the same frequency band. Such a network is known as a *heterogeneous network* or *HetNet* [29].

Heterogeneous networks suffer from two interference problems. The first (Figure 19.7) occurs if a high-power base station such as a macrocell is close to a low-power base station such as a picocell. Interference from the macrocell reduces the picocell's coverage area, so it limits the benefits that the picocell might otherwise provide. The second problem

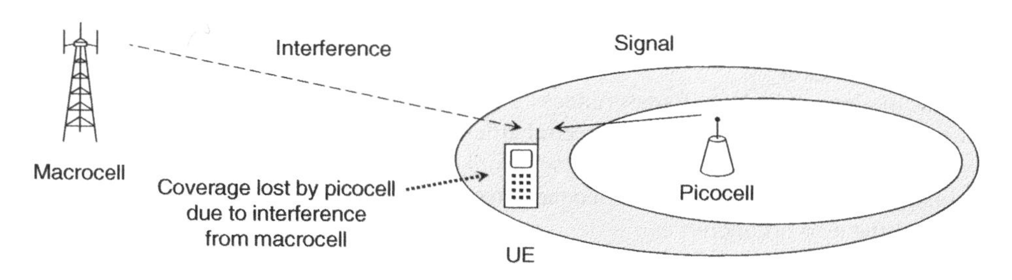

Figure 19.7 Reduction of a picocell's coverage area due to interference from a nearby macrocell. Source: TS 36.300. Reproduced by permission of ETSI

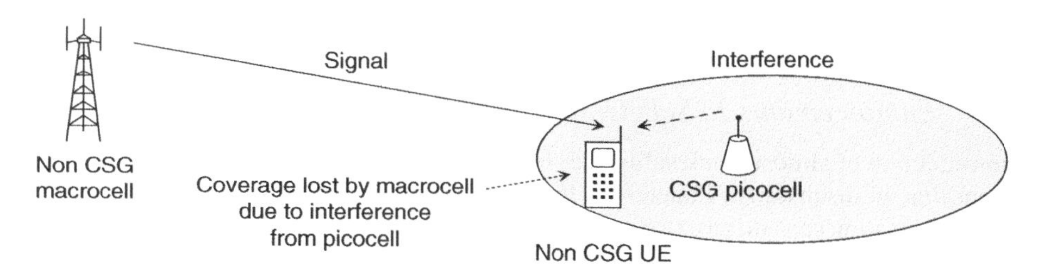

Figure 19.8 Reduction of a non-CSG macrocell's coverage area due to interference from a nearby CSG picocell. Source: TS 36.300. Reproduced by permission of ETSI

(Figure 19.8) occurs if the picocell belongs to a closed subscriber group (CSG). If a nearby mobile does not belong to the closed subscriber group, then it may be unable to communicate with the macrocell because of interference from the picocell. Note the distinction between these scenarios: in the first scenario, the macrocell is the aggressor and the picocell is the victim, while the reverse is true in the second scenario.

19.5.2 Enhanced Inter-Cell Interference Coordination

The interference issues can be reduced using a technique known as *time domain interference coordination* or *enhanced inter-cell interference coordination* (EICIC). When using this technique, the aggressor cell configures certain subframes as *almost blank subframes* (ABSs), so as to reduce the levels of interference in the victim cell.

Within an almost blank subframe, the aggressor cell transmits the minimum information that is required for backwards compatibility with Releases 8 and 9, such as the cell-specific reference signals, PCFICH and PHICH. It does not transmit information that it could schedule in other subframes, such as the majority of scheduling commands on the PDCCH and data transmissions on the PDSCH. The victim cell can then schedule vulnerable mobiles within the almost blank subframes, where the interference is lower. The almost blank subframes appear with a cell-specific pattern that has a period of 40 subframes in FDD mode and a period that depends on the TDD configuration in TDD mode. The network operator can configure almost blank subframes in two ways: either manually or by using the self-optimization techniques that we discuss below.

The use of almost blank subframes reduces the number of subframes that are available to the aggressor cell, so reduces its average spectral efficiency. However, it also allows the aggressor and victim cells to share the same frequency band, so it increases the bandwidth available to each one. In a suitable network, the second effect has a greater impact than the first, so the capacity of the network increases.

Almost blank subframes risk causing an additional problem: the interference in the victim cell can fluctuate from one subframe to the next and can disrupt the mobiles' measurements. The network can solve the problem by the use of *measurement resource restriction patterns*. These patterns limit a mobile's measurements of a victim cell to either the almost blank subframes or the ones remaining, and let the mobile make a consistent set of measurements for cell reselection, handover and the reporting of channel state information.

19.5.3 Enhancements to Self-Optimizing Networks

The introduction of almost blank subframes brings two enhancements to the self-optimization functions that we discussed in Chapter 17. In the procedure for inter-cell interference coordination, the base station can add an extra information element, *ABS information*, to its X2-AP Load Information message. The field lists the almost blank subframes that its cells have configured, and a neighbour can use these to schedule mobiles that are vulnerable to interference.

In the procedure for mobility load balancing, a neighbouring cell can add an extra information element, *ABS Status*, to its X2-AP Resource Status Update message. The field reports the neighbour's resource usage within subframes that the original base station has previously configured as almost blank subframes. The original base station can use the information to adjust the number of ABSs that it has configured; for example, by adding more ABSs if the neighbour's resource usage is high.

19.6 Traffic Offload Techniques

19.6.1 Local IP Access

The increase in mobile data traffic that we saw in Chapter 1 has been threatening to overwhelm operators' networks and has given them an incentive to offload their traffic elsewhere. The Release 10 specifications introduce four ways of achieving this. In *local IP access* (LIPA) [30, 31], a home eNB contains a *local gateway* (LGW) that acts in a similar way to a PDN gateway. The mobile can use the local gateway to communicate with a local device such as a printer: the traffic does not have to travel through the evolved packet core. The architecture is shown in Figure 19.9.

To help support local IP access, the home eNB sends the local gateway's IP address to the MME as part of any S1-AP Uplink NAS Transport or Initial UE Message. If the user requests connectivity to an access point name for which LIPA is permitted, then the MME can select the local gateway in place of the usual PDN gateway and can indicate this selection to the home eNB. The local traffic can now travel directly between the home eNB and its embedded local gateway, and does not have to pass through the serving gateway.

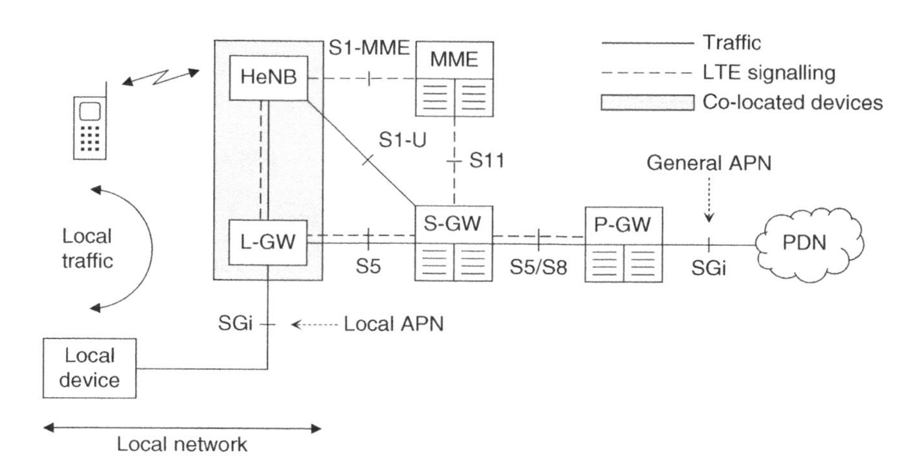

Figure 19.9 Example network architecture for LIPA. Source: TR 23.829. Reproduced by permission of ETSI

The only difficulty arises if data arrive on the downlink while the user is in RRC_IDLE. If this happens, the local gateway sends the first downlink packet over the S5 interface to the serving gateway, which triggers the usual paging procedure and moves the mobile into RRC_CONNECTED. The local gateway can then deliver subsequent downlink packets directly to the home eNB.

19.6.2 Selective IP Traffic Offload

Using *selective IP traffic offload* (SIPTO) [32], the MME can use its knowledge of the mobile's location to select a serving gateway and a PDN gateway that are located nearby. In the basic Release 10 implementation, the serving and PDN gateways are both in the operator's core network. The traffic continues to travel through the evolved packet core but takes a shorter route than it would otherwise do.

In Release 12, the PDN gateway can be replaced by a local gateway that lies in a local access network. There are two possible architectures. Firstly, the local gateway can be co-located with a home eNB, in an extension of the LIPA architecture that we introduced above (Figure 19.10). Secondly, the local gateway can be co-located with a serving gateway that typically lies in a corporate picocell network (Figure 19.11). Both architectures are intended for internet access alone and allow the operator to offload internet traffic from the evolved packet core. Traffic for other packet data networks, such as the IP multimedia subsystem, is unaffected.

19.6.3 Multi-Access PDN Connectivity

Using *multi-access PDN connectivity* (MAPCON) [33], a network operator can connect a mobile to different access point names at the same time using different radio access technologies, specifically one 3GPP network and one non-3GPP network. For example (Figure 19.12), an operator might connect a residential user to the IP multimedia subsystem using an LTE

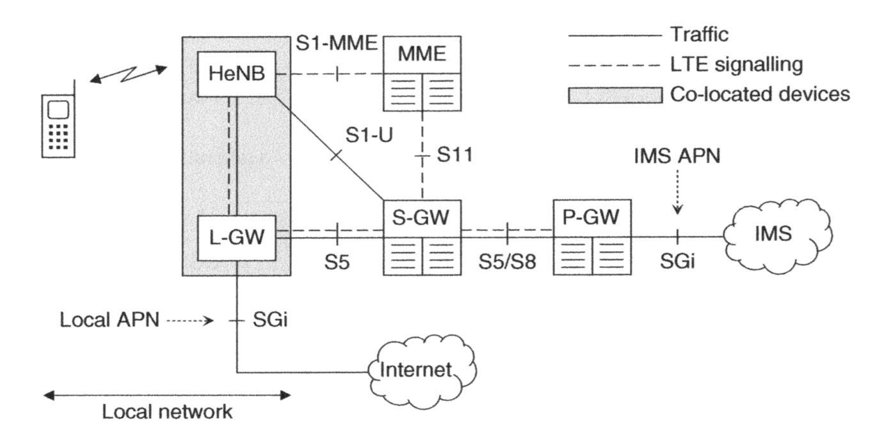

Figure 19.10 Example architecture for SIPTO at the local network, with the local gateway co-located with a home eNB. Source: TR 23.829. Reproduced by permission of ETSI

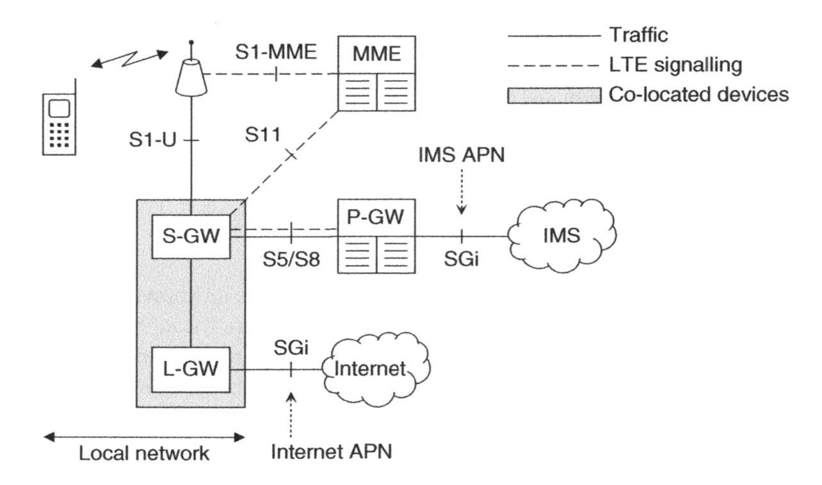

Figure 19.11 Example architecture for SIPTO at the local network, with the local gateway co-located with a serving gateway. Source: TR 23.829. Reproduced by permission of ETSI

macrocell, so as to exploit the LTE network's quality of service guarantees. At the same time, the operator might connect the user to the internet using a domestic wireless local area network, so as to offload internet traffic from the macrocell. The operator can also transfer individual APNs from one radio access technology to another; for example, if the user moves outside the wireless network's coverage area.

To help support MAPCON, the access network discovery and selection function (ANDSF) provides the mobile with a prioritized list of *service-based intersystem routing policy* (ISRP) *rules*. Each rule contains an access point name, a prioritized list of radio access technologies

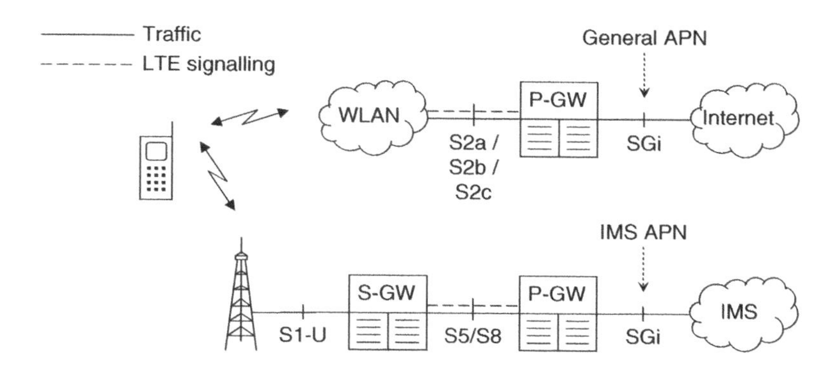

Figure 19.12 Example network architecture for MAPCON

and network or cell identities, and optional information describing the rule's validity in terms of the mobile's location, the date and the time of day.

On requesting a new PDN connection, the mobile examines the intersystem routing policy rules that it downloaded from the ANDSF and decides whether it should connect using LTE or using a non-3GPP technology. Later on, the mobile can move one or more APNs from one radio access technology to another using the reselection procedures that we described in Chapter 16.

19.6.4 IP Flow Mobility

IP flow mobility (IFOM) [34] is similar to MAPCON but more complex. Using IFOM, a network operator can handle different IP traffic flows for the same access point name using different radio access technologies and can transfer individual flows from one technology to another. In a typical scenario (Figure 19.13), a user might connect through a general APN to the network operator's servers and to the internet. The operator can then use a wireless local area network for the internet and for any other best effort traffic, while reserving the LTE access network for real time traffic such as streaming video from its own servers. In the case of wireless local area networks, IP flow mobility is also known as *seamless WLAN offload*.

To help support IFOM, the ANDSF provides the mobile with a prioritized list of *flow-based intersystem routing policy rules*. These are similar to the service-based ISRP rules that MAPCON uses, except that each rule identifies an individual traffic flow using routing filters that contain information such as its source and destination IP addresses and port numbers.

IFOM requires the host-based non-3GPP mobility architecture described in Chapter 16. The mobile has to support dual-stack mobile IPv6 (DSMIPv6) [35, 36], with extensions that let it define individual traffic flows by means of routing filters [37, 38]. To implement IFOM, the mobile begins by connecting to a PDN gateway through LTE and through a non-3GPP access network. It can then assign a traffic flow to a particular radio access technology by composing a DSMIPv6 *Binding Update* message that contains the routing filters and by sending the message across the access network to the PDN gateway. The PDN gateway updates its routing information and directs future downlink packets to the mobile over the requested network.

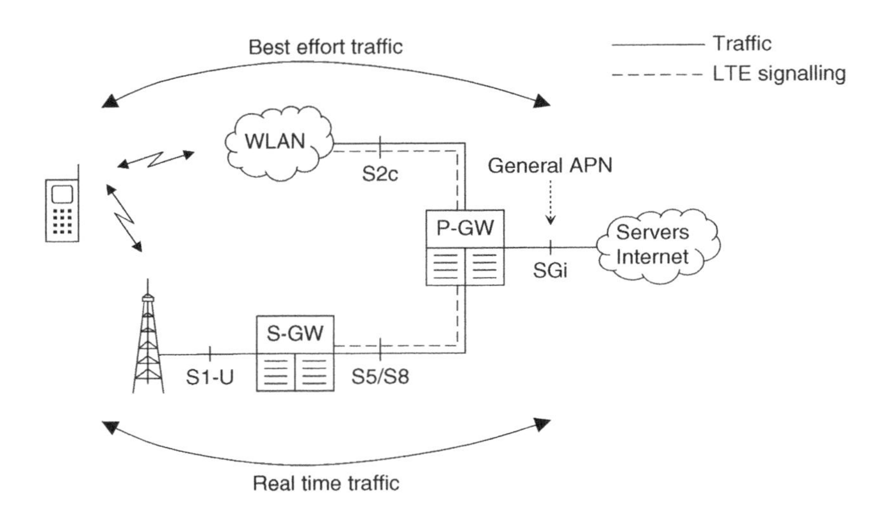

Figure 19.13 Example network architecture for IFOM

19.7 Overload Control for Machine-Type Communications

Like other mobile communication systems, LTE was designed for use by people. However, there is an increasing market for *machine to machine* (M2M) communications, which the 3GPP specifications describe as *machine-type communications* (MTC) [39, 40]. In MTC, an automated device communicates with an external server to support applications such as vehicle and goods tracking, wireless utility metering and consumer electronics.

The number of machine-type devices can be large, with the potential for thousands of devices per cell. Each individual device might only generate a small amount of traffic, so their traffic load is usually not a problem. There are, however, two other difficulties. Firstly, each device can generate just as many signalling messages as a traditional mobile, so their total signalling load can be large. Secondly, traffic and signalling messages can arrive in bursts, due for example to network reselections after a failure of the home PLMN.

Release 10 adds several features that help prevent machine-type devices from overloading the network [41, 42]. The starting point is that a machine-type device can be associated with a set of non-access stratum configuration parameters, which are stored in the USIM and can be downloaded from a device management server [43, 44]. The most important parameter gives the device a *low access priority*, on the basis that most machine-type applications can tolerate long delays.

If a device has a low access priority, then it indicates this to the network in its RRC connection requests and its non-access stratum signalling. If the network is overloaded then it can respond in several ways; for example, it might reject the message and supply the device with a backoff timer, such that the device will wait until the timer expires before communicating again. The MME can also ask the serving gateway to reduce the number of low-priority downlink data notifications by a specified percentage. The serving gateway responds by randomly discarding incoming packets for idle low-priority devices so as to help alleviate the overload condition.

The overload control techniques from Release 10 apply to mobiles that have completed the random access procedure, but they still leave that procedure vulnerable. Release 11 addresses the problem using a technique known as *extended access barring* [45, 46], in which the operator can selectively restrict the rate of access attempts by MTC devices using information broadcast from a new system information block, SIB 14.

References

1. 3GPP TR 36.912 (2012) Feasibility Study for Further Advancements for E-UTRA (LTE-Advanced), Release 11, September 2012.
2. 3rd Generation Partnership Project (2013) FTP Directory, ftp://ftp.3gpp.org/Information/WORK_PLAN/Description_Releases/ (accessed 15 October 2013).
3. 4G Americas (2012) 4G Mobile Broadband Evolution: Release 10, Release 11 and Beyond: HSPA+, SAE/LTE and LTE-Advanced, October 2012.
4. 3GPP TS 36.300 (2013) Evolved Universal Terrestrial Radio Access (E-UTRA) and Evolved Universal Terrestrial Radio Access Network (E-UTRAN); Overall Description; Stage 2, Release 11, Sections 5.5, 6.4, 7.5, September 2013.
5. 3GPP TS 36.101 (2013) User Equipment (UE) Radio Transmission and Reception, Release 11, Sections 5.5A, 5.6A, September 2013.
6. 3GPP TS 36.306 (2013) User Equipment (UE) Radio Access Capabilities, Release 11, Sections 4.1, 4.3.5.2, September 2013.
7. 3GPP TS 36.331 (2013) Radio Resource Control (RRC); Protocol Specification, Release 11, Section 6.3.6 (UE-EUTRA-Capability), September 2013.
8. 3GPP TS 36.212 (2013) Multiplexing and Channel Coding, Release 11, Section 5.3.3.1, June 2013.
9. 3GPP TS 36.213 (2013) Physical Layer Procedures, Release 11, Sections 7.1, 8.0, 9.1.1, September 2013.
10. 3GPP TS 36.306 (2013) User Equipment (UE) Radio Access Capabilities, Release 11, Sections 4.3.4.12, 4.3.4.13, 4.3.4.14, September 2013.
11. 3GPP TS 36.211 (2013) Physical Channels and Modulation, Release 11, Section 5.4.2A, September 2013.
12. 3GPP TS 36.212 (2013) Multiplexing and Channel Coding, Release 11, Sections 5.2.2.6, 5.2.3.1, June 2013.
13. 3GPP TS 36.213 (2013) Physical Layer Procedures, Release 11, Sections 7.3, 10, September 2013.
14. 3GPP TS 36.321 (2013) Medium Access Control (MAC) Protocol Specification, Release 11, Sections 5.13, 6.1.3.8, July 2013.
15. 3GPP TS 36.211 (2013) Physical Channels and Modulation, Release 11, Sections 6.10.3, 6.10.5, September 2013.
16. 3GPP TS 36.211 (2013) Physical Channels and Modulation, Release 11, Section 6.3.4.2.3, 6.3.4.4, September 2013.
17. 3GPP TS 36.212 (2013) Multiplexing and Channel Coding, Release 11, Section 5.3.3.1.5C, June 2013.
18. 3GPP TS 36.213 (2013) Physical Layer Procedures, Release 11, Section 7.1, September 2013.
19. 3GPP TS 36.306 (2013) User Equipment (UE) Radio Access Capabilities, Release 11, Section 4.3.4.6, September 2013.
20. 3GPP TS 36.213 (2013) Physical Layer Procedures, Release 11, Section 8.0, September 2013.
21. 3GPP TS 36.212 (2013) Multiplexing and Channel Coding, Release 11, Section 5.3.3.1.8, June 2013.
22. 3GPP TS 36.211 (2013) Physical Channels and Modulation, Release 11, Sections 5.2.1, 5.3.2A, 5.3.3A, September 2013.
23. 3GPP TS 36.106 (2013) Evolved Universal Terrestrial Radio Access (E-UTRA); FDD Repeater Radio Transmission and Reception, Release 11, March 2013.
24. 3GPP TS 23.401 (2013) General Packet Radio Service (GPRS) Enhancements for Evolved Universal Terrestrial Radio Access Network (E-UTRAN) Access, Release 11, Sections 4.3.20, 4.4.10, September 2013.
25. 3GPP TS 36.300 (2013) Evolved Universal Terrestrial Radio Access (E-UTRA) and Evolved Universal Terrestrial Radio Access Network (E-UTRAN); Overall Description; Stage 2, Release 11, Section 4.7, September 2013.
26. 3GPP TR 36.806 (2010) Relay Architectures for E-UTRA (LTE-Advanced), Release 9, Section 4.2, April 2010.
27. 3GPP TS 36.216 (2012) Evolved Universal Terrestrial Radio Access (E-UTRA); Physical Layer for Relaying Operation, Release 11, September 2012.
28. 3GPP TS 36.331 (2013) Radio Resource Control (RRC); Protocol Specification, Release 11, Section 5.9, September 2013.
29. 3GPP TS 36.300 (2013) Evolved Universal Terrestrial Radio Access (E-UTRA) and Evolved Universal Terrestrial Radio Access Network (E-UTRAN); Overall Description; Stage 2, Release 11, Section 16.1.5, Annex K, September 2013.

30. 3GPP TS 23.401 (2013) General Packet Radio Service (GPRS) Enhancements for Evolved Universal Terrestrial Radio Access Network (E-UTRAN) Access, Release 11, Section 4.3.16, September 2013.
31. 3GPP TS 36.300 (2013) Evolved Universal Terrestrial Radio Access (E-UTRA) and Evolved Universal Terrestrial Radio Access Network (E-UTRAN); Overall Description; Stage 2, Release 11, Section 4.6.5, September 2013.
32. 3GPP TS 23.401 (2013) General Packet Radio Service (GPRS) Enhancements for Evolved Universal Terrestrial Radio Access Network (E-UTRAN) Access, Release 12, Sections 4.3.15, 4.3.15a, September 2013.
33. 3GPP TR 23.861 (2012) Multi Access PDN Connectivity and IP Flow Mobility, Release 12, Annex A, November 2012.
34. 3GPP TS 23.261 (2012) IP flow Mobility and Seamless Wireless Local Area Network (WLAN) Offload, Release 11, September 2012.
35. 3GPP TS 24.303 (2013) Mobility Management Based on Dual-Stack Mobile IPv6, Release 11, June 2013.
36. IETF RFC 5555 (2009) Mobile IPv6 Support for Dual Stack Hosts and Routers, June 2009.
37. IETF RFC 5648 (2009) Multiple Care-of Addresses Registration, October 2009.
38. IETF RFC 6089 (2011) Flow Bindings in Mobile IPv6 and Network Mobility (NEMO) Basic Support, January 2011.
39. 3GPP TS 22.368 (2012) Service Requirements for Machine-Type Communications (MTC), Release 11, September 2012.
40. 3GPP TR 37.868 (2011) Study on RAN Improvements for Machine-Type Communications, Release 11, Annex B, October 2011.
41. 3GPP TR 23.888 (2012) System Improvements for Machine-Type Communications (MTC), Release 11, Sections 7.1, 8.1, September 2012.
42. 3GPP TS 23.401 (2013) General Packet Radio Service (GPRS) Enhancements for Evolved Universal Terrestrial Radio Access Network (E-UTRAN) Access, Release 11, Section 4.3.17, September 2013.
43. 3GPP TS 31.102 (2013) Characteristics of the Universal Subscriber Identity Module (USIM) Application, Release 11, Section 4.2.94, September 2013.
44. 3GPP TS 24.368 (2012) Non-Access Stratum (NAS) Configuration Management Object (MO), Release 11, September 2012.
45. 3GPP TS 31.102 (2013) Characteristics of the Universal Subscriber Identity Module (USIM) Application, Release 11, Sections 4.2.15, 4.2.94, September 2013.
46. 3GPP TS 22.011 (2013) Service Accessibility, Release 11, Section 4, March 2013.

20

Releases 11 and 12

In this chapter, we address the two most recent releases of the 3GPP specifications, namely Releases 11 and 12. Release 11 was frozen in June 2013, and we begin this chapter by discussing its main features. These include coordinated multipoint transmission and reception, the enhanced physical downlink control channel, avoidance of interference between different radio access technologies, and enhancements to machine-type communications and mobile data applications.

The scheduled freeze date for Release 12 was September 2014, so the specifications were not yet finalized at the time of writing. We first cover the main features of that release, which include device-to-device communications, dynamic adaptation of the TDD configuration, further enhancements to machine-type communications and mobile data applications, and improved inter-operation with wireless local area networks. We also discuss two other issues that 3GPP studied during that release, namely improvements to small cells and heterogeneous networks and elevation beamforming. Some details of Release 12 will have changed by the time the specifications are finalized, but we note the remaining uncertainties where appropriate.

Details of Releases 11 and 12 can be found in the corresponding 3GPP release summaries [1]. There is also information about the latest 3GPP releases in a regularly updated white paper by 4G Americas [2], which will ultimately contain information about Releases 13, 14 and beyond.

20.1 Coordinated Multipoint Transmission and Reception

20.1.1 Objectives

If a mobile moves away from a base station's antennas and towards the edge of a cell, then it receives a weaker signal from the serving cell and more interference from other cells that are nearby. Taken together, these two effects reduce the mobile's data rate and degrade the user's service. In a technique known as *coordinated multipoint transmission and reception* (CoMP) [3], nearby antennas cooperate so as to increase the power received by a mobile at the cell edge, reduce its interference and increase its achievable data rate. In contrast, CoMP does not have much effect on the average data rate within the cell.

An Introduction to LTE: LTE, LTE-Advanced, SAE, VoLTE and 4G Mobile Communications, Second Edition.
Christopher Cox.

20.1.2 Scenarios

In designing the support for CoMP in LTE, 3GPP considered four scenarios that are shown in Figure 20.1. Scenario 1 is a homogeneous network in which the cooperating base station antennas control different sectors at a single site. Scenario 2 is similar, but the antennas are at different sites. Scenario 3 is a heterogeneous network containing macrocells and picocells, while, in scenario 4, the picocells are replaced by *remote radio heads* (RRHs) that have the same physical cell identity as the parent.

These scenarios bring out a few relevant issues. Firstly, it is often easier for the different cells to cooperate if their antennas are at the same site (scenario 1) than if they are at different sites (scenarios 2 and 3). A network operator can, however, configure different sites in two ways, as shown in Figure 20.2. The most obvious configuration is to use different eNBs, which cooperate either across the X2 interface or by using proprietary techniques. However, the sites can also be configured using remote radio heads that communicate with a centralized eNB across a high-speed digital communication link. This second option makes the implementation of inter-site CoMP more straightforward than it might otherwise be.

Secondly, the cooperating antennas can be in different cells or in the same cell. The latter possibility appears in scenario 4 as shown in Figure 20.1, which also illustrates another application of remote radio heads.

To describe these architectural options, the 3GPP specifications define a *point* as a set of geographically co-located antennas, such that different points can cooperate using CoMP techniques. Different sectors within the same site are treated as different points, as in scenario 1, while different radio heads within the same sector can also be treated as different points, as in scenario 4.

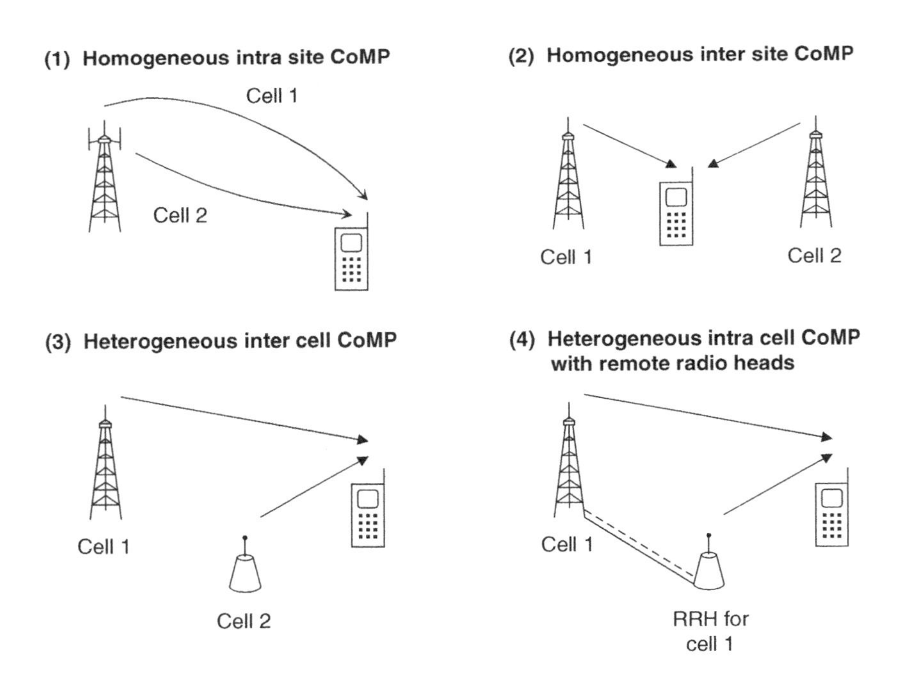

Figure 20.1 Scenarios for coordinated multipoint transmission and reception in Release 11

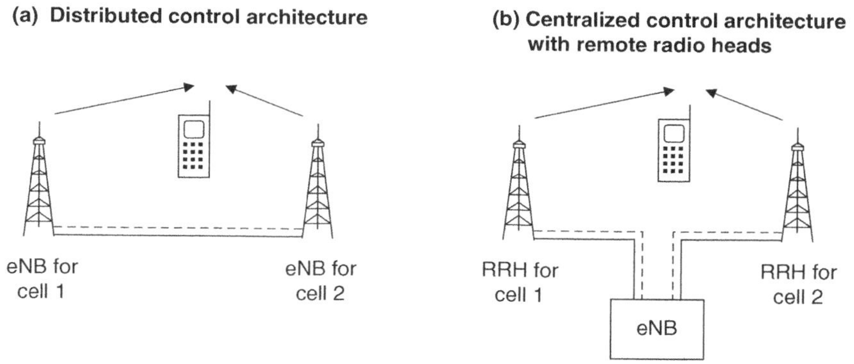

Figure 20.2 Distributed and centralized control architectures for the radio access network

20.1.3 CoMP Techniques

Downlink CoMP can be implemented using a variety of techniques, which are shown in Figure 20.3. In *coordinated scheduling/beamforming* (CS/CB), data are available for transmission at only one point, which is chosen using the normal handover procedures and is only occasionally changed. Nearby points coordinate their scheduling decisions so as to minimize the interference that they are sending to the target mobile; for example, by transmitting on different resource blocks. They can also coordinate their beamforming decisions; for example, by directing nulls of their antenna patterns towards the target mobile.

In *joint processing* (JP), data are available for transmission at multiple points. There are two varieties. In *dynamic point selection* (DPS), the network actually transmits from only one point at a time, with the selection potentially changing from one subframe to the next. In *joint transmission* (JT), the network can simultaneously transmit from more than one point. In turn, there

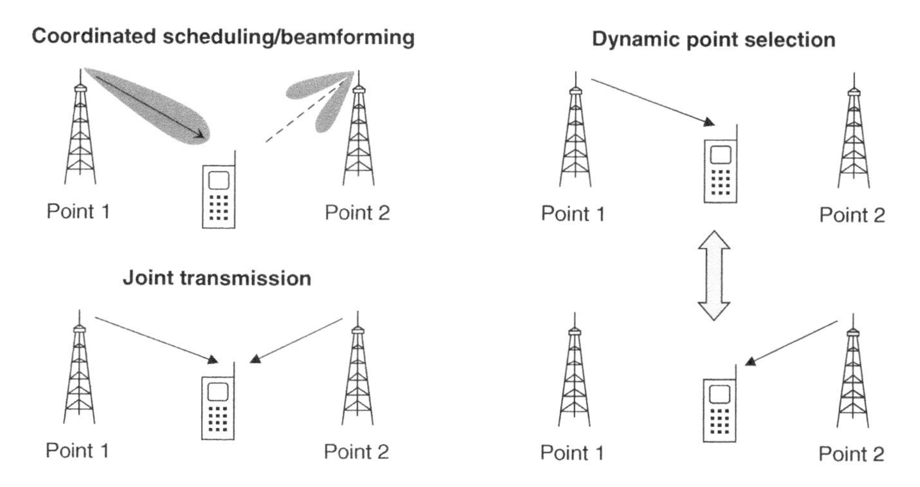

Figure 20.3 Coordinated multipoint transmission techniques

are two varieties of joint transmission. In *coherent joint transmission*, the network knows the phase relationship between the different ray paths and can ensure that the transmissions reach the mobile with the same phase angle, in a way similar to closed loop transmit diversity or beamforming. In *non-coherent joint transmission*, the network does not have any such knowledge, so the phase relationship between the received transmissions is completely arbitrary, as in multicast/broadcast over a single frequency network (MBSFN).

Once again, these options require some new terminology. The *CoMP cooperating set* is the set of points that are participating in any of these techniques. A point's participation can be either direct, in which case it has data available for transmission, or indirect, in which case it has no data but still participates by means of scheduling and beamforming. A point that actually transmits in a particular subframe is known as a *CoMP transmission point*.

Uplink CoMP uses similar techniques but is more straightforward because the data are available at multiple reception points already. In coordinated scheduling/beamforming, nearby points coordinate their uplink scheduling and beamforming decisions so as to minimize the interference that they receive from other mobiles. In dynamic point selection, the network receives data at multiple points but only selects data from one point at a time. In *joint reception* (JR), the network receives data at multiple points and combines them to improve the quality of the received signal. Whichever technique is used, the CoMP cooperating sets on the uplink and downlink can be different from each other, and the CoMP reception points can be different from the CoMP transmission points.

20.1.4 Standardization

The Release 11 specifications support uplink and downlink CoMP using coordinated scheduling/beamforming, dynamic point selection and non-coherent joint transmission and reception [4–7]. The main enhancement is the introduction of a new channel state information reporting technique to support downlink CoMP. In this technique, the base station configures the mobile with a *CoMP measurement set* that contains up to four *CSI processes*. Each process contains a CSI reference signal configuration, similar to the one in Release 10, which defines the resource elements on which the mobile should measure the CSI reference signal power. However, the CSI process also contains a separate *CSI interference measurement configuration*, on which the mobile measures the corresponding interference level.

For coordinated scheduling/beamforming, the CSI processes all refer to a single CoMP transmission point. Amongst them, however, the different CSI interference measurement configurations define different sets of resource elements on which neighbouring points may or may not be transmitting, leading to different levels of interference. The mobile reports the channel quality indicator for each of its CSI processes, and the base station compares the results to help it schedule its downlink transmissions. For joint processing, the CSI processes refer to different transmission points. The mobile reports the channel quality indicator for each of its CSI processes as before, and the base station compares the results to help it decide which transmission point(s) to use.

The radio access network also configures the mobile with up to four PDSCH parameter sets. Each of these is associated with one or more of the points in the downlink cooperating set, and defines the resource element mapping of the PDSCH using the maximum size of the downlink control region and any zero-power CSI reference signal configurations. The network then places the mobile in a new transmission mode, number 10, and schedules it using a new DCI format, 2D. This format is similar to format 2C (eight layer spatial multiplexing), but it directs the mobile to receive the PDSCH using one of the parameter sets defined earlier. For dynamic point selection, the network simply changes the parameter set from one subframe to the next, while for non-coherent joint transmission, the network must also ensure that the chosen CoMP transmission points share the same parameter set.

The Release 11 specifications leave three main issues outstanding. Firstly, there are no additional RRC measurements to help the base station decide which points should be in the CoMP measurement set. Secondly, there is no support for coherent joint transmission, which would require additional feedback to describe the phase relationship between the incoming transmissions. Thirdly, there is no attempt to standardize anything within the network, so the S1 and X2 interfaces are unaffected. Because of this, CoMP is best implemented in different sectors of one site or using remote radio heads, but not using different eNBs.

20.1.5 Performance

3GPP have carried out simulations of the different CoMP techniques in the four scenarios that we described earlier. There is much variability in the results, depending on issues such as the chosen CoMP technique and the antenna configuration, but a typical set is for joint transmission and reception in a heterogeneous network without enhanced inter-cell interference coordination [8]. Here, the data rate at the cell edge rises by 24% in the downlink and 40% in the uplink, while the cell capacity rises by 3% in the downlink and 14% in the uplink. As expected, CoMP has more impact on mobiles at the cell edge than elsewhere and it also has more impact on the uplink than the downlink.

20.2 Enhanced Physical Downlink Control Channel

The physical downlink control channel (PDCCH) works well in most situations but has a few limitations. Firstly, the channel is restricted to the first three or four control symbols in a subframe, so it can limit the capacity of cells that contain a large number of low-data rate devices. Secondly, the PDSCH and PUSCH support multiple-user MIMO, so they can communicate with even more devices as the number of base station antennas increases, while the PDCCH cannot. Thirdly, the PDCCH transmissions to individual mobiles occupy the whole of the frequency band and cannot benefit from frequency domain inter-cell interference coordination.

Release 11 addresses these limitations by means of a new physical channel known as the *enhanced physical downlink control channel* (EPDCCH) [9–11]. The EPDCCH carries the same information as the PDCCH but is transmitted in the downlink data region. Within each subframe, the base station assigns individual resource block pairs to either the PDSCH or the EPDCCH, so that the EPDCCH has an adjustable capacity and can benefit from interference coordination.

The EPDCCH is transmitted on four new antenna ports, numbered from 107 to 110. These are associated with reference signals that occupy the same resource elements as the ones on ports 7 to 10, but carry slightly different information. The base station precodes both the EPDCCH and its reference signals by means of a mobile-specific precoding matrix, so that the precoding process is transparent to the mobile. The channel supports simultaneous MU-MIMO communications to four different mobiles, with each mobile receiving a single layer on a single antenna port.

Before using the EPDCCH, the base station schedules the target mobile on the PDCCH in the usual way and waits for it to feed back some channel state information. The base station also tells the mobile which subframes and resource blocks to inspect for EPDCCH scheduling messages, by means of mobile-specific RRC signalling. In each subframe that contains the EPDCCH, the mobile searches for scheduling messages that are directed to one of its radio network temporary identities, with a match triggering data transmission or reception in the usual way. In the case of downlink reception, the EPDCCH and PDSCH arrive in parallel, so the mobile buffers the whole subframe, processes the EPDCCH and finally processes the PDSCH if required.

20.3 Interference Avoidance for in Device Coexistence

Most LTE devices support other communication technologies such as WiFi, Bluetooth and GPS. If an LTE device is also transmitting and receiving using one of those other technologies, then the resulting *in device coexistence* (IDC) interference can sometimes be severe [12]. In LTE band 7, for example, the uplink lies immediately above a WiFi band at 2.4 GHz, which is reserved for industrial, scientific and medical equipment. If a device is using both frequency bands, then its LTE transmissions can leak into its WiFi receiver and can cause interference there.

In Release 11, the network can help solve these interference problems using the signalling procedure shown in Figure 20.4 [13]. The message sequence starts with the usual capability enquiry procedure (1), in which the mobile indicates its ability to report IDC interference problems. The base station replies by including an *IDC configuration* in the usual RRC connection reconfiguration procedure (2), which indicates whether it is willing to accept such a report.

Later, the mobile may have an interference problem that is unable to resolve by itself. It reacts by transmitting an RRC *In Device Coex Indication* message (3), which indicates the LTE carrier frequencies that are causing or suffering from interference. Optionally, the mobile

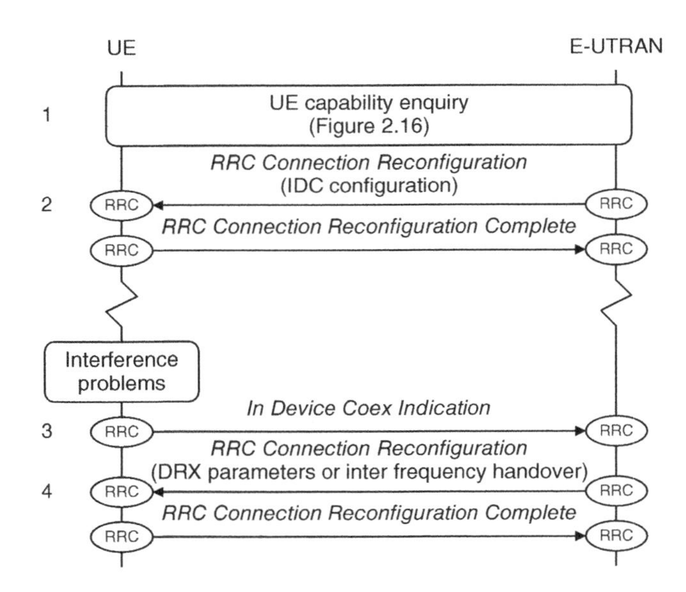

Figure 20.4 In device coexistence indication procedure. Source: TS 36.331. Reproduced by permission of ETSI

can also indicate that it could resolve the problem by alternating between LTE and the other technology and can suggest a suitable LTE discontinuous reception pattern.

The base station can respond to the mobile's report in two main ways (4), either by handing the mobile over to another LTE carrier that has no interference problems or by putting the mobile into a state of discontinuous reception. As an alternative, the base station can apply DRX as a temporary solution while collecting the measurement reports that it requires before an inter frequency handover.

To supplement the main procedure, the mobile can refuse to transmit in occasional uplink subframes so as to protect the other communication technology from interference. The base station limits the occurrence of such *autonomous denial* subframes as part of step (2).

20.4 Machine-Type Communications

20.4.1 Device Triggering

In the last chapter, we saw that Release 10 introduced several overload control techniques for machine-type communications (MTC). Release 11 takes the support of MTC further by addressing device triggering and identification.

Machine-type communications were originally developed for 2G and 3G systems. These systems do not guarantee always-on connectivity; instead, a device attaches to the core network and connects to a packet data network using two distinct procedures and does not necessarily have a PDN connection or an IP address. However, an application server may still wish to contact a device and trigger it into action, so it requires another way to do so. The most common technique is the use of mobile-terminated SMS so, despite its own support for always-on connectivity, it is useful for LTE to handle this technique as well.

Figure 20.5 shows the architecture for device triggering in LTE Release 11 [14]. The application server (AS) is owned by a third-party service provider. It can communicate with the device

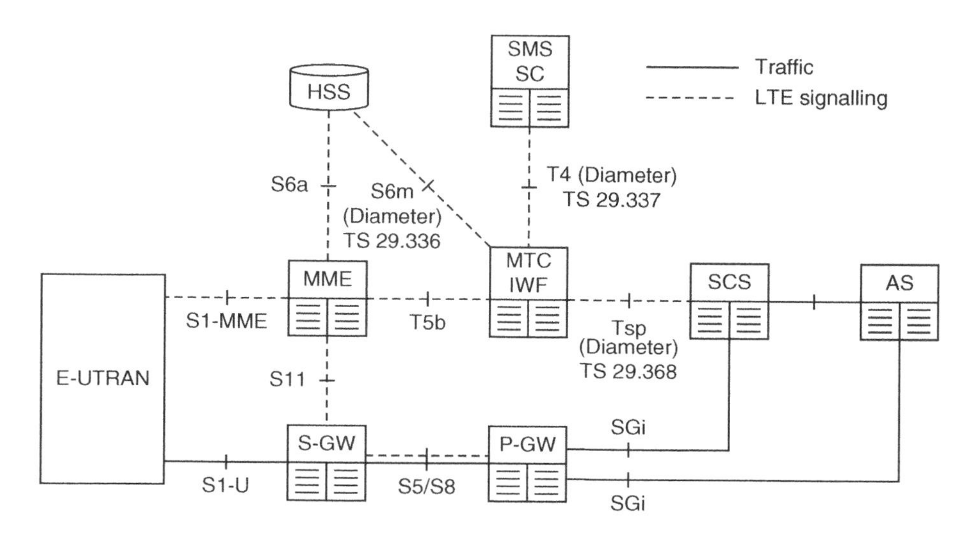

Figure 20.5 Architecture for machine type communications in LTE. Source: TS 23.682. Reproduced by permission of ETSI

in two ways, either directly over the SGi interface or indirectly through a *service capability server* (SCS). In turn, the SCS can either reach the device directly or send a device trigger request over the Tsp interface [15] to the evolved packet core. On receiving such a trigger request, the *machine-type communications interworking function* (MTC-IWF) looks up the user's subscription details in the home subscriber server [16], decides the delivery mechanism that it will use and triggers the device over the control plane of LTE.

Release 11 only supports one delivery mechanism, in which the MTC-IWF contacts the SMS service centre over the T4 interface [17] and asks it to trigger the device using a mobile-terminated SMS. The service centre can then deliver the SMS using any of the techniques that we will discuss in Chapters 21 and 22. The MTC-IWF should eventually be able to contact the MME directly over the T5b interface, but that interface is yet to be specified.

20.4.2 Numbering, Addressing and Identification

Traditionally, the short message service identifies a device using a phone number in the form of a *mobile subscriber ISDN number* (MSISDN). Unfortunately, the likely proliferation of machine-type devices is leading to a shortage of MSISDNs. To help deal with the problem, the Release 11 specifications introduce a new identity known as the *external identifier* [18] for use on the Tsp, T4, S6m and SGi interfaces from earlier.

The external identifier is a network access identifier (NAI) with the format `<local identifier>@<domain identifier>`. The local identifier maps to the device's IMSI; indeed, it may actually be the IMSI, but in that case would only be made available to trusted service providers and applications. Similarly, the domain identifier maps to the network operator.

As far as IP addresses are concerned, IP version 6 is considered to be the main addressing mechanism for machine-type devices [19]. Although it is still supported, IP version 4 is considered to be a temporary solution and is deprecated.

20.5 Mobile Data Applications

LTE was originally designed for traditional data applications such as web browsing and file transfers. In these applications, there is a clear distinction between an active state in which the device is transmitting or receiving and an idle state in which the user is viewing downloaded content. The base station can detect transitions between these states by observing the flow of traffic and can respond by switching the device between the LTE states of RRC_CONNECTED and RRC_IDLE.

When running newer applications such as Facebook and Twitter, the device has to send and receive frequent, small packets such as status reports and keep-alive indications, even if there is no interaction with the user [20]. This lessens the distinction between the active and idle states and makes it harder for the base station to respond in the appropriate way. If, for example, a device is moved to RRC_IDLE in the same way as for a web-browsing application, then it will have to change the RRC state whenever it transmits or receives, causing excessive signalling. If it remains in RRC_CONNECTED as if the user were still active, then its power consumption will be excessively high.

Release 11 introduces a lightweight solution in which the device can help the base station manage it in the appropriate way. On detecting a period of user inactivity, the device sends an RRC *UE Assistance Information* message to the base station, in which it requests a move to a low-power state [21]. The base station can respond in various ways, depending on its own configuration and the device's traffic profile. However, it might typically move the device into discontinuous reception with a long DRX cycle to offer it a compromise between the basic RRC_CONNECTED and RRC_IDLE states.

20.6 New Features in Release 12

20.6.1 Proximity Services and Device to Device Communications

In 3GPP networks, devices have traditionally communicated with the base station alone and have been unable to communicate directly with each other. That capability does, however, form part of other wireless communication technologies, such as Terrestrial Trunked Radio (TETRA), WiFi and Bluetooth. As part of Release 12, 3GPP started adding the capability to LTE under the name of *proximity services* (ProSe) or *device-to-device* (D2D) *communications*.

For applications such as voice, video and file exchange, proximity services benefit users by increasing their data rates and reducing their power consumption, and benefit the network by offloading traffic from the base station. However, the most important single application is public safety networks, also known as *critical communications* [22]. In 2011, the US Federal Communications Commission selected LTE as the preferred technology for a national network to be used by US emergency services. Such a network must have a large percentage coverage area and a high reliability. Proximity services can help the system to meet these requirements, by allowing two public safety users to communicate directly if they are outside the coverage area of the radio access network or if the network has failed.

Although proximity services require significant changes to the specifications, the public safety application prompted 3GPP to introduce support for their most important features in Release 12. The description that follows is based on the specifications from December 2013, which included the requirements and high-level architecture of proximity services but had not yet defined their architectural details or implementation [23–27].

Figure 20.6 shows the high-level architecture for proximity services. The ProSe application server is a third-party device, which runs applications that make use of proximity services.

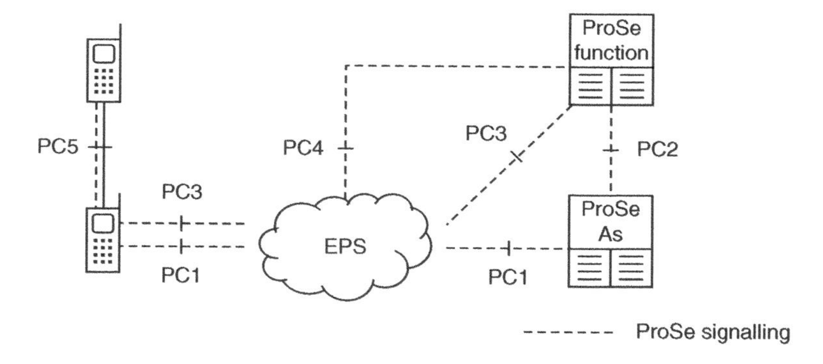

Figure 20.6 Architecture for LTE proximity services

It communicates with a mobile application over the PC1 interface, which currently lies outside the 3GPP specifications. The ProSe function is an LTE device, which manages the mobile's proximity service functions. It communicates with the mobile's LTE protocol stack over the PC3 interface, which currently runs across the user plane of LTE. By using this architecture, two mobiles can establish a direct communication path across the interface denoted as PC5.

Proximity services have to be authorized by the network operator and have two phases, discovery and communication. In the discovery phase, two mobiles find out that they are nearby. Two discovery techniques are scheduled for inclusion in Release 12. The first is network-based discovery, in which the ProSe function acts as a location service client over the PC4 interface, measures the mobiles' locations and establishes that they are nearby. The second is direct discovery over the LTE air interface, in which one mobile transmits a *ProSe code* that identifies the ProSe application(s) in which it is interested, and the other mobile detects the code. For applications other than public safety, the LTE network manages the discovery process, so both mobiles must lie within the coverage area of LTE. They can, however, be in different cells.

In the communication phase, the network establishes a direct communication path between the two mobiles. The only technique currently scheduled for Release 12 is direct communication using WiFi. This is triggered by the discovery process described above, when the ProSe function tells the mobiles about each other's identities, supplies them with the necessary parameters such as secure keys and helps them set up a direct communication path. The communication phase could also use the LTE air interface, in which case the radio access network would control the mobiles' use of resources subject to their feedback about the quality of the radio channel.

Public safety communications require a further level of authorization from the network operator but are intended to have several additional features. Two public safety mobiles can discover each other autonomously, so they can lie outside the coverage area of LTE. One discovery solution is scheduled for Release 12, in which the ProSe function pre-configures the mobiles by giving them permission to form a peer-to-peer WiFi network and supplying them with the parameters that they will require. A public safety mobile can act as a relay to support various new communication paths, with mobile-to-network relays scheduled for Release 12, and mobile-to-mobile relays also under consideration. Finally, public safety applications support not only traditional one-to-one communications, but also broadcast communications and communications within a multicast group.

20.6.2 Dynamic Adaptation of the TDD Configuration

In TDD mode, a cell assigns subframes to the uplink or downlink by means of its TDD configuration. In Releases 8 to 11, the TDD configuration is chosen using the long-term balance between the traffic levels on the uplink and downlink and usually remains unchanged. In Release 12, the specifications are enhanced to support dynamic changes of the TDD configuration [28]. There are two reasons for this, which relate to the progressive introduction of smaller cells: a small cell contains fewer mobiles than a larger one, so its traffic levels fluctuate more, while small cells tend to lie in small clusters, in which the interference is easier to manage.

To implement dynamic adaptation, a cell actually uses three separate TDD configurations. Firstly, a cell continues to advertise a static configuration using its system information. This is the most uplink-oriented configuration that the cell will ever use: its uplink subframes may be reallocated to the downlink as part of the adaptation process, but its downlink subframes are stable. Legacy mobiles use the static configuration throughout, while Release 12 mobiles also use it for measurements.

The next issue is the hybrid ARQ process timing for Release 12 mobiles; in other words, the intervals between scheduling messages, data transmissions and hybrid ARQ acknowledgements. The uplink timing simply follows the static configuration from the system information. However, the downlink timing follows a second, semi-static TDD configuration. This is the most downlink-oriented configuration that the cell will ever use and is sent to the mobiles by means of RRC signalling.

Finally, the cell informs Release 12 mobiles about its dynamic TDD configuration using downlink control information on the PDCCH or EPDCCH. This indicates the subframes that are currently assigned to the downlink, so a Release 12 mobile can inspect these for scheduling messages.

To illustrate the end result, Figure 20.7 shows an example where the system information is advertising TDD configuration 0, the cell is currently using configuration 1, and Release 12 mobiles are following the downlink hybrid ARQ timing from configuration 2. Legacy mobiles use configuration 0 throughout and can receive downlink data in subframes 1 and 6 in the usual way. Data in subframes 0 and 5 would usually trigger uplink acknowledgements in subframes 4 and 9, but those subframes are currently assigned to the downlink and are unavailable. There is therefore a loss of capacity for legacy devices. Release 12 mobiles use configuration 1,

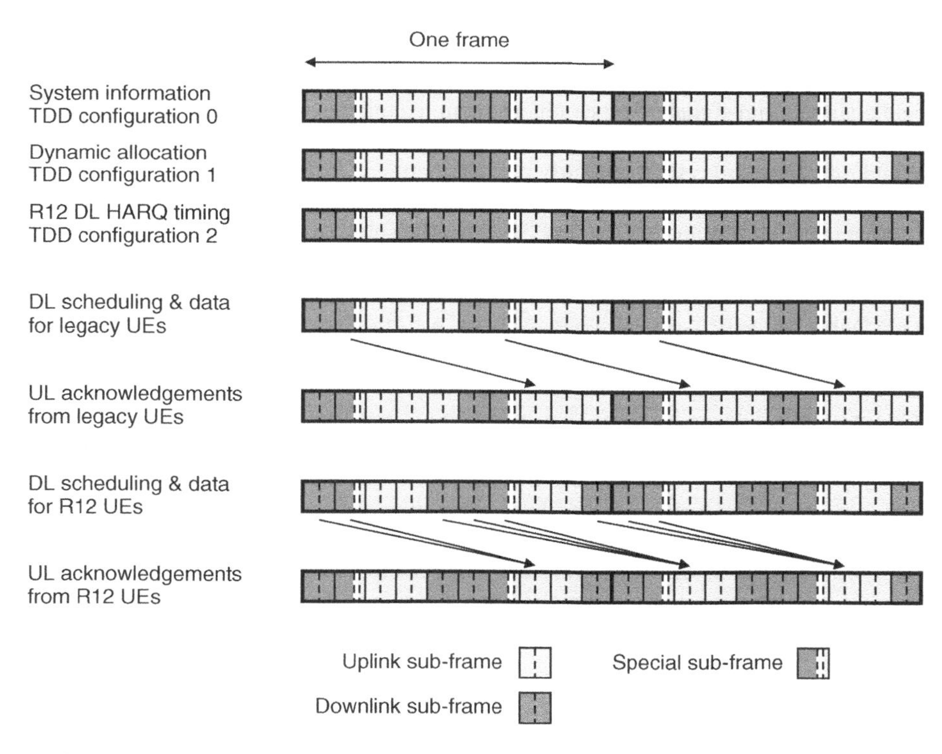

Figure 20.7 Example downlink transmission timings when using dynamic adaptation of the TDD configuration

except that their downlink timing follows that of configuration 2 and remains stable even if the dynamic configuration changes.

To help it manage the resulting interference, a cell circulates information to its neighbours about its current interference levels and about the TDD configuration that it is intending to use, by means of X2 signalling. The neighbouring cells can then use this information to maximize their capacity; for example, by using the same configuration or by scheduling vulnerable mobiles in subframes that are unaffected by interference.

20.6.3 Enhancements for Machine-Type Communications and Mobile Data

There has previously been little incentive for the manufacturers of machine-type devices to support LTE, because GSM devices are cheap and the coverage of GSM networks is excellent. This situation is the opposite of what network operators want, because the spectral efficiency of GSM devices is poor and their existence may hinder the eventual shutdown of GSM networks. As part of Release 12, 3GPP started the specification of an LTE device for machine-type applications, with the aim that its cost should be comparable with that of a GPRS modem [29]. The device is supported by means of a new UE category, whose likely capabilities include one downlink receive antenna, a maximum transport block size of 1000 bits and a maximum downlink allocation of six resource blocks.

Machine-type devices are often deployed indoors, where the coverage may be poor. To solve the problem, 3GPP started work to improve the coverage of low-data rate devices, with the aim of handling signals that are 15 dB weaker than before. Because the data rates are low, these techniques have to be applied not only to the PDSCH and PUSCH but also to most of the other physical channels and signals. Some enhancement techniques, such as repetition, come at the expense of spectral efficiency, so they are only applied to the new UE category introduced above and only to mobiles that explicitly request them.

3GPP also started work to reduce the power consumption of machine-type devices and other devices that use mobile data applications [30, 31]. The solution scheduled for Release 12 is the introduction of a new power saving state, in which the mobile transmits periodic tracking area updates as it does in RRC_IDLE but does not monitor the base station for paging messages and is not reachable by the network until it reawakens. A further possibility is the introduction of an extended discontinuous reception cycle, in either RRC_IDLE, RRC_CONNECTED or both.

Earlier, we saw that a Release 11 mobile could request a move into a low-power state that is suitable for mobile data applications such as Facebook and Twitter. The technique is simple but has some disadvantages; for example, it is only usable by Release 11 mobiles and depends on the implementation of both the mobile and the network. 3GPP started work on a more powerful solution in Release 12. During the S1 release procedure, the base station tells the MME about the mobile's current parameter settings such as the duration of its DRX cycle, which would otherwise be lost. The MME can then send the parameters to the next base station during the next service request procedure, so that the base station can immediately handle the mobile in the optimal way.

20.6.4 Traffic Offloading Enhancements

In Chapters 16 and 19, we saw that Release 12 has improved LTE's ability to inter-operate with wireless local area networks and carry out selective IP traffic offload. In addition, the access network discovery and selection function (ANDSF) has been enhanced to align it with the WiFi Alliance specifications for *Passpoint*, otherwise known as *Hotspot 2.0* [32, 33]. Using a set of *WLAN selection policy* data, the ANDSF can supply the mobile with information such as the maximum desired utilization of the wireless LAN air interface and the required capacity of the backhaul. The mobile compares this with information broadcast by a Passpoint-compliant access point to ensure that the access point has enough capacity for it to handle the mobile. The ANDSF can also identify access points using information other than the WiFi service set identifier, in particular, the access point's realm and *organizational unique identifier* (OUI).

20.7 Release 12 Studies

20.7.1 Enhancements to Small Cells and Heterogeneous Networks

3GPP carried out several studies during Release 12, amongst which was a study of ways to improve the performance of small cells and heterogeneous networks [34–36]. There are two ways to increase the data rate in a small cell. Firstly, the received signal power in a small cell is higher than usual, so the mobiles can benefit from the use of higher order modulation schemes. One new modulation scheme, downlink 256-QAM, is likely to form part of the Release 12 specifications. Secondly, the low delay spread results in a higher coherence bandwidth than usual, while the mobiles' low speed results in a higher coherence time. These issues suggest the introduction of a new mobile-specific reference signal that occupies fewer resource elements than usual, thereby increasing the capacity of the PDSCH and PUSCH.

In our earlier discussion of self-optimizing networks, we saw that a cell can switch off when not in use so as to minimize its power consumption. In Release 11, a cell can do this every few seconds, a timescale dictated by the need for measurements and for handover signalling. If a cell could switch on and off more often, then it could reduce its power consumption further and could reduce the interference that it generated elsewhere. Timescales of tens of milliseconds are achievable with little impact on the specifications, for example by ensuring that dormant cells continue to broadcast occasional synchronization signals, and these are likely to form part of the Release 12 specifications. The ultimate goal would be a cell that could switch on and off every subframe, although the impact on the specifications would be more severe.

Dual connectivity is the ability of a mobile to communicate simultaneously with two base stations, namely a *master eNB* (MeNB) and a *slave eNB* (SeNB), which are typically a macrocell and a picocell using different carrier frequencies. Dual connectivity has three main motivations. The most important is to reduce the number of handover failures in a heterogeneous network. Handovers are difficult for a mobile that is moving out of a picocell because it may not have time to discover a surrounding macrocell before losing its original signal. By maintaining the RRC signalling within the macrocell, the robustness of the handover can be improved. In addition, the network's signalling load can be reduced by minimizing the total number of handovers, while the network's capacity and the user's throughput can both be increased.

3GPP also studied the introduction of a new carrier type, also known as a lean carrier. The new carrier does not transmit information such as the cell-specific reference signals or the legacy PDCCH so it cannot be used by legacy mobiles or as a stand-alone cell; instead, it is intended for use as a secondary cell during carrier aggregation or as a slave during dual connectivity.

20.7.2 Elevation Beamforming and Full Dimension MIMO

In Releases 8 to 11, a base station's antennas lie in a horizontal plane, so the resulting beams are restricted to different azimuth angles. A cell's performance could potentially be improved using two-dimensional antenna arrays, in which the antennas also lie in a vertical plane and the beams are also at different elevations. Such antennas can be used for user-specific elevation beamforming. They can also be used for a more complex technique known as *massive MIMO* or *full dimension MIMO* (FD-MIMO), in which the base station has a two-dimensional array that contains many more antennas than in previous deployments. 3GPP studied the relevant channel models and performance benefits as part of Release 12 [37], with any specifications likely to form part of future releases.

References

1. 3rd Generation Partnership Project (2013) FTP Directory, ftp://ftp.3gpp.org/Information/WORK_PLAN/Description_Releases/ (accessed 15 October 2013).
2. 4G Americas (2012) 4G Mobile Broadband Evolution: Release 10, Release 11 and Beyond – HSPA, SAE/LTE and LTE-Advanced, October 2012.
3. 3GPP TR 36.819 (2013) Coordinated Multi-point Operation for LTE Physical Layer Aspects, Release 11, September 2013.
4. 3GPP TS 36.211 (2013) Physical Channels and Modulation, Release 11, Section 5.5.1.5, 6.10.3.1, 6.10.5.1, September 2013.
5. 3GPP TS 36.212 (2013) Multiplexing and Channel Coding, Release 11, Section 5.3.3.1.5D, June 2013.
6. 3GPP TS 36.213 (2013) Physical Layer Procedures, Release 11, Section 7.1.9, 7.1.10, 7.2.5, September 2013.
7. 3GPP TS 36.331 (2013) Radio Resource Control (RRC); Protocol Specification, Release 11, Section 6.3.2 (CSI-IM-Config, CSI-Process, CSI-RS-ConfigNZP, CSI-RS-ConfigZP, PDSCH-Config), September 2013.
8. 3GPP TR 36.819 (2013) Coordinated Multi-point Operation for LTE Physical Layer Aspects, Release 11, Sections 7.3.1.1, 7.3.2, September 2013.
9. 3GPP TS 36.211 (2013) Physical Channels and Modulation, Release 11, Sections 6.2.4A, 6.8A, 6.10.3A, September 2013.
10. 3GPP TS 36.213 (2013) Physical Layer Procedures, Release 11, Section 9.1.4, September 2013.
11. 3GPP TS 36.331 (2013) Radio Resource Control (RRC); Protocol Specification, Release 11, Section 6.3.2 (EPDCCH-Config), September 2013.
12. 3GPP TR 36.816 (2012) Study on Signalling and Procedure for Interference Avoidance for In-device Coexistence, Release 11, January 2012.
13. 3GPP TS 36.331 (2013) Radio Resource Control (RRC); Protocol Specification, Release 11, Section 5.6.9, September 2013.
14. 3GPP TS 23.682 (2013) Architecture Enhancements to Facilitate Communications With Packet Data Networks and Applications, Release 11, September 2013.
15. 3GPP TS 29.368 (2013) Tsp Interface Protocol Between the MTC Interworking Function (MTC-IWF) and Service Capability Server (SCS), Release 11, September 2013.
16. 3GPP TS 29.336 (2012) Home Subscriber Server (HSS) Diameter Interfaces for Interworking With Packet Data Networks and Applications, Release 11, December 2012.
17. 3GPP TS 29.337 (2013) Diameter-based T4 Interface for Communications with Packet Data Networks and Applications, Release 11, June 2013.
18. 3GPP TS 23.003 (2013) Numbering, Addressing and Identification, Release 11, Section 19.7, September 2013.
19. 3GPP TS 23.221 (2013) Architectural Requirements, Release 11, Section 5.1, June 2013.
20. 3GPP TR 36.822 (2012) LTE Radio Access Network (RAN) Enhancements for Diverse Data Applications, Release 11, September 2012.
21. 3GPP TS 36.331 (2013) Radio Resource Control (RRC); Protocol Specification, Release 11, Section 5.6.10, September 2013.
22. 3rd Generation Partnership Project (2013) Delivering Public Safety Communications with LTE, http://www.3gpp.org/news-events/3gpp-news/1455-public-safety (accessed 26 November 2013).

23. 3GPP TR 22.803 (2013) Feasibility Study for Proximity Services (ProSe), Release 12, June 2013.
24. 3GPP TS 22.278 (2012) Service Requirements for the Evolved Packet System (EPS), Release 12, Section 7A, September 2012.
25. 3GPP TR 23.703 (2013) Study on Architecture Enhancements to Support Proximity Services (ProSe), Release 12, October 2013.
26. 3GPP TS 23.303 (2014) Architecture Enhancements to Support Proximity Services (ProSe), Release 12, January 2014.
27. 3GPP TR 36.843 (2013) Feasibility Study on LTE Device to Device Proximity Services – Radio Aspects, Release 12, October 2013.
28. 3GPP TR 36.828 (2012) Further Enhancements to LTE Time Division Duplex (TDD) for Downlink-Uplink (DL-UL) Interference Management and Traffic Adaptation, Release 11, June 2012.
29. 3GPP TR 36.888 (2013) Study on Provision of Low-cost Machine-Type Communications (MTC) User Equipments (UEs) Based on LTE, Release 12, June 2013.
30. 3GPP TR 23.887 (2013) Machine-Type and Other Mobile Data Applications Communications Enhancements, Release 12, October 2013.
31. 3GPP TR 37.869 (2013) Study on Enhancements to Machine-Type Communications (MTC) and other Mobile Data Applications; Radio Access Network (RAN) aspects, Release 12, September 2013.
32. WiFi Alliance (2013) Wi-Fi CERTIFIED Passpoint&, http://www.wi-fi.org/discover-and-learn/wi-fi-certified-passpoint (accessed 15 October, 2013).
33. 3GPP TR 23.865 (2013) Study on Wireless Local Area Network (WLAN) Network Selection for 3GPP Terminals, Release 12, September 2013.
34. 3GPP TR 36.932 (2013) Scenarios and Requirements for Small Cell Enhancements for E-UTRA and E-UTRAN, Release 12, March 2013.
35. 3GPP TR 36.872 (2013) Small Cell Enhancements for E-UTRA and E-UTRAN – Physical Layer Aspects, Release 12, September 2013.
36. 3GPP TR 36.842 (2013) Study on Small Cell Enhancements for E-UTRA and E-UTRAN – Higher Layer Aspects, Release 12, June 2013.
37. 3GPP TR 36.873 (2013) 3D Channel Model for LTE, Release 12, September 2013.

21

Circuit Switched Fallback

As we explained in Chapter 1, LTE is designed as a data pipe: a system that delivers information to and from the user, but does not concern itself with the overlying application. For most data services, such as web browsing and emails, the applications are separate from the delivery system and are supplied by third parties, so this approach works well. However, for voice and text messages, the applications have previously been supplied by the operator's circuit switched network and have been tightly integrated into the delivery system. This is a very different principle from the one adopted for LTE.

The 3GPP specifications support two main approaches for the delivery of voice over LTE. The main short-term solution is circuit switched fallback, in which the mobile accesses the circuit switched domain of a legacy 2G or 3G network by moving to a 2G or 3G cell. The main long-term solution is the IP multimedia subsystem, an external network that handles the signalling functions needed for voice over IP. There are also three other approaches: the delivery of voice over IP services by a third party, the use of a dual radio device and another 2G/3G inter-working technique known as voice over LTE by generic access. Each approach can be adapted for the delivery of text messages, using either SMS or a proprietary messaging application.

In this chapter, we review the market for voice and text messaging and introduce each of the five approaches. We then give a more detailed account of circuit switched fallback, a technique that is summarized in a 3GPP stage 2 specification, TS 23.272 [1]. In the next chapter, we will cover the IP multimedia subsystem.

21.1 Delivery of Voice and Text Messages over LTE

21.1.1 The Market for Voice and SMS

To illustrate the importance of voice and SMS to mobile network operators, Figure 21.1 shows the revenue that Western European operators have been earning from voice, messaging services such as SMS, and other data applications. The information is from market research by Analysys Mason and uses operators' data up to the end of 2012 and forecasts thereafter.

There is a striking contrast between the information in Figure 21.1 and the information about worldwide network traffic that we presented in Figures 1.5 and 1.6: data applications supply

---

An Introduction to LTE: LTE, LTE-Advanced, SAE, VoLTE and 4G Mobile Communications, Second Edition.
Christopher Cox.

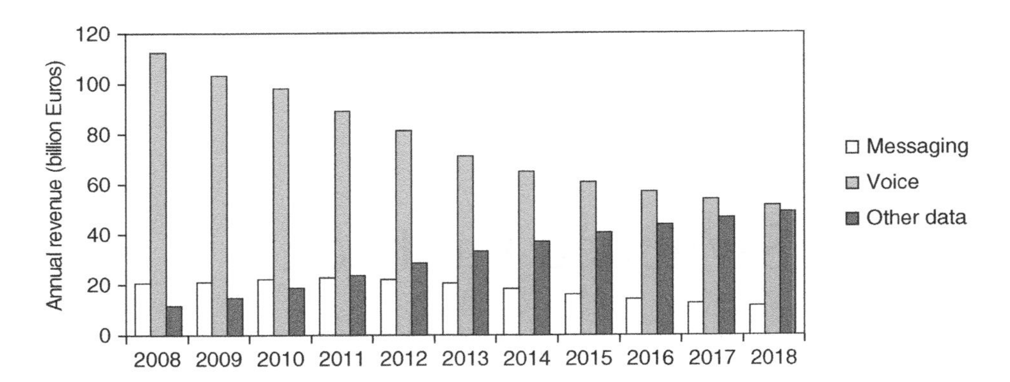

Figure 21.1 Revenue earned by network operators in Western Europe from voice, messaging and other data services, with operators' data up to 2013 and forecasts thereafter. Data supplied by Analysys Mason

most of the operators' traffic, but voice supplies most of their revenue. Note that the imbalance between voice and data traffic is even more extreme in Western Europe than elsewhere; for example, data comprised about 95% of Western European traffic in 2013 compared with about 85% worldwide.

Voice data rates are low (64 kbps or less in the circuit switched domains of 2G and 3G) and now only make up a small percentage of the total network traffic. However, voice applications provide many valuable features to the user, notably supplementary services such as voicemail and call forwarding, communication with fixed phones on the public switched telephone network and the ability to place emergency calls. Because of this, operators can still charge a large premium for voice services. This premium is falling, but, despite this, voice still makes a disproportionate contribution to operator revenue. A similar but more extreme situation applies in the case of messaging services, which make a negligible contribution to network traffic. In contrast, mobile data services often require a high data rate, yet do not provide the extra value that could justify correspondingly high charges.

21.1.2 Third Party Voice over IP

The simplest technique is to offer a voice over IP (VoIP) service through a third-party supplier such as Skype, using the same principles as any other IP-based application. Figure 21.2 shows the basic architecture, although the details will differ from one supplier to another.

In this architecture, the user sets up a call by exchanging VoIP signalling messages with an external server, and ultimately with another VoIP device. From LTE's point of view, these signalling messages look just like any other kind of data, and are transported through EPS bearers in exactly the same way. The service provider may also support media gateways, which convert the VoIP packets to and from the information flows that are used by traditional circuit switched networks. If it does so, then the user will also be able to communicate with a 2G/3G mobile phone or with a land line.

This approach requires little investment by the network operator and delivers cheap or free calls to the user. It does, however, have several disadvantages. Firstly, the network operator no longer owns the voice service. Secondly, the user's quality of service is limited by what the

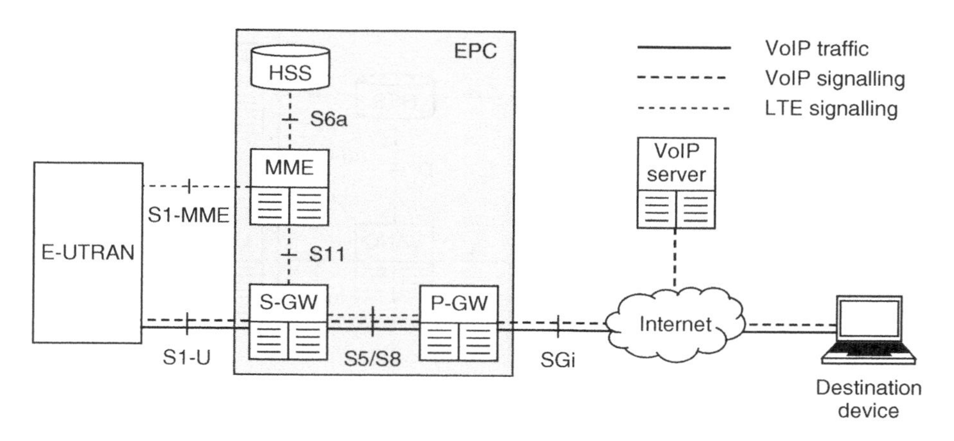

Figure 21.2 Architecture of a generic third party VoIP system

Internet can provide. Thirdly, the quality of service will fall if the mobile has to move to a 2G or 3G cell, or the call may be dropped altogether. Finally, Skype does not support emergency calls at the time of writing, so cannot yet be used to replace a traditional voice service. Because of these issues, the network operator can generate little revenue from this approach.

21.1.3 The IP Multimedia Subsystem

The main long-term approach for voice over LTE is the IP multimedia subsystem (IMS). As we saw in Chapter 1, the IMS is an external network containing the signalling functions that manage VoIP calls for an LTE device. When viewed in this way, the IMS is very like the third-party VoIP servers that we have just been discussing, but it brings two main advantages. Firstly, the IMS is owned by the network operator, not by a third-party service provider. Secondly, it is more powerful than any third-party system; for example, it guarantees the quality of service of a voice call, supports handovers to 2G or 3G cells and includes full support for emergency calls. Because of these advantages, it is likely to generate far more revenue for the network operator.

The IMS is, however, a complex system, whose rollout has been taking place a few years after the initial rollouts of LTE. This has led to a need for interim approaches to cover a period in which LTE networks are available but the IMS is not.

21.1.4 VoLGA

One interim approach that attracted attention in the early days of LTE was *Voice over LTE via generic access* (VoLGA). The technique is defined in three main specifications [2–4] with a useful technical introduction in Reference [5].

The technique is based on earlier 3GPP specifications [6] through which a mobile can reach the 2G/3G core network through a generic access network such as a wireless local area network. As shown in Figure 21.3, the VoLGA architecture exploits this approach by connecting

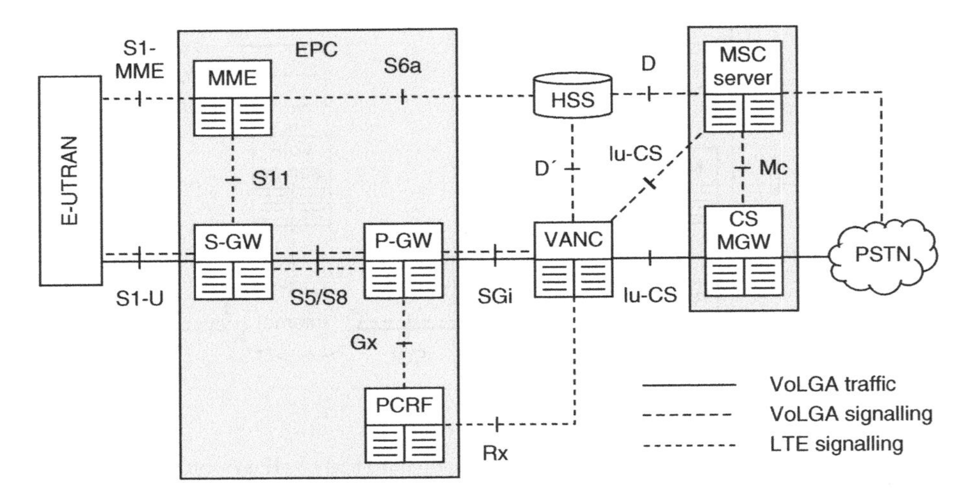

Figure 21.3 VoLGA architecture

the PDN gateway to the 2G/3G circuit switched domain through a device called a *VoLGA access network controller* (VANC). The VANC has three main functions: it authenticates the mobile over the D′ interface, relays 2G/3G signalling messages between the mobile and the circuit switched domain, and converts the user's voice traffic between VoIP and circuit switching.

The VoLGA architecture brings two other benefits. Firstly, the VANC can use the Rx interface to request a suitable quality of service for a voice call, and the evolved packet core can respond by activating a dedicated bearer. Secondly, the VANC can convert an LTE VoIP call to a 2G/3G circuit switched call if the user moves outside the coverage area of LTE. These packet- to circuit-switched handovers re-use an IMS technique known as single radio voice call continuity (SRVCC), which we will discuss in the next chapter.

The VoLGA specifications are written by an industrial collaboration known as the *VoLGA forum* [7], which was set up in March 2009. Although there was some early interest, most vendors and network operators decided to focus on circuit switched fallback and the IP multimedia subsystem, and the technique has never yet made its way into the 3GPP specifications. The VoLGA specifications have not been updated since 2010, and it seems unlikely that the technique will be widely adopted.

21.1.5 Dual Radio Devices

A second interim approach is to use mobiles with two completely separate transceivers, one for data communications over LTE and another for voice communications over a legacy 2G/3G network. This approach has two advantages: it is straightforward to design and it gives continuous voice and data connectivity to the user. However, it also has a couple of disadvantages: the devices are bulky and the battery life is greatly reduced.

This approach has been adopted by operators of legacy cdma2000 networks, such as Verizon in the United States, because the LTE and cdma2000 transceivers can run independently and because the other forms of inter-operation are rather complex. It is not suitable for the operators of legacy 3GPP networks because the 3GPP specifications do not yet allow a mobile to communicate using two 3GPP radio access technologies at the same time.

21.1.6 Circuit Switched Fallback

The main interim approach, used by most early rollouts of LTE, is *circuit switched fallback* (CSFB). Using this approach, the network transfers an LTE mobile to a legacy 2G/3G cell, so that it can place a voice call through the 2G/3G circuit switched domain in the traditional way. In the rest of this chapter, we will discuss the architecture of circuit switched fallback and the procedures for registration, mobility management, call setup and the delivery of SMS messages.

21.2 System Architecture

21.2.1 Architecture of the 2G/3G Circuit Switched Domain

In Chapter 15, we discussed the architecture that is used for inter-operation between LTE and the packet switched domain of 2G and 3G. Figure 21.4 builds upon that architecture, by introducing the 2G/3G circuit switched domain [8].

When making a voice call, the mobile digitizes its voice information with a sample rate and resolution that are usually 8 kHz and 8 bits respectively, giving a raw bit rate of 64 kbps. It then compresses the information to a lower data rate by means of a *codec*. The main codec in UMTS is the *adaptive multi rate* (AMR) codec [9], which supports eight compressed data rates that lie between 4.75 and 12.2 kbps.

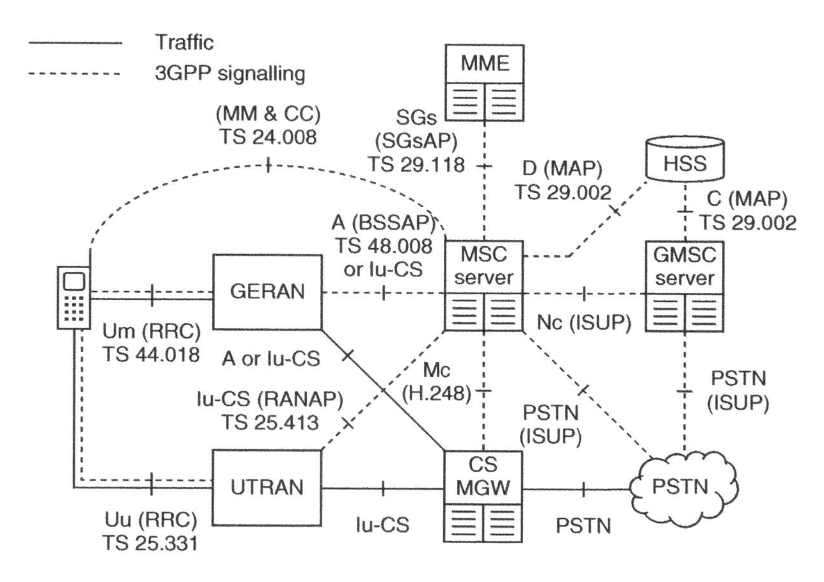

Figure 21.4 Circuit switched fallback architecture. Source: TS 23.272. Reproduced by permission of ETSI

The circuit switched media gateway (CS-MGW) carries a user's voice traffic, converts the traffic between different coding schemes and transport protocols, and acts as a user plane interface with other networks. It is similar to the serving and PDN gateways from LTE. The mobile switching centre (MSC) server looks after a set of mobiles, contains a local copy of their subscription data and controls them by means of signalling messages, so is very like an MME. A gateway MSC (GMSC) server supports all the MSC server's functions and can also receive signalling messages from another network in the case of incoming calls.

The architecture shown was first introduced in Release 4 of the 3GPP specifications. In earlier releases, the local copy of the subscription data was stored in another device, the visitor location register (VLR). The MSC server's other functions were amalgamated with those of the media gateway, so as to form the mobile switching centre (MSC).

The network uses several signalling protocols. An MSC server controls a mobile's high-level behaviour using two non-access stratum protocols [10]: the *mobility management* (MM) protocol handles internal bookkeeping within the circuit switched domain, while the *call control* (CC) protocol contains the application level signalling that manages phone calls. MSC servers communicate with the public switched telephone network and with each other using standard telephony signalling protocols. The most common are the *integrated services digital network* (ISDN) *user part* (ISUP) [11–14], which forms part of a protocol stack known as *signalling system 7* (SS7), and the *bearer independent call control* (BICC) protocol [15], an evolution of ISUP that supports any underlying transport protocol. Inside the circuit switched domain, MSC servers communicate with the home subscriber server using the mobile application part (MAP) [16] and control the media gateways using an ITU protocol known as H.248 or the *gateway control protocol* (GCP) [17]. This last protocol was originally developed in collaboration with the internet engineering task force and also goes by the IETF name of *media gateway control* (MEGACO).

The circuit switched domain keeps its own record of a mobile's state, independently of the packet switched domain. From the viewpoint of the 2G/3G radio access network, a mobile is only in RRC_IDLE if it is simultaneously idle in both the circuit switched and the packet switched domains of the core network. The circuit switched domain is organized into *location areas* (LAs), each of which comprises one or more packet switched routing areas. It identifies each user by means of a *temporary mobile subscriber identity* (TMSI).

Optionally, an MSC server can also communicate with an SGSN over the Gs interface, using a signalling protocol known as the *base station subsystem application part plus* (BSSAP+) [18]. If the Gs interface is present, then the network is said to be in *network mode of operation* (NMO) I, and the mobile carries out combined routing and location area updates by way of the SGSN. If the interface is absent, then the network is in NMO II, and the mobile carries out its routing and location area updates independently.

21.2.2 Circuit Switched Fallback Architecture

To support circuit switched fallback, the network architecture is enhanced by adding a new signalling interface, denoted SGs, between the MME and the MSC server. The messages on this interface are written using the *SGs application protocol* (SGsAP) [19]. Using these messages, an MME registers a mobile with the 2G/3G circuit switched domain and makes the network's circuit switched services available to the user.

We can then use this architecture for two distinct types of service. The mobile handles voice calls, video calls and supplementary services by moving to a 2G or 3G cell, under the control of the network. In contrast, the mobile can send and receive SMS messages within an LTE cell using a service known as *SMS over SGs*, in which the messages are relayed over the S1-MME and SGs interfaces to the MSC server. Network operators and device manufacturers can choose to support SMS over SGs alone, which can be useful for machine-type devices that have no need for voice services or for 2G or 3G

21.3 Attach Procedure

21.3.1 Combined EPS/IMSI Attach Procedure

A mobile registers for circuit switched fallback as part of the attach procedure from Chapter 11, using the message sequence shown in Figure 21.5. In its attach request (1), the mobile asks to register with an MSC server by setting the *EPS attach type* to a combined EPS/IMSI attach, and asks either for the full set of CS fallback services or for SMS alone using the *additional update type*. Amongst the other information elements, the mobile indicates which voice codecs it supports and signals two other parameters that we will discuss below, namely the voice domain preference and the UE usage setting.

The mobile sends the attach request to the base station in the usual way, and the base station selects an MME, forwards the attach request and indicates the tracking area in which it lies.

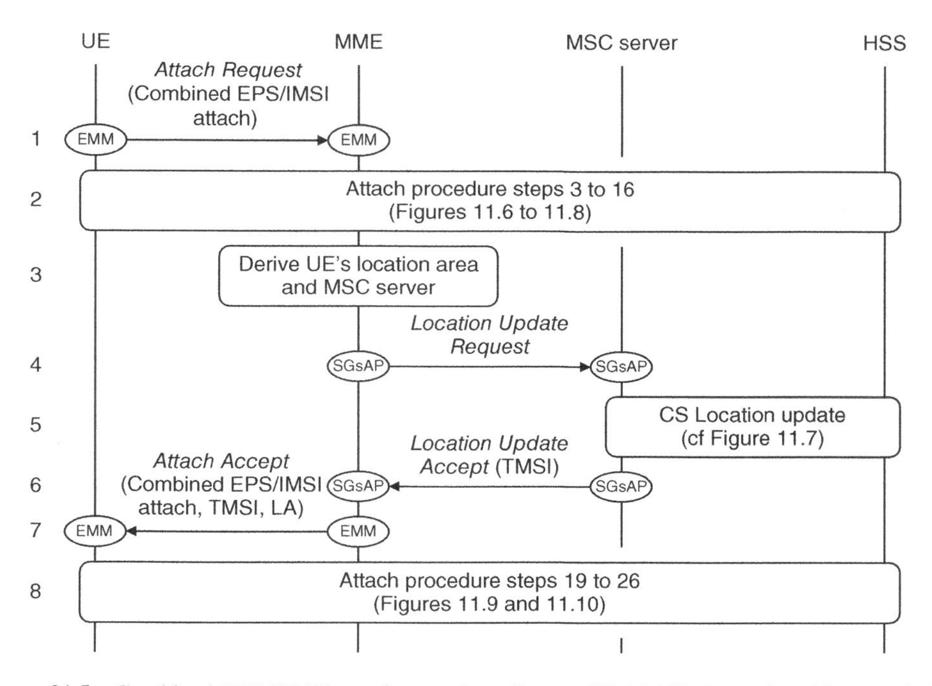

Figure 21.5 Combined EPS / IMSI attach procedure. Source: TS 23.272. Reproduced by permission of ETSI

In response, the MME runs steps 3 to 16 of the attach procedure (2), which cover the steps required for identification and security, location updating and default bearer creation.

Using the tracking area that it received from the base station, the MME identifies the 2G/3G location area that the mobile is in and selects an MSC server that controls the location area (3). It then sends the MSC server an SGsAP *Location Update Request* (4), in which it asks the MSC server to register the mobile for circuit switched services and specifies the mobile's location area and IMSI. The MSC server registers the mobile with the home subscriber server (5), acknowledges the MME's request (6) and provides the mobile with a temporary mobile subscriber identity. As a result of these SGs messages, the MME and MSC server have created a logical relationship known as an *SGs association*, through which they know that the mobile is registered for circuit switched fallback and know each other's identities.

The MME can now accept the mobile's attach request (7). If the procedure was successful, then the MME supplies the mobile with its location area identity and TMSI, uses the *EPS attach result* to indicate a combined EPS/IMSI attach and uses the *additional update result* to indicate registration for the full set of CSFB services or for SMS alone. The procedure concludes with the remaining steps from the normal LTE attach procedure (8).

21.3.2 Voice Domain Preference and UE Usage Setting

A mobile's registration is influenced using two internal parameters, namely the *voice domain preference* and the *UE usage setting* [20]. The voice domain preference indicates which voice techniques the device supports and, if applicable, which it prefers. There are four possible values: CS voice only, IMS PS voice only, IMS PS voice preferred/CS voice as secondary and CS voice preferred/IMS PS voice as secondary. The UE usage setting has two possible values: voice centric and data centric. It determines whether the mobile prefers a 2G or 3G voice service or the high data rates of LTE, in the event that the LTE network does not support voice at all. The two parameters are stored by the mobile equipment, and the network operator can also update the voice domain preference using a device management server [21].

To illustrate how these settings are used, Figure 21.6 shows how a mobile is expected to behave if its voice domain preference is set to CS voice only. The mobile first tries to register

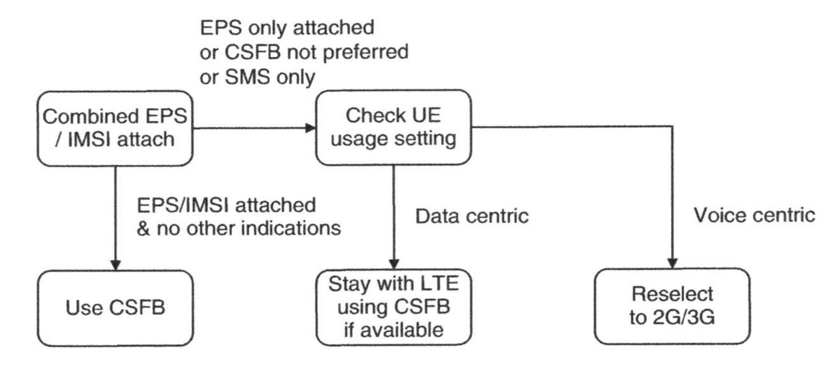

Figure 21.6 Behaviour of a mobile whose voice domain preference is CS voice only. Source: TS 23.221. Reproduced by permission of ETSI

for LTE and circuit switched fallback using the combined attach procedure. If the network supports circuit switched fallback, then it accepts the attach request and the procedure ends.

If the network does not support circuit switched fallback, then the MME accepts the mobile's attach request but sets the EPS attach result to *EPS only*. The mobile's behaviour then depends on the UE usage setting: a data centric device (such as a tablet) stays attached to LTE, while a voice centric device (such as a smartphone) detaches from LTE and tries to re-attach through 2G or 3G. The situation is the same if the MME signals a combined EPS/IMSI attach with an additional update result of *SMS only*, to indicate that it only supports SMS over SGs, not voice.

A final possibility is that the MME signals a combined EPS/IMSI attach but sets the additional update result to *CS fallback not preferred*. In that situation, a data centric device stays attached to LTE but continues to use circuit switched fallback, while a voice centric device detaches from LTE and tries to re-attach through 2G or 3G as before.

21.4 Mobility Management

21.4.1 Combined Tracking Area/Location Area Update Procedure

If a mobile moves into a tracking area in which it was not previously registered, then it runs the tracking area update procedure from Chapter 14. In its tracking area update request, the mobile sets the *EPS update type* to combined TA/LA updating, to indicate that it would like to carry out a location area update as well.

As in the attach procedure, the MME reads the tracking area that it received from the base station, identifies the corresponding location area, discovers whether it has to change the MSC server and sends the MSC server an SGsAP Location Update Request. As a result of the procedure, the mobile's location is updated in both the evolved packet core and the 2G/3G circuit switched domain.

21.4.2 Alignment of Tracking Areas and Location Areas

In the procedures for a combined attach and a combined tracking/location area update, the MME received a tracking area identity from the base station, looked up the corresponding location area and identified an MSC server that could control the mobile. For the process to work correctly, the MME has to contain a reliable mapping from tracking areas to location areas.

This leads to the following rule. To make circuit switched fallback work in the manner intended, network operators should avoid situations in which one tracking area spans multiple location areas. Instead, they should align their tracking and location area boundaries as closely as they can, so that one location area is divided into one or more tracking areas. The same advice applies to any mobile-specific tracking area lists, and also applies to the networks that are run by an operator's roaming partners.

Figure 21.7 shows the effect. If the tracking and location areas are perfectly aligned (Figure 21.7a), then a change of location area is always associated with a change of tracking area. As a result, the MME always knows the location area that the mobile is in and can keep the mobile registered with the correct MSC server.

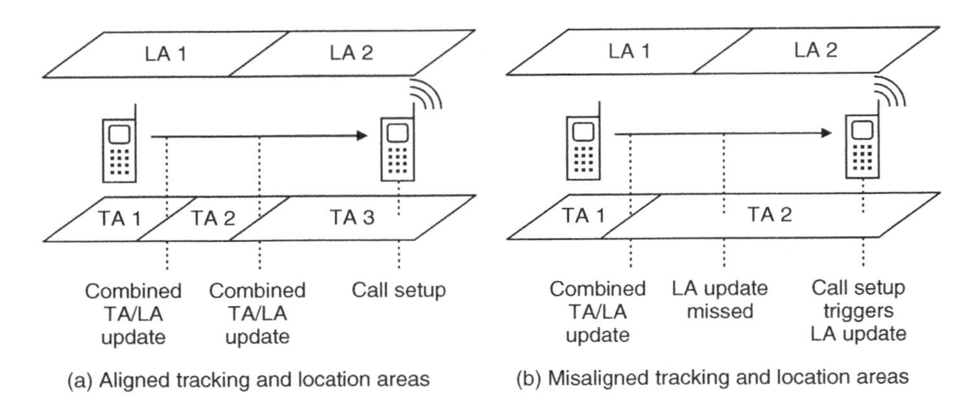

Figure 21.7 Alignment of tracking areas and location areas

If a tracking area spans more than one location area (Figure 21.7b), then a mobile can move from one location area to another without triggering the combined tracking/location area update procedure. If the mobile tries to start a call later on, it moves to a 2G/3G cell, reads the new cell's system information and only now discovers its new location area. This discovery triggers the location update procedure that should have taken place earlier, and delays the process of call setup. The delays are longer if the new location area is controlled by a different MSC server from before, because of the extra signalling that is needed to transfer control of the mobile. The same situation applies if the MME uses a mobile-specific tracking area list that spans more than one location area.

It is impossible to align the tracking and location areas perfectly, even if the 2G, 3G and LTE cells are controlled from the same sites, because the cells are operating on different carrier frequencies and have different coverage areas. Nevertheless, network operators can significantly improve the performance of circuit switched fallback by following the above rule as closely as they reasonably can.

21.4.3 Cell Reselection to UMTS or GSM

If an idle mobile moves outside the coverage area of LTE and into a region covered by 2G or 3G alone, then it carries out a reselection to a 2G or 3G cell. The mobile then reads the new cell's system information, which contains the network mode of operation and the new cell's routing and location areas. In the discussion that follows, we assume that the location area has remained unchanged.

In network mode of operation I, the mobile sends a Routing Area Update Request to an SGSN in the manner described in Chapter 15, but sets the update type to a combined routing/location area update. The SGSN retrieves the mobile's subscription data from the MME as before, and the MME responds by tearing down its end of the SGs association. At the same time, the SGSN sends a BSSAP+ *Location Update Request* to the MSC server, which responds by tearing down the other end of the SGs association.

The mobile has now left the scope of circuit switched fallback and is handling voice calls in the normal manner for UMTS or GSM. If the mobile moves back into the coverage area of

LTE later on, then it runs the combined tracking/location area update procedure from before, and the MME responds by re-establishing the SGs association.

In network mode of operation II, the mobile still sends a Routing Area Update request to the SGSN, but it sets the update type to a routing area update alone. As before, the SGSN retrieves the mobile's subscription data from the MME, which tears down its end of the SGs association. This time, however, there is no communication with the MSC server, which is unaware that the mobile has moved to a 2G/3G cell. We will resolve this issue during mobile-terminated call setup later on.

If the network wishes to activate idle mode signalling reduction (ISR) for a mobile that supports circuit switched fallback, then it can do so during the tracking area update procedure but not during the routing area update, which helps to ensure that the SGs association remains in place. Once ISR is active, the MSC server always contacts the mobile across the SGs interface to the MME: it never does so directly.

21.5 Call Setup

21.5.1 Mobile-Originated Call Setup using RRC Connection Release

To set up a call using circuit switched fallback, we can use either of the techniques that we introduced in Chapter 15 for transferring a mobile from LTE to UMTS or GSM. Figure 21.8 shows the more common technique, RRC connection release with redirection to another carrier. To keep things simple, the figure assumes that the mobile starts in RRC_CONNECTED, the call is mobile-originated and the target radio access technology is UMTS.

In step (1), the user dials a phone number and the mobile responds by sending an EMM *Extended Service Request* to the MME. In general terms, the mobile's message is an extension of an EMM Service Request; here, it indicates that the mobile would like to place a call using circuit switched fallback and includes the reason for the request, in this case a mobile-originated call. In response (2, 3), the MME sends the base station a *UE Context Modification Request*, which includes a CS fallback indicator that tells the base station to transfer the mobile to a 2G or 3G cell.

In this message sequence, the base station decides to transfer the mobile by releasing its RRC signalling connection (4), so it runs the S1 release procedure from Chapter 14 (5). In its RRC Connection Release message, the base station redirects the mobile to a 3G carrier and, from Release 9, can optionally include the system information of nearby 3G cells.

The mobile moves to the requested carrier frequency, selects a 3G cell (6) and runs the 3G procedure for RRC connection establishment (7). If the mobile did not receive the new cell's system information earlier then it has to read the system information as part of this process, which delays the call setup procedure. Furthermore, if the mobile finds itself in an unexpected location area because of misalignment of the location and tracking area boundaries, then it has to run the 3G location area update procedure (8). This causes additional delays, which are greater if the location update requires a change of MSC server.

The mobile now runs the usual 3G procedure for mobile-originated call setup (9), which we will discuss shortly. At the same time, the mobile can also run the procedures for a 3G routing area update and a 3G service request (10, 11). The second of these re-activates the data bearers so that the user can send and receive data during the call, albeit after a break in communications and with a lower data rate than before.

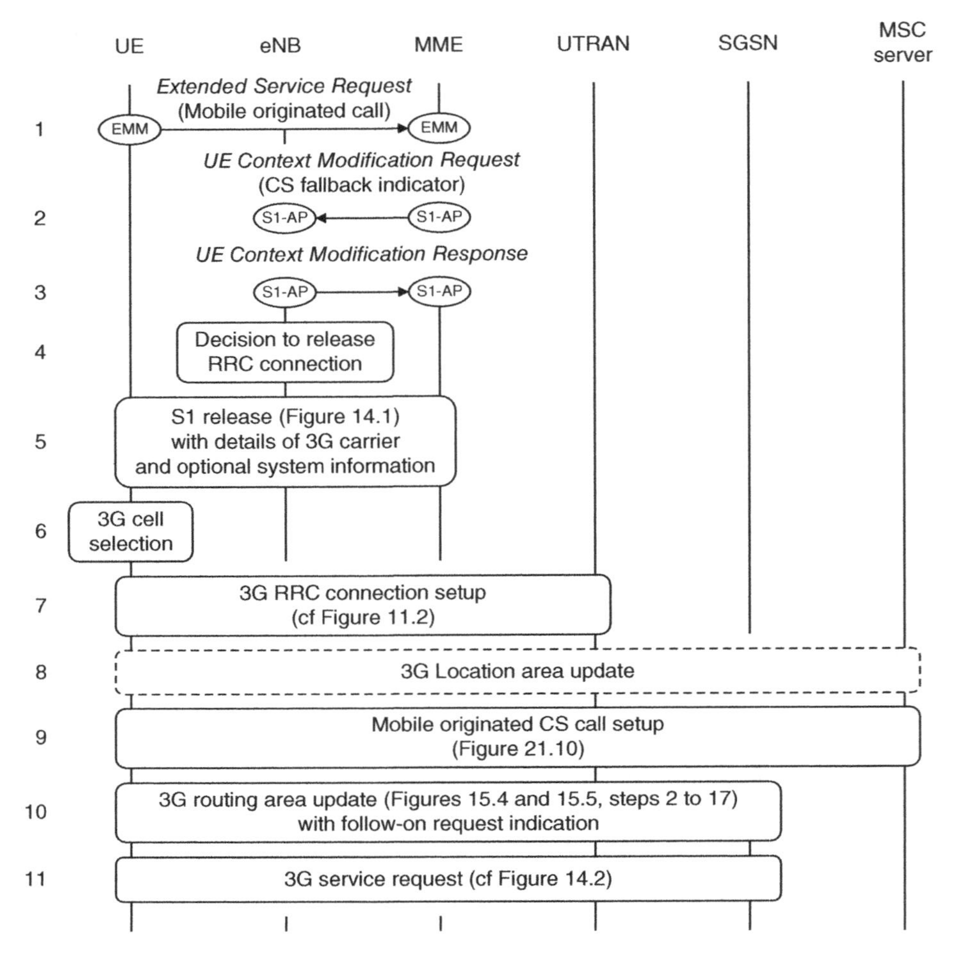

Figure 21.8 Mobile originated call setup from a mobile in RRC_CONNECTED, using CS fallback and RRC connection release with redirection to a 3G carrier. Source: TS 23.272. Reproduced by permission of ETSI

If the mobile starts in RRC_IDLE, then it begins by running the usual procedures for random access and RRC connection establishment. In place of steps 2 and 3, the MME sends the base station an S1-AP Initial Context Setup Request that includes a CS fallback indicator, and the base station moves the mobile into ECM_CONNECTED using the usual steps from the service request procedure. The call setup procedure can then continue from step 4, with the omission of the concluding 3G service request.

If the target radio access technology is GSM, then the mobile can only transfer data during the call if both mobile and network support GSM *dual transfer mode* (DTM). If this is not the case, then the MME contacts the serving and PDN gateways after the S1 release procedure (step 5), so as to place the mobile's EPS bearers into a *suspended* state in which the PDN gateway discards the mobile's incoming data packets [22]. On arrival in the target cell, the mobile sends a suspend request to the SGSN so as to prevent it from paging the mobile [23], and the 3G service request is once again omitted.

21.5.2 Mobile Originated Call Setup using Handover

Figure 21.9 shows what happens if the network implements circuit switched fallback using a packet switched handover. Once again, we assume that the mobile starts in RRC_CONNECTED, the call is mobile-originated and the target radio access technology is UMTS.

As before, the mobile sends an Extended Service Request to the MME (1), which tells the base station to transfer the mobile to 2G or 3G (2, 3). This time, however, the base station decides to transfer the mobile using a packet switched handover (4). The base station does not yet know which target cell is the best, so it instructs the mobile to measure the nearby 3G cells and return a measurement report (5). It can then start the inter-system handover procedure from Chapter 15 using the steps for handover preparation and execution (6).

The mobile arrives in the target cell, carries out a location area update if required (7) and runs the call setup procedure that we discuss below (8). At the same time, the network can run the steps for handover completion (9) by redirecting the downlink data path and tearing down the mobile's resources in the eNB.

This procedure has the same advantages and disadvantages as the original handover procedure from Chapter 15: there is no interruption to the data stream, but there is a risk that

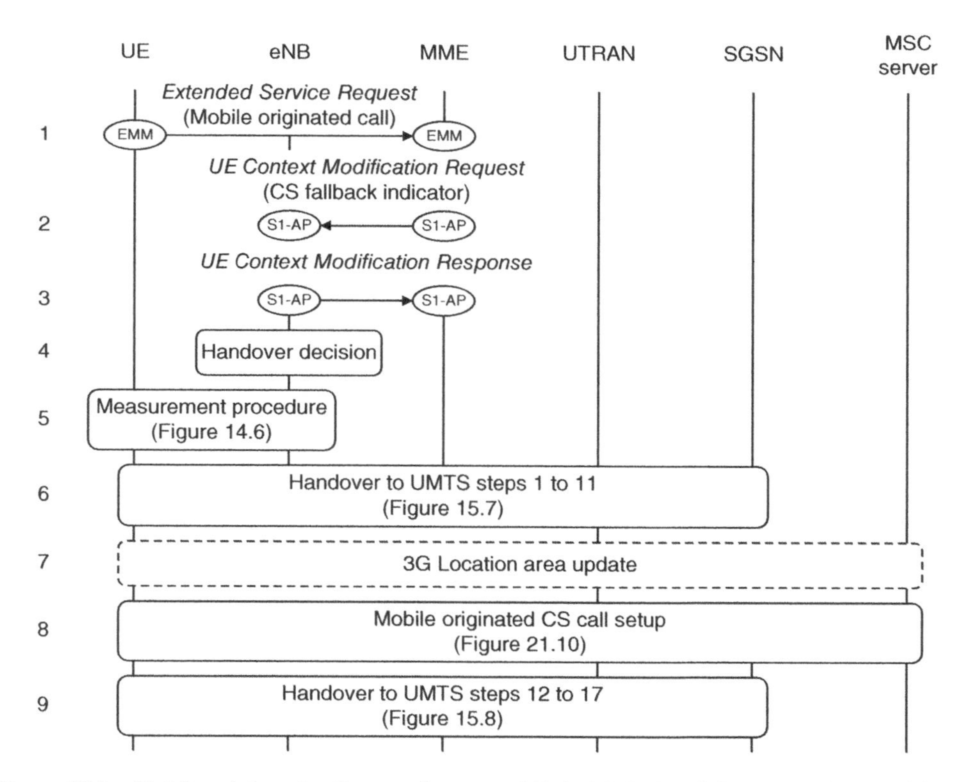

Figure 21.9 Mobile originated call setup, from a mobile in RRC_CONNECTED, using CS fallback and handover to a 3G cell. Source: TS 23.272. Reproduced by permission of ETSI

the handover will fail. The risk of failure is, however, particularly serious in the case of circuit switched fallback because a handover failure leads to a dropped call, not just a break in the data transfer. Because of this problem, the procedure is not often used, at least in early implementations.

The procedure can also be used for a mobile that starts in RRC_IDLE, but there is little incentive to do so as the mobile does not have any active data streams. If the target radio access technology is GSM and either the mobile or the network do not support dual transfer mode, then the mobile's EPS bearers are suspended using similar steps to the ones we saw earlier.

21.5.3 Signalling Messages in the Circuit Switched Domain

In the procedures shown earlier, we skipped over the call setup messages in the 2G/3G circuit switched domain. Figure 21.10 shows the messages that are appropriate for a mobile-originated call [24].

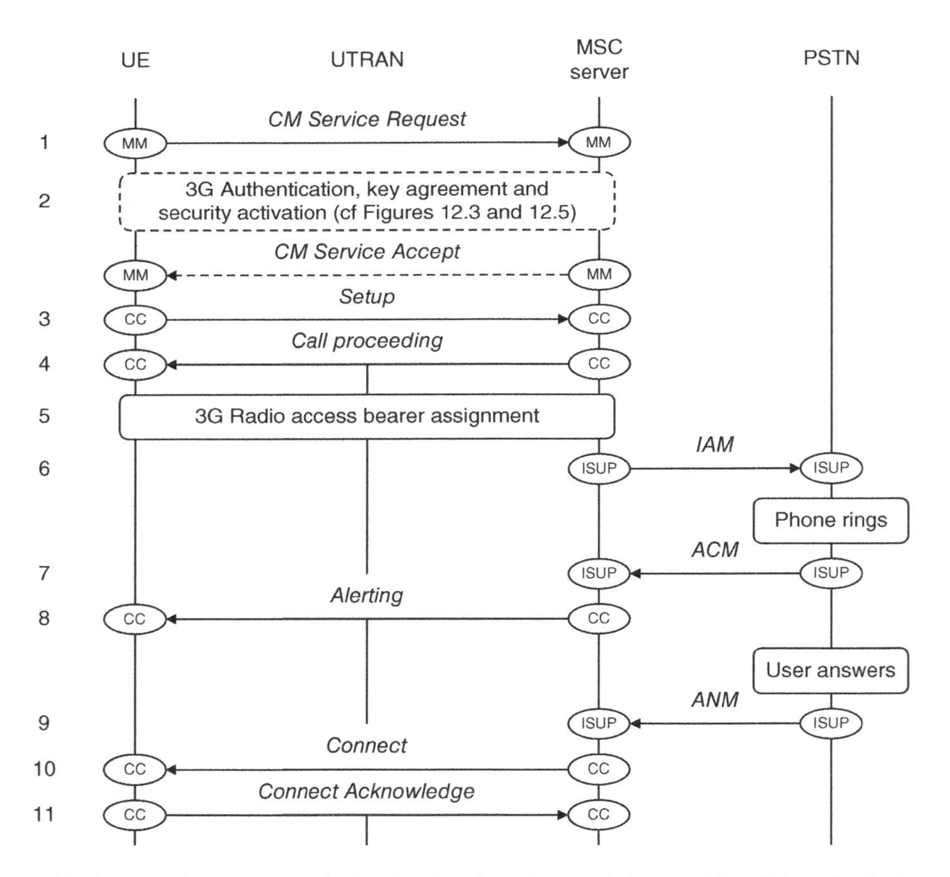

Figure 21.10 Signalling messages in the circuit switched network for a mobile originated call. Source: TS 23.018. Reproduced by permission of ETSI

After arriving in the target cell and completing any location area update, the mobile sends a *CM* (connection management) *Service Request* to the MSC server to request a move to the circuit switched connected state (1). The MSC server either accepts the mobile's request right away using a *CM Service Accept*, or authenticates the mobile and activates the access stratum security procedures (2). In the latter case, the mobile interprets completion of the 2G/3G security mode command procedure as an implicit acceptance of the service request.

The mobile now sends a *Setup* message to the MSC server (3), which is written using the 2G/3G call control protocol. As part of the message, the mobile states its relevant capabilities such as the codecs that it supports and includes the dialled number. The MSC server acknowledges the message (4), sets up radio access bearers to carry the call (5) and sends an ISUP *Initial Address Message* (IAM) to the destination device (6).

The phone rings, which triggers an *Address Complete Message* (ACM) to the MSC server (7) and an *Alerting* message to the originating mobile (8). Eventually the user answers, which triggers an *Answer Message* (ANM) to the MSC server (9) and a *Connect* message to the mobile (10). After the mobile's acknowledgement (11), the call can proceed.

21.5.4 Mobile-Terminated Call Setup

Figure 21.11 shows the signalling messages for a mobile-terminated call, in the case where the target mobile is in RRC_CONNECTED. To start the procedure, the originating network creates an ISUP Initial Address Message that contains the dialled number and routes it to a gateway MSC server in the mobile's home network (1). The GMSC server looks up the MSC server at which the mobile is registered (2) and sends the message there (3). In turn, the MSC

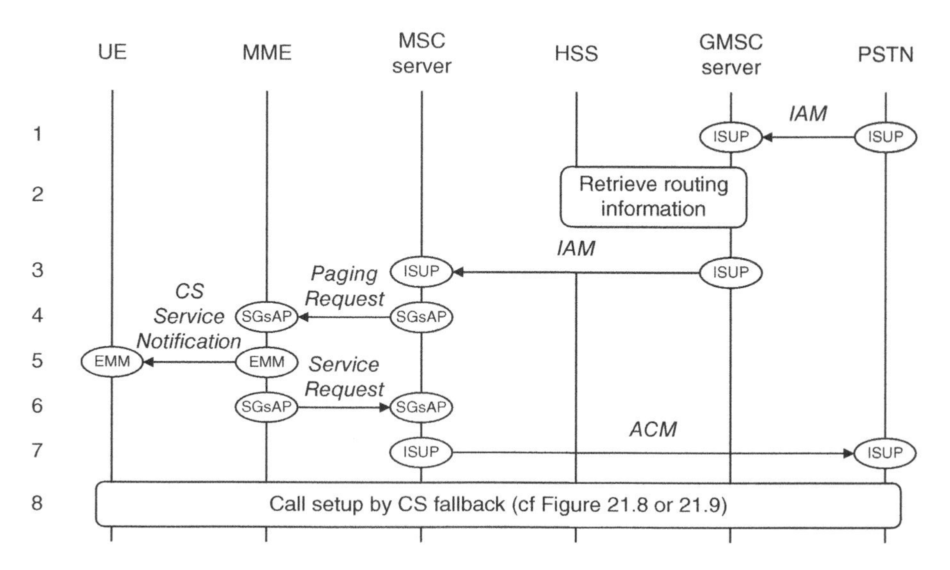

Figure 21.11 Mobile terminated call setup to a mobile in RRC_CONNECTED, using CS fallback. Source: TS 23.272. Reproduced by permission of ETSI

server looks up the mobile's SGs association and contacts the appropriate MME by means of an SGsAP *Paging Request* (4). As part of this message, the MSC server indicates that the incoming service is a voice call and states the originating device's *calling line identification* (CLI).

If the target mobile is in RRC_CONNECTED, then the MME contacts it directly by means of a *CS Service Notification* that includes the calling line information (5). As soon as it has delivered the message, the MME sends an acknowledgement back to the MSC server (6), which informs the calling party that the user is being alerted (7). The user either accepts or rejects the call, and the mobile replies to the MME using an Extended Service Request that indicates the user's response. If the user has accepted the call then the message sequence can continue using either of the procedures that we discussed earlier (8). Some of the circuit switched signalling messages are in the opposite direction from before, but their overall effect is the same.

If the target mobile is in RRC_IDLE, then the MME pages the mobile in place of message 5 and delays its acknowledgement to the MSC server until the mobile replies. The MME cannot deliver the calling line information as part of the paging procedure; instead, this information reaches the mobile later as part of circuit switched signalling. Otherwise, the procedure is unchanged.

If a mobile-terminated call arrives while idle mode signalling reduction is active, then the MSC server sends its SGsAP Paging Request to the MME in the usual way. The MME contacts the SGSN across the S3 interface, both devices page the mobile, and the mobile responds through an LTE cell to the MME or through a 2G/3G cell to the MSC server. There is no direct communication between the MSC server and the SGSN.

If a mobile-terminated call arrives after a reselection to a 2G/3G cell in network mode of operation II, then the MSC server may be unaware that the mobile has moved. In this scenario, the MSC server sends its SGsAP Paging Request to the MME in the usual way, but the MME replies with an SGsAP *Paging Reject*. The MSC server reacts by paging the mobile across the 2G and 3G radio access networks, after which the mobile responds. The message sequence then continues with the usual procedures for 2G or 3G.

21.5.5 Returning to LTE

At the end of a call, the 3G circuit switched domain releases the mobile's signalling connection using similar messages to those in the S1 release procedure from Chapter 14. If there is no further action, then the mobile stays in a 3G cell, in the state of RRC_CONNECTED if it has an active packet data session and RRC_IDLE otherwise, until an eventual reselection or handover back to LTE. A similar process takes place in the case of 2G.

There are, however, ways for the mobile to move back to LTE as part of call termination, which are collectively known as *fast return to LTE*. When releasing the mobile's signalling connection, the circuit switched domain can tell a Release 10 radio access network that the connection was originally set up as a result of circuit switched fallback. The radio access network can use this knowledge to return the mobile to LTE right away, using RRC connection release with redirection or a packet switched handover. The mobile can also ask the radio access network to release any RRC signalling connection so as to allow a mobile-initiated cell reselection. However, this last possibility depends on the mobile's implementation.

21.6 SMS over SGs

21.6.1 System Architecture

We can also use the circuit switched fallback architecture to deliver SMS messages, as shown in Figure 21.12 [25]. In this architecture, the *SMS service centre* (SMS-SC) is a standard SMS device that receives a mobile-originated SMS, stores it and forwards it as a mobile-terminated SMS to the destination. It communicates with an MSC server through two devices that respectively handle the mobile-originated and mobile-terminated cases, namely the *SMS interworking MSC* (SMS-IWMSC) and the *SMS gateway MSC* (SMS-GMSC). Thankfully, the mobile does not have to move to a 2G or 3G cell before sending or receiving a message; instead, the network transports SMS messages by embedding them into lower-level signalling messages on the Uu, S1-MME and SGs interfaces. The technique is known as *SMS over SGs*.

The Release 11 specifications introduce a variant known as *SMS in MME*, in which the MSC server's SMS functions are embedded into the MME itself [26]. The SMS devices communicate with the MME and HSS using a Diameter application [27], either directly or through an *interworking function* (IWF) that converts their legacy MAP messages into Diameter commands. The architecture is useful for handling the machine-type communication devices that we discussed in Chapter 20, because a network operator may wish to keep these devices away from its legacy MSC servers so as to prevent the MSC servers from being overloaded.

21.6.2 SMS Delivery

To illustrate the delivery of an SMS message, Figure 21.13 summarizes the procedure for a mobile-terminated SMS. The steps are almost identical to those in the normal SMS procedure except for the use of a different delivery path. To start the procedure, the service centre sends a message to the SMS gateway MSC (1), which looks up the MSC server at which the mobile is registered (2) and forwards the message there (3). In turn, the MSC server contacts the MME and indicates the nature of the incoming service (4). If the mobile is in RRC_IDLE, then the MME sends it a paging message and the mobile responds in the usual way (5). The MME can

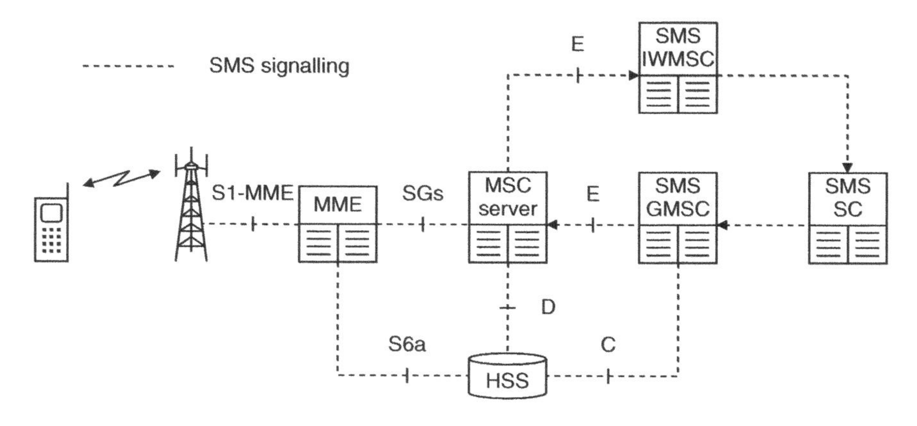

Figure 21.12 Architecture for SMS over SGs

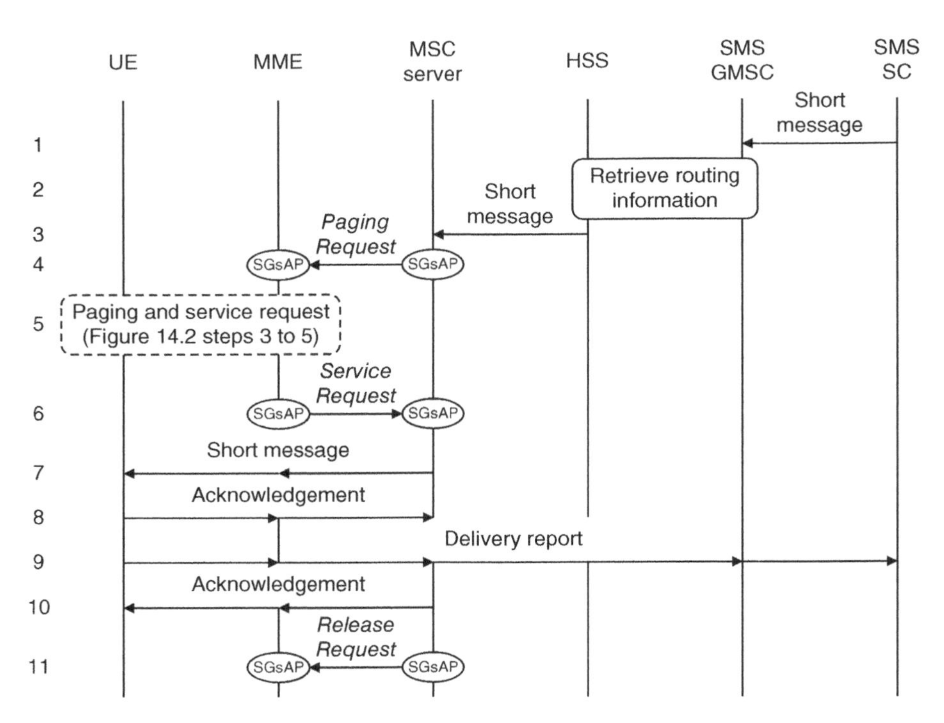

Figure 21.13 Message sequence for a mobile terminated SMS, using SMS over SGs. Source: TS 23.272 and TS 23.040. Reproduced by permission of ETSI

then send an acknowledgement to the MSC server (6), and the two devices deliver the message by embedding it into an SGsAP *Downlink UnitData* message and an EMM *Downlink NAS Transport* (7).

On receiving the message, the mobile sends an SMS acknowledgement back to the MSC server, which is delivered by embedding it into an EMM *Uplink NAS Transport* message and an SGsAP *Uplink Unitdata* (8). The mobile also sends an SMS delivery report to the service centre (9), which triggers a further SMS acknowledgement from the MSC server (10). If there are no more messages to deliver, then the MSC server indicates the end of the procedure by sending a final confirmation to the MME (11).

21.7 Circuit Switched Fallback to cdma2000 1xRTT

The 3GPP specifications also support circuit switched fallback to cdma2000 1xRTT networks [28]. Figure 21.14 shows the architecture.

The technique uses similar principles to the ones we saw in Chapter 15 for inter-operation with packet switched cdma2000 HRPD networks. After attaching to LTE, the mobile preregisters with a cdma2000 1xRTT MSC, using 1xRTT signalling messages that are transported across the S102 interface [29]. To initiate a voice call, the mobile sends an extended service request to the MME in the manner described earlier, and the MME tells the base station to transfer the mobile for circuit switched fallback.

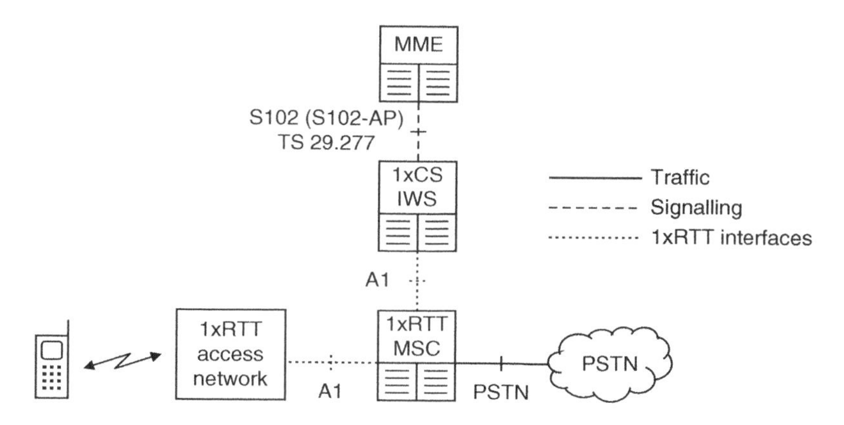

Figure 21.14 Architecture for circuit switched fallback to cdma2000 1xRTT. Source: TS 23.272. Reproduced by permission of ETSI

The Release 8 specifications only support one transfer mechanism: the base station releases the mobile's RRC connection and redirects it to a cdma2000 1xRTT carrier. The mobile's packet switched bearers are suspended for the duration of the call by re-using the GSM suspend procedure, on the assumption that the mobile cannot simultaneously communicate using 1xRTT and HRPD. From Release 9, the base station can also hand the mobile over to a specific 1xRTT cell in a procedure known as *enhanced CS fallback to 1xRTT*. If the mobile does support simultaneous voice and data communications using 1xRTT and HRPD, then the network can hand over the mobile's data bearers as well.

If an incoming call arrives for the mobile, then the 1xRTT MSC sends the mobile a paging message across the S102 interface. This triggers an extended service request from the mobile, and the call setup procedure continues as before. The mobile can also send and receive SMS messages, by tunnelling them to and from the 1xRTT MSC over S102.

The Release 9 specifications also introduce support for mobiles that have one transmitter but two receivers. Such mobiles can simultaneously camp on an LTE cell and a cdma2000 1xRTT cell, but can only be active using one radio access technology at a time. As such, their behaviour is intermediate between the single radio devices used for conventional circuit switched fallback and the dual radio devices that we noted earlier.

21.8 Performance of Circuit Switched Fallback

Using circuit switched fallback, network operators can provide a voice service to their LTE users without the need to deploy the IP multimedia subsystem. The technique does, however, have several disadvantages. Firstly, the mobile can only use the technique if it is simultaneously in the coverage areas of LTE and 2G or 3G. Secondly, the technique forces network operators to align the boundaries of their location areas, tracking areas and tracking area lists. Thirdly, there can be noticeable delays in the call setup procedure, particularly if the mobile has to read the new cell's system information or carry out a location update. Finally, the user's data rates are lower than usual throughout the call, while the RRC connection release procedure interrupts the data streams while the call is being set up.

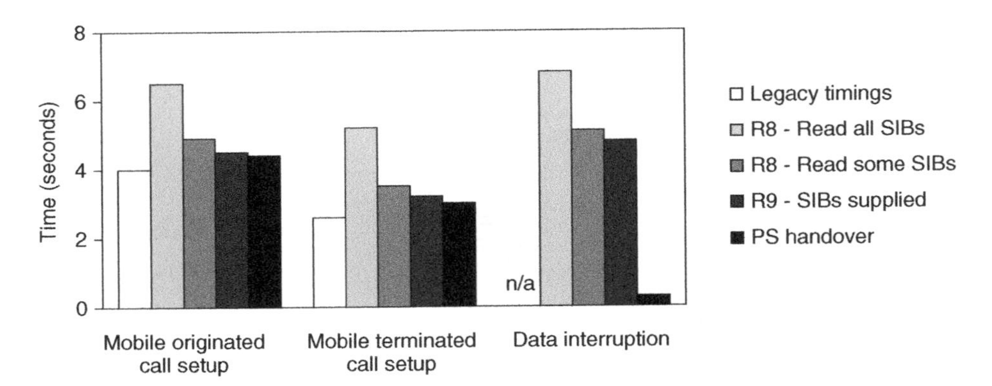

Figure 21.15 Measurements of the call setup delay and data interruption time when using circuit switched fallback

Nevertheless, the performance of circuit switched fallback is better than that might be feared. To illustrate this, Figure 21.15 shows measurements by Qualcomm of the call setup delay and data interruption time in the case of circuit switched fallback to a 3G carrier [30]. According to these results, the additional call setup delay is about 2.5 s in a basic Release 8 implementation of RRC connection release with redirection. This figure falls to 0.5 to 1 s if the base station tells the mobile about the system information of nearby cells in advance or if it hands the mobile over to a specific 3G cell. The delay can also be reduced if the mobile reads only the most important system information at the outset and ignores other information such as the target cell's neighbour list, but that capability is specific to a particular mobile's implementation. There is an additional delay of 1 to 2 s if the mobile has to do a location update after arriving in the target cell or of 4 to 5 s if that process leads to a change of the MSC server. The data interruption time is several seconds for any implementation of RRC connection release with redirection, and the only way to reduce that is by using handovers.

References

1. 3GPP TS 23.272 (2012) Circuit Switched (CS) Fallback in Evolved Packet System (EPS); Stage 2, Release 11, September 2012.
2. VoLGA Forum (2010) Voice over LTE via Generic Access; Requirements Specification; Phase 2, Version 2.0.0, April 2010.
3. VoLGA Forum (2010) Voice over LTE via Generic Access; Stage 2 Specification; Phase 2, Version 2.0.0, June 2010.
4. VoLGA Forum (2010) Voice over LTE via Generic Access; Stage 3 Specification; Phase 2, Version 2.0.0, June 2010.
5. Sauter M. (2009) Voice Over LTE via Generic Access (VoLGA) – A Whitepaper, http://cm-networks.de/volga-a-whitepaper.pdf (accessed 15 October 2013).
6. 3GPP TS 43.318 (2012) Generic Access Network (GAN); Stage 2, Release 11, September 2012.
7. VoLGA Forum (2010) VoLGA Forum – Start, http://www.volga-forum.com (accessed 15 October 2013).
8. 3GPP TS 23.002 (2013) Network Architecture, Release 11, Sections 4.1.2, 6.4.1, June 2013.
9. 3GPP TS 26.071 (2012) AMR Speech CODEC; General Description, Release 11, September 2012.
10. 3GPP TS 24.008 (2013) Mobile Radio Interface Layer 3 Specification; Core Network Protocols; Stage 3, Release 11, September 2013.

11. ITU-T Recommendation Q.761 (1999) Signalling System No. 7 – ISDN User Part Functional Description.
12. ITU-T Recommendation Q.762 (1999) Signalling System No. 7 – ISDN User Part General Functions of Messages and Signals.
13. ITU-T Recommendation Q.763 (1999) Signalling System No. 7 – ISDN User Part Formats and Codes.
14. ITU-T Recommendation Q.764 (1999) Signalling System No. 7 – ISDN User Part Signalling Procedures.
15. ITU-T Recommendation Q.1901 (2000) Bearer Independent Call Control Protocol.
16. 3GPP TS 29.002 (2013) Mobile Application Part (MAP) Specification, Release 11, September 2013.
17. ITU-T Recommendation H.248.1 (2013) Gateway Control Protocol: Version 3.
18. 3GPP TS 29.018 (2013) Serving GPRS Support Node (SGSN) – Visitors Location Register (VLR); Gs Interface Layer 3 Specification, Release 11, March 2013.
19. 3GPP TS 29.118 (2013) Mobility Management Entity (MME) – Visitor Location Register (VLR) SGs Interface Specification, Release 11, September 2013.
20. 3GPP TS 23.221 (2013) Architectural Requirements, Release 11, Section 7.2a, Annex A, June 2013.
21. 3GPP TS 24.167 (2012) 3GPP IMS Management Object (MO), Release 11, Section 5.27, December 2012.
22. 3GPP TS 29.274 (2013) Evolved General Packet Radio Service (GPRS) Tunnelling Protocol for Control Plane (GTPv2-C); Stage 3, Release 11, Section 7.4, September 2013.
23. 3GPP TS 23.060 (2013) General Packet Radio Service (GPRS); Service Description; Stage 2, Release 11, Section 16.2.1, September 2013.
24. 3GPP TS 23.018 (2013) Basic Call Handling; Technical Realization, Release 11, Section 5.1, March 2013.
25. 3GPP TS 23.272 (2012) Circuit Switched (CS) Fallback in Evolved Packet System (EPS); Stage 2, Release 11, Section 8, September 2012.
26. 3GPP TS 23.272 (2012) Circuit Switched (CS) Fallback in Evolved Packet System (EPS); Stage 2, Release 11, Annex C, September 2012.
27. 3GPP TS 29.338 (2013) Diameter Based Protocols to Support Short Message Service (SMS) Capable Mobile Management Entities (MMEs), Release 11, September 2013.
28. 3GPP TS 23.272 (2012) Circuit Switched (CS) Fallback in Evolved Packet System (EPS); Stage 2, Release 11, Annex B, September 2012.
29. 3GPP TS 29.277 (2012) Optimised Handover Procedures and Protocol Between EUTRAN Access and Non-3GPP Accesses (S102); Stage 3, Release 11, December 2012.
30. Qualcomm (2012) Circuit-Switched Fallback. The First Phase of Voice Evolution for Mobile LTE Devices, http://www.qualcomm.com/media/documents/files/circuit-switched-fallback-the-first-phase-of-voice-evolution-for-mobile-lte-devices.pdf (accessed 15 October 2013).

22

VoLTE and the IP Multimedia Subsystem

In this chapter, we discuss the likely long-term solution for LTE voice calls, namely the delivery of voice over IP streams that are controlled by the IP multimedia subsystem (IMS). The IMS is not part of LTE: instead it is a separate network whose relationship with LTE is the same as that of the Internet. Despite this, it is valuable to cover the IMS as part of this book, because LTE voice calls are important and because the IMS illustrates several aspects of LTE's operation. The delivery of voice calls over LTE and the IMS is often known as Voice over LTE (VoLTE).

In the course of the chapter, we review the history of the IMS and VoLTE, and discuss the architecture and protocols of the IMS and the procedures for registration and call setup. We then cover three aspects of interoperation between the IMS and the 2G/3G circuit switched domain, namely access domain selection, single radio voice call continuity and IMS centralized services, and conclude by discussing IMS emergency calls and SMS. There are some excellent accounts of the IP multimedia subsystem for those who need further detail, notably References [1–3].

22.1 Introduction

22.1.1 The IP Multimedia Subsystem

The IP multimedia subsystem (IMS) was originally designed for the management and delivery of real-time multimedia services over the 3G packet switched domain. It was first defined in 3GPP Release 5, which was frozen in 2002. The specifications attracted a great deal of interest, and were later enhanced to support other access technologies such as wireless local area networks (Release 6) and fixed networks based on *digital subscriber line* (DSL) or cable technology (Release 7).

Although there were some early implementations in fixed networks, mobile operators initially concluded that they had no viable business case for the IMS. The 3G circuit switched domain could handle voice and video calls, so the IMS would be limited to peripheral services such as *push to talk over cellular* (PoC), instant messaging and presence. Network operators

An Introduction to LTE: LTE, LTE-Advanced, SAE, VoLTE and 4G Mobile Communications, Second Edition.
Christopher Cox.

could not justify the expense of rolling out the IMS for these services alone, and very few chose to do so. Subsequently, however, LTE was designed without a circuit switched core network and with the intention that LTE voice calls should be transported using voice over IP. This is an ideal application for the IP multimedia subsystem and has led to a resurgence of interest in the technology.

The IP multimedia subsystem is specified by 3GPP in the same way as LTE, UMTS and GSM. TS 23.218 [4] is a useful introduction, while TS 23.228 [5] is the main stage 2 specification. There is also an overview of the signalling procedures in TS 24.228 [6], but that document has not been updated since Release 5, so the contents are out of date and should be read with caution. We will see several other specifications in the course of the chapter.

22.1.2 VoLTE

The IMS specifications are complex and have a large number of implementation options. After the introduction of LTE, equipment manufacturers and network operators became concerned that this complexity would delay the IMS further, and would make it hard to introduce a fully interoperable voice service. They also feared that the IMS might be overtaken by other approaches, such as VoLGA and third-party voice over IP.

These concerns led to the formation of an industrial initiative, *One Voice*, in 2009. The aim of One Voice was to define a *profile* for voice over LTE using the IMS, in other words a minimum set of functionality that manufacturers would be invited to follow in the interests of interoperability. The initiative proved popular, and in 2010 it was adopted by the industry's main trade association, the *GSM Association*, under the name *Voice over LTE* (VoLTE) [7]. The term VoLTE therefore hides some complexity: it refers to the delivery of VoIP calls over LTE and the IMS, in a way that complies with the GSM Association's VoLTE specifications.

VoLTE is defined in three main documents. The most important is the profile for voice over LTE using the IMS [8], while the others contain guidelines for roaming and interoperation across the two networks [9, 10]. Altogether, these documents prescribe several issues that the 3GPP specifications leave as implementation options. As an example, the 3GPP specifications allow a device to support IP version 4, version 6 or both, while the VoLTE specifications only allow the last of these.

To keep the chapter concise, we will follow the VoLTE specifications almost entirely, and we will say little about the IMS's other services, access technologies and implementation options. For reference, Table 22.1 lists the main restrictions that the VoLTE specifications impose, and compares them with the options that are available in the underlying 3GPP specifications.

22.1.3 Rich Communication Services

The VoLTE specifications are a promising approach to voice over LTE, but they have little to say about other services or access technologies. These issues are addressed by the GSM Association's specifications for *Rich Communication Services* (RCS), also known as the *Rich Communication Suite* [11–13].

RCS defines a set of services that network operators can deliver using the IP multimedia subsystem, namely voice and video calls, instant messaging, one-to-one and group chat, file transfer, presence and geolocation. The services can be delivered across access technologies that include LTE, HSPA and circuit switching, and are built on a framework that lets RCS devices discover each other's existence and capabilities. Devices and networks can conform

Table 22.1 Key differences between the GSMA specifications for VoLTE and the 3GPP specifications for LTE and the IMS

Feature	VoLTE requirements	3GPP requirements
Supported UE IP versions	Both IPv4 and IPv6	Both IPv4 and IPv6 IPv4 alone IPv6 alone
Supported UE voice domain preferences	IMS only IMS preferred/CS secondary	IMS only IMS preferred/CS secondary CS preferred/IMS secondary CS only
Support of MMTel voice service	Mandatory	Optional
Support of AMR voice codec	Mandatory	Mandatory if MMTel voice is supported
Support of MMTel supplementary services	Some mandatory services	All services optional
Support of emergency calls	Mandatory	Optional
Support of SMS	Mandatory	Optional
Support of SRVCC	Mandatory if CS is supported	Optional
IMS access point name	IMS well-known APN	Any APN
Location of PGW and PCSCF	Visited network	Home or visited network
PCSCF discovery technique	During default bearer activation	Four techniques supported
Transport of SIP signalling	Default bearer (QCI 5)	Any bearer and QCI
Transport of VoIP traffic	One dedicated bearer (QCI 1)	Any bearer and QCI
VoIP SDF establishment	Triggered by PCSCF	Triggered by PCSCF or UE

with the RCS specifications by supporting the framework and at least one service. Users access RCS through an application known as *Joyn*, which offers them the same set of services irrespective of their network operator, access technology or location in a home or visited network.

Rich Communication Services enhanced (RCS-e) is a variant of RCS, which a group of network operators introduced in 2011 to reduce the cost of RCS networks and devices. RCS-e offers only some of the services of RCS, and uses simpler mechanisms for discovering the existence and capabilities of other devices. The RCS-e specifications were folded into those of RCS from RCS version 5.0, which was published in 2012.

RCS has several similarities with VoLTE: it defines the provision of services by the IMS, and refers to VoLTE specifications for the case of voice over LTE. However there are also some differences, the most important being that RCS does not require a device to support any particular service or access technology. Because of its size and complexity, we will not consider RCS further.

22.2 Hardware Architecture of the IMS

22.2.1 High-Level Architecture

Figure 22.1 shows the most important components of the IP multimedia subsystem [14, 15]. The IMS is mainly concerned with signalling, and we will discuss its individual signalling components in the following sections. The IMS can also manipulate the user's traffic by means

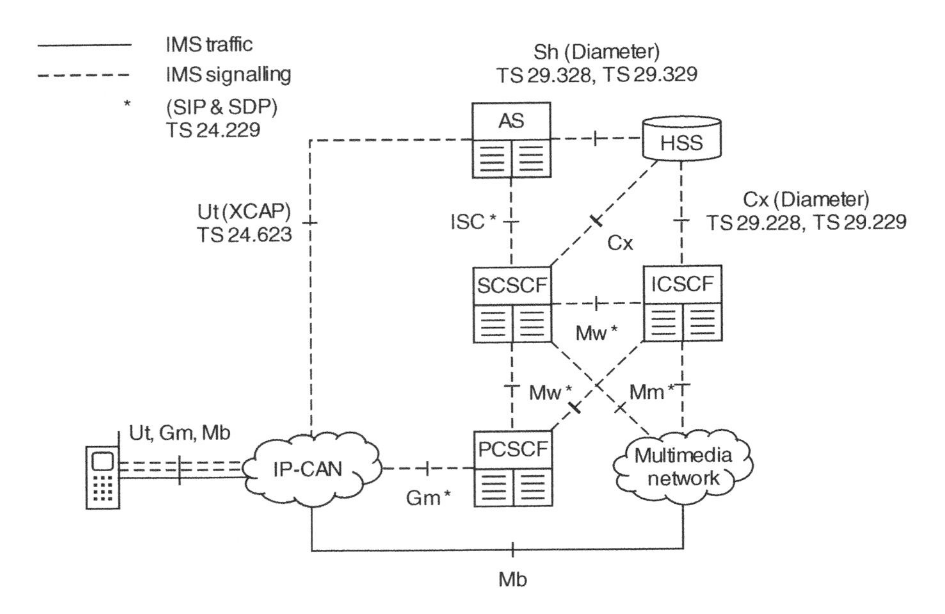

Figure 22.1 Main architectural elements of the IP multimedia subsystem. Source: TS 23.228. Reproduced by permission of ETSI

of other components that we will introduce later on, but this is not always required, so the traffic can actually bypass the IMS altogether.

The user reaches the IMS across an *IP connectivity access network* (IP-CAN) such as LTE, 3G or wireless LAN. Most of the IMS specifications are independent of the access network, but there are a few access-specific features as well [16, 17]. In turn, the IMS can communicate with other IP multimedia networks, for example with IMSs that are owned by other network operators. The most important IMS signalling protocol is the *session initiation protocol* (SIP), but we will introduce other several other protocols later in this chapter.

22.2.2 Call Session Control Functions

The most important components of the IMS are known as *call session control functions* (CSCFs). These are of three types. Each user is registered with a *serving CSCF* (S-CSCF), which controls the mobile and gives it access to services such as voice calls. The serving CSCF is similar to the MME, but with one important distinction: the serving CSCF is always in the mobile's home network, which helps to ensure that the user receives a consistent set of IMS services, even while roaming.

The *proxy CSCF* (PCSCF) is the mobile's first point of contact with the IMS, with the two devices communicating across a signalling interface known as Gm. The proxy CSCF secures signalling messages across the IP connectivity access network by means of encryption and integrity protection, and relays those messages between the mobile and the serving CSCF. It also controls the quality of service of media flows across the access network, for example by acting as an LTE application function (AF) towards the evolved packet core.

The *interrogating CSCF* (I-CSCF) is a point of contact for signalling messages that arrive from other IP multimedia networks. On receiving such a message, the interrogating CSCF asks the home subscriber server (HSS) for the serving CSCF that is controlling the target mobile, and forwards the signalling message to the serving CSCF.

22.2.3 Application Servers

An *application server* (AS) supplies the user with services such as multimedia telephony, voicemail and SMS. The services are invoked by the user's serving CSCF, but users can also manipulate their application server data, such as voicemail preferences or supplementary service configurations, across a separate signalling interface known as Ut. An application server should not be confused with the LTE application function (AF), a device outside the evolved packet core whose role in the IMS is filled by the proxy CSCF.

Most application servers are stand-alone devices, but there are two special types that act as interfaces to other application environments. The *open service access* (OSA) *service capability server* (SCS) gives access to OSA application servers across the OSA *application programming interface* (API), while the *IP multimedia service switching function* (IM-SSF) gives access to the service environment for *customized applications for mobile network enhanced logic* (CAMEL).

22.2.4 Home Subscriber Server

The home subscriber server (HSS) is a central database that contains the user's IMS subscription data. The IMS generally uses the same database as the 3GPP access network to support interoperation between the IMS and the 2G/3G circuit switched domain. However the two databases could potentially be different, which would allow different operators to manage a particular user's access network and IMS.

In the IMS, the home subscriber server's functions can be distributed across more than one physical device. When using that architecture, the IMS contains an extra device, the *subscription locator function* (SLF), which returns the name of the HSS where a particular user's details are stored.

22.2.5 User Equipment

The user equipment contains application level software that communicates with the IMS across the signalling interfaces noted earlier. The VoLTE specifications insist that the UE should support both IP version 4 and IP version 6.

The universal integrated circuit card (UICC) generally contains an application known as the *IP multimedia services identity module* (ISIM) [18, 19], which interacts with the IMS in the same way that the USIM interacts with LTE. The ISIM contains the home network operator's IMS domain name and one *IP multimedia private identity* (IMPI), an identity similar to the IMSI that the network operator uses for internal bookkeeping.

The ISIM also contains one or more instances of the *IP multimedia public identity* (IMPU), an identity similar to an email address or phone number that identifies the user to the outside world. An IMPU is usually a SIP *uniform resource identifier* (URI), which identifies the user, network operator and application using a format such as `sip:username@domain` [20]. An IMPU can also support traditional phone numbers in two ways: a SIP URI that includes a phone number using a format such as `sip:+358-555-1234567@domain;user=phone`; and a Tel URI that describes a stand-alone phone number using the format `tel:+358-555-1234567` [21].

Adding some more detail, a user's public identities can be grouped together into *implicit registration sets*, each of which contains a set of public identities that are simultaneously activated when the user registers with the IMS [22]. The ISIM only has to store one public identity from each implicit registration set: the others can be stored by the home subscriber server. Furthermore, each subscription can be associated with more than one ISIM, each of which contains a different private identity. Finally, a user can also reach the IMS using a USIM alone, using procedures that derive an IP multimedia private identity from the user's IMSI.

22.2.6 Relationship with LTE

Figure 22.2 shows the relationship between LTE and the IP multimedia subsystem, for a roaming scenario that is compliant with the VoLTE specifications. If the mobile is roaming, then it reaches the IMS through a PDN gateway in the visited network [23]. This allows the user to make a local phone call without the media travelling all the way back to the home network, in a technique known as *optimal routing*. The proxy CSCF also lies in the visited network, which gives that network visibility of the IMS signalling. To ensure that a mobile can access the IMS while roaming, the VoLTE specifications insist that every network operator should refer to the IMS using the *IMS well-known access point name*, which is *IMS*.

If the visited network operator has not yet implemented the IMS, then the mobile can reach a proxy CSCF in the home IMS using a PDN gateway in the home network's evolved packet core. This architecture does not comply with the VoLTE specifications and cannot be used for voice calls, but it can be used for other IMS services such as SMS.

The evolved packet core does not understand the SIP signalling messages that travel between the mobile and the IMS, so it transports them in the LTE user plane through a default EPS bearer with quality of service class indicator (QCI) 5. The bearer is set up before the mobile registers with the IMS and is torn down after it deregisters. The evolved packet core also transports the mobile's voice traffic using a dedicated EPS bearer with QCI 1, which is set up at the start of a call and torn down at the end. The VoLTE specifications assume that the mobile only supports one such bearer, so the network bundles multiple voice streams into the same bearer and gives them the same allocation and retention priorities. Finally, the EPC handles any non-real-time streams, such as picture files, using dedicated EPS bearers with QCI 8 or 9.

Users manipulate their application server data through a general access point name that is controlled by the home network operator. The Ut interface therefore travels through a PDN gateway in the home network, typically using the same bearer as the one used for home-routed access to the Internet. The mobile now has two IP addresses: one for the Internet and the other for the IMS.

Different networks exchange voice traffic using the IP packet exchange (IPX) [24], an enhanced version of the GPRS roaming exchange (GRX) that can also guarantee the quality of service of a media stream. By transporting the voice media using LTE dedicated bearers and the IPX, the IMS can guarantee the end-to-end quality of service that the user will receive.

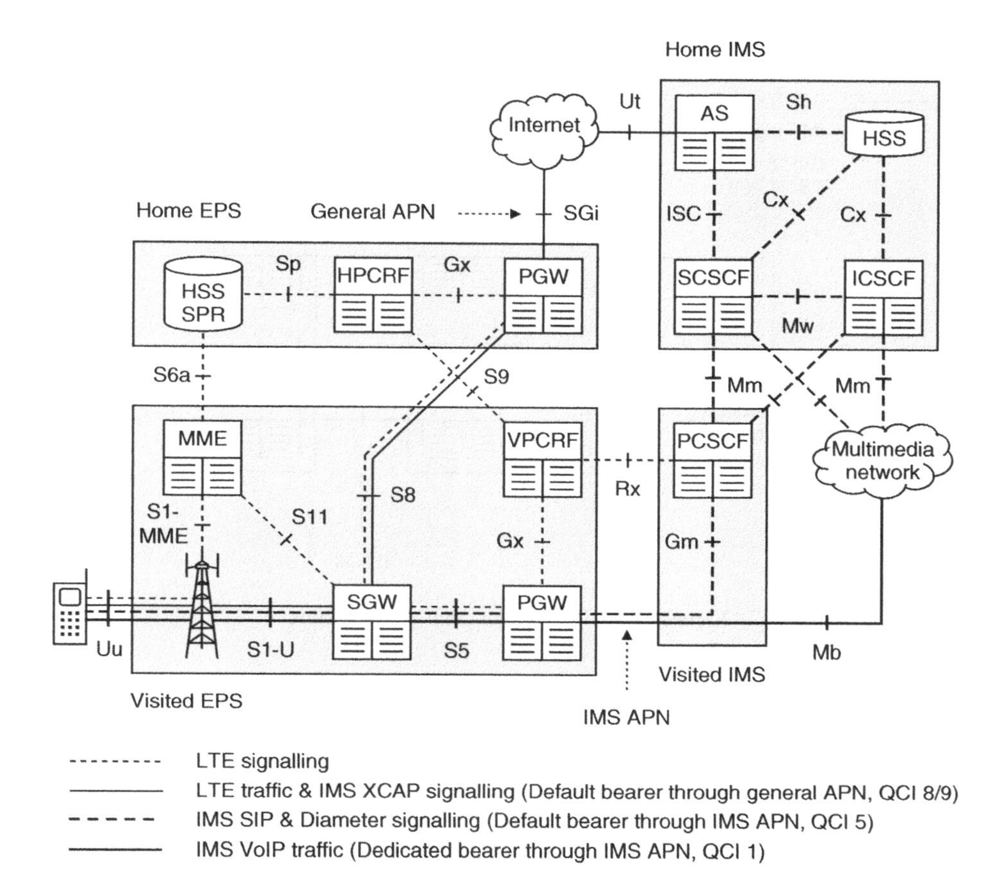

Figure 22.2 System architecture for a roaming VoLTE mobile

22.2.7 *Border Control Functions*

The IMS contains several components that we have not yet introduced, the first being the border control functions shown in Figure 22.3 [25, 26]. The *interconnection border control function* (IBCF) is the point of contact for SIP signalling communications with other networks, so, in the case of incoming signalling, it lies between the interrogating CSCF and the outside world. The *transition gateway* (TrGW) is controlled by the IBCF and is the point of contact for IMS media. If the other network is another IP multimedia subsystem, then the communications take place across the *inter-IMS network-to-network interface* (II-NNI).

One role of these devices is to help with media routing. By including an IBCF in the signalling path and a TrGW in the traffic path, the IP multimedia subsystem can force a user's voice traffic to travel through the visited IMS, the home IMS or both. The first choice is particularly useful for roaming users, because it allows the visited IMS to see the user's traffic but ensures that the traffic does not have to travel back to the home IMS. Other roles include screening SIP messages, transcoding, and interworking between networks that use IP versions 4 and 6.

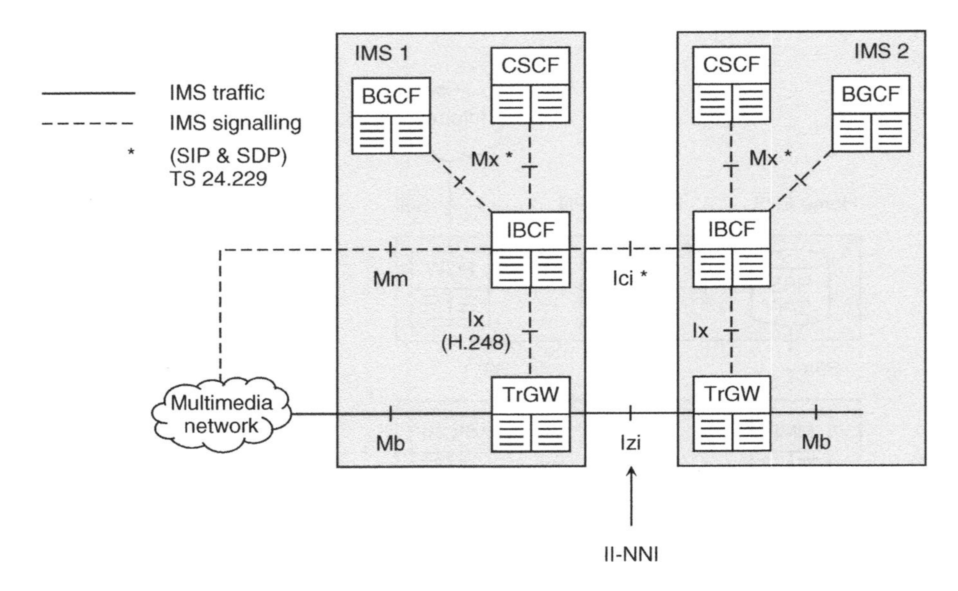

Figure 22.3 IMS border control architecture

22.2.8 Media Gateway Functions

In addition to its communication with other IP networks, the IP multimedia subsystem can also communicate with the public switched telephone network (PSTN) and with the circuit switched domains of 2G/3G network operators. Figure 22.4 shows the devices that are used.

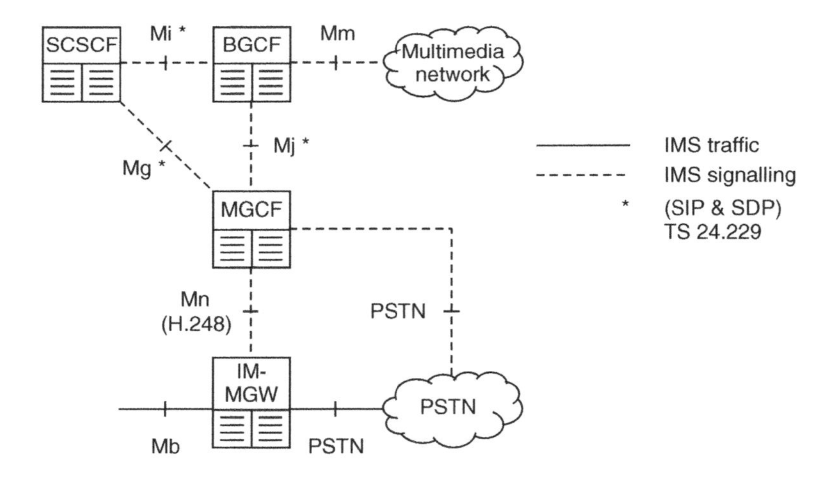

Figure 22.4 IMS media gateway architecture

The *IMS media gateway* (IM-MGW) is a user plane interface between the IMS and an external circuit switched network that handles tasks such as transcoding. It has the same functions as the 2G/3G circuit switched media gateway (CS-MGW), but is a different logical device and is part of a different network. The IM-MGW is controlled by a *media gateway control function* (MGCF), which also translates signalling messages between the ones used by the IMS and the ones used for circuit switching.

The *breakout gateway control function* (BGCF) determines the next hop for routing an outgoing signalling message that is destined for a circuit switched network. It can do this either by choosing a suitable MGCF in the same network or by delegating the choice to another BGCF. The last choice is useful if the user is roaming, because it allows a BGCF in the home network to request an MGCF in the visited network, so that the traffic can break out to the public telephone network there.

22.2.9 Multimedia Resource Functions

The last two sets of devices have dealt exclusively with one-to-one communications. However, the IMS can also act as a source and mixing point for IP multimedia streams, using the devices shown in Figure 22.5.

The *multimedia resource function processor* (MRFP) manages the user plane of a conference call by mixing media streams, playing back tones and announcements, and transcoding. It is controlled by the *multimedia resource function controller* (MRFC) and, together, the two devices comprise the *multimedia resource function* (MRF). The *media resource broker* (MRB) chooses the multimedia resource function that will handle a particular media stream, using the requirements of the application and the capabilities of each device.

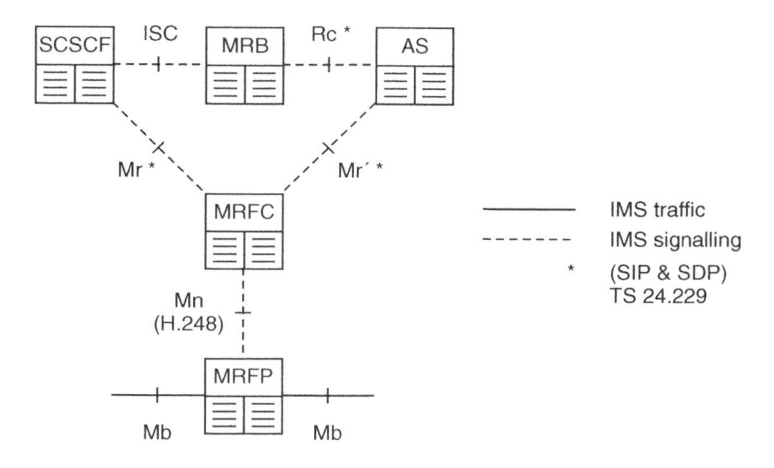

Figure 22.5 IMS multimedia resource function architecture

22.2.10 Security Architecture

Figure 22.6 shows the VoLTE security architecture [27]. The IMS uses security mechanisms similar to those of LTE, but the two architectures are completely independent. This allows the mobile to reach the IMS over an insecure access network such as wireless LAN, and to rely on the security mechanisms of the IMS alone.

In the case of network access security, the IMS re-uses the authentication and key agreement procedure from UMTS. The procedure is similar to that of the equivalent LTE procedure from Chapter 12, the main difference being that the keys CK and IK are used directly for ciphering and integrity protection instead of being the starting point for a hierarchy of additional keys. The procedure is based on a user-specific key, K, which is stored in the home subscriber server and securely distributed to the user within the ISIM, and which is different from the LTE key in the USIM.

During the authentication and key agreement procedure, the serving CSCF passes the values of CK and IK to the proxy CSCF. The proxy CSCF then sets up a security association with the mobile, so as to apply optional ciphering and mandatory integrity protection to the SIP signalling messages that the two devices exchange. The procedures are implemented using the Internet Protocol Security (IPSec) Encapsulating Security Payload (ESP) in transport mode.

Network domain security in the IMS is identical to network domain security in the evolved packet core. Two security domains first authenticate each other and establish a security association using Internet Key Exchange version 2, and then use ESP in tunnel mode to secure the information that they exchange. The IMS does not by itself secure the user plane traffic, but two devices can negotiate end-to-end application layer security during call setup.

22.2.11 Charging Architecture

The IP multimedia subsystem uses the same architecture for charging and billing as LTE [28, 29]. Any network element in the IMS can send charging data records to the offline charging system, while the serving CSCF, application server and multimedia resource function controller can also communicate with the online charging system. The charging system

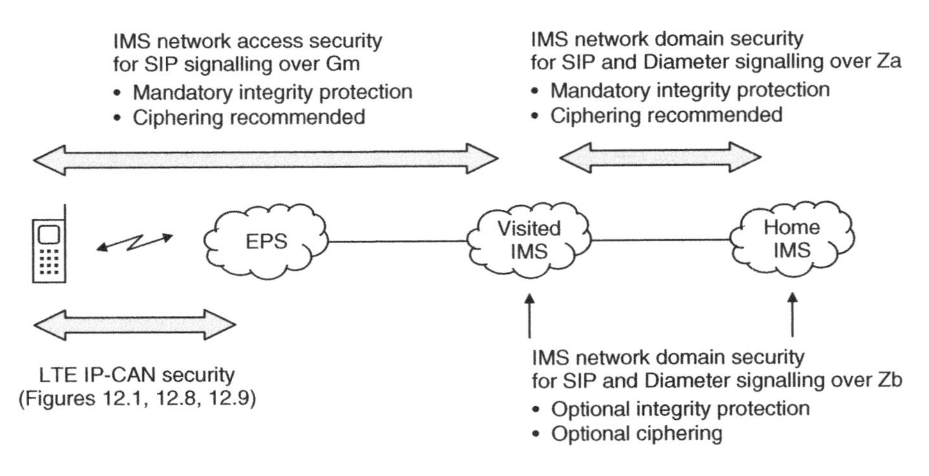

Figure 22.6 IMS security architecture. Source: TS 33.203. Reproduced by permission of ETSI

therefore receives all the information that it needs to bill the user, even if the home IMS has no direct visibility of the user's traffic.

22.3 Signalling Protocols

22.3.1 Session Initiation Protocol

The most important IMS signalling protocol is the session initiation protocol (SIP). SIP is used on most of the IMS's signalling interfaces, notably those between the mobile, the CSCFs and the application servers. It is very different from the protocols used by LTE, so needs a few words of explanation.

SIP was developed by the Internet Engineering Task Force (IETF) for the control of real-time packet switched multimedia, and is based on the hypertext transfer protocol (HTTP). RFC 3261 [30] defines the basic SIP protocol, but there are extensions in several other IETF specifications, notably RFC 3455 [31], which defines SIP extensions for use in the IP multimedia subsystem. The most important 3GPP specification is TS 24.229 [32], which defines the usage of SIP within the IMS.

Unlike the protocols that we have previously discussed, SIP is text based rather than binary, which makes the signalling messages long but easy to read. Like HTTP, SIP is a client–server protocol: a client sends a request to a server, which replies with a response. Unlike HTTP, however, an individual device can function both as a client and as a server. By default, SIP messages are transported using UDP rather than TCP. There is no guarantee that the messages are delivered reliably, so SIP includes its own mechanisms for acknowledgements and retransmissions.

A client's request is a simple text expression. Example requests are *REGISTER*, which establishes signalling communications between the mobile and the IMS, and *INVITE*, which sets up a call. The server's response contains a three-digit numerical code and a short text description. The first digit of the response code can have six values: 1 indicates a provisional response; 2 indicates a successful response; 3 asks the client to take further action and 4 to 6 indicate various types of error. Requests and responses carry additional information using *header* fields, which have a similar role to the information elements that we have previously seen, while individual options within a header are known as *tags*. A SIP message can also carry embedded content, such as media descriptions that are written using the session description protocol (SDP).

Introducing some more terminology, a *transaction* comprises a SIP request, some optional provisional responses and a final, non-provisional response. A *dialog* is a sequence of related transactions between two parties that is often associated with a *session*, an exchange of media such as a VoIP call. As its name implies, the main aim of SIP is to set up and manage these media sessions.

There are also names for the different functional components. A *user agent* (UA) is one of the endpoints in a SIP dialog, a role most often taken by the mobile. A *registrar server* stores the match between a user's public identity and the IP address where that user can be found. A *proxy server* takes part in an existing dialog by forwarding SIP requests and responses, by adding new headers and by modifying some of the existing headers, while a *back-to-back user agent* (B2BUA) can also initiate a new dialog. The IMS network elements might be implemented using a mix of proxy server and B2BUA functionality, while the serving CSCF also takes on the role of the registrar server.

22.3.2 Session Description Protocol

By itself, SIP does not say anything about the media that a session will use. That task is left to the *session description protocol* (SDP) [33]. The original version of SDP just defined a media stream, using session information such as the device's IP address and media information such as the media types, data rates and codecs. However, the protocol was later enhanced by means of an *offer-answer model* that allows two or more parties to negotiate the media and codecs that they would like to use [34]. The IMS uses SDP in exactly this way, with the SDP offers and answers being transported using the SIP requests and responses that we have already seen.

22.3.3 Other Signalling Protocols

The IMS also uses three other signalling protocols. The first is the *extensible markup language* (XML) *configuration access protocol* (XCAP) [35, 36], which is used on the Ut interface between a mobile and an application server. XCAP allows a client to manipulate data that are stored using XML format on a server, by mapping the XML data to HTTP uniform resource identifiers (URIs) that can be accessed using HTTP. Devices can then use XCAP to read and modify information stored on application servers, such as their voicemail settings and supplementary service configurations.

The home subscriber server and subscription locator function communicate with CSCFs and application servers using Diameter applications [37–40], in the same way as in LTE. Finally, the transition gateway, media gateway and media resource function processor are controlled using the H.248 protocol [41], in the same way as the circuit switched media gateway.

22.4 Service Provision in the IMS

22.4.1 Service Profiles

Each user is associated with an IMS *user profile* [42] that is stored by the home subscriber server. As shown in Figure 22.7, the user profile contains a set of *service profiles*, which define the behaviour of individual services such as multimedia telephony and SMS. Each service profile is associated with one or more public identities, and is activated if any of those public identities become active during the registration procedure. It is also associated with various sets of *initial filter criteria* (iFC), which define how the serving CSCF interacts with application servers so as to invoke the service.

Looking more deeply into the hierarchy, each set of initial filter criteria contains a priority number, the details of the corresponding application server and a set of *service point triggers* (SPTs). Finally, each service point trigger defines the conditions under which the serving CSCF will contact the application server, using information such as the SIP URI of the originating device, the SIP request, the presence or contents of any SIP headers or embedded SDP, and the session case (originating or terminating).

When a serving CSCF receives the first SIP request in a new dialog, it examines the user's initial filter criteria in priority order and compares the request with each of the service point triggers. If there is a match, then the serving CSCF forwards the request to the application server. The server can then act accordingly, for example by processing an INVITE request in accordance with a user's supplementary service configuration. The server can also ensure that

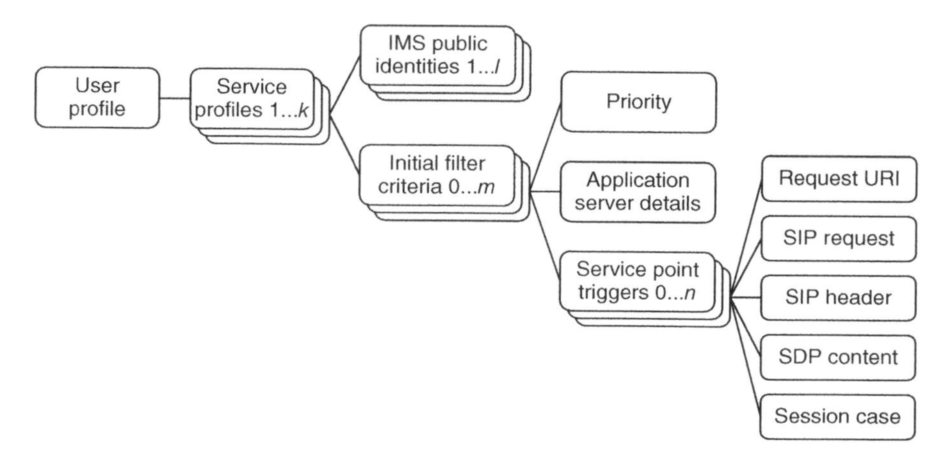

Figure 22.7 Principal contents of the IMS user profile

it receives all subsequent requests and responses in the dialog, by adding SIP header fields that place itself in the message routing path. The whole process is very like the action of CAMEL in the 2G/3G circuit switched domain.

22.4.2 *Media Feature Tags*

A SIP user agent can declare its capabilities using *media feature tags* [43, 44], which the IMS uses in two ways. During registration, a user agent can include a media feature tag as part of its REGISTER request. This declares the device's capabilities to the serving CSCF and can trigger registration with an application server through its use as a service point trigger. Later on, a user agent can include a media feature tag in a SIP INVITE request. This declares the device's capabilities as before, indicates a wish to communicate with another device that shares those capabilities and can trigger further interactions with the application server. We will see some examples shortly.

The 3GPP specifications develop the concept by defining two types of media feature tags that are specific to the IMS [45, 46]. The *IMS communication service identifier* (ICSI) indicates support for an IMS service, and triggers interactions with an IMS application server in the usual way. The service could be used by several applications, so the *IMS application reference identifier* (IARI) indicates support for a particular application, and triggers interactions with the corresponding software inside the server.

22.4.3 *The Multimedia Telephony Service for IMS*

The most important IMS service is the *multimedia telephony service* (MMTel) *for IMS* (MTSI) [47, 48]. This service encompasses traditional voice communications, and defines supplementary services such as call forwarding and call barring that replicate the services in a traditional circuit switched network. It also supports other capabilities such as UDP-based video and fax transmissions, as well as text and file sharing during a voice or video call.

All these capabilities are optional in the basic 3GPP service, but the VoLTE specifications mandate support for MMTel voice calls and for some of the MMTel supplementary services.

An MMTel voice device can support any number of codecs, but the list must include the adaptive multi-rate (AMR) codec that we introduced in Chapter 21 [49]. Optionally, the device can also support wideband voice communications. If it does, then it is mandatory for the device to support the *wideband adaptive multi-rate* (AMR-WB) codec [50], which uses a sampling rate of 16 000 samples per second and compressed data rates that are roughly twice those of AMR.

The IMS transports voice packets using IP, UDP and the *real-time transport protocol* (RTP). RTP supports the delivery of real-time media over an IP network, by carrying out tasks such as labelling a packet using its sequence number and transmission time [51]. It also has various profiles that define how particular applications should be handled, the relevant profile for IMS voice being the RTP *audio visual profile* (RTP/AVP) [52]. Finally, the protocol defines how the coded bits should be mapped to the RTP payload, the mappings for AMR and AMR-WB being defined in Reference [53].

Two devices can report information about the quality of a media stream by means of control messages, which are written using the *RTP control protocol* (RTCP) and transported on the media path. In VoLTE, a device can use RTCP to help trigger AMR rate adaptation [54]. If a cell is congested, then the base station can inform a mobile by setting an *explicit congestion notification* (ECN) field in the downlink IP packet header. The mobile can then send an RTCP packet that contains a field known as a *codec mode request* to the other device, and both devices can switch to a slower AMR mode.

22.5 VoLTE Registration Procedure

22.5.1 Introduction

The VoLTE registration procedure establishes SIP signalling communications between a mobile and a serving CSCF in the IP multimedia subsystem. During the procedure, the mobile sends its IP address and private identity to the serving CSCF, and quotes one of its public identities. The serving CSCF contacts the home subscriber server, retrieves the other public identities from the corresponding implicit registration set and sets up a mapping between each of these fields. The user can then receive incoming calls that are directed towards any of those public identities and can also make outgoing calls.

There are four stages, which are summarized in Figure 22.8. In the first stage, the mobile attaches to the evolved packet core and sets up a connection through a default EPS bearer to the IMS well-known access point name, either during the attach procedure itself or later on. The mobile then registers itself with a serving CSCF, which carries out a third-party registration with the mobile's application servers. Finally, the mobile subscribes to future notifications about its registration state to support the possibility of network-initiated deregistration. We will now discuss these stages in turn.

22.5.2 LTE Procedures

The mobile starts by running the LTE attach procedure from Chapter 11. If the subscription data identify the IMS access point name as the default and the mobile does not request any

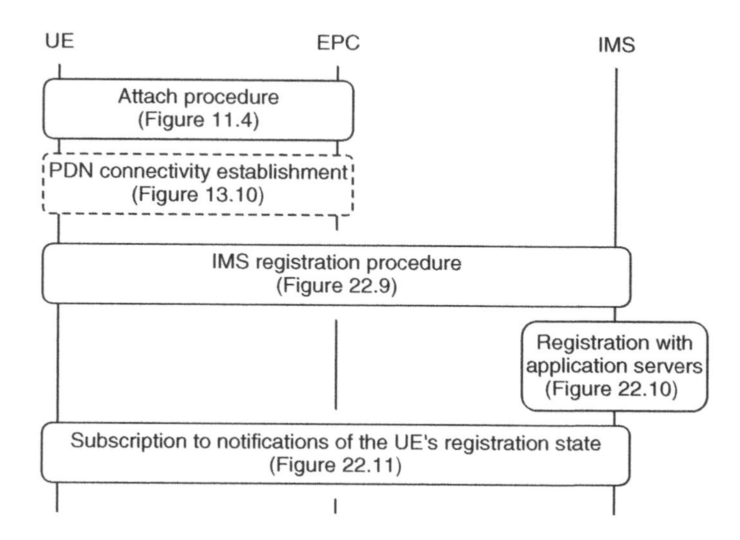

Figure 22.8 Overview of the VoLTE registration procedure

other APN of its own, then the MME connects the mobile to the IMS APN using a default EPS bearer with QCI 5. The MME also supplies the mobile with the IP address of a proxy CSCF as part of its Activate Default EPS Bearer Context Request, for use during the IMS registration procedure. In addition, the MME tells the mobile whether it supports IMS voice calls as part of its message Attach Accept, and tells the home subscriber server as part of its Update Location Request. (The VoLTE specifications forbid the mobile from explicitly requesting the IMS access point name during the attach procedure [55], which prevents the procedure from failing if the home network operator does not support the IMS.)

If the mobile is roaming and both the home and visited networks support the IMS, then the PDN gateway and proxy CSCF both lie in the visited network, and the MME declares support for IMS voice. If the home network supports the IMS but the visited network does not, then the PDN gateway and proxy CSCF both lie in the home network, and the MME denies support for IMS voice. In this latter state, the mobile can use the home IMS for other services such as SMS but will be unable to make an IMS voice call, and will react using the access domain selection procedures that we discuss later in the chapter.

If the mobile connects to a different APN during the attach procedure, then it can still connect to the IMS later on using the PDN connectivity establishment procedure from Chapter 13. In its PDN connectivity request, the mobile specifies the IMS access point name, asks for a bearer that will be suitable for SIP signalling messages and asks for the IP address of a proxy CSCF. The MME sets up a default bearer with QCI 5 and returns the proxy CSCF's IP address, as before.

22.5.3 Contents of the REGISTER Request

The mobile can now communicate with the IP multimedia subsystem, so it can write its SIP REGISTER request. Table 22.2 shows the contents of a typical request [56]. To keep the

Table 22.2 Example of a VoLTE REGISTER request

```
REGISTER sip:registrar.home1.net SIP/2.0
Via: SIP/2.0/UDP [5555::aaa:bbb:ccc:ddd]; branch=z9hG4bKnasiuen8
Max-Forwards: 70
P-Access-Network-Info: 3GPP-E-UTRAN-FDD;
   utran-cell-id-3gpp=234151D0FCE11
From: <sip:beatrice@home1.net>;tag=2hiue
To: <sip:beatrice@home1.net>
Call-ID: E05133BD26DD
CSeq: 1 REGISTER
Require: sec-agree
Proxy-Require: sec-agree
Supported: path
Contact: <sip:[5555::aaa:bbb:ccc:ddd]>;
   +sip.instance="<urn:gsma:imei:90420156-025763-0>";
   +g.3gpp.icsi-ref="urn%3Aurn-7%3A3gpp-service.ims.icsi.mmtel";
   +g.3gpp.smsip;expires=600000
Authorization: Digest username="beatrice_private@home1.net",
   +realm="registrar.home1.net", nonce="",
   uri="sip:registrar.home1.net", response=""
Security-Client: ipsec-3gpp; alg=hmac-sha-1-96;
   spi-c=23456789; spi-s=12345678; port-c=1234; port-s=5678
Content-Length: 0
```

discussion concise, we will not go through all the header fields, but will instead focus on the most important fields.

The first line states the SIP request and identifies the registrar server to which the request should be delivered. The `From:` header identifies the device that is sending the request, while the `To:` header contains the public identity that the user would like to register. In this message, the `From:` and `To:` headers are the same, but they are different in the case of a third-party registration. The mobile reads its public identity from the ISIM and identifies the registrar server using the home network operator's domain name.

The `Contact:` header contains the IP address that the mobile received during PDN connectivity establishment. It indicates support for SMS using the media feature tag `g.3gpp.smsip`, and for the multimedia telephony service using the IMS communication service identifier `urn%3Aurn-7%3A3gpp-service.ims.icsi.mmtel`. The strange looking `%3A` fields are replacements for colon characters, which the media feature tag format does not support. The header also states the international mobile equipment identity and the registration expiry time, measured in seconds.

Among the other headers, the `Authorization:` header carries authentication data, notably the private identity that the mobile is registering, while the `Security-Client:` header states the security algorithms that the device supports. `Call-ID:` is a unique identity for the dialog, while `CSeq:` identifies the individual transaction within the dialog. The `Via:` header is initialized using the mobile's IP address, and will be used later on to route the network's response back to the mobile.

22.5.4 IMS Registration Procedure

The mobile can now begin the IMS registration procedure [57–59], which is shown in Figure 22.9. There are two stages. In the first stage, the mobile sends its REGISTER request to the IMS, which responds with an authentication challenge. In the second stage, the mobile sends a second REGISTER request that contains its reply to the network's challenge, and the IMS accepts the mobile's registration.

To save space, the figure omits the usual protocol markers from the individual messages. All communications with the home subscriber server use the Diameter protocol, while all the other messages use SIP. We assume that the mobile is roaming, so that the mobile and proxy

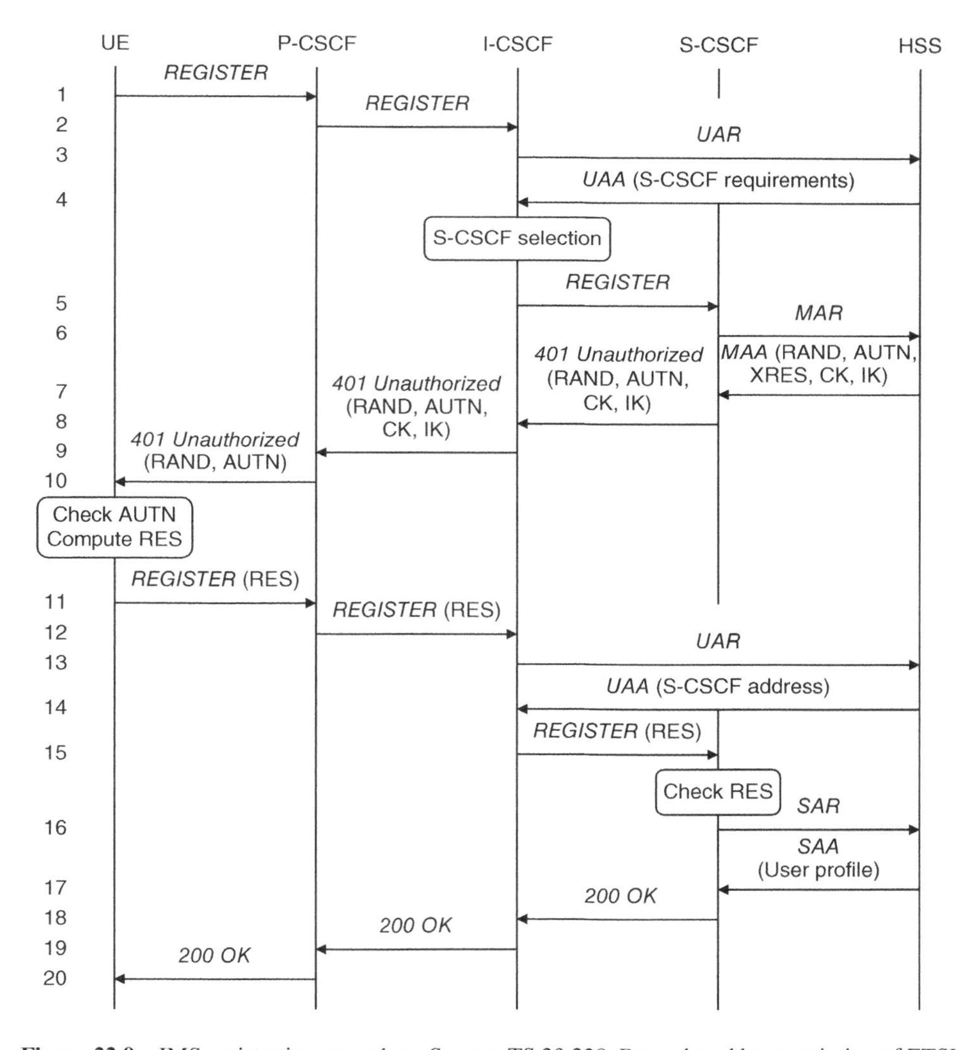

Figure 22.9 IMS registration procedure. Source: TS 23.228. Reproduced by permission of ETSI

CSCF are in the visited network while the other devices are in the home network. The figure omits the networks' interconnection border control functions, which lie between the visited network's proxy CSCF and the home network's interrogating CSCF.

To begin the procedure, the mobile sends its REGISTER request to the proxy CSCF (1), using the IP address that it discovered earlier. In turn, the proxy CSCF forwards the request to an interrogating CSCF (2). (If the mobile is roaming, then this second step will take place through the networks' interconnection border control functions, and will use a domain name server to find an entry point into the home network.) The interrogating CSCF now has to find a serving CSCF that will register the mobile. To do this, it sends a Diameter *User Authorization Request* (UAR) to the home subscriber server, and states the user's public and private identities (3). In an unusual looking reply, the HSS returns the capabilities that the serving CSCF should have in order to look after that user (4). By inspecting an internal list of devices and their capabilities, the interrogating CSCF can select a suitable serving CSCF and can forward the REGISTER request there (5).

The serving CSCF now has to authenticate the user. To do this, it sends a Diameter *Multimedia Authentication Request* (MAR) to the home subscriber server (6), and the HSS replies with an authentication vector that contains the quantities RAND, AUTN, XRES, CK and IK from Chapter 12 (7). The serving CSCF retains XRES, but sends the other quantities to the proxy and interrogating CSCFs in a *401 Unauthorized* error response (8, 9). The proxy CSCF retains CK and IK for use in encryption and integrity protection, and sends RAND and AUTN to the mobile (10).

The mobile checks the network's authentication token, calculates its response, RES, and sends RES to the proxy CSCF in a second REGISTER request (11). The proxy CSCF forwards that request to an interrogating CSCF as before (12). The interrogating CSCF sends another User Authorization Request to the home subscriber server (13), but the HSS now knows about the serving CSCF and can return its IP address (14). In turn, the interrogating CSCF sends the REGISTER request on (15).

The serving CSCF can now check the mobile's response to the authentication challenge. If the response is correct, the serving CSCF tells the home subscriber server that the mobile is now registered using a Diameter *Server Assignment Request* (SAR) (16), and the HSS replies with the mobile's user profile (17). To complete the procedure, the serving CSCF sends a *200 OK* success response back to the mobile (18–20), which includes all the public identities from the implicit registration set. The mobile is now registered with the IMS.

22.5.5 Routing of SIP Requests and Responses

The SIP requests and responses in Figure 22.9 are routed by means of headers. When the first REGISTER request travels through the network (steps 1 to 5), the proxy and interrogating CSCFs add their identities to the `Via:` header that we saw earlier by means of their IP addresses or URIs. The serving CSCF copies the information into its 401 unauthorized response, and the network uses the resulting header to route its response back to the mobile (steps 8 to 10). In this way, the `Via:` header is used to route all the responses to a SIP request.

SIP uses four other headers to route subsequent requests. In message 2, the proxy CSCF adds its identity to a new header, `Path:`, which asks the serving CSCF to include it in all future downlink requests throughout the lifetime of the mobile's registration. In message 18, the serving CSCF adds its identity to a Service-`Route:` header, which instructs the mobile to include it in all future uplink requests. A server can also add its identity to a `Record-Route:` header in any other initial SIP request, to request inclusion in the signalling path for all subsequent requests in the resulting dialog. The information from all these fields is used to populate a final header, `Route:`, which we will see being used to route those requests later on.

22.5.6 Third-Party Registration with Application Servers

As part of the IMS registration procedure, the home subscriber server sent the mobile's user profile to the serving CSCF, which extracted the public identities in the implicit registration set and sent them back to the mobile. The serving CSCF can now identify the services that are associated with those public identities and can register the mobile with the corresponding application servers in the manner shown in Figure 22.10 [60]. To start the procedure, the serving CSCF examines the REGISTER request and compares it with the service point triggers from the user profile. If there is a match, for example if the mobile declared support for the corresponding media feature tag or ICSI, then the serving CSCF writes a third-party REGISTER request on the mobile's behalf and sends it to the application server (1). The request is similar to the one we saw earlier, except that the `From:` header identifies the serving CSCF while the `To:` header identifies the mobile.

The application server can now retrieve any application-specific user data from the home subscriber server (2). It can do this either using a Diameter *User Data Request* (UDR), which simply requests the data, or using a *Subscribe Notifications Request* (SNR), which also asks for notifications if the data are updated. After the home subscriber server's reply (3), the application server sends an acknowledgement to the serving CSCF (4).

22.5.7 Subscription for Network-Initiated Deregistration

The basic SIP protocol allows a user to deregister from the network by sending it a REGISTER request with an expiry time of zero, but it does not support network-initiated deregistration. Such support is provided using two extensions to SIP. The first extension defines two SIP requests, *SUBSCRIBE* and *NOTIFY* [61], through which a device can ask to be notified if an

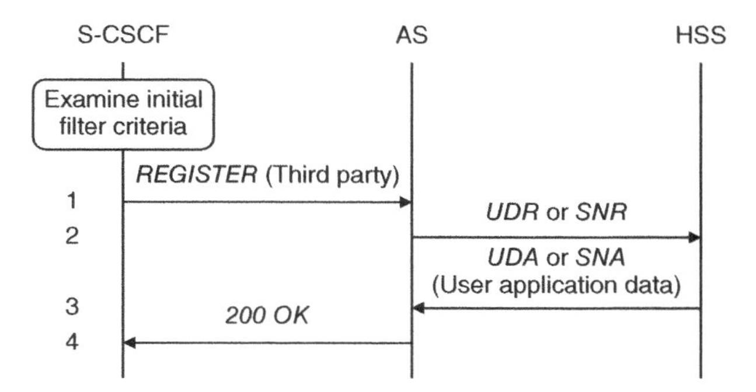

Figure 22.10 Third party registration of the mobile with application servers. Source: TS 29.328. Reproduced by permission of ETSI

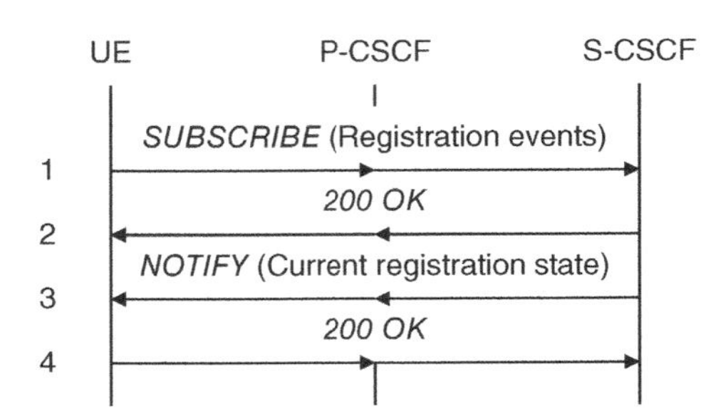

Figure 22.11 Subscription for future notifications of the mobile's registration state. Source: TS 24.228. Reproduced by permission of ETSI

event occurs. The second extension defines a registration event package [62], in which a device asks to be notified about any changes in its registration state.

After the registration procedure has completed, the mobile subscribes to the registration event package in the manner shown in Figure 22.11. The mobile sends a SUBSCRIBE request to the proxy CSCF in the usual way, but the request contains a `Route:` header that identifies the mobile's serving CSCF, so the proxy CSCF can forward the request directly there (1). After its response (2), the serving CSCF immediately sends the mobile a NOTIFY request that confirms the user's registration (3, 4). If the serving CSCF has to deregister the mobile, it simply sends the mobile another NOTIFY request later on. The proxy CSCF subscribes to the mobile's registration event package in a similar way, and any application servers can do so if they require.

22.6 Call Setup and Release

22.6.1 Contents of the INVITE Request

To set up an IMS voice call, a mobile creates a SIP INVITE request and sends it through the IMS to the destination device. Table 22.3 shows the contents of a typical INVITE [63]. As before, the first line identifies the SIP request and the public identity of the destination device. The `Contact:` header states the source device's IP address and confirms support for the multimedia telephony service, while the `Accept-Contact:` header indicates a wish to communicate with another multimedia telephony device.

The `Route:` header ensures that the request travels through the originating mobile's proxy and serving CSCFs. The mobile populates the first of these using the IP address that it received during PDN connectivity establishment, and the second using the identity that it received during IMS registration. The `Supported:` header confirms support for two SIP extensions that are mandated by VoLTE, namely preconditions and reliable acknowledgements of provisional responses. Finally, the `Content-Type:` and `Content-Length:` headers state that the message includes an embedded SDP media offer with the stated length.

Table 22.4 shows the typical contents of the embedded SDP. The offer begins with the session information, in which the `o=` line states the IP address to be used for SDP signalling and the `c=` line states the IP address for media communications. The media information begins with an `m=` line, which requests the establishment of an audio stream using the RTP audio visual profile on port 49152, and declares support for three codecs numbered 97, 98 and 99. The

Table 22.3 Example of a VoLTE INVITE request

```
INVITE sip:benedick@home2.net SIP/2.0
Via: SIP/2.0/UDP [5555::aaa:bbb:ccc:ddd]; branch=z9hG4bKnashds7
Max-Forwards: 70
Route: <sip:[5555::ccc:ddd:eee:fff]:7531;lr>,
   <sip:orig@scscf1.home1.net;lr>
P-Preferred-Service: urn:urn-7:3gpp-service.ims.icsi.mmtel
P-Access-Network-Info: 3GPP-E-UTRAN-FDD;
   utran-cell-id-3gpp=234151D0FCE11
Privacy: none
From: <sip:beatrice@home1.net>;tag=171828
To: <sip:benedick@home2.net>
Call-ID: cb03a0s09a2sdfglkj490333
Cseq: 127 INVITE
Require: sec-agree
Proxy-Require: sec-agree
Supported: precondition, 100rel
Contact: <sip:[5555::aaa:bbb:ccc:ddd]>;
   +g.3gpp.icsi-ref="urn%3Aurn-7%3A3gpp-service.ims.icsi.mmtel"
Accept-Contact: *;+g.3gpp.icsi-ref=
   "urn%3Aurn-7%3A3gpp-service.ims.icsi.mmtel"
Allow: INVITE, ACK, CANCEL, BYE, PRACK, UPDATE, REFER,
   MESSAGE, OPTIONS
Accept: application/sdp, application/3gpp-ims+xml
Security-Verify: ipsec-3gpp; alg=hmac-sha-1-96;
   spi-c=98765432; spi-s=87654321; port-c=8642; port-s=7531
Content-Type: application/sdp
Content-Length: (...)
```

device could ask for additional media streams, such as video, by adding other sets of media information to the end of the offer.

The `b=AS:` line states the bit rate to be reserved for RTP traffic, in kbps, while `b=RS:` and `b=RR:` state the bit rates to be reserved for RTCP sender and receiver reports, in bps. The `a=curr:qos` and `a=des:qos` lines support the SIP precondition extension [64] by stating that the device has no quality of service guarantees at present, and that it is only willing to communicate once it receives a suitable quality of service from its own IMS. Finally, the `a=rtpmap` lines identify codecs 97 and 98 with two slightly different formats of the AMR codec, known as bandwidth efficient and octet aligned. They also declare support for *dual-tone multi-frequency* (DTMF) signalling using codec 99.

22.6.2 Initial INVITE Request and Response

The mobile can now send its INVITE request to the IP multimedia subsystem. The SIP and SDP negotiations can take place in various different ways, so Figures 22.12 and 22.13 show the signalling messages that might typically be exchanged between two LTE mobiles that are registered with different network operators [65, 66].

Table 22.4 Example of a VoLTE SDP offer

```
v=0
o=- 2987933615 2987933615 IN IP6 5555::aaa:bbb:ccc:ddd
s=-
c=IN IP6 5555::aaa:bbb:ccc:ddd
t=0 0
m=audio 49152 RTP/AVP 97 98 99
b=AS:38
b=RS:0
b=RR:4000
a=curr:qos local none
a=curr:qos remote none
a=des:qos mandatory local sendrecv
a=des:qos optional remote sendrecv
a=inactive
a=rtpmap:97 AMR/8000/1
a=fmtp:97 mode-change-capability=2; max-red=220
a=rtpmap:98 AMR/8000/1
a=fmtp:98 mode-change-capability=2; max-red=220; octet-align=1
a=rtpmap:99 telephone-event/8000/1
a=fmtp:99 0-15
a=ptime:20
a=maxptime:240
```

The originating mobile sends its INVITE request and SDP offer to its proxy CSCF, which sends the request to the serving CSCF (1, 2). The serving CSCF looks up the destination mobile's domain name and sends the request to an interrogating CSCF in the destination network by way of the interconnection border control functions (3). In turn, the interrogating CSCF looks up the destination mobile's registration details in the home subscriber server and forwards the request to the mobile's serving CSCF (4), which sends the request to the proxy CSCF and to the mobile (5, 6). In the case of an INVITE request, each CSCF also sends a *100 Trying* provisional response back to its predecessor to confirm that the request has arrived successfully and does not have to be retransmitted.

Along the way, the proxy and serving CSCFs inspect the SDP offer to check that it complies with the networks' policies and with the users' subscriptions. If they find anything unacceptable (such as too high a bit rate), then they reply with the error response *488 Not Acceptable Here* and include an SDP media description that they are willing to accept.

The serving CSCFs also inspect the INVITE request and compare it with the service point triggers of their respective mobiles. In this example, they find a match with the multimedia telephony service, so they forward the request to the mobiles' MMTel application servers, which process the request through tasks such as number translation or call forwarding. In common with the other devices shown in the figure, the application servers also add themselves to the request's `Record-Route:` and `Via:` headers, so that they receive all subsequent requests and responses in the dialog.

If the destination mobile is in ECM-IDLE, then the evolved packet core cannot deliver the INVITE request right away. Instead, the request travels as far as the serving gateway, which asks the MME to contact the mobile using the paging procedure from Chapter 14. The mobile responds by running the service request procedure, which takes it into the state of ECM-CONNECTED. The serving gateway can then deliver the INVITE request to the mobile.

If the destination mobile is not switched on, then it will not be registered with a serving CSCF at all. In this situation, however, the interrogating CSCF can still select a serving CSCF to handle the INVITE request, in the same way that it did during registration. On receiving the request, the serving CSCF reads the mobile's user profile from the HSS, compares the request with the service point triggers and forwards the request to a voicemail application server.

When the INVITE request finally arrives, the destination mobile inspects the SDP offer. If the offer contains any media streams that it does not support, such as video in the case of a voice-only device, then the mobile modifies the SDP by setting the port numbers of those media streams to zero. It then selects one codec for each media stream that it does support, declares any additional support for DTMF tones, embeds its SDP answer into a *183 Session Progress* provisional response and sends the response back to the originating mobile (7).

Along the way, each proxy CSCF inspects the SDP answer and calculates the quality of service parameters that the media will require. It then asks the corresponding PCRF to set up a new service data flow, using the procedure for network-initiated SDF establishment from Chapter 13. The PDN gateway sets up a new dedicated bearer with QCI 1 for the mobile's first voice call, or modifies an existing bearer if there is already an ongoing voice call. In this way, the SDP parameters make their way down to the evolved packet core and give the mobile the quality of service that it requires.

22.6.3 Acceptance of the Initial INVITE

At the point we have just reached, the destination mobile has sent an SDP answer back to the source, which it has embedded into a SIP provisional response. The SDP answer contains important information, so the destination mobile asks for an acknowledgement of its provisional response by including the header `Require: 100rel` in message 7 [67]. The source mobile sends the acknowledgement by means of a *PRACK* request (8), and the destination mobile responds with a 200 OK (9). The mobiles can match this response to the PRACK rather than the initial INVITE by examining the SIP headers.

On completion of the dedicated bearer activation procedure, the source mobile knows that the LTE access network can give it the quality of service that it requires. It informs the destination device using a second SDP offer, in which its local quality of service has changed to `a=curr:qos local sendrecv`. It embeds the offer into a SIP *UPDATE* request (10), this being an extension to SIP that modifies the parameters of a session [68], and the destination mobile responds in the usual way (11). If its own resource allocation process has completed, then the destination mobile can indicate this within its SIP response in the manner shown. Otherwise, it does so by sending an UPDATE request of its own later on.

The message sequence may require further stages of media negotiation. If, for example, the destination device does not support any of the codecs that were originally offered, then the negotiation can continue with additional SDP offers and answers, which are embedded into additional SIP requests and responses. Alternatively, the bearer activation process may already be complete when the source mobile sends its provisional response acknowledgement

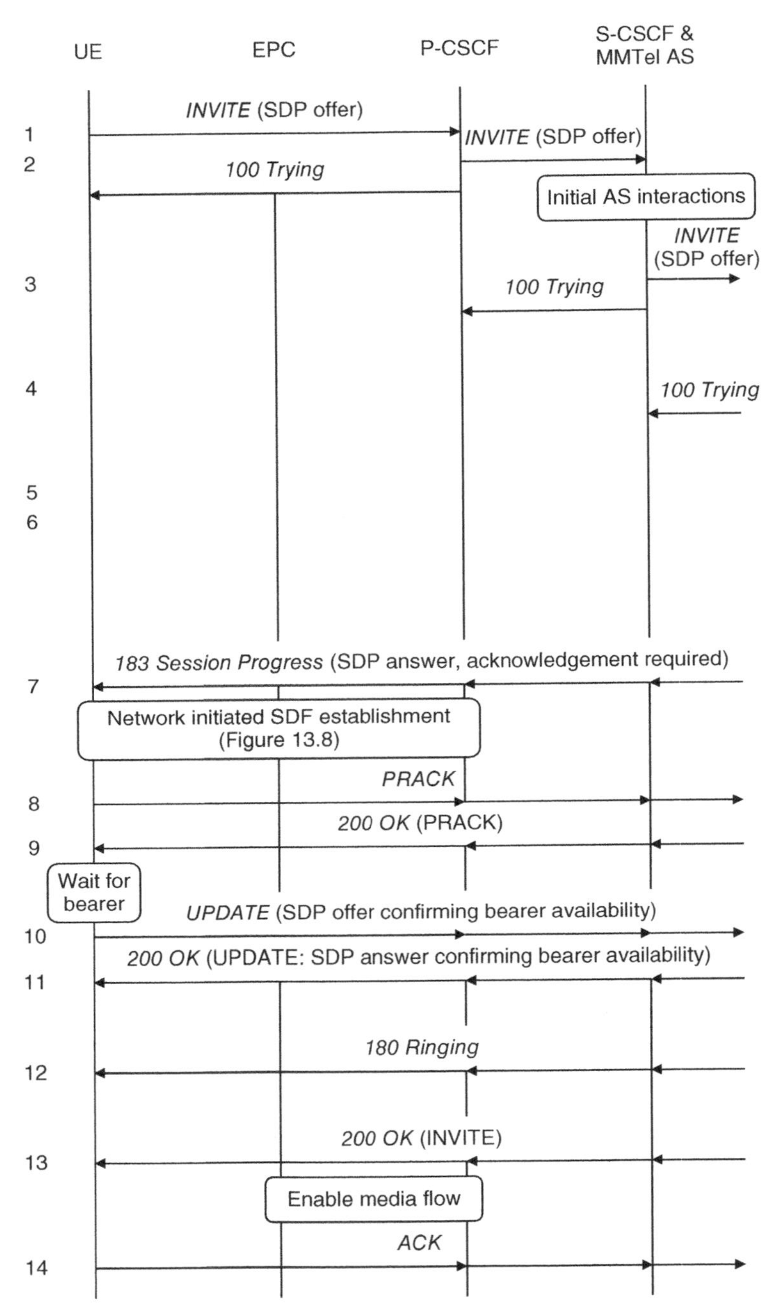

Figure 22.12 IMS session initiation procedure. (1) Messages in the originating network. Source: TS 23.228. Reproduced by permission of ETSI

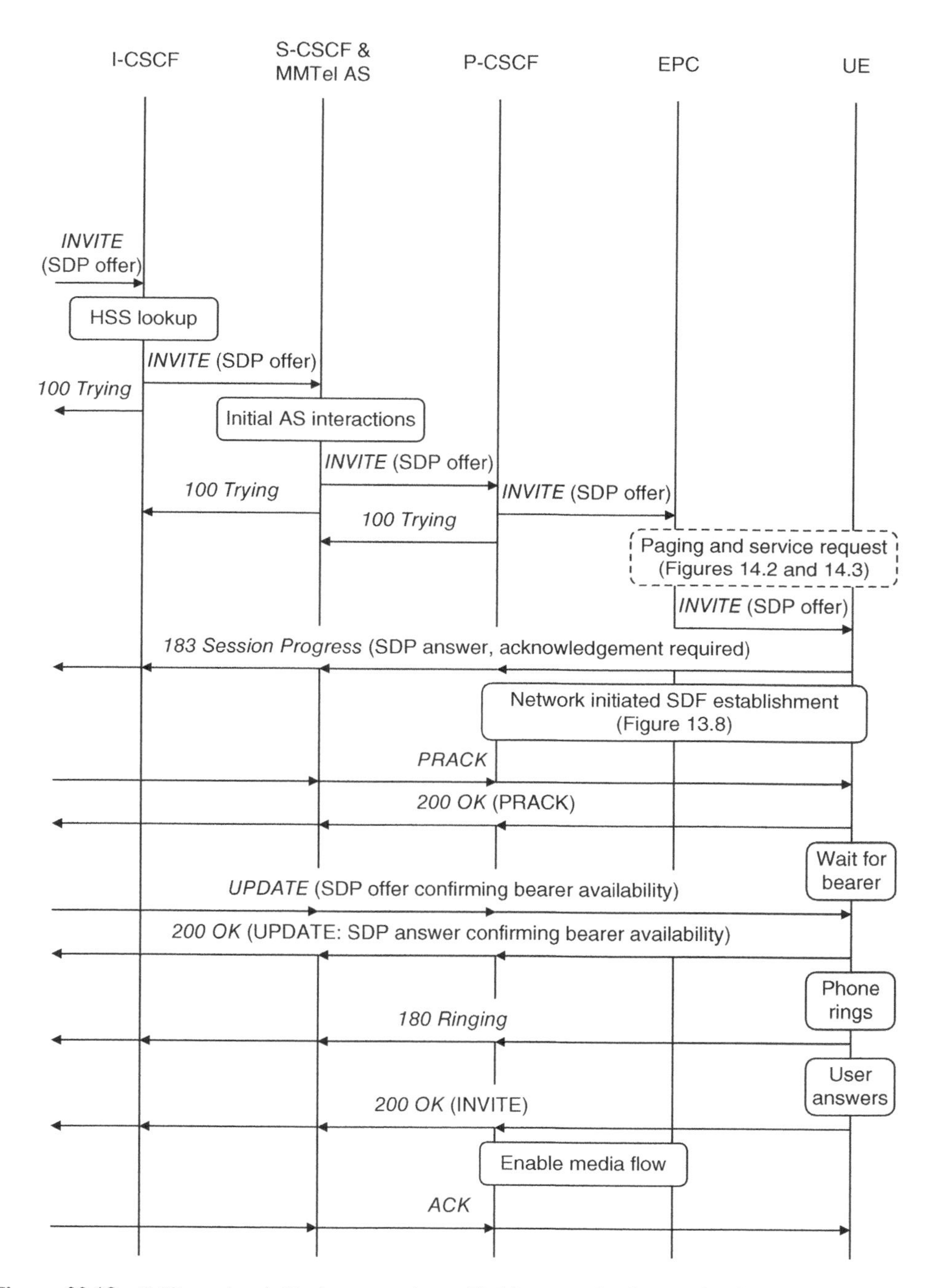

Figure 22.13 IMS session initiation procedure. (2) Messages in the terminating network. Source: TS 23.228. Reproduced by permission of ETSI

in message 8. If so, it can inform the destination device right away and does not have to send message 10.

Once all these steps are complete, the phone rings and the destination mobile returns a *180 Ringing* provisional response back to the source (12). There is no need to request an acknowledgement here, because there is no longer any embedded SDP information. Once the user answers, the destination mobile returns a 200 OK success response to the initial INVITE (13). On receiving this response, the proxy CSCFs know that the SIP and SDP signalling has completed, so they ask their respective PCRFs to enable the media flows, and the PCRFs tell the PDN gateways to open the gates for the users' traffic. To conclude the procedure, the source device sends a final SIP request known as *ACK* (14), which indicates that the session can begin.

22.6.4 Establishment of a Call to a Circuit Switched Network

If the source or destination device is in a circuit switched network, the message sequence is slightly different but the underlying principles remain the same [69]. First, let us consider what happens if the destination device is in the public switched telephone network. The mobile sends a SIP INVITE request to the IMS as before, but identifies the destination device using a phone number that is expressed as a Tel URI. The serving CSCF cannot handle this directly, so it converts the phone number to a domain name [70] and looks up the result in a domain name server, in an attempt to translate the number to a routable SIP URI.

If this process fails, then the serving gateway sends the INVITE request to a breakout gateway control function. By inspecting the phone number, the BGCF can either select a media gateway control function in the home IMS or, if the user is roaming, can delegate the choice to a BGCF in the visited IMS. The MGCF now acts as the destination SIP user agent, and the message sequence continues in a similar way to the previous one. In addition, the MGCF configures its media gateway, translates the SIP INVITE into an ISUP Initial Address Message and forwards that message to the destination network after message 11. It then waits for an ISUP Address Complete Message before sending its 180 Ringing response, and waits for an ISUP Answer Message before sending the final 200 OK.

Now consider what happens if a device in the public telephone network calls a SIP user. Using the dialled number, the public telephone network routes the ISUP Initial Address Message to a media gateway control function in the user's home IMS. The MGCF translates the message into a SIP INVITE request and forwards it to an interrogating CSCF. From this point, the message sequence can continue as before, except that the MGCF acts as a SIP user agent in place of the source mobile.

22.6.5 Call Release

To end a call [71, 72], one device transmits a SIP *BYE* request and the other device responds with a 200 OK. When using an LTE access network, the proxy CSCFs ask their respective PCRFs to release the service data flows, and the PDN gateways either modify or tear down the dedicated bearers. At that point, the call ends.

If an LTE mobile moves out of radio coverage, then the base station notices that it can no longer communicate with the mobile, so it initiates the S1 release procedure from Chapter 14. The MME tears down the mobile's guaranteed bit rate bearers [73], including the dedicated bearer that it is using for the voice call. In turn, the PCRF tells the proxy CSCF that the service data flow has been torn down, and the proxy CSCF transmits a SIP BYE request so as to end the call. The IMS can then tear down all the resources that were associated with the call, and can ensure that the user is no longer charged.

22.7 Access Domain Selection

22.7.1 Mobile-Originated Calls

So far we have assumed that the mobile only supports voice over the IMS. If the mobile also supports circuit switching, then it selects the access domain for outgoing voice calls using two parameters, namely the network's indication of whether it supports IMS voice and the mobile's voice domain preference [74, 75]. The VoLTE specifications insist that the second of these should be set to IMS PS voice preferred/CS voice as secondary.

The LTE attach procedure can use two alternative techniques, which are shown in Figure 22.14. (The mobile's choice is influenced by its preferred access domain for SMS, in the manner noted at the end of the chapter.) In the first alternative, the mobile attaches for LTE alone using the EPS attach procedure from Chapter 11. If the network supports IMS voice, then the mobile registers with the IMS as described earlier. If not, the mobile registers for circuit switched fallback using a combined tracking/location area update and continues as described in Chapter 21. In the second alternative, the mobile uses the combined EPS/IMSI attach procedure from the beginning. The outcome is similar, but if the network supports both IMS voice and circuit switched fallback, then the mobile ends up registered with both the IMS and the 2G/3G circuit switched domain. On the basis of its voice domain preference, the mobile then chooses the IMS for its outgoing voice calls.

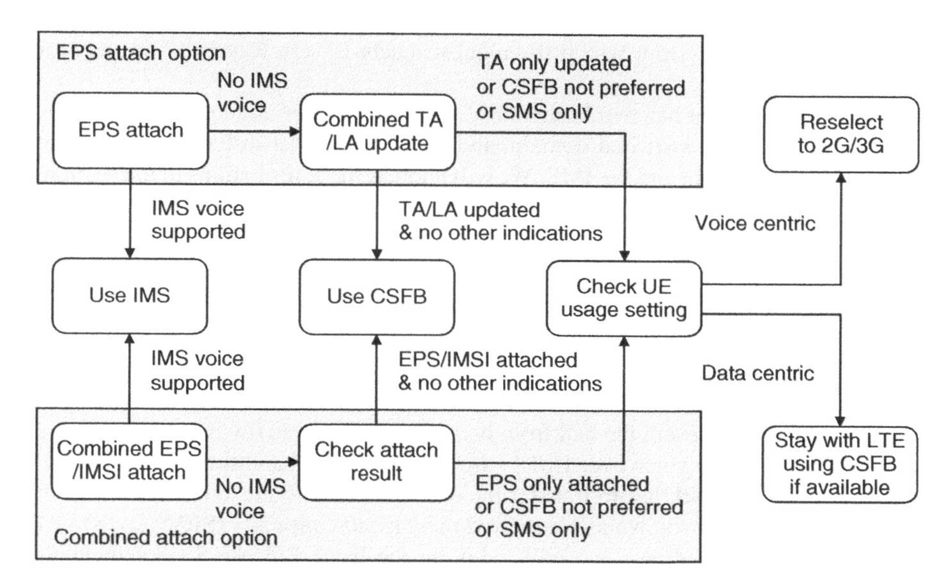

Figure 22.14 Behaviour of a mobile whose voice domain preference is IMS voice preferred / CS voice as secondary. Source: TS 23.221. Reproduced by permission of ETSI

If the mobile moves outside the coverage area of LTE and into a 2G or 3G cell, then the network operator can continue to offer it voice services in two ways. In a 3G cell, the latency may be low enough for a technique known as *voice over HSPA*, in which the network delivers VoIP calls using the 3G packet switched domain. If the network has implemented this technique, then the SGSN indicates support for IMS voice in its Routing Area Update Accept and informs the home subscriber server. The mobile can then continue to use the IMS.

In a 2G cell or a basic 3G cell, the only viable option is traditional circuit switching, so the SGSN denies support for IMS voice and informs the home subscriber server. The mobile attaches to the 2G/3G circuit switched domain if it has not already done so and chooses circuit switching for its outgoing voice calls. Once again, the mobile is simultaneously registered with the IMS and the circuit switched domain.

22.7.2 Mobile-Terminated Calls

Access domain selection for a mobile-terminated call is more complicated [76–80]. If an incoming call arrives at the circuit switched domain, then the gateway MSC looks up the mobile's location in the home subscriber server. The HSS returns an MGCF if the mobile is in a region of the network that supports IMS voice and the mobile's MSC server otherwise. The call can then be handled using the normal procedures for the chosen access domain.

If an incoming call arrives at the IMS, then the incoming INVITE request arrives at the mobile's serving CSCF in the usual way. The serving CSCF forwards the request to a new application server, known as the *service centralization and continuity application server* (SCC-AS), which manages interactions between the IMS and the circuit switched domain. The SCC-AS interrogates the home subscriber server to discover whether the mobile is in a region of the network that supports IMS voice. If it is, then the SCC-AS routes the INVITE request through the IMS in the usual way. If not, then it passes the request to a breakout gateway control function, from which the request reaches a gateway MSC in the mobile's circuit switched domain.

Access domain selection has two limitations: it cannot transfer an ongoing voice call from LTE to the 2G/3G circuit switched domain, and it offers the user different services through the circuit switched domain and the IMS. We will address these limitations in the sections that follow.

22.8 Single Radio Voice Call Continuity

22.8.1 Introduction

If the network hands over a mobile from LTE to a 2G/3G cell during a voice call, then it has to take additional steps to prevent the call from being dropped. It can do this in two ways. If the network has implemented voice over HSPA, then it can simply hand the mobile over to a 3G cell. Otherwise, the network has to transfer the call from LTE to the 2G/3G circuit switched domain, using a technique known as *single radio voice call continuity* (SRVCC) [81–84].

In the discussion that follows, we will focus on the basic Release 8 implementation of SRVCC. This allows the network to hand over a single ongoing voice call from LTE to the

2G/3G circuit switched domain, and is a requirement for VoLTE networks and mobiles that support circuit switched voice. Later on, we will review some enhancements to SRVCC in later releases, which VoLTE does not mandate.

22.8.2 SRVCC Architecture

Figure 22.15 shows the architecture for SRVCC in 3GPP Release 8. There are two important features. Firstly, the circuit switched domain contains one or more MSC servers that are enhanced for SRVCC and support a new signalling interface known as Sv. Using this interface, the MME can hand the mobile over from LTE to the circuit switched domain using a signalling protocol that is based on GTPv2-C [85]. The mobile can be controlled from a different MSC server, so the network operator does not have to enhance all its MSC servers in this way. Secondly, the IMS manages the process using the service centralization and continuity application server (SCC-AS) that we introduced earlier.

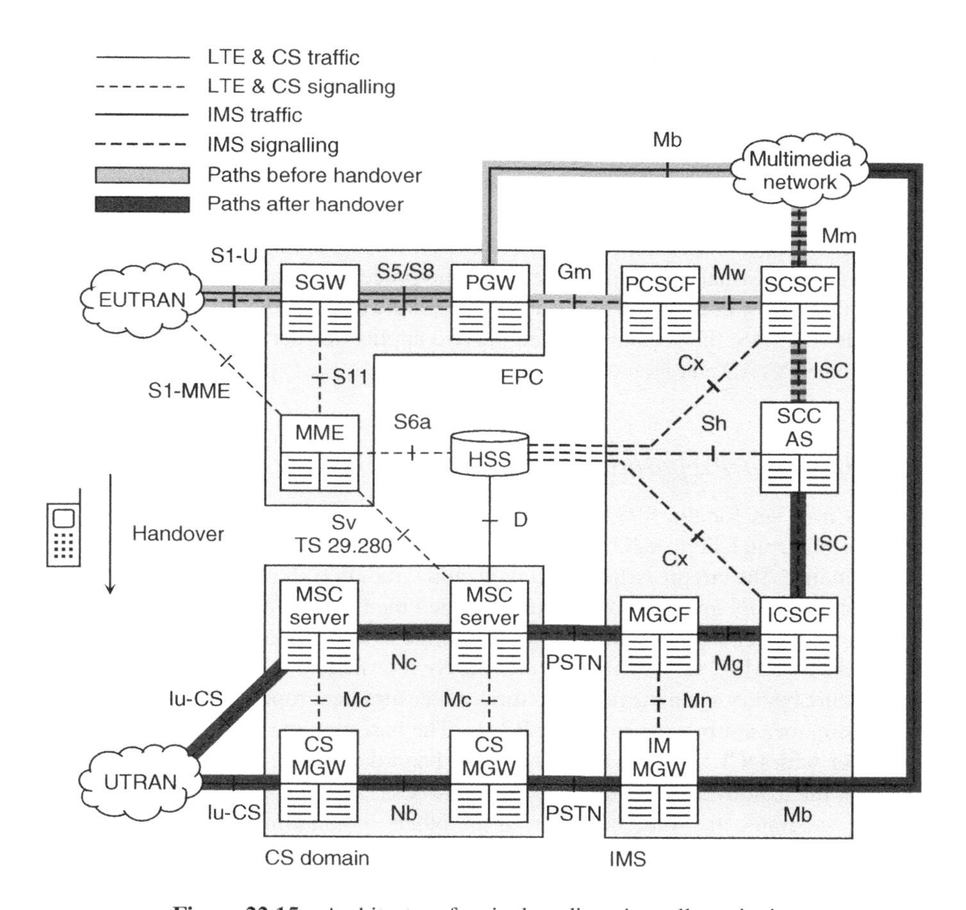

Figure 22.15 Architecture for single radio voice call continuity

The SRVCC procedures use two important parameters [86]. The *session transfer number single radio* (STN-SR) is a phone number that identifies the SCC-AS and is configured as a SIP identity known as a *public service identity*. The *correlation mobile subscriber ISDN number* (C-MSISDN) is one of the user's existing phone numbers, which is uniquely associated with the user's IMSI and IMPI and which identifies the user.

In Release 8, the SRVCC handover procedure hands over a single voice bearer from LTE to the 2G/3G circuit switched domain. The other bearers are handed over to the packet switched domain if they can be supported there, or are dropped otherwise. The MME identifies voice bearers by assuming that they have a QoS class indicator of 1, so the PCRF has to reserve QCI 1 for use with voice bearers alone.

22.8.3 Attach, Registration and Call Setup Procedures

SRVCC requires a few preparatory steps before the handover can begin. In the LTE attach procedure, the MME retrieves the mobile's STN-SR and C-MSISDN from the home subscriber server as part of the subscription data, and informs the base station that SRVCC is supported. In the IMS registration procedure, the serving CSCF retrieves the mobile's SRVCC service profile and registers the mobile with the correct SCC-AS. The SCC-AS can then retrieve the relevant application data from the HSS, in particular the mobile's C-MSISDN.

In the call setup procedure, the INVITE request arrives at the originating mobile's SCC-AS using the normal routing mechanisms for application servers. Acting as a back-to-back user agent, the SCC-AS creates a third-party INVITE request, which it sends to the destination network. The request arrives at the destination SCC-AS, which creates another INVITE request and sends it to the destination mobile. The call setup procedure can then continue in the usual way, except that it now contains as many as three separate dialogs, the first between the source device and its SCC-AS, the second between the two application servers and the third between the destination SCC-AS and its mobile.

22.8.4 Handover Preparation

The stage is now set for the SRVCC handover procedure [87, 88]. This has three aims: it hands the mobile from LTE to a 2G or 3G cell, it transfers a single voice stream from the packet switched domain to the circuit switched domain, and it redirects the destination device's traffic from the mobile's PDN gateway to a circuit switched media gateway. Figure 22.16 shows the first stage of the procedure. We assume that the mobile is moving to a 3G cell, in which it will be controlled by an MSC server that supports the Sv interface.

The procedure begins when the mobile returns a measurement report that describes a specific target cell using measurement event B1 or B2 (1). The base station realizes that the mobile has a voice bearer with QCI 1, so it asks the MME to hand the mobile over to the target cell and indicates that the handover is for the purpose of SRVCC (2).

The MME separates the voice bearer from the others by examining its QCI, in a function known as *bearer splitting*. It then selects a suitable MSC server and asks that device to accept the voice bearer using an *SRVCC PS to CS Request* (3). In the message, the MME identifies

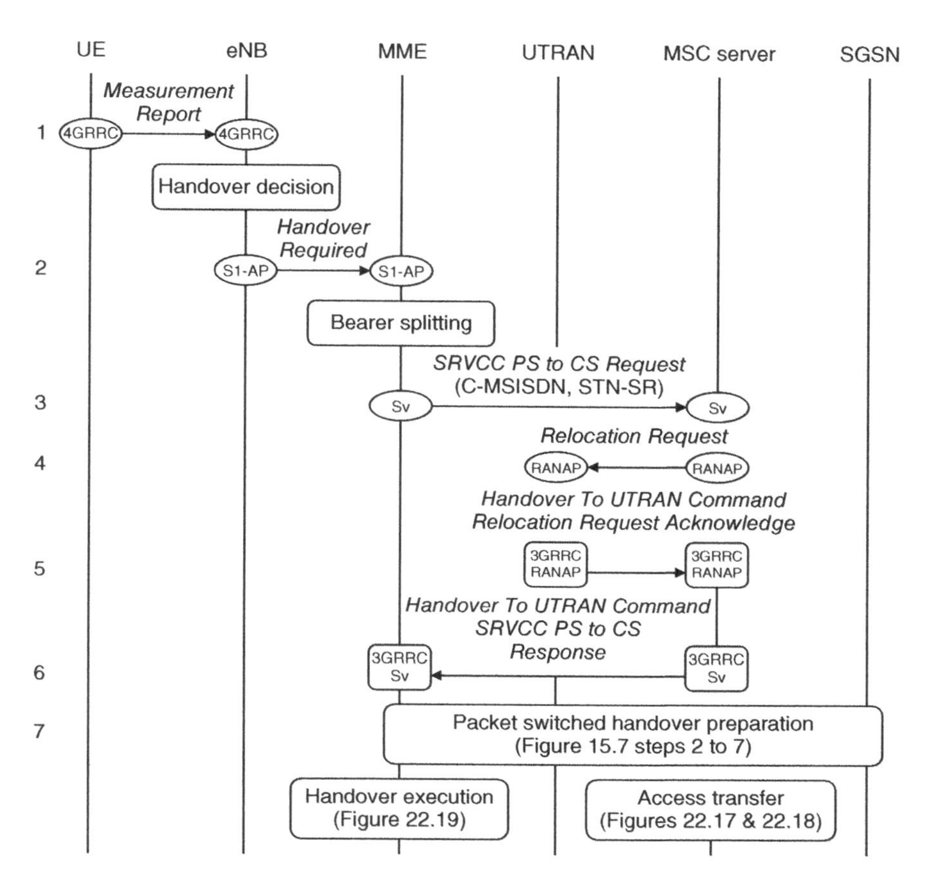

Figure 22.16 Handover from LTE to UMTS using SRVCC. (1) Handover preparation. Source: TS 23.216. Reproduced by permission of ETSI

the target cell and includes the mobile's STN-SR and C-MSISDN. The MSC server asks the 3G radio network controller to accept the voice bearer (4), and the RNC's reply includes a 3G Handover to UTRAN Command that instructs the mobile how to communicate with the target cell (5). On receiving that reply, the MSC server returns an acknowledgement to the MME (6). At the same time, the MME selects a suitable SGSN and asks that device to accept the mobile's other bearers by preparing a packet switched handover (7).

22.8.5 Updating the Remote Leg

The network can now carry out two tasks in parallel, namely changing the destination device's traffic path and handing over the mobile. Figure 22.17 shows how the first of these begins. After returning its acknowledgement to the MME, the MSC server starts the configuration of its media gateway, and creates an ISUP Initial Address Message in which the originating

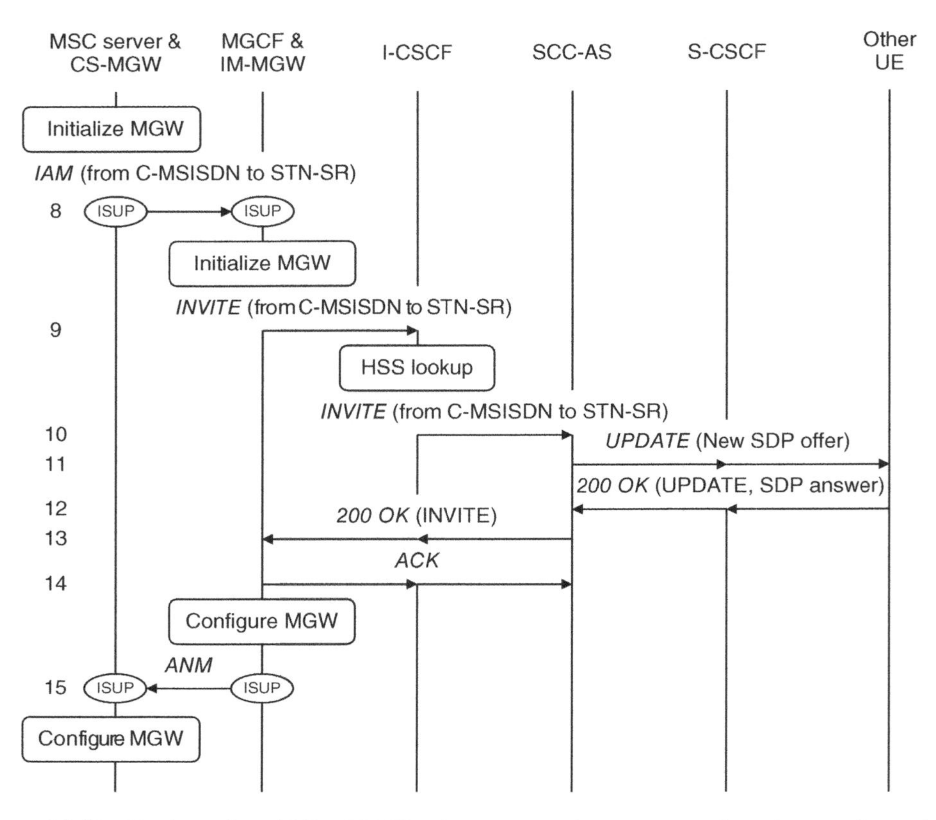

Figure 22.17 Handover from LTE to UMTS using SRVCC. (2) Access transfer and remote leg update. Source: TS 23.237. Reproduced by permission of ETSI

number is the C-MSISDN and the terminating number is the STN-SR (8). The message is routed to a media gateway control function in the IMS, because that is the network in which the STN-SR is registered.

The MGCF configures its own media gateway and converts the incoming message to a SIP INVITE request (9). In the message, it identifies the source and destination devices using Tel URIs that contain the C-MSISDN and STN-SR, and identifies the traffic path using the IP address of the media gateway. It then sends the request to an interrogating CSCF, which looks up the STN-SR in the home subscriber server and sends the request to the corresponding SCC-AS (10).

The SCC-AS inspects the C-MSISDN, remembers the call setup procedure from earlier and realizes that the mobile already has an ongoing call. (If there is more than one ongoing call, then the SCC-AS chooses the most recent.) Acting as a back-to-back user agent, it composes a SIP UPDATE request with an embedded SDP offer that tells the destination device to redirect its traffic path to the media gateway. The SCC-AS can then send the request to the mobile's serving CSCF, which forwards it to the destination (11). The destination device redirects its traffic path as requested and responds (12).

The SCC-AS can then send a 200 OK response to the original INVITE (13), which triggers a SIP ACK request in the usual way (14). At the same time, the MGCF responds to the MSC server using an ISUP Answer Message (15) and the two devices complete the configuration of their media gateways.

22.8.6 Releasing the Source Leg

The call is now being controlled through the circuit switched domain, so the SCC-AS can release the source leg from the IMS. Figure 22.18 shows the steps that are required. To start the process, the SCC-AS composes a SIP BYE request and sends the request through the serving and proxy CSCFs to the mobile (16). If the service data flow has not already been released as part of the handover procedure below, then the proxy CSCF asks the PCRF to release it here, which triggers the release of the dedicated bearer. After the mobile's reply (17), the tasks of the IMS are complete.

22.8.7 Handover Execution and Completion

At the same time as the preceding IMS messages, the MME can hand the mobile over to the target cell. As shown in Figure 22.19, the MME waits for the SGSN and the MSC server to reply in steps 6 and 7, and merges their responses into a single 3G Handover to UTRAN Command. It then tells the base station to hand the mobile over to the target cell, and the base station relays that instruction to the mobile (18, 19). The mobile extracts the embedded 3G message, synchronizes with the target cell and contacts the radio network controller (20). The RNC tells the MSC server that the handover is complete (21) and the MSC server tells the MME (22, 23). The MME tears down the voice bearer if the PDN gateway has not already done so, so any remaining voice calls are dropped. At the same time, the network can complete the normal steps for a packet switched handover (24).

At the end of the procedure, the network has handed over one of the mobile's voice calls from LTE to the 3G circuit switched domain. The mobile retains its non-GBR bearers, including the default IMS bearer that carries its SIP signalling messages, but these messages are now decoupled from the voice call and cannot be used to control it. Instead, the mobile stays in the 2G/3G radio access network until the call ends, even if it returns to the coverage area of LTE. If the mobile has handed over to a 2G cell and there is no support for dual transfer mode, then the non-GBR bearers are suspended, but the effect on the call is the same.

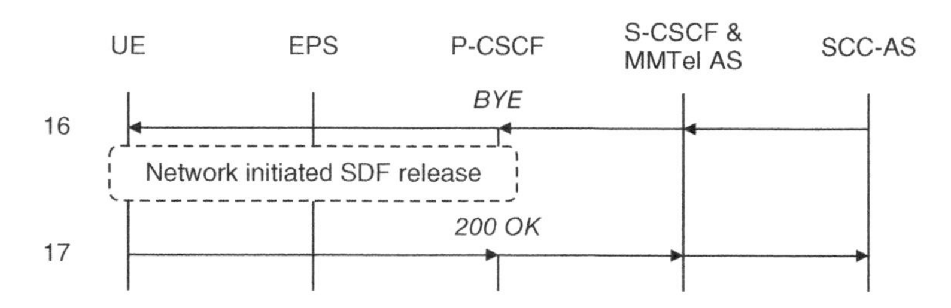

Figure 22.18 Handover from LTE to UMTS using SRVCC. (3) Source leg access release

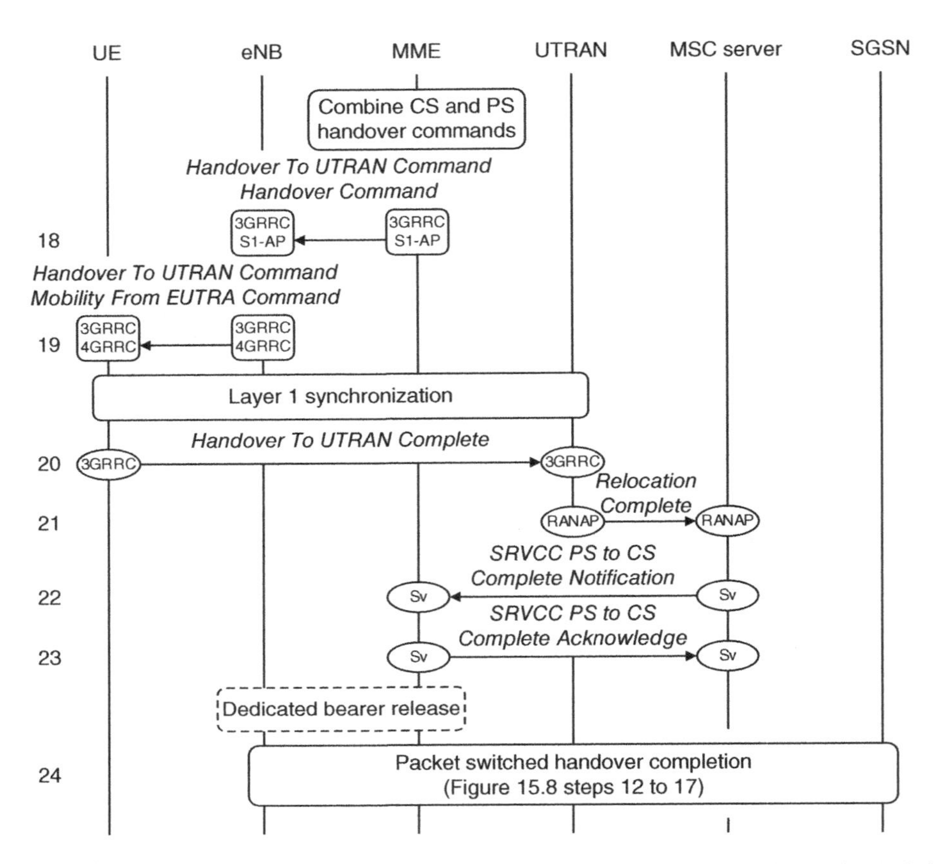

Figure 22.19 Handover from LTE to UMTS using SRVCC. (4) Handover execution and completion. Source: TS 23.216. Reproduced by permission of ETSI

22.8.8 *Evolution of SRVCC*

SRVCC has gone through several enhancements since it was first introduced. Apart from the support of emergency calls, which we cover later, none of these is mandated by VoLTE, so we will only discuss them briefly.

In the SRVCC procedure that we discussed earlier, the MME was able to hand over the mobile before the IMS signalling completed. Because of this feature, SRVCC should not increase the delay between the measurement report and the handover command, so it should not increase the likelihood of a dropped call. It can, however, cause a significant gap in the voice communications, because of the time required for the IMS to contact the destination device. This gap can be particularly long if the mobile is roaming, because the SCC-AS is located in the mobile's home network.

To deal with the problem, the Release 10 specifications allow the SCC-AS to delegate the handover to two new functions in the visited IMS, the *access transfer control function* (ATCF)

for signalling and the *access transfer gateway* (ATGW) for traffic [89]. During the IMS registration procedure, the access transfer control function chooses an STN-SR of its own and passes it to the home network's SCC-AS. The SCC-AS can then inform the home subscriber server about the new STN-SR, and the home subscriber server can inform the MME. During the call setup procedure, the access transfer control function inserts the access transfer gateway into the traffic path from the beginning. During the handover, the MME uses the new STN-SR instead of the old one, so the INVITE message arrives at the access transfer control function instead of the SCC-AS. In place of messages 11 and 12, the access transfer control function simply has to redirect the local traffic path at the access transfer gateway, and does not have to contact the destination device at all.

The Release 8 specifications also support a technique known as *dual radio voice call continuity* (DRVCC). In this technique, the mobile contacts the MSC server over the destination radio access network in place of message 3, so the Sv interface is not required. The technique is useful for handovers between a wireless local area network and a 2G/3G cell, or between LTE and cdma2000. The Release 11 specifications allow a mobile to hand over a video call, in a technique known as *single radio video call continuity* (vSRVCC).

The remaining enhancements all require the architecture for IMS centralized services that we introduce next. These include handover in the alerting phase of the call (Release 10), the preservation of mid-call supplementary services such as conference calls (Release 10) and the ability to hand a mobile back to an LTE cell before the call ends (Release 11).

22.9 IMS Centralized Services

Using *IMS Centralized Services* (ICS) [90, 91], a mobile can reach the IMS through a circuit switched access network, so can receive the same set of IMS services through a 2G or 3G cell that it does through LTE. ICS is an integral part of the rich communication services that we discussed earlier, but is not required by VoLTE, so we will only cover it briefly.

Figure 22.20 shows the architecture for ICS. The communications are managed by the service centralization and continuity application server that we introduced earlier, and take place using three options for *service control signalling*. In the first option, the mobile is controlled by an MSC server that supports two new interfaces known as I2 and I3. Using these interfaces, the MSC server converts the mobile's circuit switched signalling messages into the corresponding SIP and SDP, so that the mobile can register with a serving CSCF and access the supplementary services of the multimedia telephony application server.

In the second option, the mobile is itself enhanced to support IMS centralized services and can access the IMS across a packet switched core network. In this option, the mobile communicates with the IMS in the usual way, with its enhancements serving to couple the mobile's SIP signalling messages to the circuit switched media.

In the third option, the mobile is enhanced to support ICS as before, but has no packet switched access to the IMS. In this option, the mobile communicates with the SCC-AS using a new interface known as I1. This interface carries a binary signalling protocol [92] whose messages are transported over the circuit switched core network, typically by embedding them into *unstructured supplementary service data* (USSD).

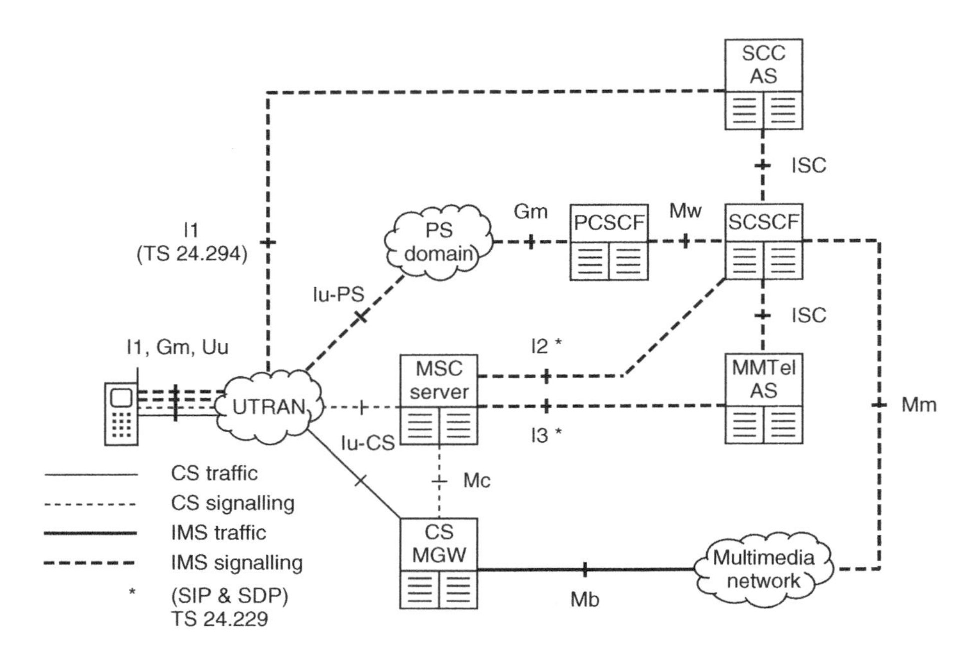

Figure 22.20 Architecture for IMS centralized services. Source: TS 23.292. Reproduced by permission of ETSI

22.10 IMS Emergency Calls

22.10.1 Emergency Call Architecture

Emergency calls have slightly different requirements from normal voice calls, including the need to measure the mobile's location and the need to support an emergency call from a mobile in a limited service state. The IMS supports emergency calls from Release 9 [93], and such support is mandated by the VoLTE specifications. Figure 22.21 shows the architecture used.

The evolved packet core dedicates a separate access point name to emergency calls. The VoLTE specifications do not prescribe any particular APN, but recommend the name *SOS*. The mobile's first point of contact with the IMS is a proxy CSCF, but this can be a different device from the one used for normal voice calls. The call is set up by an *emergency CSCF* in the visited network and is relayed to an external *public safety answering point* (PSAP), if necessary through a media gateway control function.

To complicate things, the mobile also carries out an emergency registration with a serving CSCF in its home network, which is separate from its normal IMS registration. This ensures that the mobile can receive a call back from the public safety answering point if required.

The *location retrieval function* (LRF) provides details of the mobile's location by acting as a gateway mobile location centre towards the evolved packet core. The E-CSCF uses the information to choose a suitable public safety answering point, while the PSAP uses the information to pinpoint the mobile. Finally, the *emergency access transfer function* (EATF) takes the role of the service centralization and continuity application server, by re-routing the destination leg if the mobile has to be handed over to a 2G or 3G cell.

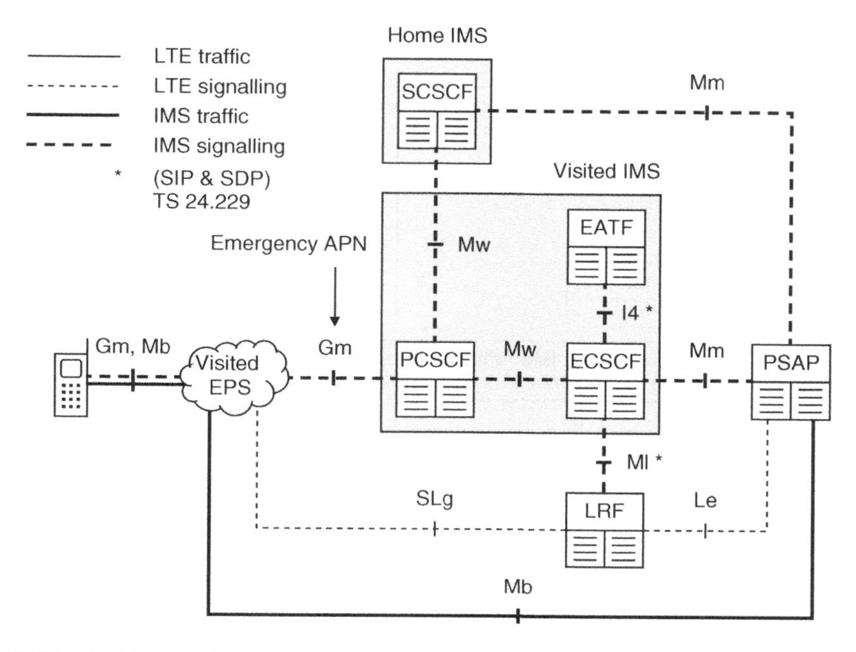

Figure 22.21 Architecture for IMS emergency calls. Source: TS 23.167. Reproduced by permission of ETSI

22.10.2 Emergency Call Setup Procedure

Figure 22.22 summarizes the most common form of the emergency call setup procedure [94–96]. During the usual attach procedure, the MME tells the mobile whether it supports IMS emergency calls and can supply the mobile with a list of local emergency numbers. (If the IMS does not support emergency calls, then the mobile will place such calls using circuit switched fallback.) The mobile then registers with the IMS in the usual way (1).

Later, the user dials an emergency number, and the mobile detects the situation using a list of standard emergency numbers and any list that it received from the MME. In response, the mobile runs the PDN connectivity procedure from Chapter 13 (2). The mobile does not specify any access point name, but instead indicates the emergency situation using an information element known as the *request type*. The network assigns the mobile a new IP address, sets up a high priority bearer to the emergency APN and supplies the IP address of the proxy CSCF.

The mobile then registers with a serving CSCF in its home network (3), but indicates that this is a separate emergency registration by including an `sos` tag in its `Contact:` header. The mobile registers using a SIP URI, but must choose this from an implicit registration set that also contains a Tel URI, to ensure that the mobile can receive a callback from the public telephone network later on.

Once these steps are complete, the mobile can place the call (4). To do this, it creates a SIP INVITE request in the usual way, supplies some basic location data using its E-UTRAN cell global identity and addresses the request to the emergency URI. The proxy CSCF detects the emergency URI and sends the request to the emergency CSCF. The emergency CSCF places

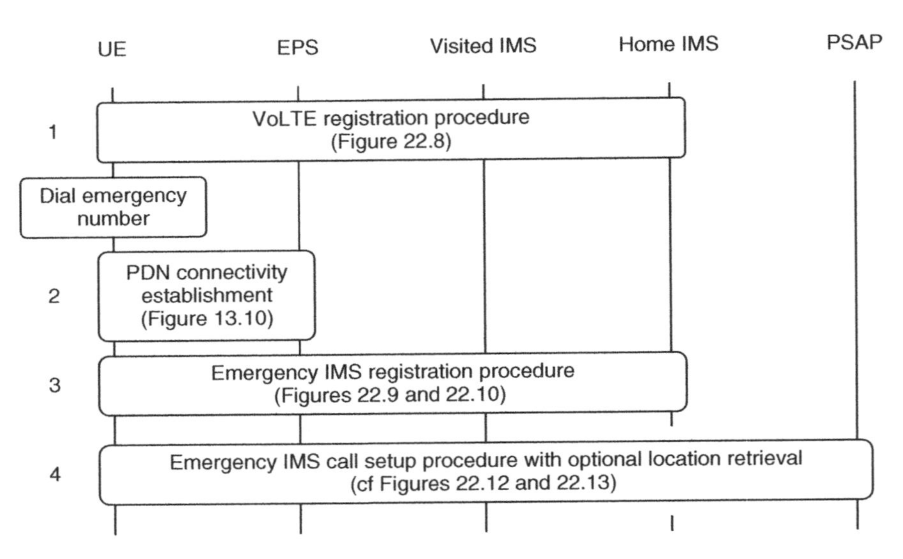

Figure 22.22 Procedure for initiation of an IMS emergency call. Source: TS 23.167. Reproduced by permission of ETSI

the emergency access transfer function in the signalling path and can also ask the location retrieval function for more details of the mobile's location. It then uses the location data to choose a public safety answering point and forwards the INVITE message there. The call setup procedure can now complete as before.

If the mobile is in a network that has no roaming agreement with its home network operator, or if the mobile does not have a UICC installed, then it will be camping on a cell in a limited service state and will not yet be attached to LTE. In this situation, the mobile has skipped step 1, so it runs the attach procedure from Chapter 11 in place of step 2, and sets the attach type to an EPS emergency attach. The MME completes the attach procedure in the usual way, but does not have to authenticate the mobile and can switch off the usual integrity protection algorithms for the sole purpose of the emergency call. Step 3 is omitted, so the user cannot receive a return call later on, but otherwise the call setup procedure continues as shown.

22.11 Delivery of SMS Messages over the IMS

22.11.1 SMS Architecture

The IP multimedia subsystem can deliver SMS messages using the architecture shown in Figure 22.23 [97, 98], which is known as *SMS over IP*. The most important component is the *IP short message gateway* (IP-SM-GW), which acts as an application server towards the IMS, and acts as an MSC server or SGSN towards the SMS interworking and gateway MSCs. It is mandatory for a VoLTE mobile to support SMS over IP, while the network must support either SMS over IP or SMS over SGs.

During the registration procedure, the mobile declares its support for SMS by including the media feature tag `g.3gpp.smsip` in its REGISTER request. The serving CSCF registers the

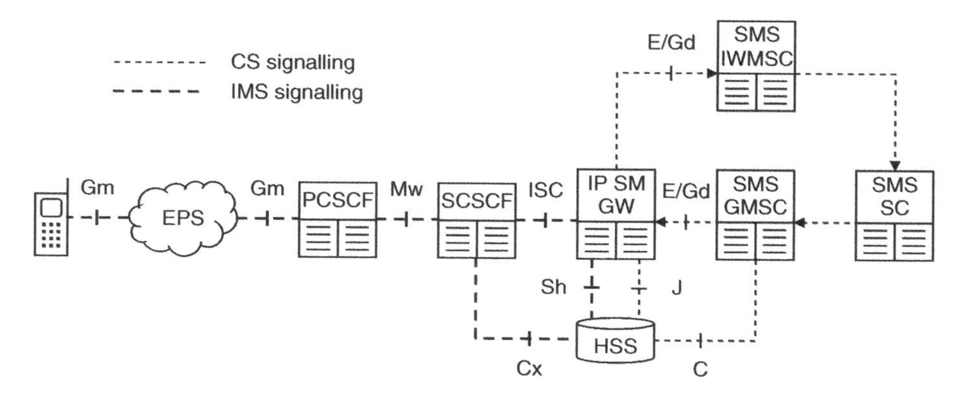

Figure 22.23 Architecture used to deliver SMS messages over the IP multimedia subsystem. Source: TS 23.204. Reproduced by permission of ETSI

mobile with the IP short message gateway, which informs the home subscriber server that it will accept incoming short messages for the mobile. The IMS can then transport an SMS message between the mobile and the IP-SM-GW by embedding it into a SIP *MESSAGE* request, in a similar way to the embedded SDP information that we saw earlier. The procedures are almost identical to the ones for SMS over SGs, so we will not discuss them further.

The IMS can also support two other types of messaging [99]. Using *page mode messaging*, the IMS transports a message by embedding it into a SIP MESSAGE request and can store a transmitted message before forwarding it to the destination user. The technique is similar to SMS, but does not use the SMS protocols or the legacy SMS devices from Figure 22.23. Using *session mode messaging*, the IMS sets up a session between two registered devices, and then transports the message in the IMS user plane using the *message session relay protocol* (MSRP). Neither of these techniques is required by VoLTE.

22.11.2 Access Domain Selection

The access domain selection functions for SMS are similar to those for voice calls, but are independent. The mobile selects the access domain for mobile-originated SMS using a parameter known as the *SMS over IP networks indication*, which is stored by the mobile equipment and can be updated from a device management server [100]. The choice of access domain influences the choice of attach procedure from Figure 22.14, as the mobile can carry out an EPS only attach if it is configured to use SMS over IP, but has to carry out a combined EPS/IMSI attach before using SMS over SGs.

The IP short message gateway maintains a prioritized list of access domains for mobile-terminated SMS [101]. If the IMS has the highest priority, then the gateway first tries to deliver a mobile-terminated SMS across that network. If the procedure fails, then it interrogates the home subscriber server to discover an MSC server and/or SGSN where the mobile is registered, and delivers the message across the circuit or packet switched domain of 2G or 3G.

References

1. Camarillo, G. and Garcia-Martin, M.-A. (2008) *The 3G IP Multimedia Subsystem (IMS): Merging the Internet and the Cellular Worlds*, 3rd edn, John Wiley & Sons, Ltd, Chichester.
2. Noldus, R., Olsson, U., Mulligan, C. *et al.* (2011) *IMS Application Developer's Handbook: Creating and Deploying Innovative IMS Applications*, Academic Press.
3. Poikselkä, M., Holma, H., Hongisto, J. *et al.* (2012) *Voice over LTE (VoLTE)*, John Wiley & Sons, Ltd, Chichester.
4. 3GPP TS 23.218 (2013) IP Multimedia (IM) Session Handling; IM Call Model; Stage 2, Release 11, June 2013.
5. 3GPP TS 23.228 (2013) IP Multimedia Subsystem (IMS); Stage 2, Release 11, September 2013.
6. 3GPP TS 24.228 (2006) Signalling Flows for the IP Multimedia Call Control Based on Session Initiation Protocol (SIP) and Session Description Protocol (SDP); Stage 3, Release 5, October 2006.
7. GSM Association (2013) GSMA VoLTE Initiative, http://www.gsma.com/technicalprojects/volte (accessed 15 October 2013).
8. GSM Association IR.92 (2013) IMS Profile for Voice and SMS, Version 7.1, September 2013.
9. GSM Association IR.88 (2013) LTE and EPC Roaming Guidelines, Version 10.0, July 2013
10. GSM Association IR.65 (2013) IMS Roaming and Interworking Guidelines, Version 12.0, February 2013.
11. GSM Association (2013) Rich Communications, http://www.gsma.com/rcs/ (accessed 15 October 2013).
12. GSM Association RCC.07 (2013) Rich Communication Suite 5.1 Advanced Communications Services and Client Specification, Version 3.0, September 2013.
13. GSM Association IR.90 (2013) RCS Interworking Guidelines, Version 5.0, May 2013.
14. 3GPP TS 23.228 (2013) IP Multimedia Subsystem (IMS); Stage 2, Release 11, Section 4, September 2013.
15. 3GPP TS 23.002 (2013) Network Architecture, Release 11, Sections 4a.7, 5.5, 6a.7, June 2013.
16. 3GPP TS 23.228 (2013) IP Multimedia Subsystem (IMS); Stage 2, Release 11, Annexes E, L, N, September 2013.
17. 3GPP TS 24.229 (2013) IP Multimedia Call Control Protocol Based on Session Initiation Protocol (SIP) and Session Description Protocol (SDP); Stage 3, Release 11, Annexes B, D, E, H, L, M, O, Q, R, S, September 2013.
18. 3GPP TS 23.003 (2013) Numbering, Addressing and Identification, Release 11, Section 13, September 2013.
19. 3GPP TS 31.103 (2012) Characteristics of the IP Multimedia Services Identity Module (ISIM) Application, Release 11, December 2012.
20. IETF RFC 3986 (2005) Uniform Resource Identifier (URI): Generic Syntax, January 2005.
21. IETF RFC 3966 (2004) The tel URI for Telephone Numbers, December 2004.
22. 3GPP TS 23.228 (2013) IP Multimedia Subsystem (IMS); Stage 2, Release 11, Section 5.2.1a, September 2013.
23. GSM Association IR.65 (2013) IMS Roaming and Interworking Guidelines, Version 12.0, Section 2, February 2013.
24. GSM Association IR.34 (2013) Guidelines for IPX Provider Networks, Version 9.1, May 2013.
25. 3GPP TS 29.162 (2012) Interworking between the IM CN Subsystem and IP Networks, Release 11, December 2012.
26. 3GPP TS 29.165 (2013) Inter-IMS Network to Network Interface (NNI), Release 11, September 2013.
27. 3GPP TS 33.203 (2012) Access Security for IP-Based Services, Release 11, June 2012.
28. 3GPP TS 32.240 (2013) Charging Architecture and Principles, Release 11, March 2013.
29. 3GPP TS 32.260 (2013) IP Multimedia Subsystem (IMS) Charging, Release 11, September 2013.
30. IETF RFC 3261 (2002) SIP: Session Initiation Protocol, June 2002.
31. IETF RFC 3455 (2003) Private Header (P-Header) Extensions to the Session Initiation Protocol (SIP) for the 3rd-Generation Partnership Project (3GPP), January 2003.
32. 3GPP TS 24.229 (2013) IP Multimedia Call Control Protocol Based on Session Initiation Protocol (SIP) and Session Description Protocol (SDP); Stage 3, Release 11, September 2013.
33. IETF RFC 4566 (2006) SDP: Session Description Protocol, July 2006.
34. IETF RFC 3264 (2002) An Offer/Answer Model with the Session Description Protocol (SDP), June 2002.
35. IETF RFC 4825 (2007) The Extensible Markup Language (XML) Configuration Access Protocol (XCAP), May 2007.
36. 3GPP TS 24.623 (2012) Extensible Markup Language (XML) Configuration Access Protocol (XCAP) over the Ut Interface for Manipulating Supplementary Services, Release 11, December 2012.
37. 3GPP TS 29.228 (2013) IP Multimedia (IM) Subsystem Cx and Dx Interfaces; Signalling Flows and Message Contents, Release 11, September 2013.

38. 3GPP TS 29.229 (2013) Cx and Dx Interfaces Based on the Diameter Protocol; Protocol Details, Release 11, June 2013.
39. 3GPP TS 29.328 (2013) IP Multimedia (IM) Subsystem Sh Interface; Signalling Flows and Message Contents, Release 11, September 2013.
40. 3GPP TS 29.329 (2013) Sh Interface Based on the Diameter Protocol; Protocol Details, Release 11, June 2013.
41. ITU-T Recommendation H.248.1 (2013) Gateway Control Protocol: Version 3.
42. 3GPP TS 29.228 (2013) IP Multimedia (IM) Subsystem Cx and Dx interfaces; Signalling Flows and Message Contents, Release 11, Annex B, September 2013.
43. IETF RFC 3840 (2004) Indicating User Agent Capabilities in the Session Initiation Protocol (SIP), August 2004.
44. Internet Assigned Numbers Authority (2013) Media Feature Tags, https://www.iana.org/assignments/media-feature-tags/media-feature-tags.xhtml (accessed 15 October 2013).
45. 3GPP TS 23.228 (2013) IP Multimedia Subsystem (IMS); Stage 2, Release 11, Section 4.13, September 2013.
46. 3rd Generation Partnership Project (2013) Uniform Resource Identifier (URI) List, http://www.3gpp.org/Uniform-Resource-Name-URN-list (accessed 15 October 2013).
47. 3GPP TS 26.114 (2013) IP Multimedia Subsystem (IMS); Multimedia Telephony; Media Handling and Interaction, Release 11, September 2013.
48. 3GPP TS 24.173 (2013) IMS Multimedia Telephony Communication Service and Supplementary Services; Stage 3, Release 11, March 2013.
49. 3GPP TS 26.071 (2012) AMR Speech CODEC; General Description, Release 11, September 2012.
50. 3GPP TS 26.171 (2012) Adaptive Multi-Rate – Wideband (AMR-WB) Speech Codec; General Description, Release 11, September 2012.
51. IETF RFC 3550 (2003) RTP: A Transport Protocol for Real-Time Applications, July 2003.
52. IETF RFC 3551 (2003) RTP Profile for Audio and Video Conferences with Minimal Control, July 2003.
53. IETF RFC 4867 (2007) RTP Payload Format and File Storage Format for the Adaptive Multi-Rate (AMR) and Adaptive Multi-Rate Wideband (AMR-WB) Audio Codecs, April 2007.
54. 3GPP TS 26.114 (2013) IP Multimedia Subsystem (IMS); Multimedia Telephony; Media Handling and Interaction, Release 11, Section 10, September 2013.
55. GSM Association IR.92 (2013) IMS Profile for Voice and SMS, Version 7.1, Section 4.3.1, September 2013.
56. 3GPP TS 24.229 (2013) IP Multimedia Call Control Protocol Based on Session Initiation Protocol (SIP) and Session Description Protocol (SDP); Stage 3, Release 11, Section 5.1.1, September 2013.
57. 3GPP TS 23.228 (2013) IP Multimedia Subsystem (IMS); Stage 2, Release 11, Section 5.2.2.3, September 2013.
58. 3GPP TS 29.228 (2013) IP Multimedia (IM) Subsystem Cx and Dx Interfaces; Signalling Flows and Message Contents, Release 11, Annex A.4.1, September 2013.
59. 3GPP TS 33.203 (2012) Access Security for IP-Based Services, Release 11, Section 6.1.1, June 2012.
60. 3GPP TS 29.328 (2013) IP Multimedia (IM) Subsystem Sh Interface; Signalling Flows and Message Contents, Release 11, Annex B.1.1, September 2013.
61. IETF RFC 3265 (2002) Session Initiation Protocol (SIP)-Specific Event Notification, June 2002.
62. IETF RFC 3680 (2004) A Session Initiation Protocol (SIP) Event Package for Registrations, March 2004.
63. 3GPP TS 24.229 (2013) IP Multimedia Call Control Protocol Based on Session Initiation Protocol (SIP) and Session Description Protocol (SDP); Stage 3, Release 11, Sections 5.1.2A, 5.1.3, 6.1, September 2013.
64. IETF RFC 3312 (2002) Integration of Resource Management and Session Initiation Protocol (SIP), October 2002.
65. 3GPP TS 23.228 (2013) IP Multimedia Subsystem (IMS); Stage 2, Release 11, Sections 5.5.1, 5.6.1, 5.7.1, 5.11.3, September 2013.
66. 3GPP TS 29.213 (2013) Policy and Charging Control Signalling Flows and Quality of Service (QoS) Parameter Mapping, Release 11, Section 6.2, Annex B.2, B.3.2, September 2013.
67. IETF RFC 3262 (2002) Reliability of Provisional Responses in the Session Initiation Protocol (SIP), June 2002.
68. IETF RFC 3311 (2002) The Session Initiation Protocol (SIP) UPDATE Method, September 2002.
69. 3GPP TS 23.228 (2013) IP Multimedia Subsystem (IMS); Stage 2, Release 11, Sections 4.3.5, 5.5.3, 5.5.4, 5.6.3, 5.7.3, 5.19, September 2013.
70. IETF RFC 3761 (2004) The E.164 to Uniform Resource Identifiers (URI) Dynamic Delegation Discovery System (DDDS) Application (ENUM), April 2004.
71. 3GPP TS 23.228 (2013) IP Multimedia Subsystem (IMS); Stage 2, Release 11, Sections 5.10, September 2013.
72. 3GPP TS 29.213 (2013) Policy and Charging Control Signalling Flows and Quality of Service (QoS) Parameter Mapping, Release 11, Section 6.2, Annex B.4, September 2013.

73. 3GPP TS 23.401 (2013) General Packet Radio Service (GPRS) Enhancements for Evolved Universal Terrestrial Radio Access Network (E-UTRAN) Access, Release 11, Section 5.4.4.2, September 2013.
74. 3GPP TS 23.221 (2013) Architectural Requirements, Release 11, Section 7.2a, Annex A.2, June 2013.
75. 3GPP TS 24.301 (2013) Non-Access-Stratum (NAS) Protocol for Evolved Packet System (EPS), Release 11, Section 4.3, March 2013.
76. 3GPP TS 23.221 (2013) Architectural Requirements, Release 11, Sections 7.2, 7.2b, 7.3, June 2013.
77. GSM Association IR.64 (2013) IMS Service Centralization and Continuity Guidelines, Version 6.0, Section 3, February 2013.
78. 3GPP TS 29.272 (2013) Mobility Management Entity (MME) and Serving GPRS Support Node (SGSN) Related Interfaces Based on Diameter Protocol, Release 11, Sections 5.2.1.1, 5.2.2.1, September 2013.
79. 3GPP TS 29.328 (2013) IP Multimedia (IM) Subsystem Sh Interface; Signalling Flows and Message Contents, Release 11, Section 7.6.18, Annex E, September 2013.
80. 3GPP TS 23.292 (2013) IP Multimedia Subsystem (IMS) Centralized Services; Stage 2, Release 11, Section 7.3.2.1.3, June 2013.
81. 3GPP TS 23.216 (2013) Single Radio Voice Call Continuity (SRVCC); Stage 2, Release 11, June 2013.
82. 3GPP TS 23.237 (2013) IP Multimedia Subsystem (IMS) Service Continuity; Stage 2, Release 11, September 2013.
83. 3GPP TS 24.237 (2013) IP Multimedia Subsystem (IMS) Service Continuity; Stage 3, Release 11, September 2013.
84. GSM Association IR.64 (2013) IMS Service Centralization and Continuity Guidelines, Version 6.0, Section 4, February 2013.
85. 3GPP TS 29.280 (2013) 3GPP Sv Interface (MME to MSC, and SGSN to MSC) for SRVCC, Release 11, September 2013.
86. 3GPP TS 23.003 (2013) Numbering, Addressing and Identification, Release 11, Section 18, September 2013.
87. 3GPP TS 23.216 (2013) Single Radio Voice Call Continuity (SRVCC); Stage 2, Release 11, Section 6.2.2.2, June 2013.
88. 3GPP TS 24.237 (2013) IP Multimedia Subsystem (IMS) Service Continuity; Stage 3, Release 11, Sections 6.2.1.3, 6.2.2.3, 6.3.1.5, 6.3.1.6, 6.3.2.1.4, September 2013.
89. 3GPP TS 24.237 (2013) IP Multimedia Subsystem (IMS) Service Continuity; Stage 3, Release 11, Sections 6.1.2, 6.2.1.4, 6.2.2.5, 6.3.2.1.9, September 2013.
90. 3GPP TS 23.292 (2013) IP Multimedia Subsystem (IMS) Centralized Services; Stage 2, Release 11, June 2013.
91. GSM Association IR.64 (2013) IMS Service Centralization and Continuity Guidelines, Version 6.0, Section 2, February 2013.
92. 3GPP TS 24.294 (2013) IP Multimedia Subsystem (IMS) Centralized Services (ICS) Protocol via I1 Interface, Release 11, September 2013.
93. 3GPP TS 23.167 (2013) IP Multimedia Subsystem (IMS) Emergency Sessions, Release 11, September 2013.
94. 3GPP TS 23.167 (2013) IP Multimedia Subsystem (IMS) Emergency Sessions, Release 11, Section 7, September 2013.
95. 3GPP TS 24.229 (2013) IP Multimedia Call Control Protocol Based on Session Initiation Protocol (SIP) and Session Description Protocol (SDP); Stage 3, Release 11, Sections 5.1.6, September 2013.
96. 3GPP TS 24.237 (2013) IP Multimedia Subsystem (IMS) Service Continuity; Stage 3, Release 11, Section 6c, September 2013.
97. 3GPP TS 23.204 (2013) Support of Short Message Service (SMS) over Generic 3GPP Internet Protocol (IP) Access; Stage 2, Release 11, September 2013.
98. 3GPP TS 24.341 (2012) Support of SMS over IP Networks; Stage 3, Release 11, December 2012.
99. 3GPP TS 24.247 (2012) Messaging Service using the IP Multimedia (IM) Core Network (CN) Subsystem; Stage 3, Release 11, December 2012.
100. 3GPP TS 24.167 (2012) 3GPP IMS Management Object (MO), Release 11, Section 5.28, December 2012.
101. 3GPP TS 23.204 (2013) Support of Short Message Service (SMS) over Generic 3GPP Internet Protocol (IP) Access; Stage 2, Release 11, Section 6.5a, September 2013.

23

Performance of LTE and LTE-Advanced

In a mobile telecommunication system, two main factors limit a cell's performance: coverage and capacity. Coverage is more important in rural areas, since a mobile far from the base station may not receive a signal that is strong enough for it to recover the transmitted information. Capacity is more important in urban areas, since every cell is limited by a maximum data rate. We consider these issues in this final chapter.

We begin by examining the peak data rate of an LTE mobile, and review the issues that might prevent it from reaching that data rate. We then discuss the techniques used for link budget estimation and propagation modelling in LTE and use the results to estimate the coverage of an LTE cell. We develop the techniques to produce an initial estimate of the capacity of an LTE cell and compare this with more robust estimates that are obtained from simulations. We conclude by examining the typical data rates and cell capacities for voice over IP.

23.1 Peak Data Rates of LTE and LTE-Advanced

23.1.1 Increase of the Peak Data Rate

Figure 23.1 shows how the peak data rate of LTE has increased since its introduction in Release 8 and compares it with the peak data rate of WCDMA from Release 99. The data are taken from the most powerful UE capabilities available in FDD mode at each release [1–3]. The vertical axis is logarithmic, which is consistent with the large increases in peak data rate that have been achieved since the introduction of 3G systems.

In Release 99, WCDMA had a peak theoretical data rate of 2 Mbps on the downlink and 1 Mbps on the uplink, although practical devices were usually limited to 384 kbps. The introduction of high-speed downlink packet access in Release 5 increased the peak downlink data rate to 14.4 Mbps, by the use of a faster coding rate and a new modulation scheme, 16-QAM. There was a similar increase for the uplink in Release 6, through the introduction of high-speed

An Introduction to LTE: LTE, LTE-Advanced, SAE, VoLTE and 4G Mobile Communications, Second Edition.
Christopher Cox.

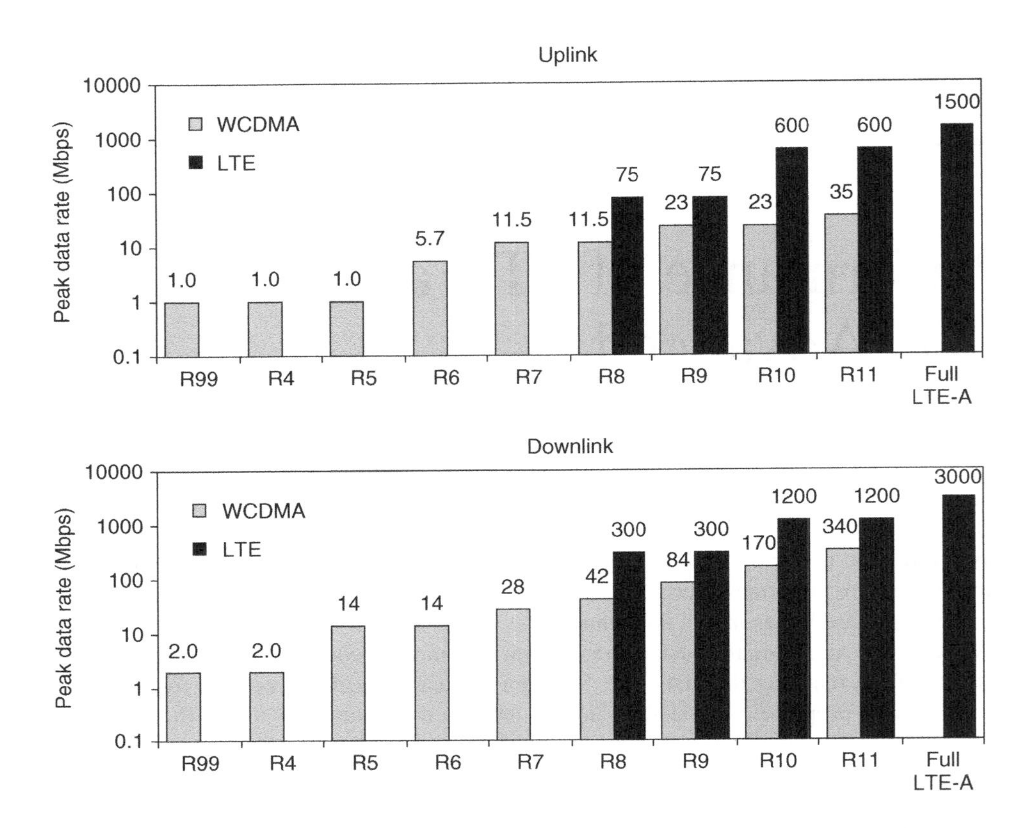

Figure 23.1 Evolution of the peak data rates of WCDMA and LTE in FDD mode

uplink packet access. Later releases have increased the peak data rate further, through the introduction of 64-QAM and spatial multiplexing, and the use of multiple carriers.

The peak data rate in LTE Release 8 is 299.6 Mbps in the downlink and 75.4 Mbps in the uplink (Table 2.1). These figures are easily understood. In the uplink, the mobile's greatest likely allocation is 96 resource blocks, because we need to reserve some resource blocks for the PUCCH and because we need an allocation with prime factors of 2, 3 or 5 only. Each resource block lasts for 0.5 ms and carries 72 PUSCH symbols (Figure 8.8), so it supports a symbol rate of 144 ksps. With a modulation scheme of 64-QAM, the resulting bit rate on the PUSCH is 82.9 Mbps. We can therefore support an information rate of 75.4 Mbps using a coding rate of 0.91, which is a reasonable maximum figure. (The effect of the CRC bits is small, so we can ignore them.)

In the downlink, the symbol rate per resource block depends on the number of transmit antennas and the size of the downlink control region, as shown in Table 23.1. With four transmit antennas and one control symbol, the result is 136 ksps. Using four layers, 100 resource blocks and a modulation scheme of 64-QAM, the resulting bit rate on the PDSCH is 326.4 Mbps.

Table 23.1 Symbol rate per resource block for the PDSCH (ksps)

Number of antennas	Number of control symbols			
	1	2	3	4 (1.4 MHz only)
1	150	138	126	114
2	144	132	120	108
4	136	124	112	100

Ignoring the CRC bits as before, we can support an information rate of 299.6 Mbps using a coding rate of 0.92. This figure is almost identical to the coding rate used for a channel quality indicator of 15 (Table 8.4).

The peak data rates in releases 10 and 11 are about 1200 Mbps on the downlink and 600 Mbps on the uplink. These arise through the use of two component carriers, eight layers on the downlink and four layers on the uplink, and meet the peak data rate requirements of IMT-Advanced. The peak data rates will eventually increase to 3000 and 1500 Mbps respectively, through the use of five component carriers.

23.1.2 Limitations on the Peak Data Rate

As we noted in Chapter 1, we can only reach the peak data rates shown above in special circumstances. There are five main criteria. Firstly, the cell must be transmitting and receiving in its maximum bandwidth of 20 MHz. This is likely to be an unusually large allocation, at least in the early days of LTE, with a value of 10 or 5 MHz being more common. In these lower bandwidths, the peak data rate will be a factor of 2 or 4 less.

Secondly, the mobile must have the most powerful UE capabilities that are available at each release. In Release 8, for example, category 5 mobiles are the only ones that support four layer spatial multiplexing on the downlink, or the use of 64-QAM on the uplink. If we switch to a category 3 mobile, common in early LTE devices, then the peak downlink data rate falls by another factor of 3 and the peak uplink data rate falls by another factor of 1.5.

Thirdly, the mobile should be close to the base station. If it is not, then the received signal-to-interference plus noise ratio may be low, and the receiver may be unable to handle the fast modulation schemes and coding rates that are required for a high data rate.

Fourthly, the cell should be well isolated from other nearby cells. This condition can often be achieved in femtocells and picocells, which are usually indoors and are isolated by the surrounding walls. A similar but weaker result applies to microcells, which are partially isolated from each other by the intervening buildings. In macrocells there is little such isolation, so the receiver may pick up significant interference from nearby cells. This reduces the SINR, and prevents the receiver from handling a fast modulation scheme or a high coding rate.

The final condition is that the mobile must be the only active mobile in the cell. If it is not, then the cell's capacity will be shared amongst all its mobiles, resulting in a large drop in the peak data rate that is available to each one.

23.2 Coverage of an LTE Cell

23.2.1 Uplink Link Budget

In Chapter 3, we defined the *propagation loss* or *path loss* (PL) as the ratio of the transmitted signal power P_T to the received signal power P_R:

$$PL = \frac{P_T}{P_R} \tag{23.1}$$

In a *link budget*, we estimate the largest value of P_T that the transmitter can send and the smallest value of P_R at which the receiver can recover the information. We can then use Equation 23.1 to estimate the greatest propagation loss that the system can handle. By combining this figure with the results of a propagation model, we can estimate the greatest possible distance between the transmitter and the receiver.

To illustrate the process, Table 23.2 shows an example link budget for the physical uplink shared channel (PUSCH). The numbers are typical for an LTE macrocell, but there will be wide variations from one deployment scenario to another. References [4–8] contain some other examples and parameter values.

Let us run through the terms in the link budget, beginning with the mobile transmitter. The 3GPP specifications limit the mobile's transmit power to 23 dBm (decibels relative to 1 mW), in other words 200 mW [9]. The antenna gain describes the focussing effect of the mobile's transmit antenna and is measured in decibels relative to an isotropic antenna (dBi), in other words an antenna that radiates equally in all directions. The mobile's antenna gain is usually set to 0 dBi, although it can differ by a few decibels each way. The body loss describes the effect of the user and is generally set to 3 dB for the case of voice services and 0 dB otherwise. If we bring these quantities together, we arrive at the *equivalent isotropic radiated power* (EIRP).

Table 23.2 Example link budget for the PUSCH

Quantity	Value	Units	
UE transmit power	23	dBm	a
UE antenna gain	0	dBi	b
Body loss	0	dB	c
Equivalent isotropic radiated power	23	dBm	d = a + b − c
Noise spectral density	−174	dBm Hz^{-1}	e
UE transmitter bandwidth (2 RBs)	55.6	dB Hz	f
eNB noise figure	3.5	dB	g
SINR target (0.28 bits per symbol)	−7.7	dB	h
Receiver sensitivity	−122.6	dBm	i = e + f + g + h
Cable and connector losses	0.5	dB	j
eNB antenna gain	18	dBi	k
Isotropic receive level	−140.1	dBm	l = i + j − k
Interference margin	3	dB	m
Penetration loss	12	dB	n
Slow fading margin	5	dB	o
Link losses and margins	20	dB	p = m + n + o
Maximum propagation loss	143.1	dB	q = d − l − p

The next three rows describe the noise in the base station receiver. The *noise spectral density* is the amount of thermal noise in a bandwidth of 1 Hz. It equals k*T*, where k is Boltzmann's constant and *T* is the temperature of the receiver, and is about $-174\,\text{dBm}\,\text{Hz}^{-1}$ at room temperature. We now have to bring in the bandwidth that we are actually using, in units of decibels relative to 1 Hz (dB Hz). The appropriate bandwidth in the LTE uplink is the mobile's transmission bandwidth, since this is the bandwidth into which the mobile is injecting the transmit power that we noted earlier. In Table 23.2, we have assumed a bandwidth of two resource blocks. (As discussed later, the uplink link budget is not sensitive to the exact value chosen here.) Finally, the *noise figure* accounts for the extra noise that arises because of imperfections in the analogue components of the receiver.

The *SINR target* is the minimum signal-to-interference plus noise ratio that is needed for satisfactory reception. It depends on the chosen modulation scheme and coding rate, and also on issues such as the fading characteristics of the radio channel and the mobile's speed. We can often obtain suitable figures from manufacturers' data, but we can also estimate the SINR target by adjusting the Shannon limit to account for imperfections in the receiver's digital signal processing [10]:

$$\epsilon = \text{Min}\left[\epsilon_{\text{max}},\ \eta \log_2\left(1 + \text{EAG}\frac{\text{SINR}}{\text{SINR}_{\text{eff}}}\right)\right] \tag{23.2}$$

Here, ϵ is the number of information bits per symbol, while η and SINR_{eff} are correction factors. Following the above reference, we have set η to 0.68, SINR_{eff} to 1.05 and ϵ_{max} to 3.2 so as to model transmissions in a fading channel that are limited to 4/5 rate 16-QAM. (There is a similar but a more conservative set of figures in Reference [11].) Using these figures, the SINR target in Table 23.2 corresponds to 0.28 bits per symbol, which is equivalent to a downlink channel quality indicator (CQI) between 2 and 3.

Using a bandwidth of two resource blocks, the mobile can transmit 288 symbols per subframe. At 0.28 bits per symbol, that figure equates to 80 information bits; in other words, a 24 bit CRC appended to a 56 bit transport block. The transport block is large enough for the mobile to send an RRC connection request, which requires a 48 bit RRC message and an 8 bit MAC header. If the layer 2 headers in the user plane add up to 24 bits, then the transport block also supports an uplink data rate of 32 kbps.

Amongst these figures, the only important quantity is the transport block size at the cell edge or, equivalently, the uplink data rate. If the data rate remains constant, then we can increase the uplink bandwidth and reduce the number of bits per symbol, or vice versa, without having much impact on the uplink link budget. We will illustrate this effect in Section 23.2.4.

Some care is required for the measurement of signal levels in a diversity receiver. In Table 23.2 and Equation 23.2 we have chosen to quote the SINR at each individual receive antenna. The equation therefore contains an explicit array gain, EAG, which equals 3 dB for the case of two antenna receive diversity. We could instead quote the combined SINR from both receive antennas, in which case the SINR target would be 3 dB greater and the explicit array gain would move to the link budget.

By combining the SINR target with the receiver noise, we arrive at the *receiver sensitivity*, which is the weakest signal that the base station's receiver can successfully handle. We now need to introduce the receive antenna gain, typically 18 dBi for a three-sector base station, and any cable or connector losses between the antenna and the receiver. By doing so, we arrive at the *isotropic receive level* (IRL), which is the weakest signal that the base station can successfully handle at its receive antenna.

In Table 23.2, we have assumed that the base station just has one receive amplifier, which could be sited either in a remote radio head, as in this example, or at the bottom of the mast with the use of a higher cable loss. Some base station receivers use two amplifiers, with one in both of these locations. To model such a deployment, we have to remove the cable loss and noise figure from the uplink link budget and replace them with a single combined noise figure F_{total} that is calculated using the Friis noise formula:

$$F_{total} = F_1 + \frac{F_2 - 1}{G_1} + \frac{F_3 - 1}{G_1 G_2} + \frac{F_4 - 1}{G_1 G_2 G_3} + \ldots \quad (23.3)$$

Here, F_i and G_i are the noise figure and gain of the ith stage of the receiver, and are measured in real numbers rather than decibels. The noise figure is the reciprocal of the gain for a passive component such as a cable, while for an active component such as an amplifier, the two quantities are independent.

We now have to account for issues in the air interface. If the base station is outdoors but the mobile is indoors then the signal is weakened by *penetration losses* through the building walls, which are hard to estimate but are typically in the range 10 to 20 dB. The *interference margin* accounts for interference from LTE mobiles in nearby cells that are transmitting on the same resource blocks. It is hard to estimate the uplink interference margin reliably, because it is sensitive to the locations and transmit powers of a small number of interfering mobiles, so we generally impose a value instead. Typical interference margins are a few decibels in macrocells but greater in microcells because the smaller propagation losses permit higher levels of interference there.

So far, we have assumed that the received signal power is constant. In practice, the received signal power can fluctuate as the mobile moves in and out of the shadows formed by hills and buildings. The effect is known as *slow fading* or *log-normal fading*. Sometimes slow fading does not matter as the mobile can simply experience it as a variation in the received data rate. If, however, we are interested in constant data rate services or in worst case reception conditions, then we can make the link budget more conservative by including a *slow fading margin*. We can estimate the slow fading margin in an isolated cell using Jakes' equation [12]. The margin required in a wide area network is rather less, because the mobile may be able to see more than one base station yet only requires a satisfactory signal from one of them.

The link budgets for GSM and UMTS also included a *fast fading margin*, which accounted for fluctuations in the received signal power due to destructive interference between the incoming rays. The LTE link budget sometimes requires a fast fading margin, particularly in scenarios where the delay spread and the mobile's speed are both low so that the mobile can become stuck in a wide bandwidth fade. In other scenarios, we may be able to minimize the power fluctuations by the use of time and frequency-dependent scheduling, and we have assumed that to be the situation here. By combining the losses and margins with the EIRP and IRL from earlier, we finally arrive at the maximum propagation loss that the base station can handle.

23.2.2 Downlink Link Budget

Table 23.3 shows the corresponding link budget for the physical downlink shared channel (PDSCH). Most of the terms are the same as in the uplink, so we will focus on the differences.

The 3GPP specifications do not define the base station's transmit power in the case of wide area transmission, but typical figures are from 40 to 46 dBm (10 to 40 W). The specifications do, however, constrain the transmit power in a picocell or femtocell, which limits the received

Table 23.3 Example link budget for the PDSCH

Quantity	Value	Units	
eNB transmit power	43	dBm	a
Cable and connector losses	0.5	dB	b
eNB antenna gain	18	dBi	c
Equivalent isotropic radiated power	60.5	dBm	d = a − b + c
Noise spectral density	−174	dBm Hz^{-1}	e
eNB transmitter bandwidth	69.5	dB Hz	f
UE noise figure	7	dB	g
SINR target (0.66 bits per symbol)	−4.2	dB	h
Receiver sensitivity	−101.7	dBm	i = e + f + g + h
UE antenna gain	0	dBi	j
Body loss	0	dB	k
Isotropic receive level	−101.7	dBm	l = i − j + k
Interference margin	2.1	dB	m
Penetration loss	12	dB	n
Slow fading margin	5	dB	o
Link losses and margins	19.1	dB	p = m + n + o
Maximum propagation loss	143.1	dB	q = d − l − p

signal power in scenarios where the base station is close to the user [13]. The appropriate bandwidth in the LTE downlink is the whole bandwidth of the cell, since this is the bandwidth into which the base station is injecting its power. In the example shown, we have used a bandwidth of 50 resource blocks, which corresponds to a 10 MHz cell.

The downlink SINR target can be estimated in a similar way to the uplink. Following Reference [14], we have set η to 0.78, SINR_{eff} to 0.95 and $\epsilon_{\max}$ to 4.8, for transmissions in a fading channel that use Alamouti's diversity technique and frequency-dependent scheduling, and are limited to 4/5 rate 64-QAM.

Unlike the situation in the uplink, we can also make a satisfactory estimate of the downlink interference margin as follows [15]:

$$\text{IM} = \frac{1}{1 - \gamma \frac{\text{SINR}}{\text{SIR}_{\min}}} \tag{23.4}$$

Here, $\text{SIR}_{\min}$ is the signal-to-interference ratio at the edge of the cell, which depends on the network's geometry and is typically in the range −4 to −1 dB, while SINR is the target signal-to-interference plus noise ratio at the cell edge. γ is the target cell load, in other words the percentage of resource elements that are actually used, and IM is the resulting interference margin. The SINR target in Table 23.3 corresponds to 0.66 bits per symbol, equivalent to a CQI between 4 and 5. To calculate the resulting interference margin, we have set $\text{SIR}_{\min}$ to −3 dB and γ to 50%. The resulting propagation loss is the same as that of the uplink.

23.2.3 Propagation Modelling

Propagation models relate the propagation loss to the distance between the transmitter and the receiver. Several propagation models exist and vary greatly in their complexity. A simple and frequently used example is the Okumura-Hata model, which predicts the coverage of macrocells in the frequency range 150 to 1500 MHz [16]. The model was later extended to the range 1500 to 2000 MHz, as part of a project in the *European Cooperation in Science and Technology* (COST) framework known as COST 231 [17]. As another example, the *Wireless World Initiative New Radio* (WINNER) consortium has developed propagation models for several different propagation scenarios in the frequency range 2 to 6 GHz [18].

The results of a propagation model depend upon the carrier frequency: low carrier frequencies such as 800 MHz are associated with a high coverage, while at high carrier frequencies such as 2600 MHz, the coverage is less. The main reason is that the receive antenna has an effective collecting area proportional to λ^2, where λ is the wavelength of the incoming radio waves. As the carrier frequency increases, the wavelength falls, and the power collected by the receive antenna becomes progressively less. Because of this, operators often prefer low carrier frequencies for wide area networks, while reserving high carrier frequencies for boosting network capacity in urban areas or for carrier aggregation.

We do, however, need to issue a warning about the use of propagation models. The parameters are estimated by fitting the predictions to a large number of measurements, but, in practice, the actual propagation loss can vary greatly from one environment to another. To deal with this, it is important for a network operator to measure the actual radio propagation in the region of interest by means of drive tests and to adjust the parameters in the chosen model using the results. If we use a propagation model without making these adjustments, then we get a rough estimate of the coverage of a cell, but no more.

23.2.4 Coverage Estimation

With the above warning issued, we can combine the link budget with a suitable propagation model and estimate the size of the cell. If we use the COST 231/Okumura-Hata model at a carrier frequency of 1800 MHz, then the earlier examples give a maximum range of 1.6 km. As usual, wide variations are possible.

We can go further, by calculating the maximum range for a variety of modulation and coding schemes instead of just one. The results are shown in Figure 23.2, in which the horizontal axes show the distance between the base station and the mobile, and the vertical axes show the achievable number of bits per symbol. In the downlink, the result depends only on the distance between the base station and the mobile: a nearby mobile receives a stronger signal than a distant one, so can handle a faster modulation and coding scheme. In the uplink, the result also depends on the mobile's transmission bandwidth: as the transmission bandwidth rises, the noise in the base station receiver rises, and the SINR and the achievable number of bits per symbol both fall.

By introducing the mobile's bandwidth a second time, we can convert the achievable number of bits per symbol to the achievable data rate. The results are shown in Figure 23.3, in which the vertical axes are now logarithmic. In the downlink, the data rate is directly proportional to the mobile's reception bandwidth, and is limited to about 32 Mbps in this example because we are operating in a 10 MHz bandwidth and are not using MIMO. The relationship in the uplink is more complex, because a large bandwidth increases the data rate if the number of bits per symbol remains fixed, but reduces the achievable number of bits per symbol in accordance with Figure 23.2. However, the most important result is that a mobile cannot transmit with a high data rate at the edge of the cell because it is limited by its maximum transmit power.

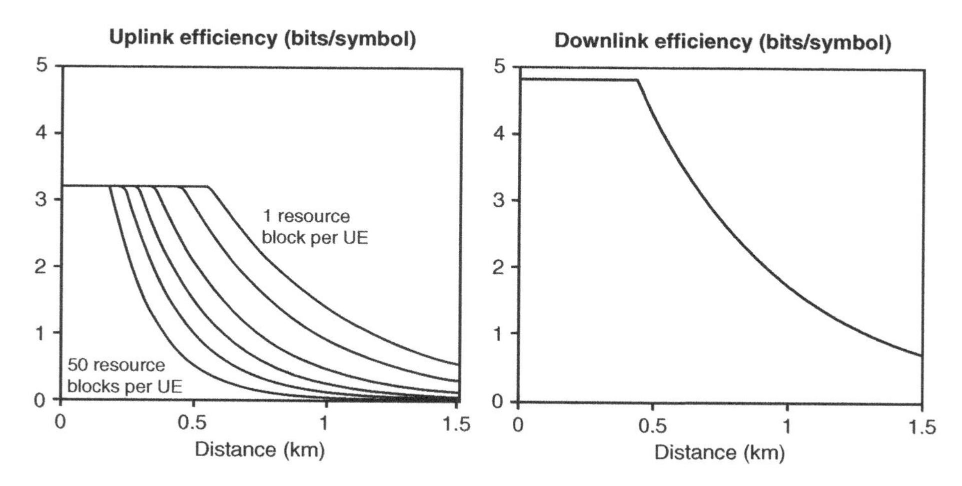

Figure 23.2 Example relationships between distance and the achievable number of bits per symbol, for allocations of 1, 2, 5, 10, 20 and 50 resource blocks per mobile

23.3 Capacity of an LTE Cell

23.3.1 Capacity Estimation

We can use the results from the previous section to make an initial estimate of the downlink cell capacity. To do this, we simply add the downlink data rates in Figure 23.3, under the assumption that the cell contains a large number of mobiles that are uniformly distributed across the cell. For services with a variable data rate, it is reasonable to assume that the scheduling algorithm gives the same number of resource blocks to each mobile, in a manner consistent with the cell load of 50% that we quoted earlier. We can also omit the slow fading margin from the

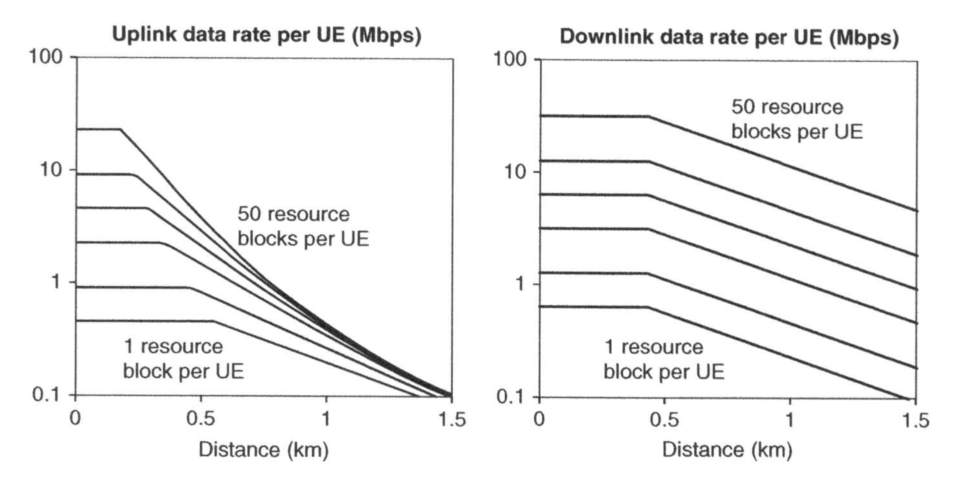

Figure 23.3 Example relationships between distance and data rate, for allocations of 1, 2, 5, 10, 20 and 50 resource blocks per mobile

link budget, on the grounds that we are now considering the mobiles' typical data rates rather than the worst case.

If we carry out the calculation in this way, then the downlink cell capacity for the example considered earlier comes to about 9.5 Mbps. If we allow the cell load to increase to 100%, then the result increases to 14 Mbps, although it does not double because of the higher levels of interference.

The downlink capacity can vary in several ways. The capacity will be greater if some of the mobiles are in better propagation conditions than the ones assumed in the link budget, for example if they are outdoors. The capacity will also be greater if the cell supports downlink MIMO, because mobiles with a high SINR will then be able to receive at a higher data rate. On the other hand, the capacity will be less if the services have a fixed data rate, because the base station will have to allocate a disproportionate number of resource blocks to distant mobiles that cannot use them efficiently. Finally, the downlink capacity will fluctuate if the cell only contains a few mobiles, which can lead to much higher data rates if there is just a single mobile located close to the base station.

23.3.2 Cell Capacity Simulations

More robust estimates of the cell capacity are typically made using simulations. To illustrate them, we will look at some simulations that have been carried out by 3GPP during system design [19, 20].

The simulations we will describe are known as 3GPP case 1 and case 3. They are both carried out for macrocell geometries, in which the receiver can pick up significant interference from nearby cells. The coverage requirements are demanding, as the base station is outdoors but the mobile phone is indoors, with a penetration loss of 20 decibels through the walls of the building. Furthermore, the carrier frequency is 2 GHz, a higher-than-average value that is associated with worse-than-average propagation losses. To compensate for these issues, the distance between the base stations is rather low for a macrocell geometry, at 500 m in case 1 and 1732 m in case 3. There are three sectors per site: we will show the capacity of each sector throughout.

Figure 23.4 shows the total sector capacity in FDD mode from a number of different simulations in the uplink and downlink, which cover baseline results from WCDMA Release 6, early results from LTE, later results from LTE and early results from LTE-Advanced. The figure shows the results for a bandwidth of 10 MHz, which is the same bandwidth that was actually used for the simulations of LTE, but implies the use of two carriers per sector for the case of WCDMA. The vertical axis is linear, which is different from Figure 23.1, but is consistent with the smaller increase in capacity since the introduction of 3G systems.

There are several points to note. Firstly, the downlink cell capacities are somewhat greater than they were in the examples from earlier, but they are still broadly comparable. The main differences are that the simulations use a cell load of 100% and permit the use of downlink MIMO, both of which serve to increase the resulting capacity. The net effect of other differences, such as the cell size and the penetration loss, is small.

Secondly, the cell size makes little difference to the results, at least in the range we are considering, since there is little difference between cases 1 and 3. Instead, the data rates are limited by the interference from neighbouring cells, which depends on the amount of overlap between them. To increase the cell capacity, we would have to move to a microcell or femtocell

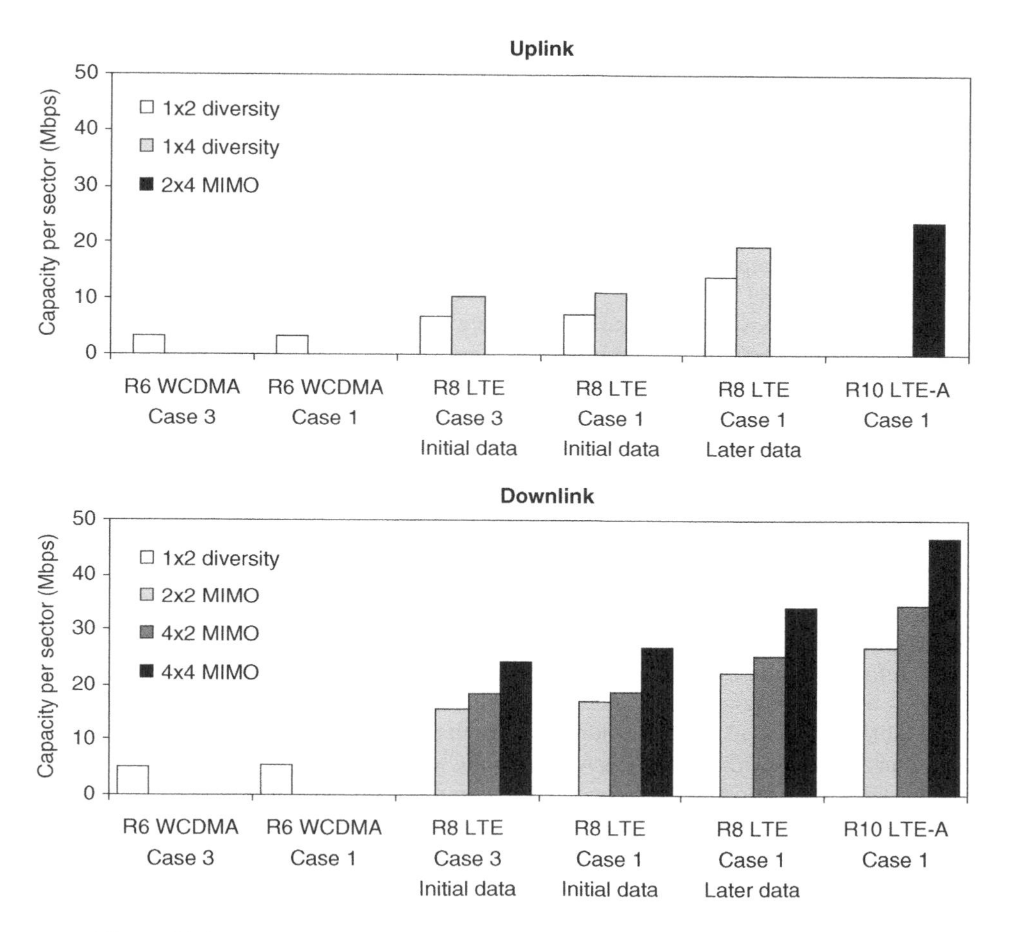

Figure 23.4 Capacity per sector of a typical LTE macrocell in a 10 MHz bandwidth

geometry, in which the greater isolation would reduce the interference and permit a higher data rate.

Thirdly, the capacity increases significantly as we move from the baseline WCDMA results to the early results for LTE. There are several reasons for this, particularly the more effective treatment of fading and inter-symbol interference through the use of OFDMA and SC-FDMA, and the introduction of higher order diversity on the uplink and spatial multiplexing on the downlink.

Finally, there is another increase in capacity as we move from the initial simulations of LTE to the later ones. Possible reasons include improvements in the receiver software, which allow the receiver to support a high data rate at lower signal-to-interference ratios than before. There is another capacity increase as we move from LTE to LTE-Advanced. This is primarily due to the introduction of single user MIMO in the uplink and a proper implementation of multiple user MIMO in the downlink.

Estimates of sector capacity should be treated with caution, since they are sensitive to issues such as the bandwidth, the antenna geometry and the amount of incoming interference.

Roughly speaking, however, we can say that the capacity of a 10 MHz LTE macrocell in these simulations is about 25 Mbps per sector in the downlink and 15 Mbps per sector in the uplink. These figures are far less than the peak data rates quoted earlier and have to be shared amongst all the mobiles in the cell.

There is a useful analogy here with the motor industry. In Release 8 LTE, we have seen that the maximum data rate on the downlink is 300 Mbps, while the typical capacity of a macrocell is somewhere around 25 Mbps per sector. Coincidentally, the top speed of a Ferrari is around 300 kph (180 mph), while the average traffic speed across Greater London is somewhere around 25 kph (15 mph) [21]. So, at the risk of stretching the analogy too far, we are about as likely to achieve 300 Mbps in a Release 8 macrocell, as we are to drive a Ferrari at 300 kph on the streets of London. A similar situation applies in Release 10, except that we can exchange the Ferrari for the Thrust SSC [22].

23.4 Performance of Voice over IP

23.4.1 AMR Codec Modes

We close this chapter by discussing how an LTE network handles voice over IP. VoIP calls most often use the adaptive multi rate (AMR) codec, which digitizes speech information using 8 bit samples at a rate of 8 ksps, giving an underlying bit rate of 64 kbps. It then collects the samples into 20 ms *frames* and compresses each one to a rate that depends on the *AMR mode*. Table 23.4 lists the different AMR modes that are available.

In the table, the payload size is the number of compressed data bits, while the frame size also includes an AMR header. The table is for the bandwidth efficient format of AMR: in the octet aligned format, the payload size is rounded up to a multiple of 8 bits before header addition, so the frame size is slightly greater. If the user is not speaking then the codec transmits occasional *silence information descriptor* (SID) frames, which give an audible indication to the other user that the line is still working.

Traditionally, listeners have assessed the quality of a voice call subjectively, using a *mean opinion score* (MOS) that ranges from 1 (bad) up to 5 (excellent). The table lists some mean opinion scores that have been gathered in this way [23]. The quality of voice calls can also be assessed automatically using algorithms whose output is designed to replicate the mean opinion scores of human listeners. An example is *perceptual objective listening quality assessment*

Table 23.4 AMR codec modes, including frame sizes for the bandwidth efficient format and example mean opinion scores

	AMR mode								
	12.2	10.2	7.95	7.4	6.7	5.9	5.15	4.75	SID
Payload size (bits)	244	204	159	148	134	118	103	95	39
Frame size (bits)	256	216	176	160	144	128	120	112	56
Example MOS	3.20	3.08	3.10	3.00	2.90	2.90		2.59	

(POLQA) [24], an ITU standard that is a successor to *perceptual evaluation of speech quality* (PESQ) [25].

The wideband AMR (AMR-WB) codec captures higher audio frequencies by using a sample rate of 16 ksps and produces compressed data rates that are roughly twice as high as before. Table 23.5 lists the corresponding frame sizes and mean opinion scores for wideband AMR. We can immediately see how much better this codec performs than AMR, even when using similar compressed data rates.

23.4.2 Transmission of AMR Frames on the Air Interface

Table 23.6 shows how the LTE air interface might typically handle some commonly used AMR modes. The table is for the modes and header sizes that are quoted in References [26, 27] and uses an example bandwidth allocation of three resource blocks.

The RTP, UDP and IP protocols add their own headers to the AMR frames from earlier, but the packet data convergence protocol compresses those headers to just a few bytes. The quoted figures assume a size of 48 bits in the case of the silence information descriptor frames and 24 bits otherwise. The air interface's layer 2 protocols then add headers of their own, with the usual sizes being 8 bits each [28–30]. There are also some spare bits to allow for variations in the header size, MAC control elements or MAC padding. The result is mapped onto one of the allowed transport block sizes for the LTE physical layer [31]. These have been designed to handle six of the above modes efficiently by the use of 16 spare bits, while the additional overheads for the other two modes are small.

The transport blocks are now mapped to 1 ms sub-frames and are transmitted with an interval of 20 ms, typically using semi-persistent scheduling. The table uses an example bandwidth allocation of three resource blocks per mobile, at which all of the quoted transport block sizes are available. With this allocation, each transport block occupies 432 resource elements on the PUSCH: the same figure is suitable for the PDSCH if the base station has two antennas and the control region contains one symbol, and changes only slightly in other configurations. The resulting number of bits per symbol is between about 0.4 and 0.8, which corresponds to a downlink channel quality indicator between 3 and 5.

If the radio conditions improve, then the base station can reduce the mobile's bandwidth allocation and/or change its modulation scheme to 16-QAM. (The format of semi-persistent scheduling commands precludes the use of 64-QAM [32].) If the conditions degrade, then

Table 23.5 Wideband AMR codec modes, including frame sizes for the bandwidth efficient format and example mean opinion scores

	AMR-WB mode									
	23.85	23.05	19.85	18.25	15.85	14.25	12.65	8.85	6.6	SID
Payload size (bits)	477	461	397	365	317	285	253	177	132	40
Frame size (bits)	488	472	408	376	328	296	264	192	144	56
Example MOS				4.06	3.94	3.93	3.78	3.34	3.07	

Table 23.6 Transmission of selected AMR and AMR-WB modes on the air interface, for an example allocation of 3 resource blocks

	AMR-WB mode			AMR mode				
	12.65	8.85	6.6	12.2	7.4	5.9	4.75	SID
AMR frame size	264	192	144	256	160	128	112	56
RTP, UDP & IP headers	24	24	24	24	24	24	24	48
PDCP, RLC & MAC headers	24	24	24	24	24	24	24	24
MAC padding / spare	16	16	16	24	16	32	16	16
Transport block size	328	256	208	328	224	208	176	144
Transport block CRC	24	24	24	24	24	24	24	24
Bits before coding	352	280	232	352	248	232	200	168
Bit rate (kbps)	17.6	14.0	11.6	17.6	12.4	11.6	10.0	
Physical channel symbols	432	432	432	432	432	432	432	432
Number of bits per symbol	0.81	0.65	0.54	0.81	0.57	0.54	0.46	0.39

the base station can increase the bandwidth allocation. If that leads to congestion, then the base station can ask the mobile to change to a slower AMR mode using the rate adaptation mechanism that we introduced in Chapter 22.

In Chapter 8, we noted that the coverage of LTE voice risks being limited by the uplink. We can see this effect in Figure 23.3, which highlights two issues: the uplink data rate at the cell edge is less than that of the downlink, and is almost unaffected by changes in the uplink transmission bandwidth. The problem can be alleviated using the technique of TTI bundling, in which the mobile repeats its uplink transmissions four times so as to increase the received signal energy, at the expense of increasing the mobile's use of resources.

Reference [33] describes simulations of the cell capacity for voice over IP, using the AMR codec at a rate of 12.2 kbps and a cell bandwidth of 10 MHz. The results suggest downlink capacities of around 634 and 578 users per sector for 3GPP cases 1 and 3, and uplink capacities of 482 and 246 users, consistent with the idea that the voice service is limited by the uplink. In contrast, the corresponding figure in a basic GSM network is around 30 users per sector, assuming that each GSM carrier is used once in a cluster of four 3-sector sites, and supports a maximum of eight voice users through the use of a full rate voice codec.

23.4.3 Transmission of AMR Frames in the Fixed Network

In the fixed network, we start with the same AMR frames as before, but this time we have to add the full RTP, UDP and IP headers without any compression. The results are shown in Table 23.7 [34]. Assuming the use of IP version 6, the resulting packet sizes lie between 536 and 744 bits, and the corresponding bit rates lie between 29.6 and 37.2 kbps. These last figures are the same ones that the session description protocol uses to request a resource allocation within the IMS, with the slightly larger figure in Table 22.4 arising through the use of the octet aligned format.

Table 23.7 Transmission of selected AMR and AMR-WB modes in the fixed network

	AMR-WB mode			AMR mode				
	12.65	8.85	6.6	12.2	7.4	5.9	4.75	SID
AMR frame size	264	192	144	256	160	128	112	56
RTP header	96	96	96	96	96	96	96	96
UDP header	64	64	64	64	64	64	64	64
IPv6 header	320	320	320	320	320	320	320	320
IP packet size	744	672	624	736	640	608	592	536
Bit rate (kbps)	37.2	33.6	31.2	36.8	32.0	30.4	29.6	

References

1. 3GPP TS 25.306 (2013) UE Radio Access Capabilities, Release 11, Section 5, September 2013.
2. 3GPP TS 36.306 (2013) User Equipment (UE) Radio Access Capabilities, Release 11, Section 4.1, September 2013.
3. 3GPP TS 36.101 (2013) User Equipment (UE) Radio Transmission and Reception, Release 11, Section 5.6A, September 2013.
4. Penttinen, J. (2011) *The LTE/SAE Deployment Handbook*, John Wiley & Sons, Ltd, Chichester.
5. Holma, H. and Toskala, A. (2011) *LTE for UMTS: Evolution to LTE-Advanced*, Chapter 10, 2nd edn, John Wiley & Sons, Ltd, Chichester.
6. Sesia, S., Toufik, I. and Baker, M. (2011) *LTE – The UMTS Long Term Evolution*, Chapter 26, 2nd edn, John Wiley & Sons, Ltd, Chichester.
7. 3GPP TR 25.814 (2006) Physical Layer Aspect for Evolved Universal Terrestrial Radio Access (UTRA), Release 7, Annex A, October 2006.
8. 3GPP TR 36.814 (2010) Further Advancements for E-UTRA Physical Layer Aspects, Release 9, Annex A, March 2010.
9. 3GPP TS 36.101 (2013) User Equipment (UE) Radio Transmission and Reception, Release 11, Section 6.2, September 2013.
10. Anas, M., Rosa, C., Calabrese, F.D. *et al.* (2008) QoS-Aware single cell admission control for UTRAN LTE uplink. IEEE 67th Vehicular Technology Conference, pp. 2487–2491.
11. 3GPP TR 36.942 (2012) Radio Frequency (RF) System Scenarios, Release 11, Annex A, September 2012.
12. Jakes, W.C. (1994) *Microwave Mobile Communications*, Chapter 2, 2nd edn, John Wiley & Sons, Ltd, Chichester.
13. 3GPP TS 36.104 (2013) Base Station (BS) Radio Transmission and Reception, Release 11, Section 6.2, September 2013.
14. Mogensen, P., Wei, N., Kovács, I.Z. *et al.* (2007) LTE Capacity compared to the Shannon Bound, IEEE 65th Vehicular Technology Conference, pp. 1234–1238.
15. Salo, J., Nur-Alam, M. and Chang, K. (2010) Practical Introduction to LTE Radio Planning, http://4g-portal.com/practical-introduction-to-lte-radio-planning (accessed 15 October 2013).
16. Hata, M. (1980) Empirical formula for propagation loss in land mobile radio services. *IEEE Transactions on Vehicular Technology*, **29**, 317–325.
17. European Cooperation in Science and Technology (1999) Digital Mobile Radio Towards Future Generation System, COST 231 Final Report, http://www.lx.it.pt/cost231/final_report.htm (accessed 15 October, 2013).
18. Wireless World Initiative New Radio (2008) WINNER II Channel Models, WINNER II Deliverable D1.1.2, Version 1.2, http://www.ist-winner.org/WINNER2-Deliverables/D1.1.2.zip (accessed 15 October 2013).
19. 3GPP TR 25.912 (2012) Feasibility Study for Evolved Universal Terrestrial Radio Access (UTRA) and Universal Terrestrial Radio Access Network (UTRAN), Release 11, Section 13.5, September 2012.
20. 3GPP TR 36.814 (2010) Further Advancements for E-UTRA Physical Layer Aspects, 3rd Generation Partnership Project, Release 9, Section 10, March 2010.

21. Department for Transport, UK (2008) Road Statistics 2008: Traffic, Speeds and Congestion, www.ukroads.org/ukroadsignals/articlespapers/roadstats08tsc.pdf (accessed 15 October, 2013).
22. Coventry Transport Museum (2012) Thrust SSC, http://www.transport-museum.com/about/ThrustSSC.aspx (accessed 15 October, 2013).
23. Rämö, A. and Toukomaa, H. (2005) On comparing speech quality of various narrow- and wideband speech codecs. Proceedings of the 8th International Symposium on Signal Processing and its Applications, pp. 603–606.
24. ITU-T Recommendation P.863 (2011) Perceptual Objective Listening Quality Assessment.
25. ITU-T Recommendation P.862 (2001) Perceptual Evaluation of Speech Quality (PESQ): An Objective Method for End-to-end Speech Quality Assessment of Narrow-band Telephone Networks and Speech Codecs.
26. 3GPP R2-084764 (2008) LS on Considerations on Transport Block Sizes for VoIP.
27. 3GPP R1-083367 (2008) Adjusting TBS Sizes to for VoIP.
28. 3GPP TS 36.323 (2013) Packet Data Convergence Protocol (PDCP) Specification, Release 11, Section 6.2.4, March 2013.
29. 3GPP TS 36.322 (2012) Radio Link Control (RLC) Protocol Specification, Release 11, Section 6.2.1.3, September 2012.
30. 3GPP TS 36.321 (2013) Medium Access Control (MAC) Protocol Specification, Release 11, Section 6.1.2, July 2013.
31. 3GPP TS 36.213 (2013) Physical Layer Procedures, Release 11, Section 7.1.7.2.1, September 2013.
32. 3GPP TS 36.213 (2013) Physical Layer Procedures, Release 11, Sections 7.1.7.1, 8.6.1, 9.2, September 2013.
33. 3GPP R1-072570 (2007) Performance Evaluation Checkpoint: VoIP Summary.
34. 3GPP TS 26.114 (2013) IP Multimedia Subsystem (IMS); Multimedia Telephony; Media handling and Interaction, Release 11, Annex K, September 2013.

Bibliography

LTE Air Interface

Dahlman, E., Parkvall, S. and Sköld, J. (2013) *4G: LTE/LTE-Advanced for Mobile Broadband*, 2nd edn, Academic Press.

Ghosh, A. and Ratasuk, R. (2011) *Essentials of LTE and LTE-A*, Cambridge University Press, Cambridge.

Ghosh, A., Zhang, J., Andrews, J.G. and Muhamed, R. (2010) *Fundamentals of LTE*, Prentice Hall.

Holma, H. and Toskala, A. (2011) *LTE for UMTS: Evolution to LTE-Advanced*, 2nd edn, John Wiley & Sons, Ltd, Chichester.

Holma, H. and Toskala, A. (2012) *LTE-Advanced: 3GPP Solution for IMT-Advanced*, John Wiley & Sons, Ltd, Chichester.

Johnson, C. (2012) *Long Term Evolution in Bullets*, 2nd edn, Createspace.

Khan, F. (2009) *LTE for 4G Mobile Broadband: Air Interface Technologies and Performance*, Cambridge University Press, Cambridge.

Rumney, M. (2012) *LTE and the Evolution to 4G Wireless: Design and Measurement Challenges*, 2nd edn, John Wiley & Sons, Ltd, Chichester.

Sesia, S., Toufik, I. and Baker, M. (2011) *LTE: The UMTS Long Term Evolution: From Theory to Practice*, 2nd edn, John Wiley & Sons, Ltd, Chichester.

Signalling and System Operation

Kreher, R. and Gaenger, K. (2010) *LTE Signaling: Troubleshooting and Optimisation*, John Wiley & Sons, Ltd, Chichester.

Olsson, M., Sultana, S., Rommer, S. *et al.* (2012) *EPC and 4G Packet Networks: Driving the Mobile Broadband Revolution*, 2nd edn, Academic Press.

White Papers

4G Americas (2014) *4G Mobile Broadband Evolution: 3GPP Release 11 & Release 12 and Beyond* (updated annually), February 2014.

4G Americas (2013) *Mobile Broadband Explosion: The 3GPP Wireless Evolution* (updated annually), August 2013.

An Introduction to LTE: LTE, LTE-Advanced, SAE, VoLTE and 4G Mobile Communications, Second Edition.
Christopher Cox.

Index

An Introduction to LTE: LTE, LTE-Advanced, SAE, VoLTE and 4G Mobile Communications, Second Edition.
Christopher Cox.

Printed and bound by CPI Group (UK) Ltd, Croydon, CR0 4YY

12/01/2025

14624503-0002